MODERN TOWING

Modern Towing

BY
JOHN S. BLANK, 3rd

Cornell Maritime Press

CENTREVILLE, MARYLAND

Copyright © 1989 by Cornell Maritime Press, Inc.

All rights reserved. No part of this book may be used or reproduced in any manner whatsoever without written permission except in the case of brief quotations embodied in critical articles and reviews. For information, address Cornell Maritime Press, Inc., Centreville, Maryland 21617.

All of the sketches in this book that are not otherwise credited are the work of Jeffrey Eldredge; all photos not attributed to others are in the collection of the author.

Library of Congress Cataloging-in-Publication Data
Blank, John S., 1911-
   Modern towing.

   Bibliography: p.
   Includes index.
   1.Tugboats. 2. Towboats. 3. Towing.
I. Title.
VM464.B42 1988   387.1'66   87-47736
ISBN 0-87033-372-0

Manufactured in the United States of America
First edition, 1989; second printing, 1994

To my lovely wife, Constance Thompson-Blank,
sailor and mother of our six children,
who was forced to take over the helm at home
whenever I was away at sea,
and to the many other good wives and families
of those mariners
whose careers take them away in sailing or towing around the world.

Contents

Preface   *xi*

PART I
The Industry—Past, Present, and Future

1. Towing, a Growing Industry   3
   The Industry; The Vessel; Power; Personnel; Management; World Towing, Past and Future; Geographic Breakdown; Deep-sea Towing Worldwide

2. Towing Nomenclature   29

3. Modern Tug Types   41
   Harbor Tugs—Ship-Assist Tugs; Tractor Tugs; Gulf of Mexico Tugs; Assist Tugs at Offshore Terminals; Coastwise Tugs; Ocean Tugs; Salvage and Rescue Tugs; Petroleum Industry Tug/Supply Vessels; Inland River Towboats; Towboats on Canadian Inland Waters and Lakes; Towboats on the Columbia River System; Tugs in the Panama Canal Zone and U.S. West Coast Ports; Power and the Harbor Tug

4. Modes of Towing   60
   Ship Assistance—Docking and Undocking; Inland or Sheltered Waters; Tug-Barge Units; Ocean and Coastwise Towing on the Hawser

5. Tug Design   88
   Pilothouses; Electronics on the Bridge, in the Pilothouse or Wheelhouse; Visibility; After Steering and Control Station; Afterdeck Engine Controls; Tow Winch Controls; Deckhouses; Stability; Hull Form; Fenders and Guards; Tumble Home and Deckhouses; The Stern—General Layout; Tow Winch Fastening, Location, and Controls; Tow Winch Types; Hawser Tension Damper; Tow Winch Power; Tow Winch Operation; Tow Winch Cable Guide; Tow Winch Line Retrieval and Warping Drum; Deck Capstans; Fairleads; Vertical and Horizontal Stern Rollers and Norman Pins; Thumb Chocks; Hydraulic Pins and Rollers; Horizontal Rollers; Cable Hold-Downs; Single Tow Arch and Sheave;

Hydraulic Cable Controls for Tug/Supply Vessels; Fairleads and Molly Goggers; Bitts (Stern, Side, and Bow as Mooring Fittings); Stern H Bitts; Forward Bitts; Stemhead; Tow Hooks; Deck Winches and Sheaves; Vents; Stowage; Outside Illumination; Outside Light Controls; Telephones and Intercom Systems; Engine Rooms; Engines; Wastewater and Sewage; Thrusters; Steering Systems; "Joystick" Steering; Rudders; Fixed Propellers; The New Tug Generation; Anchor and Cable; Painting and Rust Prevention; Dry-Docking; Zincs and Electrolysis; Tail Shaft and Outboard Bearings; Fire Equipment; Lifesaving Equipment—EPIRB and Survival Suits; Navigation Lights and Searchlights; Tug Interiors; Outside Ladders; Tug Size and Tonnage; Comparative Size of Tugs and Their Crews; Classification; Summation

PART II
Operations

6. Getting the Tug Under Way   153

Reporting Aboard; Master's Responsibility; Main Engine Warm-Up; Getting Under Way; Hand Signals; Letting Go—Departure; Feeling Out the Tug; Steering a Tug without a Tow; Turning the Tug—Advance vs. Transfer of Fixed-Propeller Tugs; Trial Runs—New Tugs; Use of Tug's Power; Coming Alongside a Stationary Object; Towing Orders; Coming Alongside a Moving Object; Communications; The Well-Equipped Tug

7. Under Way—Towing on Inland Waters   185

Ship Assisting; The Tractor Tug; Pilots and Docking Methods; Ship and Barge Work on the Hawser; Towing a Dead Ship; Whistle Signals between Pilot and Tug; Defensive Action by the Tug Captain; Ship Assisting, Great Lakes Ports; Tugs as Escorts; Big Ship Docking at Exposed Oil Terminals; Barge Work and the Tug; Getting a Barge Tow Under Way; Advantages of Pushing; Short Hawser Towing—Deep-Draft Units; Towing from a Hook; Dredge Towing; Railroad Car Float Towing; Handling Tows in Locks and Canals; Approaches to Other Locks and Canals; Moving Drill Rigs; Drawbridges; Piloting the Tug and Tow; The Tug Log; Care of the Tug; Safety

8. Under Way with the Tow and at Sea   264

Condition of the Unit to Be Towed; Inland Water Towing; The Integrated River Tow; Making and Entering a Lock; Downbound Tows—U.S. Great Lakes and Western Rivers; Push Tows on Other Inland Waters; Descriptions of Towing Practices and Maneuvers; Crew Training for Unmanned Barges; Charts and Spare Equipment; Surveys of Ocean and Coastwise Tows; Barge Retrieving Lines; Weather Considerations; Where to Connect Up to the Tow Bound for Sea—Choices; Connecting to the Drill Rig; Leaving Port for Sea with Various Units; Hawser Length on Inland Waters; Areas Commonly Used to Get on the Hawser; Hawser Territory; Connecting the Hawser in Port; Connecting Hawser When Head-and-Tail; Getting Away from Head-and-Tail Makeup; Connecting Hawser at Transition Points; Barge Pickup Line; Taking Charge of the Barge; Keeping a Good Lookout; Application of Power; Bringing the Tow out of Swift Rivers; West Coast Surge Chains; Tandem Towing; The Big Tug—Push vs. Pull; Pushing the Big Barge—Dock to Sea; Dangers in Pushing at Sea; The ITB; Skegs; Chafing Gear—Types and Rigging; Routine on the Short Sea Tow; Unforeseen Occurrences with the Tow—Sheering—Tow Attempts to Pass; Grounding and the Report of Marine Accident, Injury, or Death; Narrow Passages and Rapids; Narrow Dredged Channels and Canals; Occurrences at Sea; Approaching Port; Tow Lights and Day Signals

9. Special Types of Towing 328

Long-Distance Towing—The Tug; The Towage Market; Marine Insurance and the Ocean Tow; The Towage Contract; Planning the Towage; Preparing and Rigging the Ocean Tow; Connecting to the Ocean Tow; Departure with the Ocean Tow; At Sea with the Ocean Tow—Chafing Gear; Towing Winches at Sea; Installation and Care of the Towing Hawser; Cable Pay-out and Catenary; Communications and the Ocean Tow; Navigation and the Electronic Bridge of Seagoing Tugs; Drills at Sea; Inventory and the Computer; Eating and Exercise; Recreation and Entertainment at Sea; Routes, Current, and Weather; Going Back to Look at the Tow; Salvage and Rescue; Anchor Handling; Rig Towing at Sea; Tandem Tugs Towing Ahead; Mud Towing; Towing Toxic-Waste Barges

10. Handling Tows in Special Situations 371

Partially Loaded Barges; LASH and SeaBee Barges; Passing through Drawbridges; Shifting Personnel from Boat to Boat; Notes on Towing in Specific Places; Towing Log Rafts; Ice Conditions; Tugs Equipped for Fire Fighting; Turning an Empty Oil Barge in the Wind; Use of the Anchor When Docking on the Hawser; Anchoring the Tug with a Tow Astern; Anchoring the Barge with the Tug When Delayed En Route; Getting an Anchored Barge Under Way; Approaching Port in the Fog; Placement of Ship's Towlines; Towing "Live" Ships; Towing Barges in Rough Weather; Putting a Barge on a Mooring Buoy; Hurricane, Typhoon, and Tornado Readiness; Vessel Traffic Services; Traffic Separation Schemes; Fishing Vessels and Areas Tugs Should Avoid

PART III
Towing as a Business

11. The Shore Establishment   439

Towing Companies; Operating Costs; Management; Protection and Indemnity Coverage; Relationships; Computer Management; Operations, Dispatching, and Vessel Readiness; Tug Dispatching; Vessel Supervisors; Company Seminars; Supervisors Afloat; Profit Sharing; Sales and Marketing; Maintenance and Repairs; Surveys; Towage Rates; Supply; Feeding and Health; Health and Exercise; Safety

12. Personnel and Management—Labor Relationships   456

The Tug Master or Captain; Relief Captains and Crew; Mates; Seamen (He or She); Engineers; Cooks; Radio Operators; Manning; Tug Labor Relations and the Unions; Organized Tug Areas; Work Arrangements and Wages; Negotiations; Hiring and Training the Crew

13. Rules, Rulings, Regulations, and Other Considerations Affecting Towing   482

'72 COLREGS and the Navigation Rules; Rules Affecting Tugs; Laws Regarding Tugs and Tows; Lookouts and the Tug; Who is the Tug Master? Responsibilities—Tug Operator vs. Tug Master; Pilots and the Tug and Tow; U.S. Coast Guard's Proposed Rule Making; Use of Foreign-Flag Tugs in U.S. Harbors; Use of Foreign-Flag Tugs in U.S. Harbors; IMO Clarifications of COLREGS; Pollution and the Tug; VHF Radio and the Tug; Use of VHF Bridge-to-Bridge to Avoid Collisions

PART IV
Appendices

List of Appendices	502
Appendices	505
References	580
Index	581
About the Author	595

Preface

TOWING is an ever-growing industry. While many of the basic principles of the past remain, a whole new technology has been introduced. The tug captain now faces a more challenging and at times exciting occupation. Much, of course, depends upon where he will be operating.

Modern Towing is a compendium of towing practices gathered during 48 years of service on various types of tugs and in the office. It offers an intimate concept of many forms of towing; towing vessels; their design, construction, uses, and equipment; personnel; management; and methods of operation both on inland and ocean waters. Towing problems, projects, towage liability, and the law are discussed broadly with reference to specific treatises devoted to the tug and tow.

Modern Towing covers most of these areas and the various aspects and methods of tug handling, operations, and usage. It is not just devoted to the role of the tug master. It reaches into the engine room afloat and management ashore. It examines harbor towing's various components: shifting, shiphandling, and docking; all types of barge movements; inland river towing; ocean and coastwise towing; and towing on the Great Lakes. In some areas, the comments are based on local or U.S. practices; in others, upon worldwide usage.

A series of tables and diagrams useful to the towboat industry has been included in the appendices.

Towing is a service. As a team, the tug crew serves the owner and his customers. As a tug captain, one must thank his crew who worked so diligently with him to make each assignment a financial success. He must also thank those who recognized him, employed him, and trusted him.

My humble and extensive thanks must be given to Mr. Jack Rountree, former superintendent of Bay State Dredging & Contracting Co. of East Boston, who offered me my first berth as a tug master; to the late Kate Sutton of the Providence Steamboat Company, that fine Rhode Island towing firm, and my close friend the late Bart J. Turecamo, both of whom recognized and helped and promoted me; as well as the U.S. Navy which, in wartime, trusted me with the command of several fleet rescue tugs.

Respected employers through whom much added towing experience has been gained would include: Lavigne & Smith Dredging Company; the Marine Division of the Erie Railroad at Jersey City, New Jersey; Howard Line of New York; Seaboard Sand Company of New York; B. Turecamo Towing Company of Brooklyn; Ocean Towing Division of Meseck Towing Line of New York; Great Lakes Dredge & Dock Company, Inc.; Tecon of Dallas, Texas; Merritt, Chapman & Scott; Peter Keiwit & Company; L.A. Wells Construction Company, Cleveland, Ohio; Oil Transfer Company of New York; Harbor Towing and St. Phillips Towing of Tampa, Florida; Coyle Lines and Indian Towing Company, both of New Orleans, Louisiana; GulfCoast Transit Company of Tampa; Red Star Towing & Transportation Company of New York; Spentonbush Transportation Company of Brooklyn; Sheridan Transportation of Philadelphia; Moran Towing Line of Texas, and Interstate & Ocean Transportation of Philadelphia (now Maritrans).

I am personally indebted to former shipmate and tug captain, Jeffrey "Jeff" Eldredge of Middleborough, Massachusetts, for his fine sketches throughout. Seamanship and talent come naturally to him from his ancestors, the Eldredge captains of Cape Cod (both spellings are correct), one of whom originated the still popular annual *Eldridge Tide and Pilot Book*.

Some of the photography is my own, some generously offered by various towing companies throughout the world, and others are from the vast collection of my good friend, Steven Lang, author of the beautiful tug book, *On the Hawser*.

It is hoped that *Modern Towing* may serve its purpose to the industry and to its personnel both afloat and ashore.

PART I

The Industry—
Past, Present, and Future

1. Towing, a Growing Industry

THE INDUSTRY

THE towing industry worldwide has evolved from a declining mode of marine transportation to become a very sophisticated, modern, competitive challenge in the movement of products afloat.

In its attempt to remain competitive, the towing industry has tested many solutions to the problem of moving the greatest loads with the most efficient power, housed in the smallest possible space—the tug, manned by a minimum crew. This is the prime goal of the industry, and the *best* solution is to provide the customer with superior service. "Service" embodies the culmination of a combined effort both ashore and afloat that results in the successful operation of a good towing vessel.

THE VESSEL

Although the fictitious captain of an ultra-large crude carrier (ULCC) claims the term "vessel" only applies to something that holds liquid, and not to a ship—which is limited in that ability—through universal usage "vessel" has been accepted in both the maritime industry and its legal and regulatory appendages.

The use of "vessel" in *Modern Towing* is general, since there are many colloquialisms used throughout the world in reference to what in North America is called a tug. In the United States, towing vessels are placed into two classes—the towboat and the tugboat. To those outside the industry (even within it in certain areas), these names are synonymous. Upon the Western Rivers of the United States, however, from the passes of the Mississippi and its intersecting Gulf Intracoastal Waterway to the headwaters of the great river's tributaries, "towboat" is the name used for those flat-bottomed, powerful, pusher type towboats. Towing vessels of all sizes that can pull or tug from over their stern are referred to as tugboats with the appropriate additional designation, "harbor," "coastwise," or "ocean."

Gone are the mighty steam stern-wheel river towboats. Even the World War II-built steam propeller towboats have been scrapped, re-

The ocean tug *Smit London,* leaves Singapore with the rig *Ocean Prospector,* bound for Mobile. (Courtesy Smit Tak International Netherland BV)

placed by the steel diesel giants whose pushing knees face up to 60 or more loaded barges. Active steam tugboats are no longer in sight except in a museum or two. Their graceful and romantic being has been taken over by the unaesthetic, yet highly efficient, tugs of today.

POWER

Power is of course what towing is all about: the most efficient, judicious use of power in applying it to the unit to be moved, whether shifting, undocking, landing, or long-range towing.

In the 1930s, the average horsepower of a harbor tug was from 400 to 600. Much of it was steam produced. Ocean and coastwise tugs had from 750 to 1,000 HP, although only a few had the maximum. By the end of World War II, the U.S. Navy, Army, and War Shipping Administration had a terrific fleet of tugs whose horsepower ranged from 1,250 to 3,000. This was matched abroad by the Royal Navy's Bustler class of salvage tugs at 4,000 HP and the pride of Holland, *Zwarte Zee,* at 4,200. With the advent of bigger ships and barges, the horsepower of harbor tugs soared from 1,200 to 3,500. For coastwise tugs, a horsepower from 3,500 to 9,000 is now common. The horsepower of ocean towing and rescue vessels has risen to 26,000. The super tug has arrived!

PERSONNEL

To man this new breed of tugs has required, and continues to require, the upgrading of experienced tugmen and the training of new entrants into the towing industry. It is a great occupation with a good future, one attractive to the young aggressive worker-seafarer as well as to those who enter from industry- or labor-sponsored seamanship schools, and officers from the various national and state maritime academies.

Today's successful tug crew is a team, a highly trained group of specialists of three or four on a harbor tug, eight or ten on a coastwise vessel. Ocean tugs may have up to seven in the crew. Salvage tugs will have even more.

MANAGEMENT

The legendary tug company managers are gone, as well as nearly all the old-line family owners. Crowley, Foss, Moran, McAllister, Sause, and Turecamo may be the exceptions in the United States. Today, business administrators run the shore establishment under what remains of the original owner's family.

In the United States in particular, where each harbor once had a family-owned towing company (towboat company), the larger ones waited patiently until the smaller were ripe for absorption. In the 1980s, several huge towing organizations remain, some a part of a nonmaritime conglomerate. The European Tug Owners' Association claims that there are 141 towing companies, worldwide. Within this number, there are, in addition to deep-sea tugs, 115 designed solely for harbor work; among them are the following:

Alexandrea	Norge Salvage
Bugsier	Progmar
Cory	Safmarine
Crowley	Seaspan
Curtis Bay	Smit†
Fairplay	Svitzers
Foss	Turecamo
Lawson-Batey	Tokyo Marine
Les Abeilles	Ulstein
Maritrans*	United
McAllister	Vervicos
Moran	Wijsmuller
Neptune	Zapata-Gulf

*Formerly IOT—Sonat Marine.
†Smit operates as Smit International and/or Smit Tak

Bugsier's *Titan* off to the rescue. (Courtesy Bugsier)

It has been difficult for towing management to stay on top of the new sophistication in the fields of propulsion and electronics. A close liaison must be maintained between those ashore, who may wish they held the required sea towing experience, and those in command afloat, who probably wish management would absorb some understanding of their problems.

One outstanding towing line holds annual seminars for each classification of personnel—captains, engineers, and tankermen. Other companies are very impersonal, communicating only with their chief engineers and senior captains.

WORLD TOWING, PAST AND FUTURE

Anything that floats can be towed; it just takes the tug with the right horsepower. When considering moving something afloat, there are many people who think first of towing. This has offered the towing industry a wonderfully competitive position in the world shipping market. Today, marine towing is a stable industry; changing economic conditions of the late 1980s have caused a temporary stagnation in world towing and a surplus of tugs.

Out of the past has risen a great new fleet, improved by the lessons learned from each passing era of the towing industry. Harbor tugs no longer cruise for becalmed sailing ships off the entrance, nor tow them offshore to meet a breeze. Coastwise coal barges, three to a tow in old cut-down schooner hulls, are gone. The early lumber- and petroleum-loaded barges have held on, but these carriers have increased immensely in size. So have the vessels that tow. Both are now gigantic when compared to pre-World War II tugs and barges. Smit International's famous tug *Zwarte Zee* was the world's largest and most powerful at 4,200 HP. Their tug of the same name, built as a replacement in 1966, came out with 12,000 HP. She also has been replaced. Now there are a half dozen of these ocean towers operating with 26,000 HP.

For years, Holland led the world in deep-sea towing, with rescue tugs standing by in various strategic locations. Holland's other tugs were busily engaged in towing dredges to the East Indies or ships from one world port to another for repairs. Great Britain and Germany offered some competition. In the United States, the push towing of bulk cargoes on inland waters grew slowly. In the coastwise trade, the small steamers built for use in World War I and after bit deeply into tug and barge movements. In the period from the 1920s through 1940, many old wooden-hulled harbor and coastwise tugs converted to diesel power. Barges and scows were still mainly wooden. World War II changed all that. Hundreds of tugs of all sizes were built. Towing, both transatlantic and transpacific, became a most important part of the support system to the fighting forces. In just one campaign, the invasion of Normandy, 134 tugs took part. Eighty percent of these vessels hailed from the United States.

In the postwar lull, after surplus ships had been towed to lay-up areas, or to ship-breakers' yards, the towing industry entered a transitional era. Abroad, towing companies in the Netherlands built new and bigger tugs to replace their warworn fleets. England held on to their fleet and diminished its size. Germany and Japan reentered the ocean towing market; Italy and Greece began competing using surplus Allied tugs, many of which were sold by the United States to any interested buyers. There was little left to be towed in the Western Hemisphere.

Soon the one commodity that was so important to sustain so much—oil—changed all that. Drilling for oil in the western Gulf of Mexico and Venezuela had always been from sites onshore or in the swampy marshland. While probing to what extent these oil pockets ran offshore, drilling companies discovered a whole new series of reservoirs off the coasts of Louisiana and Texas. As a result a whole new set of floating units and attending tugs was created.

At the same time, oil wells in the Persian Gulf, Africa, and Malaysia came in, requiring more tugs. Then came the North Sea discovery. England, France, Norway, and the Low Countries all built new and powerful tugs to service this area. They were joined early on by fleets of new U.S. Gulf-built tugs. This rush to probe new areas in the late 1960s produced an unprecedented flurry of oil rigs coming off builders' ways in the United States and the Far East—all to be towed elsewhere, frequently halfway round the world, by tugs whose horsepower had risen in 20 years from 2,000 to 10,000.

Seaspan Regent is typical of a large British Columbia ocean tug. (Courtesy Seaspan International Ltd.)

In the mid-1980s the wellhead price of crude had dropped so far worldwide that capital for further exploration was lacking. The marine section of the petroleum industry was at a standstill. Many of the U.S. tugs designed and built especially for the oil industry were sold and entered into the general towing service of U.S. tug operators.

The closing of the Suez Canal had also caused concern for the safety of the lumbering ULCCs, which then sailed from northern Europe to the Persian Gulf via the Cape of Good Hope. These ships of over 200,000 tons were prone to casualty at sea. For assistance they required tugs of enormous horsepower to make any headway. In Holland, the Smit and Wijsmuller companies built huge new tugs of 20,000 to 22,000 HP, and South Africa Marine had two 26,000-HP tugs built to handle ships that required help off South Africa's long, rough coasts.

Some of these salvage and towing companies have created tows for themselves by buying valuable ships that were laid up. Most of these vessels were available due to overbuilding and a saturated market and

The supertug *John Ross,* having 26,000 HP, is towing a disabled ship. She is also indicating her inability to maneuver as specified in International Rule 27. (Courtesy South African Marine Corporation Limited)

were ten years old or less. They could be towed to a ship-breaker at a profit over the cost of towage. Some owners have made these arrangements, which have at least kept the tug in operation, in order to pay for fuel, stores, and wages.

Meanwhile, in the United States, an event occurred that brought about a new state of the towing art. Statehood for Alaska and Hawaii attracted further development in those areas. Equipment and objects too large or difficult to stow aboard a ship or destined for an area too shallow for her draft were loaded aboard a barge.

Many of Alaska's requirements for construction materials and mobile homes are barged by several towing lines. Alaska's state-run railroad is connected with the Lower Forty-eight by weekly car/barge tows. The Aleutian Islands and Alaska's fishing industry are also served by tug-barge carriers.

A typical Pacific Northwest to Alaska tow. (Courtesy Foss Maritime Co.)

GEOGRAPHIC BREAKDOWN

The North Slope

Alaska's North Slope oil fields near Prudhoe Bay have received huge barge lifts akin to floating cities of production equipment, which arrive in icebreaking convoys during the two-week period when ice loosens, usually in August.

Alert to a cost-conscious world, the modern towing vessel is laden with all the latest electronic navigational and communication equipment. Its engines in many cases burn less-expensive blends of fuel. Its entire machinery on deck and below may be controlled from the bridge

Opposite, above, Gladiator is departing on one of Crowley Maritime's annual North Slope-Prudhoe Bay convoys. The modules she is towing are up to 112 feet high. (Courtesy Crowley Maritime Corporation) *Below, King Bay* is typical of a new breed of tractor type general purpose tug being built in western Australia. The builder, Australian Shipbuilding Industries (W.A.) Pty, Ltd. terms this the Titan Class. The class has been accepted as 1A1, Det norske Veritas. At 34.15 meters (111.5 feet, LOA), this class can deliver approximately 5,200 HP through twin aft located rudderpropellers with a 360-degree azimuth. (Courtesy Australian Shipbuilding Industries [W.A.] Pty, Ltd.)

Foss Maritime's tractor tug *Andrew Foss* demonstrates her maneuverability in a 360-degree turn. (Courtesy Foss)

or pilothouse. The tugs themselves are structurally and operationally different.

The one great change is the introduction from Europe of the tractor tug, which is now a standard ship-assist craft in some U.S. West Coast ports.

As we approach the last decade of the twentieth century, the tonnage the towing industry moves within the United States and parts of Canada represents a major portion of the total of each country's self-sustaining merchant fleet. The following sections describe the products presently being moved by tug and barge in various locations.

Great Lakes Towing's tug *Oregon* is towing a bulk cargo barge out of the Milwaukee River. (Courtesy The Great Lakes Towing Company)

Great Lakes

Coal in barges is towed from Toledo to Detroit, and to other powerhouses as far east as the Niagara River. Iron ore, limestone, grain, and coal are also moved by the large, recently introduced tug-barge units. They have already replaced some of the aging Great Lakes steamships. American Oil Company of Chicago has replaced its fleet of U.S. flag tankers with newly commissioned tug-barge combinations. At least one of these notched barges and its attending tug has been assigned to the lakes. The Great Lakes Towing Company not only maintains tugs in nearly every port of the five lakes, it has also stationed a salvage tug at Sault Ste. Marie. McAllister Brothers Ltd. has a similar fleet of salvage craft at Kingston, Ontario, close to the head of the St. Lawrence River. Chicago-headquartered Great Lakes Dredge and Dock Company has tugs towing their dredges and dump scows in ports from South America to the Arabian Gulf.

The St. Lawrence River and Gulf

Modern harbor tugs at Montreal and Quebec City cover river towing, mainly ships, pulpwood, newsprint barges, and oil barges. This same type of heavy towing on the Saguenay River, the Sept-Iles area, and the Gulf of St. Lawrence is often handled by tugs from the Maritimes or New Brunswick.

The Maritimes and New Brunswick

Foundation Ltd.'s towing services have been taken over by Corys Ltd. of London, and others, including salvage within the area from the Bay of Fundy to the Saguenay. From Saint John, New Brunswick, Atlantic Towing's *Irving* tug fleet tows ships and heavy petroleum barges be-

Charlie S is one of Quebec Tugs Limited's recent harbor tugs. She has been superseded by *Donald P,* shown in chapter 3. (Courtesy Quebec Tugs Limited)

tween Maritime ports and Newfoundland. Their *Irving Miami* has regularly towed newsprint barges from the mills in Saint John or Liverpool, Nova Scotia, to Norfolk and Alexandria, Virginia, and Miami, Florida. For a profitable return, paper pulp is often loaded in a Georgia port.

Along the northeast coast of the United States, petroleum products are towed by modern tugs from the refineries on the Delaware River and along the Kills between Staten Island and New Jersey. The barges vary in capacity from 30,000 to 250,000 barrels, the tugs from 3,500 to 6,000 HP. Smaller units are towed up the upper Hudson either through the Champlain Canal to ports on Lake Champlain in Vermont or New York or westward through the Mohawk River and Erie Canal to ports on the Great Lakes, on the St. Lawrence, or in Canada.

The Upper Hudson River
From the upper Hudson, brick, cement, and stone are towed. Some cement goes to sea coastwise to Savannah, Norfolk, or Port Everglades in the Atlantic Cement Company's two 420-foot, 8,500-ton barges.

Delaware Bay and River
This is a leading towage area. Barges with various types of cargo are discharged or loaded along this waterway. Every day the lower bay receives one or more very large crude carriers (VLCCs), whose draft is over the

river's 37- to 39-foot limit. They move into Big Stone Beach Anchorage to be lightened to the permissible draft. Maritrans of Philadelphia does much of this lightering using their late-model, twin-screw, upper-and-lower-wheelhouse tugs to push, or pull partway if too rough, manned barges that hold from 150,000 to 250,000 bbl. Each month they move several million barrels upriver to the various refineries. When refined, much of this oil will again pass out the Delaware capes in barges bound for New England or New York terminals, or head southward through the Chesapeake and Delaware Canal bound for storage tanks in Baltimore or down the Bay to Norfolk.

Domestic crude from Texas has been towed into the Delaware by tugs of Bloodworth of Houston, in deep-notched barges that can be pushed at sea in moderate weather.

Other products and towing originating or terminating in the Delaware are chemicals from New Castle and liquid lead handled by Bloodworth's special tug *El Gaucho* and barge *Pampas*. Bloodworth claims that this combination can push all the way from Deepwater Point, Delaware, to South America.

The Philadelphia area is also the northern terminus of the Willis Barge Line, which operates several push towboats and a fleet of covered barges between Delaware and Florida via the Atlantic Intracoastal Waterway. Philadelphia Naval Shipyard is often the source of heavy tows of battleships and aircraft carriers, brought in for repair or storage. Tugs of the military sealift do this heavy and long-distance pulling. From Petty Island above Camden, New Jersey, Crowley Maritime through their Caribbean Division of Trailer Marine Transport (TMT) operates weekly sailings of their immense three-deck RO/RO (roll-on, roll-off) barges whose dimensions of 580 feet by 105 feet with 57-foot sides allow 374 standard 45-foot truck trailers to be loaded, all bound for the West Indies. Crowley's 9,000-HP tugs do the towing. The Port of Philadelphia has in the past, and still does on a limited scale, ship coal by barge from the Pennsylvania and Reading Railroad's coal piers. Most of this is towed or pushed by tugs of Express Marine through the Intracoastal Waterway. Farther up the Delaware River at Morrisville, steel has been barged from the mill to the West Coast using chartered equipment as required.

Chesapeake Bay and Tidewater Virginia

For years, this vast tidewater area has been served largely by tug and barge. As a spin-off from its position as the country's greatest coal-exporting area, the Hampton Roads coal piers at Lambert Point in Norfolk and in Newport News, Virginia, ship many tons of coal by barge. The largest consistent movement has been in the 500-foot *Chesapeake*,

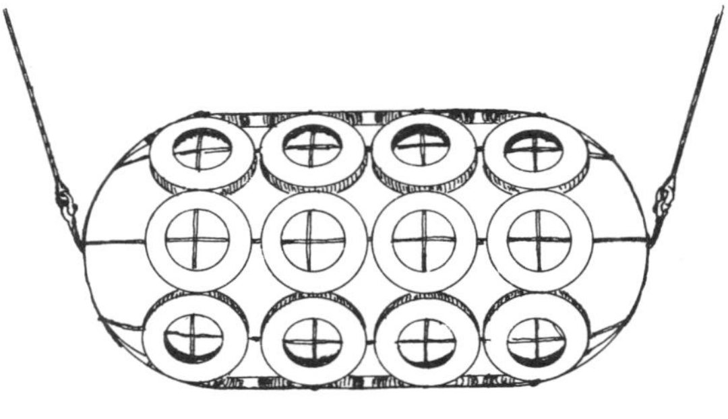

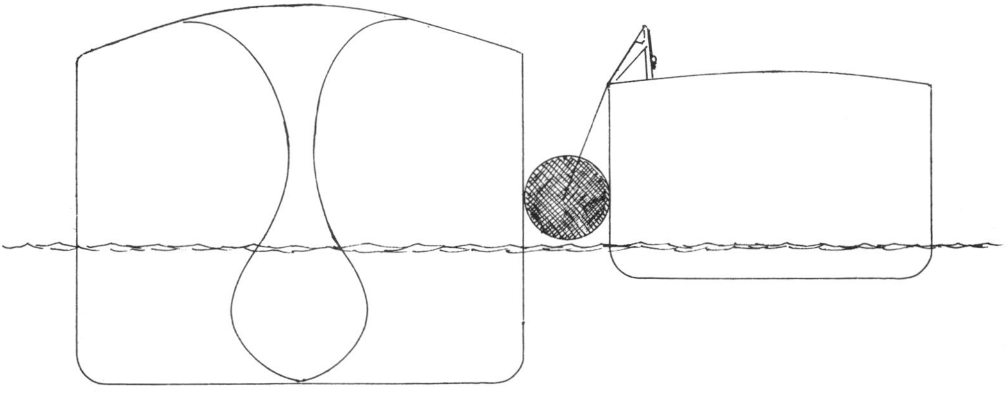

pushed twice weekly by one of the Curtis Bay Towing Company's Cape Lookout class tugs to Baltimore, and by tug-barge units to New England energy plants.

Coal, once the dominant East Coast barge cargo, has again become a vital energy fuel for some power plants. Self-propelled colliers have been replaced by tugs towing hawser-notch barges. With cargoes of coal, this towing operation exists between Hampton Roads and New England, and from Mississippi River transfer points to Florida.

Enormous bulk carriers, which because of their great draft cannot fully load alongside some coal piers, now top off in the stream in lower Delaware Bay, off the Delta of the Mississippi, and in the Canso Harbour area of Nova Scotia. New ocean barges of great capacity that have the ability to quickly transfer their cargo are towed to the U.S. anchorages. In Canada, Great Lakes ships are used for this purpose, but may in the future be replaced by the towed barge.

Norfolk, Virginia, strategically located in a pocket on the south side of Hampton Roads and at the mouth of the Elizabeth River, has access to all of the Intracoastal Waterway to the south as well as the Bay and the Atlantic. It is the transfer point for the covered-barge traffic of various tug and barge lines.

From Norfolk, a variety of cargoes are pushed or towed southward through the Intracoastal and northward up the Chesapeake Bay and its many river tributaries. Wood pulp and petroleum go up the York to West Point and on the James to City Point and Richmond.

On the York River, American Oil has a large refinery at Yorktown that is constantly loading barges for both inland and coastwise ports. The U.S. Navy has an ammunition storage area at Chetham Annex, most of the contents of which are barged by navy tugs to the Norfolk operating base.

The Potomac River is another river well served by towing. Stuart Petroleum operates a fair-sized refinery at Piney Point using crude brought in by both tanker and by their own fleet of tugs and barges. Using the latter, they deliver much of the refined products up and down the East Coast.

Opposite, above, the loaded barge is ready to depart after lightering the tanker in lower Delaware Bay. The tug is in the barge's notch and the captain has swung the barge away from the ship so as not to force the Yokohama fender up on the barge's deck. *Below,* when a barge is lightering from a tanker, the huge fenders must be placed between the ship and the barge as shown in this sketch. When not in use, they are held on skids and lowered and hoisted mechanically.

Atlantic Intracoastal Waterway

Although this waterway has a limited depth of 9 feet, considerable tonnage is towed through it between Hampton Roads and Jacksonville, Florida. Some goes on toward Port Everglades and Miami. Other than the regular barge line tows, those of the various paper companies are prevalent. Usually pulled behind conventional tugs, they consist of two closely connected barges piled high with pulpwood and are frequently encountered in North Carolina. South Carolina's section of the Intracoastal is also a busy route, with small petroleum tows and high-sided wood-chip barges, which are pushed up the Cooper River through Charleston.

Jacksonville and Florida's East Coast

For many years Jacksonville has been the shipping point for Trailer Marine Transport, the original shipment-by-barge concept of rolling regular truck trailers aboard rather than lifting them. The trailers were

Much of Puerto Rico's imports come from the United States on wheels—by barge. This is the Trailer Marine Transport barge *La Reina,* with over 360 units, pulled by a Crowley Maritime tug. (Courtesy Crowley)

unloaded in Puerto Rico also by being rolled off, and the designation RO/RO for roll-on, roll-off, as it applies to ships, was born. Originally, converted World War II U.S. Navy landing ship tank (LST) hulls were used for this purpose. Through their elevator ramps both the cargo deck below and the main deck were used. They were pulled by the battered ancient tugs of the Coppedge Towing Company at a speed of 6 knots, taking from five to seven days to travel from Florida to San Juan. Now, Coppedge is out of towing, and TMT has been taken over by Crowley Maritime. Crowley's 9,000-HP tugs pull their 730- by 105-foot, triple-deck barges *La Reina, El Conquistador, La Princesa,* or *El Rey* with 374 trailers aboard at better than 10 knots in good weather.

Jacksonville is also the home port of Gulf Atlantic Towing's several offshore tugs and barges and was the site of the Libby-Owens paper mill, which received all of its pulpwood logs via big, shallow-draft barges towed from the Bahama islands by former U.S. military tugs under a foreign flag. In addition to these towage movements, Jacksonville receives much of its petroleum products by big barge tows, the over-200,000 bbl. type, towed from Texas by Maritrans, Sabine Towing and Transportation, or Moran of Texas tugs, and huge tug-barge units of Maritrans Bulk Fleet, Amoco, Amerada Hess, and Moran of Texas.

Port Everglades and Miami

Mainly petroleum products are towed in by the same tugs that unload part or full cargoes in Jacksonville.

Florida's Gulf Coast

In tonnage towed and horsepower used, the Gulf Coast is probably the leader of all U.S. towing areas. Phosphate is the cargo westbound, coal and oil on the return trip. This traffic has allowed the Port of Tampa to expand. Two major towing lines, Gulfcoast Transit and Red Circle Towing, move 400- to 500-foot loaded barges westbound to chemical plants on the Mississippi and in Texas. Some barges are towed empty on the return. Others are loaded with coal for Tampa Electric's or Florida Power's enormous generating stations. Those that burn fuel oil are serviced by barge tows of Exxon, Sabine, Maritrans, Amoco, Allied-Dixie Carriers, or Bulk Marine. All are high-tech towing movements using the latest state of the art.

Northwest of Tampa and running around the Florida Panhandle are the minor ports of Crystal River, St. Marks, Port St. Joe, Panama City, and Pensacola. Each is served by towed barge. Some have a sufficient depth for small ships which load paper products. Pensacola is the home port for U.S. Navy training aircraft carriers. Each of these ports has or is served by one or more tugs. Much of the marine traffic to or from these northwest Florida ports is towed via the Gulf Intracoastal Waterway which runs from Apalachicola Bay, Florida, to Brownsville, Texas.

Mobile Bay and Tributaries

Mobile is a deepwater port with good towing services to assist ships in docking at Alabama state piers and the various shipyards. It also acts as a transfer point for inland towing. Some enter from the Gulf Intracoastal or descend from the Alabama and Warrior River systems and the Tennessee-Tombigbee Canal.

Pascagoula, Mississippi

At Bayou Casotte there is a large oil refinery that ships its products across the Gulf or farther using the large barges mentioned as serving Tampa, as well as those that are towed through the Gulf Intracoastal Waterway.

Lower Mississippi

Below Baton Rouge the lower Mississippi is alive with every kind of tow. Some are pulled, some taken alongside, and many pushed by the flat-bottomed river towboat or conventional harbor or coastwise tug. Grain from upriver elevators and coal from the mines of Kentucky come down in huge flotillas to be transferred to either shoreside silos or storage areas to be transferred later into ocean barges.

These river barges reload with phosphate or other products for their voyage northbound. There are also special tows of such products as liquid sulphur, asphalt, or liquid gases and toxic chemicals made up as unit tows, not unlike a unit train. They serve such places as Omaha, Nebraska, on the Missouri, St. Paul on the Mississippi, Pittsburgh on the Ohio, and towns along the banks of the Kanawha, Tennessee, and Kentucky rivers.

Louisiana's Gulf Bayou Coast

West from the mouth of the Mississippi off Grand Isle is LOOP, the Louisiana Offshore Oil Port, where VLCCs and ULCCs are off-loaded by 48-inch pipeline to shore and into ocean barges. Several deep passes and river mouths enter the Gulf of Mexico. From this nautical haven known as the bayou country come the great fleets of tugs and tug/supply vessels to service the requirements of the U.S. offshore oil industry. In the process of fulfilling the industry's towing requirements a whole new breed and concept of rough-water tugboat has emerged.

Lake Charles

From Lake Charles, Louisiana, Crowley Maritime's Caribbean division operates another of its TMT marine trailer terminals. Five specially designed and converted 100- by 400-foot barges (one by fours in towboat lingo) are loaded stern first by dockside ramp. With their triple decks fully loaded they can carry 288 trailers including tankers or reefers, all for delivery to Puerto Rico; the Dominican Republic; the Virgin, Leeward, or Windward islands; or the U.S. naval base at Guantanamo Bay, Cuba. Bagged rice from the Midwest may go from farm to consignee in the islands without leaving the original truck. Cognizant of the possibilities in ocean towing, the world's major towing lines maintain offices in both New Orleans and Houston.

Texas

Into and out of Sabine, Beaumont, Orange, Galveston, Houston, Corpus Christi, and Brownsville flows a heavy stream of tows. It is on this side of the Gulf that much local crude is refined and later shipped by tug and barge. Liquefied natural gas and toxic chemicals are also safely barged through the Gulf Intracoastal and up the Mississippi system, or coastwise.

Mexico and Atlantic Central America

This region has little barge and tug movement. Pemex, Mexico's national oil complex, has tugs and barges that move their product from Puerto Mexico-Coatzacoalcos to other domestic ports.

Panama

The Panama Canal Company has new powerful tractor tugs to assist the Panamax-size ships through the various locks. They are backed up by the older tugs and those of the dredging division, all of which have been designed and built for the canal. Privately owned tugs have also been hired to supplement the canal tug fleet and to replace those tugs that have outlived their usefulness.

Pacific Mexico

On the coast of Mexico, below the Sea of Cortez, phosphate rock is barged to the mainland using former U.S. Navy seagoing tugs.

California

From the ports of California, all sorts of cargo are barged. The cargoes are as different from those of the East Coast as are the towing modes used. Towage on the West Coast of the United States is mainly by hawser, although the largest dry-cargo, fixed tug-barge unit has been introduced in the sugar trade between Hawaii and California.

There is no intracoastal or extensive inland waterway system on the West Coast. There are few rivers—the Sacramento and San Joaquin, and the Columbia and Snake in Oregon. They are served by push towboats and covered or petroleum barges, which move general cargo upriver and rice or grain downriver.

Coastwise, crude oil is loaded from storage vessels at single anchor loading moorings (SALMs) offshore in the Santa Barbara Channel and barged to refineries in Los Angeles and the Bay Area.

Crowley Maritime Corporation, whose main offices are in San Francisco, tow covered newsprint barges from the Bay Area to Los Angeles.

All manner of cargoes—modules, mobile homes, lumber, trucks, machinery, pipe, and other building materials—go to sea regularly in single or tandem barge tows behind the powerful tugs of Crowley-Red Stack, Foss, or Sause Brothers, the three leading West Coast towing concerns. Such tows may originate in the San Francisco or the Los Angeles area and be destined for the Hawaiian Islands, Guam, Papeete, or Alaska. There is also frequent coastwise towing to Eureka and Crescent City.

Oregon, Coos Bay

Forest products and logs are towed in the huge barges of Sause Brothers to coastwise ports and throughout the Pacific islands. The Oregon coast is rough-water towing country. Each port must be served by tug and barge, which have replaced the steam schooner and small ship. Bars at such places as Newport, Bandon, Tillamook, and the Yaquina River, which have shifting sands, are very tricky. An error in judgment as to the tidal conditions can be lethal, particularly when entering the Columbia.

Columbia River

This river, navigable into the Palouse country of eastern Washington, has a traffic pattern that varies with each pool level. From Astoria to Portland and Vancouver, Washington, this traffic might be compared to that of the lower Mississippi, as some of the world's largest merchant ships mingle with both ocean and coastwise tows, as well as descending river towboats pushing four or five loaded grain barges. Ships are towed up to Portland's huge dry docks for repairs, or into the Willamette for scrapping. Above Portland, there is heavy towing traffic in specially designed covered barges that carry chemicals, steel, cement, and petroleum. These tows proceed up Columbia's gorge and through the various locks to the Snake River and on to Lewiston, Idaho.

A typical Columbia-Snake River tow. (Courtesy Tidewater Barge Lines)

By tug and barge line, Portland exports the unpowered creations of her shipyards, as well as regular sea lifts to Alaska and Hawaii, and in return receives such products as sugar, urea, and newsprint, all in huge seagoing barges, many of which were launched locally.

Puget Sound, British Columbia, and Alaska

Puget Sound is a natural marine gateway to Alaska. Alaska depends solely on shipping for all but a small portion of its requirements, which come by air or road. The modern world of towing has replaced the small

Log tows are still seen in the Pacific Northwest. (Courtesy Joseph Scaylea)

fleet of steamers once serving its many wharves, bays, and roadsteads. Regular scheduled tug-barge service departs from Seattle. Foss Alaska Line, Alaska Hydro-Train, and Arctic Freighters, all scheduled barge lines, operate to southeast Alaska. From the spring to the fall, several transportation companies tow across the rough Gulf of Alaska to the Aleutians and beyond into the shallow entrance of the Bering Sea off Nome and Kotzebue. Crowley Maritime maintains a regular tandem rail-barge tow between Seattle and Alaska Railway's Whittier terminal, and, of course, when the ice weakens off Point Barrow, they send their annual barge convoy to the North Slope. By 1987 the weight per barge of individual modules had risen to 5,000 tons and the points of origin had become diversified. Some of these tandem tows originated in Louisiana.

Canada's Dome Oil has a similar convoy that comes down the MacKenzie River delta to the Beaufort Sea. Shaver Towing of Portland has also entered into this Arctic movement, as have several other smaller Puget Sound towing companies. The barges, most of which are one by fours with a raised bow, are loaded to their extreme capabilities and resemble huge buildings.

Between southeastern Alaska and Olympia, Washington, at the foot of Puget Sound, there are on any normal day a hundred other towing operations in progress.

Northbound from the United States are railroad car-barge tows from Seattle to Port Townsend, Washington, and North Vancouver, British Columbia. Another very prominent towing movement both north- and southbound comprises the many high-sided chip barges or covered newsprint barges running between various British Columbia mills off the Strait of Georgia and to and from the wharves of the various arms of the Fraser River.

Seaspan tugs move tons of wood chips from lumber mills to forest product manufacturers' wharves. (Courtesy Seaspan)

Smit Rotterdam and *Smit London* towing *Glomar Beaufort Sea #1* (CIDS) from Japan to Alaska. (Courtesy Smit Tak)

The forest products industry within British Columbia is highly dependent on towing. The steep-sloped forests of northern Vancouver Island and many along the fjords of the province's north coast are only accessible by water.

Floating logging camps with mechanized equipment are towed to location. Cut logs are either loaded onto enormous self-dumping barges or formed into a multi-unit flat raft. Either way, they are towed south to mills or booming yards. Wood chips are barged from mills in Canada to paper plants there or to mills in the United States. At almost any time, day or night, one of these types of tows is visible on the Strait of Georgia.

With the immense oil exploration and development of natural gas, it is predicted that thousands of tons of equipment, drill rigs, and attendant plants will be towed to workings in the Arctic and the Beaufort Sea by high-tech icebreaking tugs of 11,000 HP. Once this huge field is in production, towing services of a different nature will be required. In the mid-1980s several cement island drilling systems (CIDS) had been towed from the builder's yard in Japan to locations in the Beaufort Sea. The *Molkpaq* went to the Herschel Island area, and the *Glomar Beaufort Sea* (59,000 tons at 32 feet) required two Smit International tugs totaling 48,000 HP and two Bugsier tugs totaling 40,000 HP with the help of an icebreaker to arrive at the set-down sites.

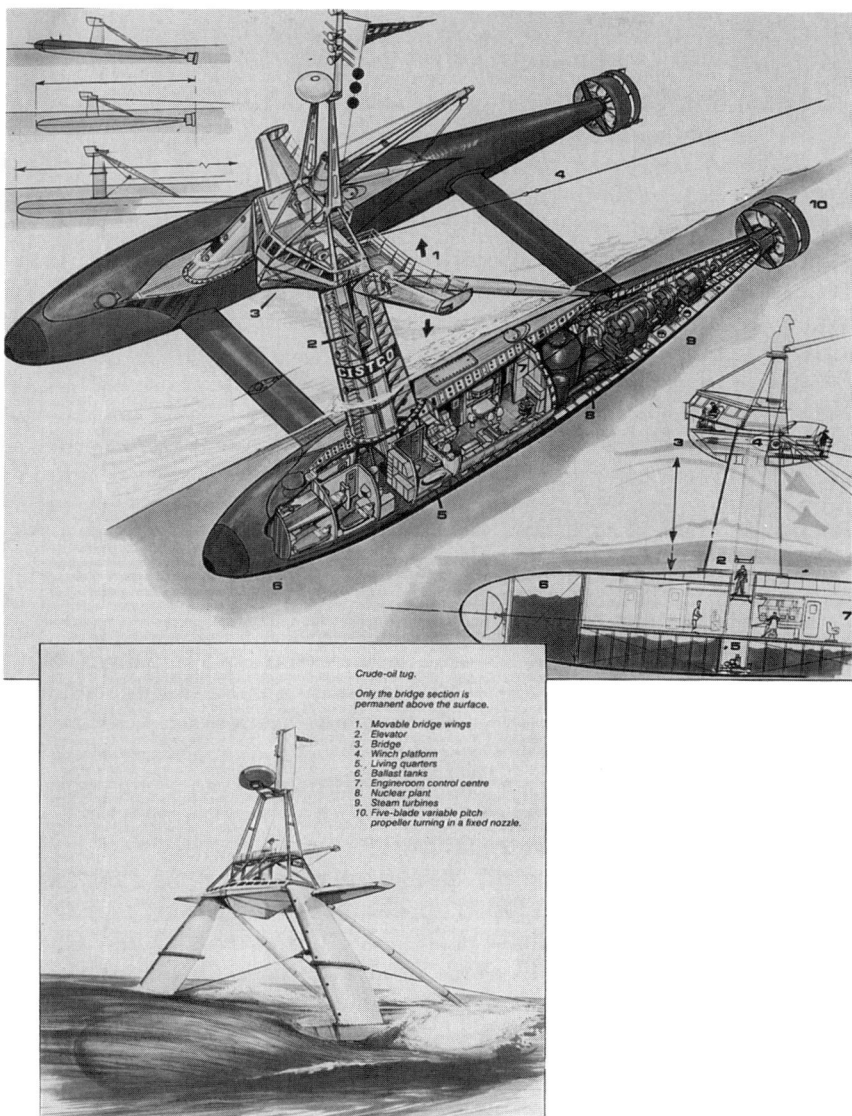

The futuristic offshore tug of the twenty-first century, as presented by Mr. R. W. Scheffer, President of Smit-International. This ocean tug would be driven by steam turbines turning variable-pitch propellers in fixed nozzles. It has a unique hawser arrangement as well as various attitudes relating to towing and desired draft (see upper left). (Courtesy Smit Tak)

DEEP-SEA TOWING WORLDWIDE

This phase of towing has changed somewhat in the 1980s. Instead of dredges, barges, and naval vessels going for scrap, the principal and most lucrative sea tows are drill rigs. They are moved by the now-elephantine tugs of Smit, Bugsier, ITC, and Les Abeilles. Distances run from a mere 1,800 to over 16,000 nautical miles. Their navigation bridges contain the most sophisticated high-tech position and communication devices. They require a whole new breed of technically trained seafarers, backed up by an equally knowledgeable operating management ashore.

In other areas towing is replacing or competing with other types of shipping. This has occurred notably in transgulf and East Coast movement of petroleum, as well as container cargo between the Southeast and Gulf to the West Indies, from the Pacific Northwest to Alaska, including railroad cars, and from the West Coast to Hawaii and the Pacific islands. Barge traffic is replacing interisland shipping throughout the Hawaiian chain. Transocean towage has been introduced through the use of the tug-barge fixed complex.

Interport container feeder barge operations are a new towing success. This has been proven in operations between Port Elizabeth, New Jersey, and Boston, as well as in the mid-Atlantic ports of Philadelphia, Baltimore, and Norfolk. These operations have been extended internationally between Seattle and Vancouver, B.C. Matson Steamship Line tows feeder container barges between Honolulu and the various Hawaiian Islands. Other towing companies may soon enter this trade between the mainland and the islands.

Just as many of the present so-called modern tugs have no resemblance to the tugs of the 1920s in appearance, horsepower, propulsion equipment, or electronics, those of the twenty-first century may reveal entirely new towing concepts. Some, which are now a little science-fictional in appearance, might develop into more realistic models by 2055.

2. Towing Nomenclature

A COMMON expression used by seafarers throughout the world is "different ships, different long splices." It refers to the difference in names of the same nautical equipment in different areas. This applies to towage equipment as well. To clarify this usage within the text, some of the most frequently mentioned items are offered and illustrated.

In many places, too, different words have the same meaning. Some examples are: vessel or craft equals tug or towboat. Tug master, captain, mate, skipper, pilot means the person on watch or handling the tug as a member of its crew. Tow wire, tow cable, or hawser signifies the line used for pulling over the stern. Tow winch or towing machine is the main afterdeck winch upon which the tow wire is reeled. Wheel means propeller. Other terms are noted in the following glossary.

Alongside. This term refers to either a tug made up to a unit to be towed or two vessels lying side by side. Also known as "on the hip."
Amidships. Refers to position of rudder as being fore and aft in line with the tug's keel and centerline; also to the middle area of a tug or vessel.
Anchor shackle. Used between chain and anchor and all towing connections.
Athwartship. Refers to anything constructed, lying, or running at right angles to a vessel's fore and aft line.
Automatic gear. An arrangement on a towing winch that allows the tow cable to pay out and heave in automatically when set at certain tensions.
Bar. That section at the entrance to ports which is more shallow than the ocean or waterway and must be carefully crossed.
Bar-tight. Solid and rigid as a steel bar.
Beaufort scale. A numerical and descriptive scale of wind velocities and sea conditions.
Beckets. Rope or wooden appliances used to hold hand steering wheels on small tugs.
Bergy bits. Small pieces of hard ice that have broken off from icebergs.

Anchor shackle

Forward or double bow bitts

The triple bow bitts installed on many large modern tugs

Side or quarter bitts

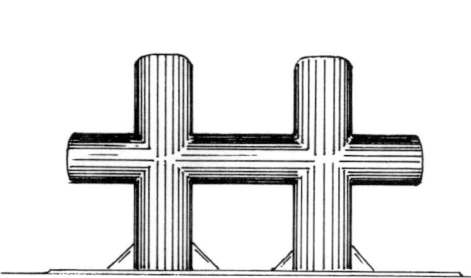

Stern, main, or H bitts

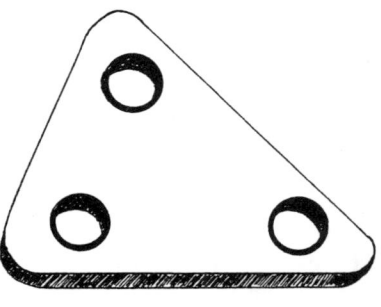

Flounder plate

Bilge keel. Finlike projections on either side of a tug's hull to decrease rolling.

Bitter end. The absolute end of an unspliced line or wire cable, or the part fastened to any object.

Bitts. Forward bitts. Any single, double, or triple arrangement for the fastening of lines on the tug's bow. *Side bitts or quarter bitts.* Bitts built into bulwark rails on either side of the tug about one-quarter of the way from the stem, and approximately even with the after H bitts or towing winch. These are used in mooring and towing alongside. (Shoulder post is British terminology.) *H bitts.* The heavy double posts joined by a cavel, forming a letter H. These are located at the tug's stern and are used for securing any heavy lines.

Bridle. A Y-shaped arrangement of line, wire, or chain. Each part or leg must be of equal length. They are joined together at a fish plate or shackle, which in turn is connected to the pendant.

Bridle leg. A single part or leg of a bridle.

Bullnose. The closed chock at the stem of a tug or ship.

Bulwark. The raised portion of the tug's hull running from the stem to the stern on either side, usually used in plural form.

Captain. The person designated as or acting as the watch officer in charge and maneuvering the tug at any particular time.

Catenary. The dip in a line or tow cable when strung between two objects such as tug and tow.

Chafing gear. Any material used to prevent wear on lines or towing cables.

Chain shackle. A metal fitting, commonly U-shaped, used to connect chain to anchor or other chain.

Chine. The point in a tug's hull where the straight side turns inward below the water. The usual location of the bilge keel.

Chock. A heavy metal casting fitted on the sides of a deck or on the bow and stern of a ship through which lines pass; may be open or closed (see sketches).

CIDS. Concrete island drill system of platforms built for and towed to arctic waters, where they rest on the bottom.

Cleat or clevis. The common deck fitting around which lines are fastened or eye splices are passed over.

Coastwise. This term refers to any area within 20 miles of the shoreline and, in towing, routes that stay within such an area.

Connecting up. The act of connecting the tow cable and other necessary parts to a unit to be towed, whether accomplished at sea or in sheltered waters.

Counter. That portion of any vessel with a rounded spoonlike stern.

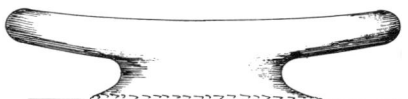
Cleat or clevis

Closed chock   Open chock

Chain shackle

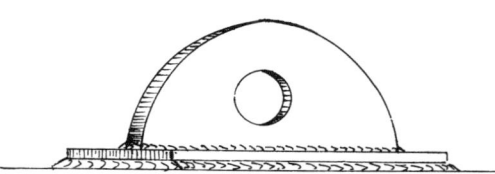
Pad eye

Rope stopper

Dead-in-the-water. Referring to a tug or tow with no way on.

Dead ship. A ship that has no power or is not using its propeller during towage.

Dip. This is an expression referring to passing one line under another.

Down-by-the-head. When any vessel's draft is greater forward than aft.

Drag. The difference in the amount of a vessel's draft when it is greater aft than forward.

Fair current. When tug and tow are favored by a current running in the direction they are proceeding.

FGMDSS. Future Global Maritime Distress and Safety System, which will be operational in 1990 after development by IMO.

Fid. A tapered piece of hardwood used in splicing cordage.

Fish plate. See Flounder plate.

Flounder plate. A heavy triangular steel plate with a bushed hole in each corner that accepts the shackle pin from each leg of a towing bridle and that of the towing pendant. It is also known as a fish plate, union plate, or spider.

Free-surface effect. Liquid in a partially filled tank or vessel that can move freely and without restriction, causing the vessel to list or roll.

Gear. Refers to that group of articles used in making up a tow.

Getting way off. Slowing a tug or tow to stop.

Getting way on. Moving a tug or tow ahead after being stopped.

Girded. A situation in which a tug capsizes due to the strain from a line at an angle of 90 degrees or more, which was not released. Girding is also incorrectly referred to as being caught in irons.

GM. This is a measure of stability related to the tug's metacentric height. It is a movable spot because of the tug's draft, loading, and weight distribution, and is closely related to the metacenter. Its position indicates the stability or lack of same and it may be figured from tables established for this purpose.

Gobrope. A line used on many European tugs to regulate the lead of a short towline. One end is usually fastened on deck and the other led through a shackle or closed fairlead to a gypsy so that its tension can be adjusted. Occasionally, it refers to the hold-down arrangement of towing hawsers on ocean tugs.

Ground tackle. The anchor(s) and chain.

Gunwale or gunnel. The sides of a skiff or dory; sometimes used in referring to bulwarks on a tug.

Gypsy. A small spool that is often an appendage of a towing winch, or that can be individually mounted vertically.

Hard over. Full rudder in either direction.

Hawser. Any line used in astern towing.

Hawser board. A wooden, steel, or aluminum plate or beam placed under the tug's hawser to prevent chafing on the stern rail when towing at sea.

Head-and-tail. When two vessels are side by side with their bows and sterns in opposite directions.

Head current. When tug or tow are heading into the current. Also known as bucking the tide.

Heave in. To pull by hand or power any line, wire, or cable.

Heave to. To hold a vessel or tow heading into the wind and sea at very slow speed and still maintain control while minimizing the effect during rough weather or during other conditions that may prevent progress on a desired course.

Heaving line. A light line that is thrown from a vessel, usually to allow a larger line to be hauled over to a ship, dock, or tug.

"Hook her up!" Used by captains and pilots as an order to go full speed.

Hooking up. An expression used in referring to connecting up to a tow.

IALA. International Association of Lighthouse Authorities.

IMCO. Inter-Governmental Maritime Consultative Organization (replaced by IMO).

IMO. International Maritime Organization.

INMARSAT. The International Maritime Satellite Organization system, made up of a consortium of 40 maritime nations, oversees a global network of geostationary satellites and coastal earth stations. It provides coverage of the earth's navigational waters through 22 earth stations.

Kort nozzle. A cylindrical ring encompassing the propeller(s) on many tugs. Tugs with fixed propeller(s) have the rudder directly behind the nozzle. Rudderpropeller tugs have the nozzle arranged to turn 360 degrees. Nozzles increase propulsion efficiency up to 30 percent as the water leaves the propeller(s) as a jet stream.

Lash up. A slang expression referring to the manner in which tugs or other vessels are secured to each other.

LCD. Liquid crystal display, used on electronic equipment.

LED. Light-emitting diode used to eliminate dials on electronic equipment.

Left-hand turn(s). The coiling or placing a line around a capstan or gypsy in an anticlockwise direction, which is usually against the lay of the line and will cause kinking.

Let go. To release and take in lines from a tug or towed unit as directed.

Light tug. A tug or towboat running without a tow, also known as running lightheaded.

List. Refers to a condition in which a vessel is deeper on one side than the other due to loading, wind, or icing.

Log. This is the tug's official record. It may be written "in the rough" when on watch or may be "smooth," as prepared from the rough log.

Making up. The placing and securing of lines from a tug or towboat in order to tow.

Master. That officer whose name is listed on the vessel's enrollment and license or registry as being the master.

Metacentric height. The center of gravity of the vessel.

Molly gogger. A line, chain, or short wire connected to a shackle on the towing cable that restricts its lead and arc at the stern.

Monkey fist. A heavy knot with a weight placed on the end of a heaving line. It is made of three interwoven strands of cordage.

Mud balls. An upwelling of clay that does not break the surface. These occur frequently in the various passes of the Mississippi River Delta.

Multipurpose tug. A tug built for all types of general towing rather than specifically to fit into a barge notch or for any other special type of towing.

Norman pin. Round steel pins that are operated hydraulically at the stern of large ocean tugs or inserted by hand into the rail on either quarter to keep the hawser leading to the tow winch and clear of the propeller(s).

One by four. A standard U.S. Pacific Coast barge used for deck or liquid cargo. Its dimensions are 100 feet in beam by 400 feet in length.

Out of shape. When a tug, towboat, or tow is in a very awkward position in attempting to land or pick up a tow, or to make a maneuver as planned.

Pad eye. A heavy deck fitting welded where needed. It usually receives a shackle attached to a chain, wire, or block.

Panamax. The maximum size of ships allowed to transit the Panama Canal: length 948 feet, beam 106 feet, draft 34.8 feet.

Parted. Any line or cable that has broken.

Pay out. To let run or slack out from the vessel any line or hawser.

Pelican hook. A fitting with a long bill that when closed is held by a slip ring. These are used to hold a chain link or eye of a wire, which may be quickly released by knocking a slip ring off the bill of the hook.

Pendant. A short length of wire or chain such as, but not limited to, that between the bridles and the tow hawser; often pronounced "pennant" aboard tugs.

Pennant. See Pendant

Pigtail. A term applied to the chain pendant, which is connected to the bridles, used on tugs on the U.S. West Coast.

Pilot. Bar pilot. A general term used in the U.S. Gulf and West Coast referring to a pilot who takes a vessel into the entrance of a harbor from the sea. He may also have to dock the vessel, anchor it, or turn it over to another pilot for transit up a river or bay. *Docking pilot.* A person, either independent or a member of a tug crew, who goes aboard another vessel or ship to take charge in the docking operation. His legal status is that of an independent contractor. *Harbor pilot.* A pilot used in a harbor for shifting berths or taking the vessel over from the bar pilot. *Riding pilot.* An independent operator, frequently a tug captain, who rides ships or tows being assisted by tugs. *River pilot.* Usually a pilot who takes over from the bar pilot in long rivers, such as the Mississippi or Columbia. *Sea pilot.* The pilot who takes a vessel or tow outside of jetties and over the bar when bound for sea. *Tug pilot.* A watch officer on a tug or towboat.

Put out. The act of paying out.

Rake. The angular underbody at the bow and stern of square-ended barges.

Rig. A drill rig is a floating structure used to drill for oil. It is usually supported by legs that are either semisubmersible or rest on pads on the ocean floor.

Rope. A term interchangeable with cordage; refers to soft lines and tow cables or hawsers.

Rope stopper. Made of soft line and used to temporarily hold larger lines when shifting from bitts to capstan, etc.

Rouse out. To haul or to get out for use something that is stowed such as spare hawsers, push cables, etc.

Safety shackle. Always used as a prudent connector in all towing arrangements.

Screw pin shackle. A handy type of shackle for onboard use.

Ship assist. A tug used for docking, undocking, or otherwise assisting a ship in a harbor.

Shock line. A piece of nylon or similar line of 10 to 12 inches or more in diameter and from 200 to 250 feet long, used between barge pendant and tug's hawser. Its elasticity absorbs some of the shock of the sea when towing. Used mainly in U.S. Gulf towing.

Shore gang. Refers to workers who service tugs and their fittings at a tug company's tie-up yard and pier.

Short stay. When the anchor is hove in and the cable out is a little more than the depth of water. Also known as straight up and down.

TOWING NOMENCLATURE

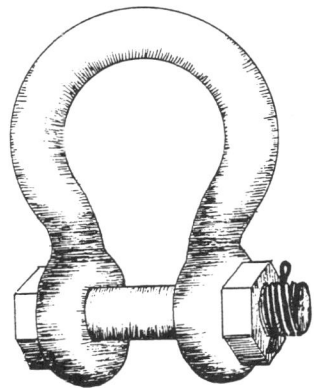

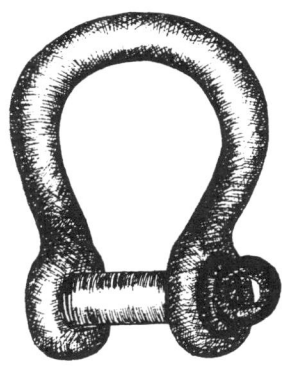

Safety shackles have a short pin. Most are long enough to receive two nuts. Some suppliers will furnish a pin with an unthreaded end. When both nuts are used and a cotter key is placed (as it always should be), the lack of exposed threads prevents thread damage and jamming of the nuts when trying to disconnect quickly from a tow.

Screw pin shackle

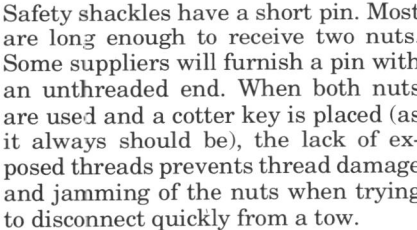

Soft or synthetic fiber rope

Steel tow hawser

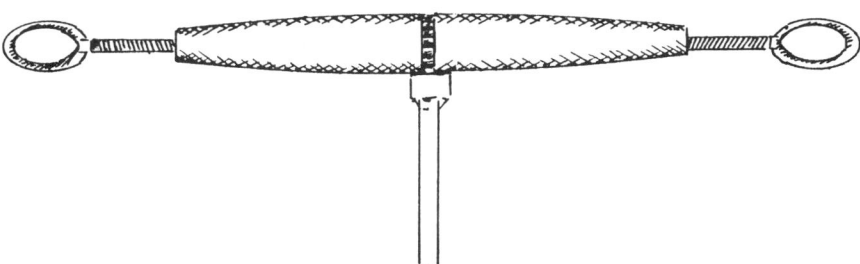

Steamboat ratchet

Shot. Refers to a length of anchor chain that is usually 15 fathoms.

Skeg. These are fixed rudder appendages, usually at a slight angle under the rake end of barges. On a tug, a skeg is a flat extension from the keel out under the propeller into which the heel pin of the rudder may fit. This applies only to single-screw tugs.

Snotter. A short piece of chain, wire, or rope used between the hold-down point and the tow hawser to restrict its movement near the stern of tugs.

Soft line. Manila, hemp, sisal, all the polyesters, nylon, or other synthetic fiber blends of cordage.

Soft or synthetic fiber rope. Rope that is usually of three strands, spliceable, and easy to handle when used as deck lines and ship towlines.

Spider. See Flounder plate.

Spuds. The long, round or square poles used as legs that reach to the bottom and hold dredges, drill boats, and other work craft in place. Such poles run through a well in the vessel's hull and may be raised or lowered mechanically.

Standing part. The middle section of a rope.

Steamboat ratchet. Used on river towboats and other types to tighten barge connectors and the push cables on some tugs.

Steel tow hawser. 6×37 or 6×41 means 6 strands with 37 or 41 wires per strand.

Stick out. Usually refers to paying out more line or hawser; also to the original connecting of the hawser.

Stop off. To tie tightly but lightly so that an object can be easily released.

Straight up and down. When the anchor cable is vertical from hawsepipe to water.

Stream. To put out and slack out a hawser to a towed unit.

Surge gear. Name used in Pacific Northwest towing for the chain bridle and pendant. Every foot of chain weighs 150 pounds.

Swedish wire rope. A retrieving wire and fiber line, strong and flexible, used on U.S. West Coast barges between the tow pendant and bridles and the barge (also called spring lay wire rope).

Tandem. More than one tug towing side by side, or one ahead of the other, or more than one unit in a single tow.

Tier. Barges grouped in a single line. There may be up to five tiers in a single river tow.

Tow hook. A simple tow hook with remote release as used on European harbor tugs.

Towing master. A person with considerable expertise, other than the tug captain, who has been placed in charge of a tug or group of tugs, usually handling a very difficult tow.

TOWING NOMENCLATURE 39

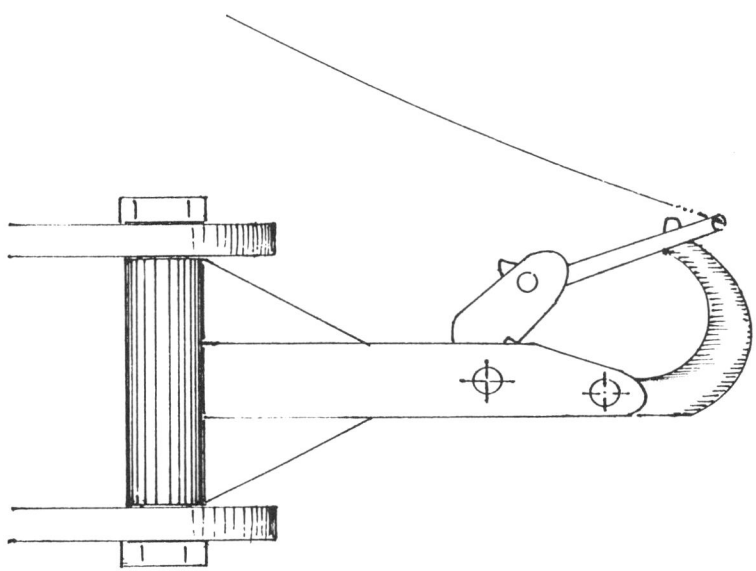

Tow hook

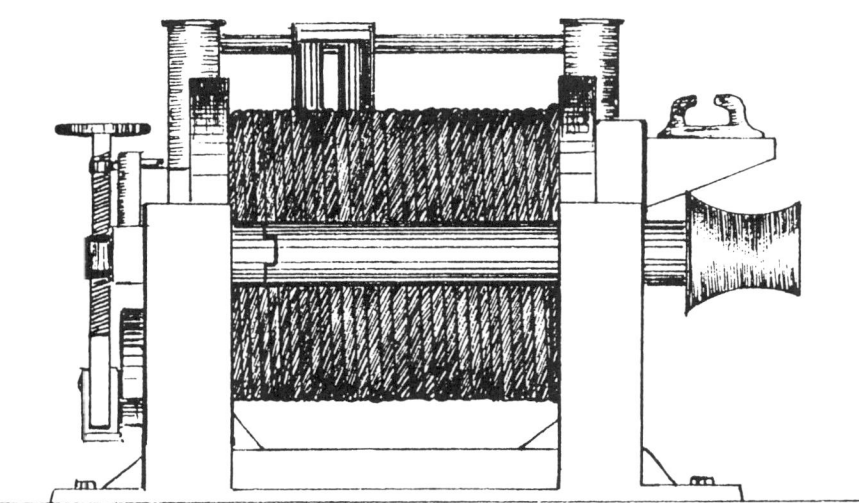

Tow winch

Tow winch. A single-drum tow winch with vertical automatic cable threader or guide. The open chock on the right gives a good lead for lines running to the gypsyhead. A hand brake is shown on the left. Power may be electric, hydraulic, or diesel.

Tractor tug. A general reference to any tug fitted with a cycloidal or rudderpropeller propulsion system. *See* Water tractor.

TSS. Traffic separation system.

Under run. To place a shackle, hook, and line under a cable and back the tug so as to clear or run toward an anchor or other object.

Vessel. Used in this text interchangeably with tug, towboat, or ship.

VHF. Very high frequency type of radiotelephone.

VTS. Vessel traffic services (or systems).

Warping drum. See Gypsy.

Water tractor. A special type of tug built with cycloidal or rudder-propeller power. The propulsion may be located near the middle of the vessel or aft. These tugs can tow from either end.

Way. Movement through the water.

Whip. To bind the end of a piece of line so it will not unravel.

Wildcat. That portion of an anchor winch into which the chain links fit so that the chain may be held for heaving in or paying out.

Williwaw. A term borrowed from the Cape Horn area to describe sudden gusts of wind that occur in narrow, fjordlike passages.

3. Modern Tug Types

FOR years, in the United States, there have been two basic types of towing vessels, tugs and towboats. The term *towboats* refers to those flat-bottomed powerhouses of various size that push their tows of from several to dozens of barges on the world's inland waterways. Of the two basic tug types, there are those that stay within the harbor, or venture out on short coastwise voyages, and those classed as seagoing, not necessarily by a classification bureau—more often by their owners or operators. Frequently they are one and the same tug as used in general harbor towing. In Canada, the United Kingdom, and northern Europe a third type was developed: the deep-sea salvage tug.

As ship dimensions, tonnage, and shore moorage facilities have increased, a whole new set of modern tug types has been required. Many have been designed for specialized towing assignments. The petroleum industry alone, dissatisfied with what was available, has introduced the radically different, high-bow, light-hull, low-tonnage tug/supply anchor-handling vessel. In many cases the modern tug differs greatly in appearance from those of the past.

HARBOR TUGS—SHIP-ASSIST TUGS

In many ports the appearance of ship-assist (docking and undocking) tugs remains the same. The tractor tug is of course a great departure from the past in appearance. It is a radical departure in shape, operation, handling, and horsepower. Harbor tugs have always been ruggedly built. Usually during their lifespan the size and deadweight of ships and other floating tonnage has increased tremendously. As a result, in some U.S. ports tugs have been rebuilt and re-engined with an increase in power. In other ports, new vessels have been built. Harbor tugs in the U.S. ports of New York, Providence, Savannah, and Houston are good examples of the modernization of the conventional harbor tug.

Later designs for original harbor tugs throughout the world offered broad squat hulls and wide decks, with strong inward-sloping bulwarks. Their hulls were heavily fendered with rubber-inserted guardrails completely down both sides and around the stern, with extra tiers at the stem. The deckhouses are set back.

Some ports on the U.S. East Coast are served by tugs that have been somewhat downgraded by their owners, most of whom have fleets in more than one port. Such an advantage allows the shifting of tugs from port to port as the need arises.

Above, Seminole Chief is a Mississippi river towboat capable of handling up to 50 "loads" in front of her. This type of vessel is from 140 to 160 feet in length and draws up to 9 feet with HP ranging from 4,200 to over 9,000. (Courtesy Jeffboat Incorporated) *Below, Sun XVIII* is a London river and harbor tug. She was the first one in that area to have pilothouse control. (Courtesy Sun Tugs Limited)

Kuanza and sister are recently Dutch-built harbor tugs used in Luanda, Angola. (Courtesy Damen Shipyards)

Bentol is a Stan 3300 class Damen tug used for short sea and ship-assist service. (Courtesy Damen)

Smit's contribution to the modern ocean towing industry is self-contained for up to three months. (Courtesy Smit Tak)

A very modern 216-foot diesel-electric supply vessel. (Courtesy Halter Marine Incorporated)

Otto Candies is a Halter-built (1985) 140-foot twin Niigata Z-peller unique tug. In addition to her two rotatable nozzled Z-pellers, she has a standard centerline propeller. As the first of "two highly innovative tugs—which can go anywhere and do anything," she has handled rigs and their anchors as well as being fitted into a deep-notched barge. (Courtesy Halter Marine)

Mathiason steam tugs undock Bull Line C-2 *Frances* from 57th Street, Brooklyn, circa 1960.

A U.S.-built Z-peller tractor tug in San Francisco. (Courtesy Steven Lang—Tug Photos & Research)

Two powerful Suderman and Young tugs turn a Keystone tanker in the Houston Ship Canal. (Courtesy Suderman & Young Towing)

TRACTOR TUGS

While this strange-looking tug has been in use for years in European harbors, it is now competing with and in some areas replacing what has been considered the standard tug. During the late 1980s, twin-screw rudderpropeller tugs were ordered for ports in Taiwan, Turkey, England, South Africa, and Belgium. There are several types of tractor propulsion. The most frequently used seems to be the Schottel rudderpropeller type. It has been copied by other designers but differs from the cycloidal Voith-Schneider propeller. Without prejudice toward other manufacturers the author believes that the operation of the Schottel propeller is fascinating and partially applicable in the handling of many tractor tugs.

GULF OF MEXICO TUGS

In the Gulf of Mexico ports, there are both the new/old, the well-rebuilt and re-engined hulls of once-proud northern steam tugs, and the ultra-modern spin-offs from the neighboring petroleum industry.

ASSIST TUGS AT OFFSHORE TERMINALS

Tugs assisting ULCCs and VLCCs* are usually built more heavily than standard harbor tugs. Many have glass in the eyebrow that rims the pilothouse, allowing the captain to see upward toward the vessel he is working. Fire fighting is often one of their added features.

COASTWISE TUGS

With the advent of diesel propulsion, tugs of all sizes have been entered into this trade when the demand was high and competition was anyone's challenge. Some were shamefully underpowered and their seaworthiness questionable. Their crews' courage was indomitable.

Economics has largely eradicated their presence. Profitable coastwise towage consists of moving the greatest amount of product in a single hull. As the size of barges increased, the size of coastwise tugs remained within the 110- to 140-foot range. Horsepower, which had hovered at under 1,000, rose to 3,500 in the 1960s. By the 1980s, many were over 9,000 HP.

With the development of the notched-stern barge, the raised forecastle head was introduced on some tugs.

This trend led to higher-sided, deep-notch barges and to the ultimate—the double or upper pilothouse coastwise tug.

---

*A VLCC is a very large crude carrier (125,000 to 200,000 gross tons); a ULCC is an ultra-large crude carrier (over 200,000 gross tons).

Quebec harbor tug exercises her fire monitor. (Courtesy Quebec Tugs/Herbert P. Bertrand)

This tug, known as an Artubar type, was built on the Great Lakes for use in the notch of five-deck trailer barges in the Florida-to-Central America trade. (Courtesy Marinette Corporation)

MODERN TUG TYPES 49

Compare this old U.S. coastwise steam tug with the modern giant, *Patriarch,* which follows. (Courtesy S. Lang)

Crowley's *Patriarch* is one of this fleet of nearly 50 ocean towing tugs, shown here with two towing cables out. (Courtesy Crowley)

Ellena Hicks is typical of the second generation high-bow tug for use in moderately notched barges. (Courtesy Watkins Photographers)

Valiant, a U.S. coastwise hawser and deep-notch tug. (Courtesy Bulkfleet Marine Corporation)

MODERN TUG TYPES 51

Turmoil is famous for her rescue and salvage duties during and after World War II. These tugs were built in Great Britain and are no longer in service.

OCEAN TUGS

The term *ocean tug* should apply only to those specifically built for this trade and not to some well-founded general harbor and coastwise tugs that have managed by some method to be awarded a Maltese Cross for ocean towing by a classification society.

The real ocean tug is with some exceptions too cumbersome for daily harbor use. It has been used as a puller for bunched tows of barges on rivers, but its real place has been offshore with the hawser out.

European towage companies have for years led the world with their ocean tugs. Prior to World War I, the United States had a few noteworthy ocean tugs, such as the great *Paul Jones* and *Cuba.* During that war, the U.S. Shipping Board, through the Emergency Fleet Corporation, produced a group of 145-foot, 900-HP, steel-hulled steam tugs of about 450 gross tons. Primarily for coastwise use, they supplemented the large railroad company tugs that hauled tandem coal barge tows along the East Coast.

In the meantime, from Europe, the great Netherlands fleets of L. Smit, led by *Zwarte Zee,* and N.V. Bureau Wijsmuller's competing ocean tugs were challenged by those of Germany, Denmark, and the French

This was the U.S. standard ocean tug built in the pre–World War I era. Horsepower averaged 800. (Courtesy S. Lang)

Between the 1930s and 1970s the Dutch developed a fleet of tugs similar to *Noordzee*. (Courtesy C. V. Hidde—L. Smit)

Les Abeilles tugs. All were powered by economical European type slow-speed oil engines (diesel).

Neither Great Britain nor the United States took up the challenge until, forced by the necessity of World War II, England introduced the 4,000-HP Bustler class ocean rescue tug and the U.S. Navy its steel 145-foot, 1,500-HP diesel electric ATA class, and wooden 165-foot, 1,500-HP steam ATRs; and the War Shipping Administration their 2,500-HP V-4 class. After the war the commercial use of the steam ATRs and the V-4s soon proved uneconomical. To overcome this, however, their power plants could have been updated.

Today, the towing companies of northern Europe continue as leaders with their 22,000-HP ocean towing vessels such as *Smit Singapore,* Bugsier *Titan* and *Wotan,* and others. These great sea tugs are 250 feet LOA with beams of 48½ feet and drafts of up to 23 feet. Many are powered by Stork-Pielstick diesels with twin propellers of four blades or controllable-pitch propellers. South Africa also has two huge 26,000-HP ocean tugs, *John Ross* and *Wolraad Woltemaad.*

SALVAGE AND RESCUE TUGS

Many of the ocean tugs described above are fitted for salvage purposes, but in today's shipping world, they are confronted with very great financial risks. An operator of big deep-sea tugs that are built for 100 percent towage and salvage, not the so-called multipurpose vessels that proliferated with the upsurge in offshore oil exploration activity, is faced with a daily conflict: Should he keep a vessel on station for relatively long periods waiting for a salvage job? Many are placed in a generally poorer-paying commercial towage environment, which for the high powered aristocrat of the seas is a loss as a consequence of the oversubscription of multipurpose tug capacity.

PETROLEUM INDUSTRY TUG/SUPPLY VESSELS

Due to the unique requirements of the oil industry in moving, mooring, and tending the huge oil rigs, a special class of tug/supply vessel with anchor-handling capabilities has been developed. These are an outgrowth of the original smaller supply vessels. All have a raised forecastle head bow with all superstructures well forward, leaving a large, long, open stern.

The latest in tug/supply vessels are fitted with tow-winch stern rollers and twin screws. Frequently, they are fitted with bow and stern thrusters. Their size runs to 225 feet and horsepower to 12,000 or over. In addition to their towing and anchor-handling facilities, they have tanks and stowage capacity to supply drill rigs with potable water, fuel, drill mud, pipe, and provisions.

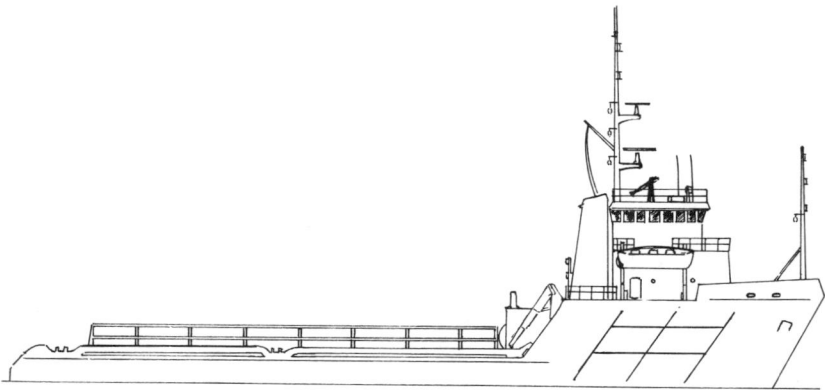

The standard U.S.-type oil industry tug/supply vessel

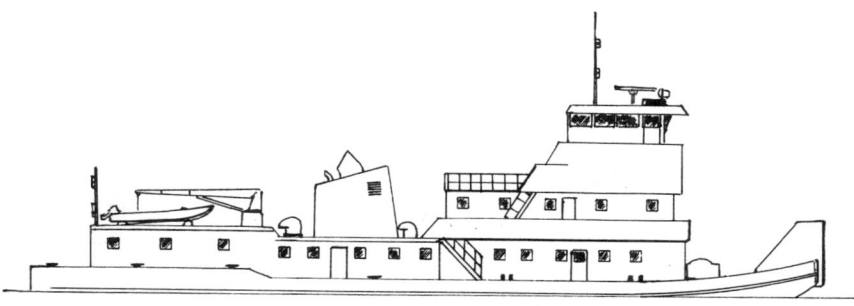

A Mississippi River line haul towboat

INLAND RIVER TOWBOATS

The term *towboat* is the preferred usage over that of *tug* when referring to the flat-hulled and large-tiered deckhouse vessels that push their tows ahead of them. Most are of a design that is an outgrowth of those used on the U.S. Mississippi-Ohio River system.

Outside of improved steering, underwater hull design, and the central console operation of deck winches, searchlight, and electronics, the basic hull and towing qualities of the vessels have changed little. While limited by draft, their horsepower and capabilities have increased so much that the pilot sitting between the rudder steering sticks of one of these vessels has the control of a floating city of thousands of tons ahead of him. This is on the so-called line haul towboat.

With a few exceptions, all are diesel powered in the 6,000- to 9,000-HP range. Their hull configuration is usually a flat underwater body that may have a slight tunnel or cathedral effect. Near the after end of the engine room, the hull curves upward to allow for up to three shafts and propellers. Behind each propeller is one main rudder, long and not any deeper than the vessel's bottom draft. Ahead and to the side of each propeller are slightly smaller flanking rudders.

In conjunction with the main rudders these flanking rudders allow some fantastic maneuvering. The Mississippi River is full of bends, a few of the hairpin variety. To negotiate these, the pilot of upbound large loaded tows will slow, and hold the tow's head where he wants it, while flanking the whole tow sideways, then slowly inching ahead until clear. Downbound, the towboat pilot will back hard holding the tow in the best water, with room to slide sideways into the bend while floating his tow around a turn until ready to pour on the power. It takes years of experience and river knowledge to become a master/pilot on the Mississippi River system.

River pilots have made many adaptations of pilothouse equipment. For years, searchlights required the use of one or more hands to turn them. Now, they may be directed electrically by fingertip touch. If the tow is caught by fog or heavy rain at night, the pilot can tell through a swing meter what the head of his tow is doing a quarter of a mile ahead. The swing meter is a gyrocompass unit temporarily placed near the center of the head tier of barges with a repeater in the tug's pilothouse. When these tows are downbound and passing around a bend in low visibility, drift buoys are streamed on a short line off the outside corners of the last barges in tow. The way they lead reveals to the pilot whether his tow is turning, flanking, or steady on a course.

Many integrated tows of three or more units in a single tier have used a thruster in the head barge to assist in negotiating bends and locks. Some thrusters are built into the barge; others are fastened on deck and can be raised and folded up like an outboard motor when not in use.

At various river ports there are tie-up fleets. In some cases, the line haul tug lands her tow and drops off barges. At other points, there are small shifting towboats that will come out and pick off barges or add some to a tow.

There are thousands of smaller push towboats owned by individuals and towing companies of the U.S. Gulf Coast that push from one to three or four petroleum or cargo barges on the Gulf Intracoastal Waterway and up the Mississippi.

TOWBOATS ON CANADIAN INLAND WATERS AND LAKES

As a natural spin-off from these towboats, a large and heavy modified version appeared in Lake Athabasca, Great Bear and Great Slave lakes,

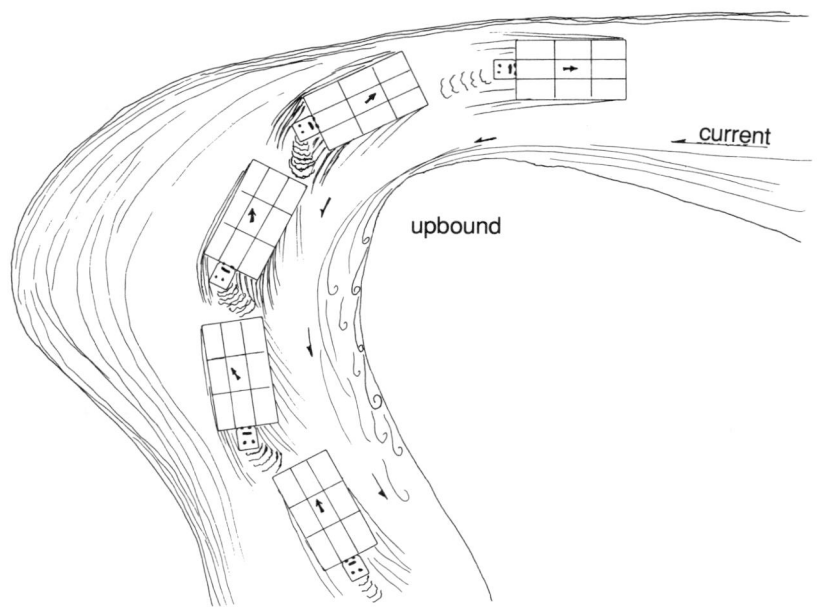

Course and rudder action used by pilots rounding bends upbound on Mississippi

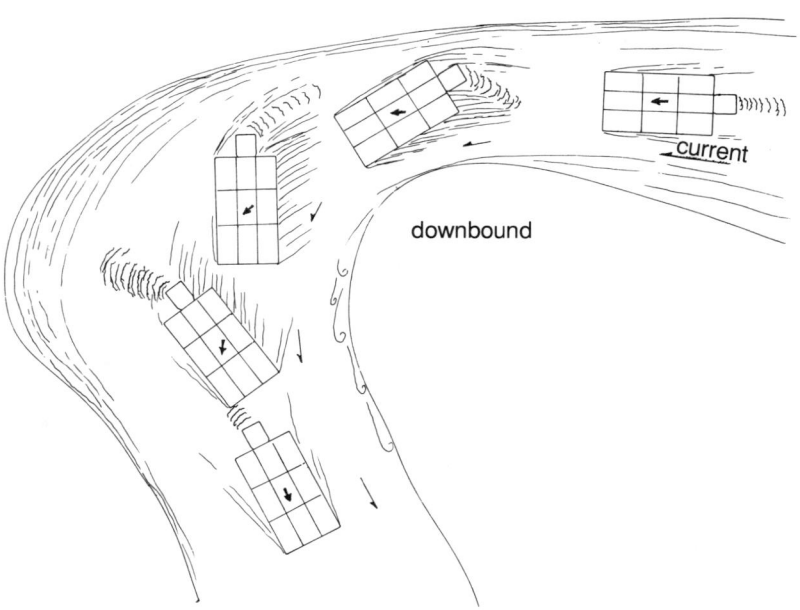

Steering and flanking a downbound Mississippi River tow

and the Mackenzie River of Canada's Northwest Territories. They can push barges on the river and across the lakes, weather permitting. They are also fitted with a long stern for hawser towing. Their bows are blunt and raised and heavily reinforced with towing knees.

Farther into the Arctic, along the U.S./Canadian North Slope and Beaufort Sea, there are specially reinforced steel-hulled pushers with knees, which take barges from ocean tows and shove them through the shallow water to a landing hard.

TOWBOATS ON THE COLUMBIA RIVER SYSTEM

Again, the Mississippi River influence in towboats is seen in the commerce on the Columbia, Willamette, and Snake River system, which extends to Lewiston, Idaho. Due to the fact that the barges used are so high and are over 100 feet in length, the towboat pilothouse is perched on steel stiltlike legs extending about three decks above the top of the main cabin.

TUGS IN THE PANAMA CANAL ZONE AND U.S. WEST COAST PORTS

In the past, most tugs were purchased from the East Coast. Presently, they are locally designed and some are built on the West Coast. Recently, these too have been challenged by the tractor tug. Owners from Los Angeles to Vancouver, British Columbia, have added one or more tractor tugs to their fleets. This is a result of the successes of the same type of tractor tug that was introduced in Europe some years ago.

Gulfport Shipbuilding's yard tug G-15 stands by to assist in picking up the barge *Barbara Vaught* as it is launched. (Courtesy Gulfport Shipbuilding & Drydock Company)

Willamette's cycloidal drive *America* works in the Portland, Oregon, harbor and the Columbia River as well as outside on the Pacific Coast. (Courtesy Reidel International—Ackroyd)

There are tugs built for shallow work, such as on the rivers of Alaska where in many places the bottom is bare at low tide. These tugs have flat bottoms and double or twin skegs, spread far enough apart so that the tug can remain upright and will not roll on its side when aground if the bottom is fairly level.

POWER AND THE HARBOR TUG

Horsepower in harbor tugs varies greatly due to location, size, and types of ships handled, and the tidal current. A 1,000-foot loaded Great Lakes ore ship can be turned in the calm waters of a Great Lakes harbor by a tug having 1,000 to 1,200 HP. The same conditions and horsepower would apply to a similar-sized ship in Baltimore or in a Puget Sound harbor. However, the same ship going through drawbridges in a tidal New England port, the Willamette at Portland, Oregon, or San Francisco Bay might require two or more tugs of from 1,500 to 3,500 HP. If the ship were bound into the Port of New York-New Jersey at Newark Bay three tugs of 3,500 HP might be required to assist this ship around Bergen Point.

This interesting 1984-built general purpose and coastal tug has many refinements. Notice the double stemhead eyelet, inward sweeping side bitts, forward deck capstan, overhead pilothouse vision, and norman pins. (Courtesy Jack Wilsky)

Tractor *America* and several others assist a tanker through one of the Willamette River bridges. Notice that both ship and tug are in ahead motion and that the tractor is using her stern as the bow, a common practice by tractor tugs. (Courtesy Reidel International/Willamette Tug & Barge)

4. Modes of Towing

SHIP ASSISTANCE—DOCKING AND UNDOCKING

FOR the short period of time involved in docking or undocking a ship, her assisting tug is frequently in a position where an incorrect move can place her in extremis. *Great care must be exercised!*

The positioning and makeup of tugs assisting ships vary greatly from port to port depending upon tidal currents, the physical makeup of piers, turning basins, or mooring slips, the required passage through drawbridges, or up winding rivers.

In the British Isles and northern European ports, harbor tugs are built so they can work off a heavy towing hook, which swivels athwartships in a nearly 180-degree arc. On a tug that has its propellers located under its afterbody, this hook is located well forward. On a tractor tug with the propeller forward or amidships, the tow hook is usually at half the distance between the propeller and the stern.

Tugs with towing hooks normally work in pairs when assisting a ship—one at each end. For full control, they pull against each other, enabling the ship to be moved ahead, astern, or breasted sideways into her berth. This method is often employed in the quiet waters of European harbor basins as well as in the ports on the Seine, Weser, and Elbe and in Canada on the St. Lawrence at Quebec City and Montreal. In this mode, the ship provides the line to the tug.

Along North America's East and Gulf coasts, ship-assist tugs are more often positioned alongside the ship at such a point as is designated by the ship's pilot. The tug's lines, usually synthetic, are secured to her H bitts forward or aft as required. Many harbor tugs are now built with or have an inverted U-bolt welded at their stemhead through which a bowline can be passed.

Opposite, above, Turecamo tugs assist in berthing *Ogden Dynachem* at Exxon Bayway, New Jersey (Courtesy Turecamo Coastal & Harbor Towing Corporation) *Below,* although both tug and ship are no longer in service, this scene illustrates what is still common practice in undocking at Montreal. The tug is *Matilda,* presently at the Hudson River Museum at Roundout, New York. (Courtesy Canadian National)

Moran tugs turning U.S. Lines' *American New York* in Kill van Kull. (Courtesy Moran Towing & Transportation Company, Inc.)

The heavy towhook arrangement on the stern of the Dublin tug *Colinmore* replaces the H bitts common on most U.S. harbor tugs.

MODES OF TOWING 63

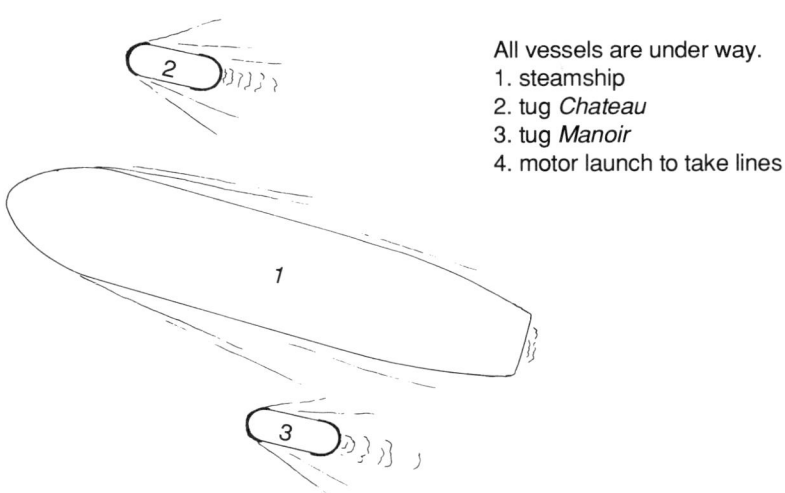

All vessels are under way.
1. steamship
2. tug *Chateau*
3. tug *Manoir*
4. motor launch to take lines

This sequence illustrates how ships are turned and docked in a tideway and the amount of time normally required as observed in the Port of Quebec. The approach is at 0900.

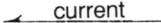

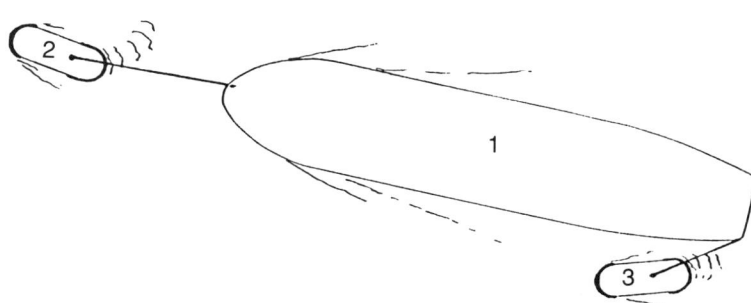

At 0903, the tugs that were approaching the ship now have lines out to it and prepare to turn with the ship's propeller stopped.

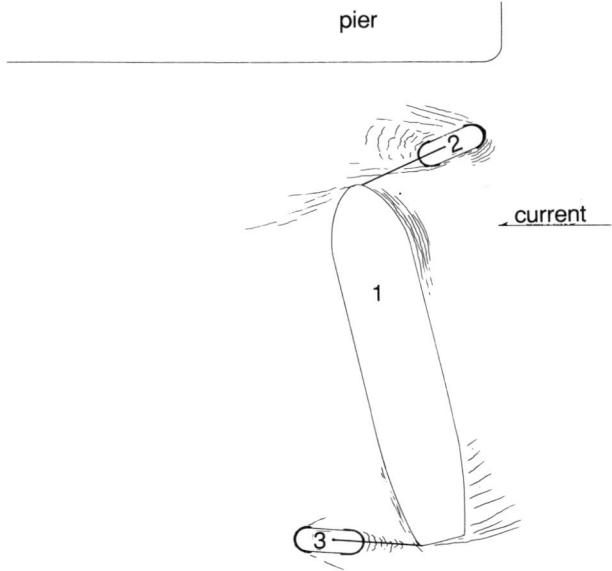

At 0905, the ship now across the current has dropped upstream near the end of the pier. Tug 2 is pulling hard to hold the bow and tug 3 pulls slowly on the ship's stern.

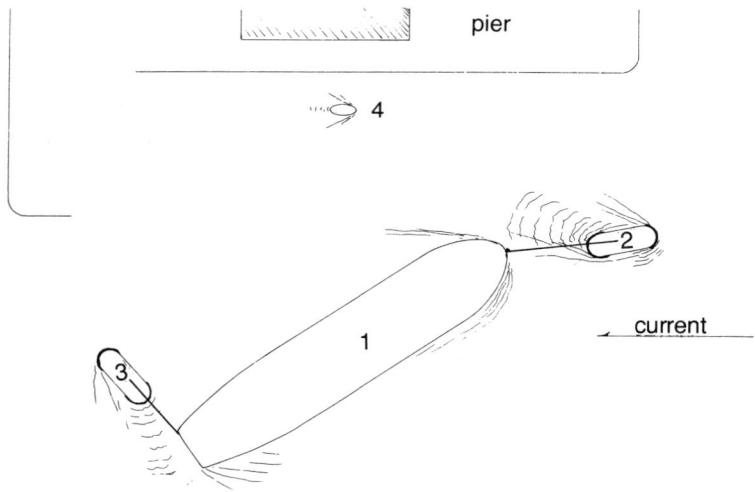

At 0906, with the ship held in the current, a launch comes out to take lines as tug 3 pulls the stern over to line up the ship with the pier.

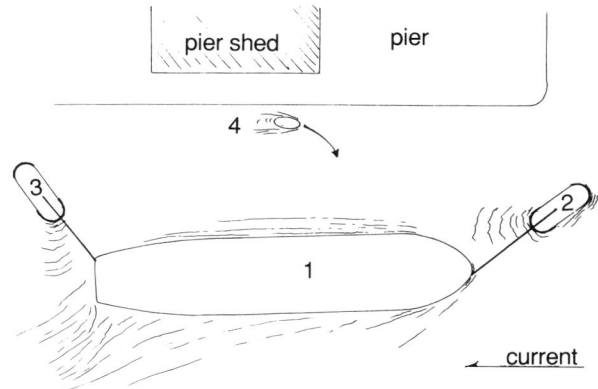

At 0907, with the ship parallel to the pier and held in the current, the tugs position themselves to pull the ship in sideways. Towing orders throughout these maneuvers have come from a pilot on the ship.

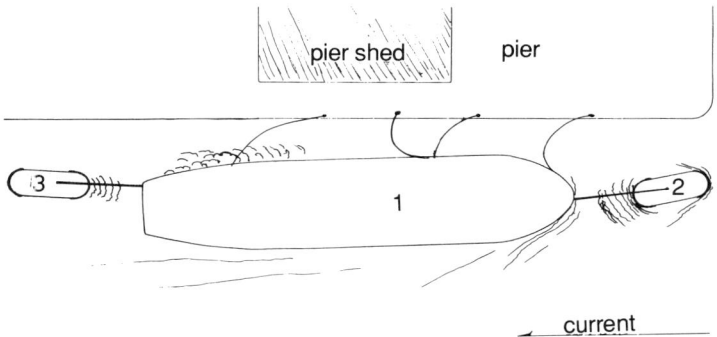

Tugs pull against each other; ship backs engine to get opposite shed on pier. Off the pier at 0908 with lines out, the ship is held as she is being heaved in.

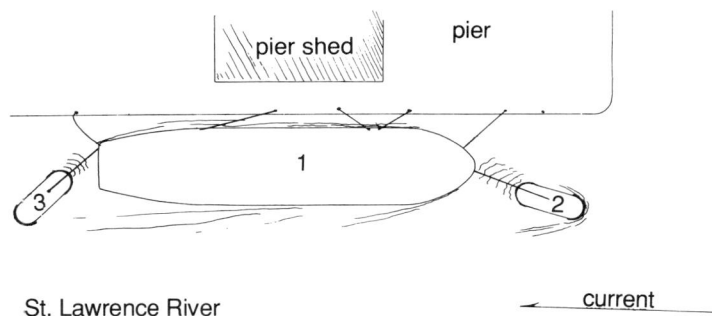

St. Lawrence River

Lines out, tugs check way of ship to berth. As the ship's lines draw her to the pier, the tugs at 0909 take a light strain to hold the ship so she will come alongside gently.

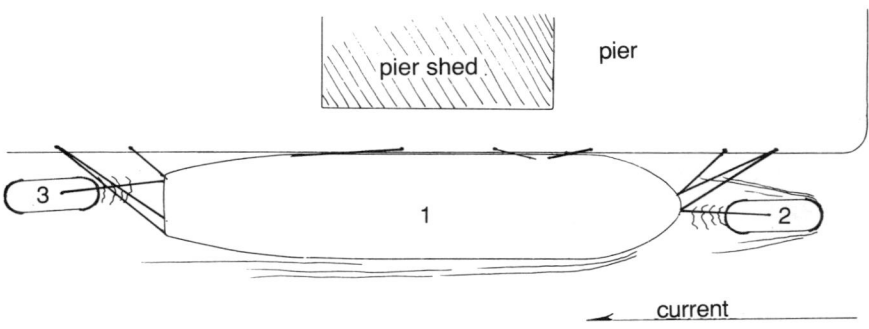

Ship in position alongside pier; tugs work engines easily to hold ship while lines are doubled. At 0910: All fast. The tugs are released, having completed another successful docking in just ten minutes.

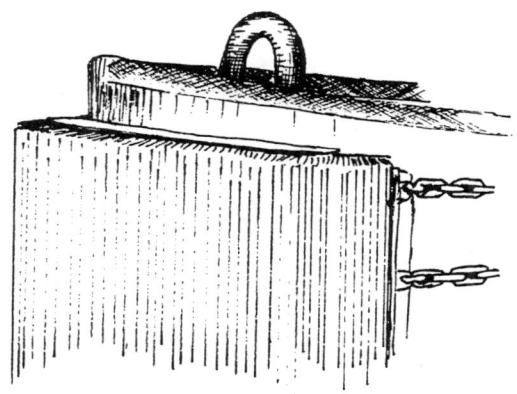

The bow U-bolt or eyelet, which has replaced the round stemhead on many tugs

Tugs doing similar work in the Pacific Northwest seldom use soft lines in shiphandling. Some are equipped with either hydraulically or air-controlled small deck winches holding suitable size and length of wire cable, which when led through fairleads and secured on the towed unit allow the tug to change positions without handling lines. Control of these winches is from the pilothouse. In less than a minute, a tug made up parallel to a ship or barge can change to a head-on position at a 90-degree angle.

In the Far East, ship assisting is done through a combination of the European tow hook method or the U.S. alongside mode. Recently in Japan, a suction type swivel cup attached to the tug's bow has been successfully used in handling large containerships without the use of any other lines.

Foss's *Wedell* illustrates the growing use of U-bolt eyelets at both ends of the tug. (Courtesy Foss)

INLAND OR SHELTERED WATERS*

Basically, there are three modes of towing used in sheltered waters. These are with the towed unit astern, alongside, or being pushed ahead.

The choice and use of any of these methods depend on many things: the area, its docks and piers, the current and weather conditions at the destination. Other factors of equal importance are the size of the unit versus that of the tug, the distance to be towed, and the confidence and ability of the person in charge of the tow.

Control of the unit to be towed is of primary consideration. It is very easy to put a towline on something and pull. It is something else to maintain complete control, whether in an open bay, a river, a harbor slip, or passing through a drawbridge or entering a lock.

*Not all sheltered waters of the United States come under the navigation regulations, Inland Rules. COLREGS '72 as amended in 1980 changed the lines of demarcation between high seas and inland waters.

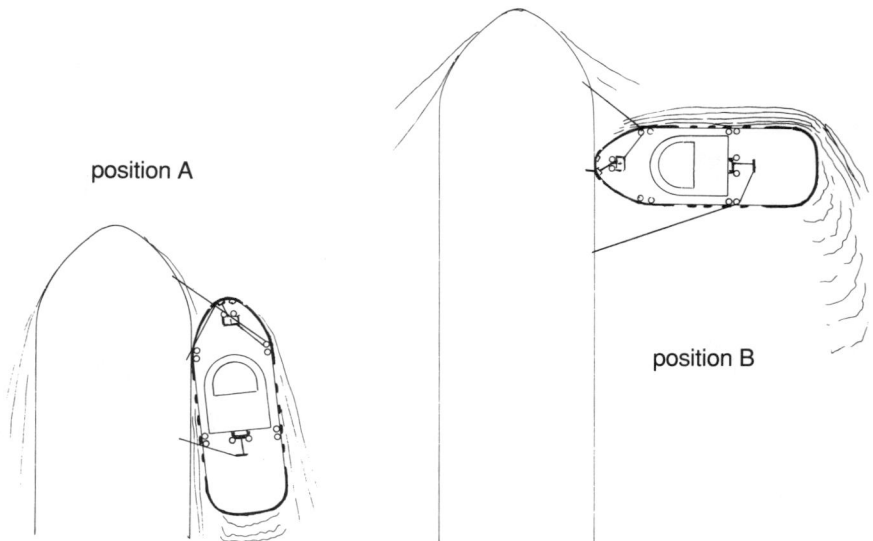

Shelley Foss, a new-generation harbor tug, has a unique line-handling arrangement. All of her deck lines come from winches that are controlled by the captain from the pilothouse. As shown in position A, she is made up with three lines, all of which are wire. At the order of the docking pilot, *Shelley* is able to change to a rigid head-on position, B, by heaving in on the towline and slackening the stern line. An adjustment to the former bowline keeps her rigidly in place as she pushes on the ship.

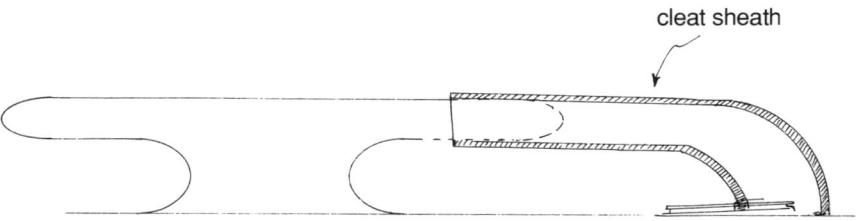

To keep lines from catching the horns of a cleat, Foss devised this removable sheath.

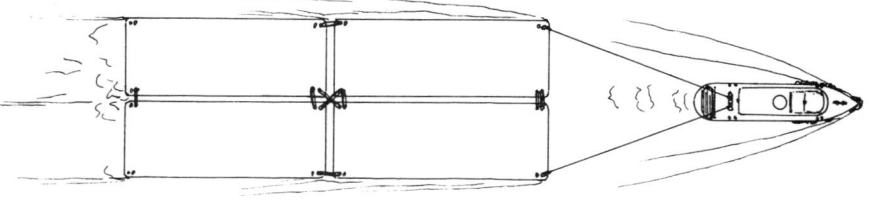

Towing a flotilla of four barges on short hawsers.

MODES OF TOWING 69

The cleat sheath used on *Shelley Foss*

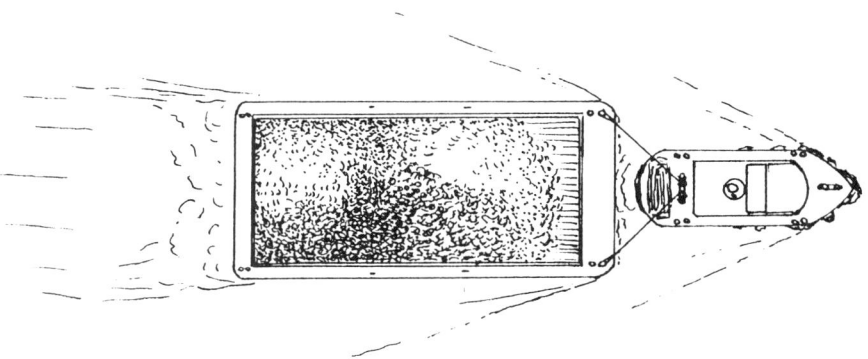

Tug towing open coal barge on gate lines

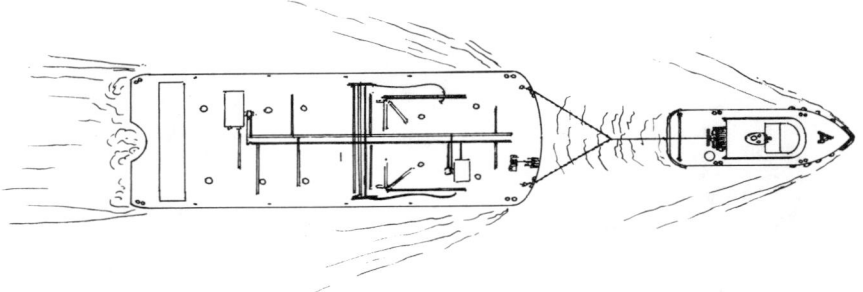

Tug towing empty oil barge on short hawser from tow winch

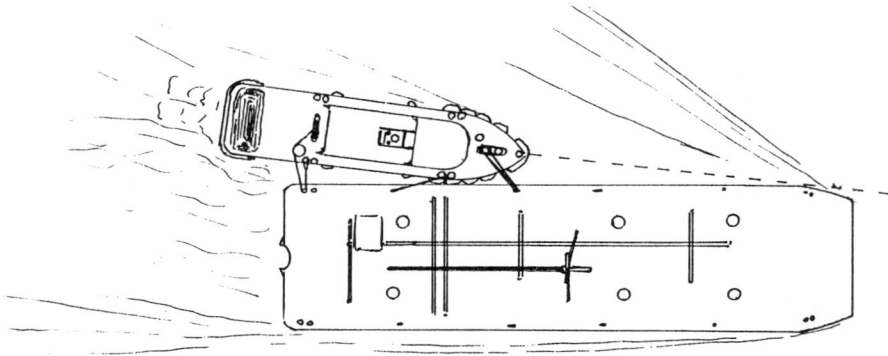

Tug made up alongside loaded oil barge showing centerline of tug pointing to the forward corner of the barge

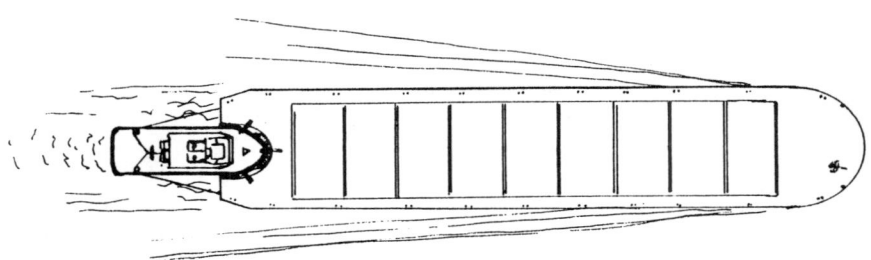

Tug pushing a moderately deep-notched ocean barge

Careful consideration of the choices, and extreme care in the execution of the maneuver, will preclude the possibility of damage, the bane of many a towboatman's otherwise successful operation.

Astern Method

Towing on a stern line or lines is the easiest and by far the most common worldwide towing mode. It is not always possible, however. Navy shipyards and marine contractors have small tugs or workboats used as tugs. These craft of from 25 to 50 feet can, under normal light wind conditions within a shipyard slip or basin, accomplish the most amazing feats in moving or towing large vessels. Much of this is done using a single line, with a crew member shifting it as the tug moves from one spot to another. By applying the minimum power required to move the unit, it is easily kept under control.

If, however, the same tug is to move a unit from the dock or basin to the harbor or waterway, a new set of considerations will have to be faced.

A handy type of tug for light towing and shifting, particularly adaptable in marine construction work. (Courtesy Atlantic Shipbuilding)

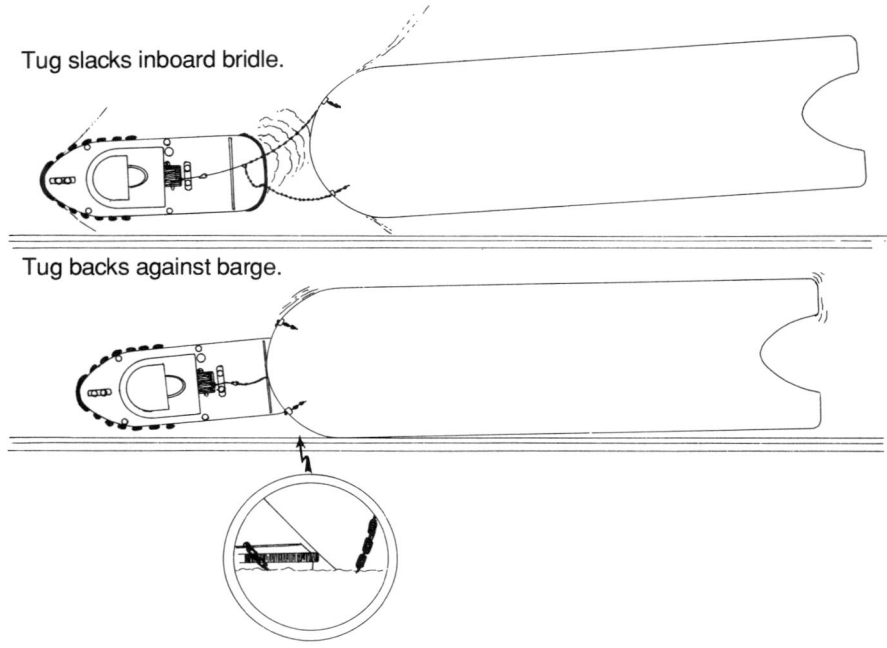

A careful landing can be made by shortening hawser until the tug's stern fender rests against the barge.

Foremost are the weather and the capability of the tug to do the job without assistance. What is the stage of the tide and current and how will they affect the tow, particularly on departure and arrival at the destination? Can and should the tow be accomplished by the astern towing method or must a combination of astern and alongside towing be used? Will it be necessary to turn the tow around on landing? Which side of the tow is to be placed alongside the dock? Should the tow land bow in or stern in? Is the location of the wharf or pier suitable for a landing on the hawser?

Remember that to land a tow on the hawser at a wooden pile pier at low tide is a risky maneuver, because square-ended units have an inclination to head between single piles. The result is bending, cracking, or loosening. The cost for such repairs by the towing company is usually much greater than the towing charges.

Another consideration when towing astern is the distance to hold between the tug's stern and the bow of the unit(s) towed. If a bridled hawser is to be used there will be less control than with two stern lines,

MODES OF TOWING   73

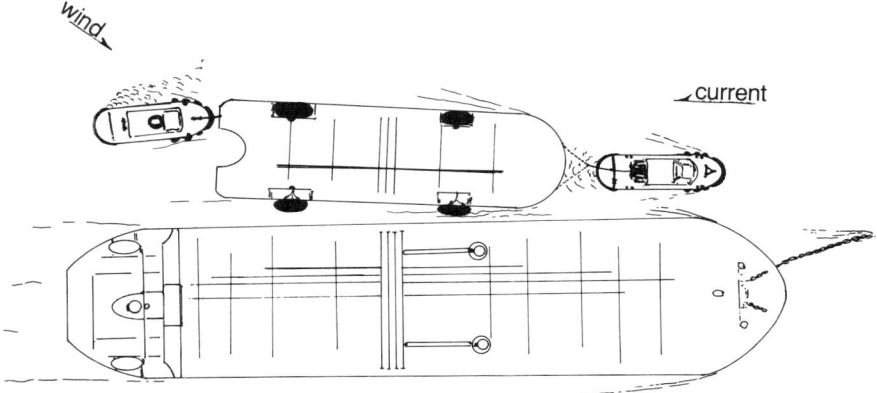

Landing a light oil barge alongside a tanker with an assisting tug. Yokohama fenders have been lowered into place. By pulling against each other, the tugs can achieve nearly perfect control and landing.

which around the New York area are termed gate hawsers.* When making a landing with either bridle or gate hawsers, greater control may be had by heaving the tow up, or backing the tug, until its stern fender rests against the tow (see chapter 8).

Landing—Barge Lines. Of equal importance when considering harbor towing on astern lines are by whom and how this towed unit's mooring lines will be put out. Must you put a man on the unit if there is no bargeman? Can he climb up from the tug's stern? Can he get around to the sides of the unit to the mooring bitts from there? When you do back up against the towed unit, is its rake so long and sharp that it will strike the tug's rail, mast, or deckhouse? This has occurred.

Using the Wind and Current. This is the safest and easiest way to proceed if on stern lines. It may be a bit rough. When landing alongside a ship, sufficient fenders should be in place.

Use of an Assisting Tug. Quite obviously, there will be times when tide, current, and wind are in different directions or the wharf location is such that to land a tow on the hawser or stern lines will require the assistance of another tug. This is a common practice when handling large empty barges, trailer barges going alongside a dock, or anchored ships, and also

---

*The term *gate hawsers* comes from Hell Gate in New York's East River, where tugs have pulled heavily loaded sand scows through using two single stern lines.

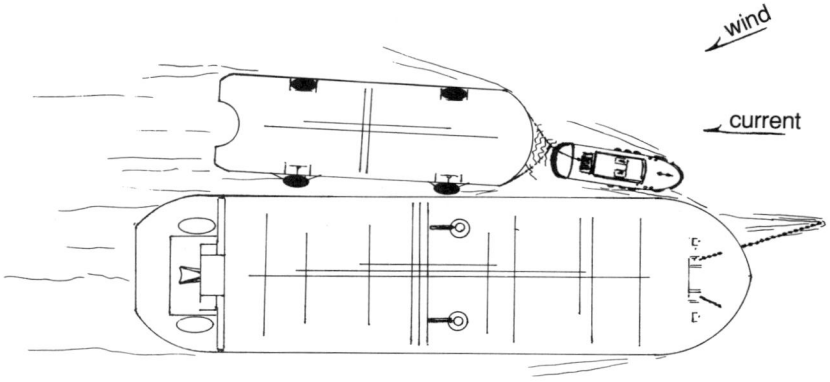

Landing an empty barge alongside a tanker without an assist tug should only be attempted when the wind and current act as a breaking force.

if the lead tug is changing from hawser to alongside or notch towing. Coordination by VHF and good maneuvering are required.

Landing a Tow on the Hawser without Assistance. Frequently, the tug captain is faced with this problem. Its solution requires careful thought and timing. If the towed unit is of light draft and is riding high, will the wind effect be more than that of the current? How much leeway will the unit make? Should the captain await better conditions? If the unit is loaded, can it be slowed sufficiently to drift into a landing with the tug's stern against it so that by backing the tug, the unit will be stopped safely? Does the tug have a proper stern fender for backing against a loaded barge? Does the shape of the barge present the possibility of its riding up over the tug and doing damage?

Alongside Towing, or "On the Hip"
By definition, the towing of a unit alongside a tug is just that and not to be confused with the term pushing as used in the navigation regulations of the International and Inland Rules of the Road. The colloquialism "on the hip," which originated along the U.S. Gulf Coast, refers to this alongside towing method. This offers control when leaving a pier or slip where it is necessary to turn upstream or downstream and also when landing a tow. When getting under way from an area where other ves-

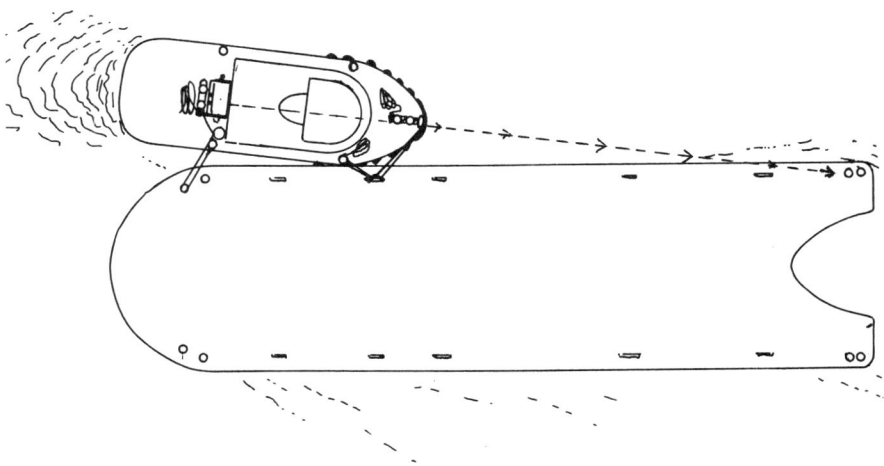

The normal towing arrangement when made up alongside to any large barge.

sels are moored or anchored, great care must be exercised. Before going on the hawser, it may be necessary to make up alongside.

Alongside towing is not without its problems. As a towing mode, it should be thought of positively, but all eventualities should be considered. To be made up alongside correctly, the tug should have her centerline toward the unit she is to be made up to. Looking forward from the tug's pilothouse, her stem should appear to be pointing directly toward, or a little inside, the most forward point or corner of the vessel to be towed.

To adjust to this position or to one that will render better towing control, the bowline should be shortened and the stern line slackened until a minimum of opposing rudder is required to hold a straight course. It should always be remembered that any alongside tow is making a course of about 10 to 20 degrees from what appears as a straight line. It must be realized that a single-screw tug in taking a tow alongside will require use of a lot of opposing rudder when first getting her tow under way. Once the tow is under way and well made up, the rudder can be eased. The tow's stern must be watched carefully. It may angle off enough to hit a wharf or vessel being passed.

A twin-screw tug using her engine against the rudders can walk an alongside tow sideways. Tractor and Z-peller tugs can move an alongside unit in many different directions.

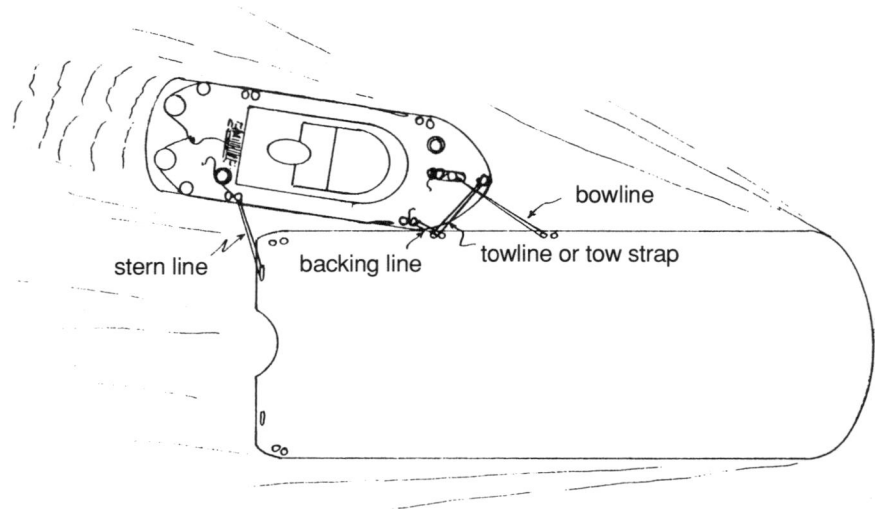

The lines used by many U.S. and Canadian tugs when made up alongside a large barge

Other conditions to consider in alongside towing are the wind, tide, and what the current will be doing where and when the tow is to be landed. Can it be held alongside all the way? Can the tug's pilot see over it? How will it tow best? Which end or side does the customer want alongside his wharf? How much windage does the tow offer? Does it have a history of being troublesome to tow? Much of this information can be acquired from someone who has previously towed the unit.

A heavy and powerful tug may be made up alongside a small barge without regard to the normal centerline and stemhead-to-barge-corner configuration. To make up in this manner, the tug's bow must be the fulcrum area. A doubled-up towline, a doubled-up backing spring, and a stern line are usually all that are necessary. This has been a method used in New York Harbor. It should always be remembered, however, that any alongside tow is making a course of about 10 to 20 degrees off the straight course line of the tug. (Handling a tow alongside is described in chapter 7.)

Push Towing

This mode of towing may be divided as follows: push towing by general harbor tugs on inland waters; push towing by huge square-bowed tow-

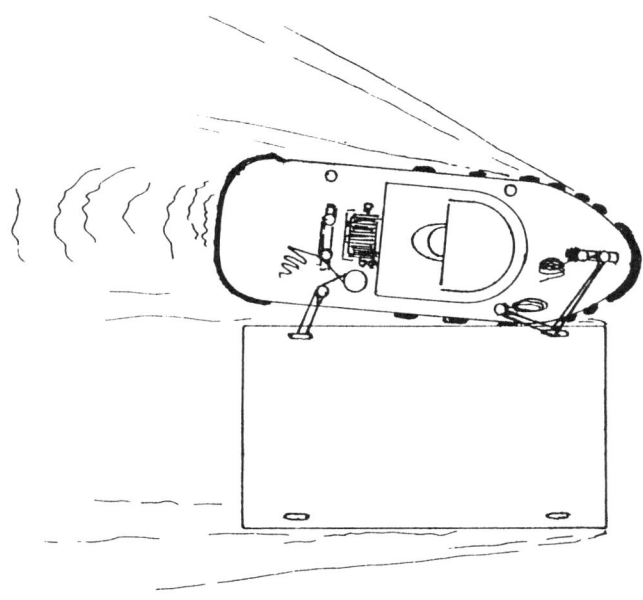

When the unit is light and the tug is long and powerful, a makeup such as this is possible.

boats on inland waterways and rivers; push towing of large notched-stern barges by specially constructed seagoing tugs; and pushing of huge barge units by rigidly attached power units, making a single shiplike "tug-barge unit," known as a TBU.

Push towing was probably started on the western rivers of the United States by early stern-wheel river steamers that pushed flatboats or barges of firewood to fuel their boilers, and later pushed barges of cotton, coal, grain, and farm produce. By the time the huge, now historically preserved, stern-wheel towboat *Sprague* had reached her zenith, diesel river towboats were pushing standard-sized river barges, three abreast in three or four tiers, up and down the Mississippi and Ohio rivers. Today, a 10,000- or 12,000-HP towboat will handle up to 50 loaded jumbo river barges. They are held tightly together by hard rigging: a combination of short wires, cross wires, and steamboat ratchets.

Inland River Push Towing—Mississippi River and Tributaries. This is an entirely different type of towboating. As a floating island, either up-

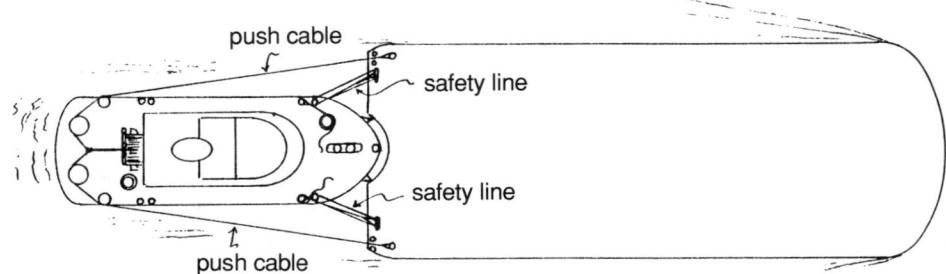

Pushing in a shallow notch with safety lines

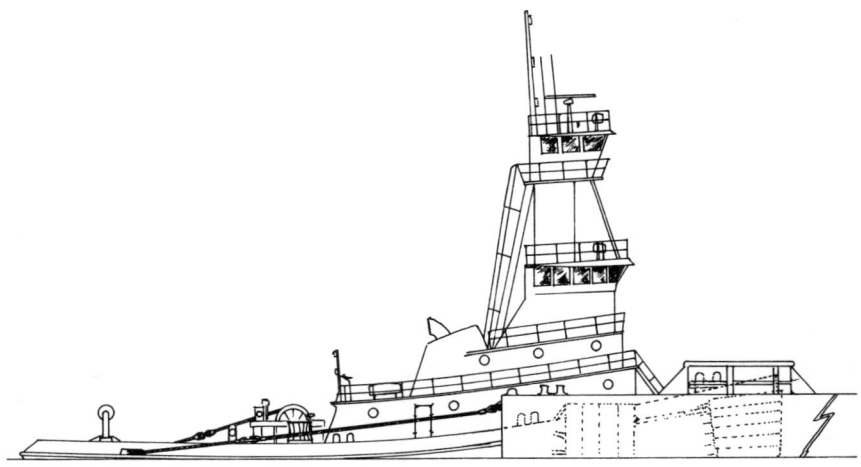

A tug in a moderately deep-notched barge would appear as above. The dotted lines represent the tug's rail and vertical and horizontal rubber fenders.

bound or down, such a tow takes up a large portion of the river. Great care, courage, patience, and local knowledge are the hallmarks of the masters/pilots of these towboats: care in meeting and passing other tows, in descending through the many unfendered bridges, and in approaching lock walls and dams; courage in running at night in rain squalls with radar blacked out—probing ahead with a searchlight for ranges or crossing buoys; patience in holding a tow in midstream for fuel, stores, crew change, or to drop or add on another barge; knowledge through experience in knowing just how to approach certain bends—to float a little, flank, or back hard, or a combination of all if the river is running. To bring such a tow from St. Louis to New Orleans without mishap is the mark of excellence in piloting.

Push Towing on Gulf Intracoastal Waterway. Push towing between Brownsville, Texas, and Port St. Joe, Florida, and tributary waters including the lower Mississippi is a modification of river towboating. The pusher type towboats are usually smaller. Not all are of the square-bow, towing-knee type. Standard coast and harbor tugs with tows will be met on these waterways. Most of them will have the tow "on the head" as they say, meaning that they are pushing. In the Gulf Intracoastal, the pushing mode offers a safe and efficient method. Due to the narrow channels, bends, bridges, river crosscurrents, and sweeping winds, tows normally consist of from one to four units in a single tier.

How push towing spread from the Mississippi and Gulf to the North is not recorded. As a mode its use still comes to an abrupt end at Chicago where the Illinois Waterway joins Lake Michigan.

From the Gulf, it spread eastward in the 1940s to Florida's west coast and across to the Atlantic Intracoastal system. Earlier it had been introduced on the Hudson River and from New York eastward, and throughout the New York State canal system onto the Great Lakes and the St. Lawrence in single units only.

Pushing. This is perhaps the easiest, safest, and most efficient mode of towing. On inland waters, it offers an ease in handling, maneuvering, and docking, particularly with deeply loaded liquid cargo barges. Much better control is present in passing other tows, passage through drawbridges, and in entering locks.

Originally, the conventional tug with model bow and elliptical stern made up to square-ended barges by running a towing line from the tug's stem to a cleat in the middle of the barge's stern. To gain the ability to steer and turn as one unit, a line from each quarter of the tug, usually a fixed length of wire rope secured on one side with a soft line on the other side, led to the tug's afterdeck capstan.

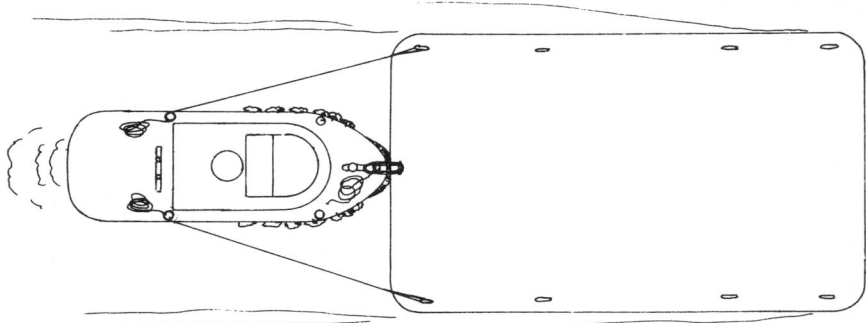

This is the original pushing makeup of a harbor tug to a scow. From this idea, barges have increased greatly in size, and the notch for the tug has become deeper and deeper until on some barges the tug fits in tightly as far as the after end of her deckhouse.

Push Towing of Notched-Stern Barges. This came about when petroleum and other liquid-carrying barges were developed to make use of the maximum dimensions of the New York State barge canal locks, particularly in the Erie Canal section between Waterford and Oswego, New York.

When in the 1930s operators realized they might carry more cargo through this New York canal system, a change in barge profile was made. Instead of the square stern, a longer barge with a deep notch aft was introduced. Now the tug's bow as far back as the wheelhouse fits into the barge snugly. As a unit the tug and barge filled 290 feet of the lock's 300 feet. It was found that with the tug snug in this notch, maneuvering of the tow was improved. The pivot point that had been near the barge's bow had moved back to approximately just forward of the middle of the barge. The deeper the tug was fitted into the barge, the greater the control and ease of steering.

Most of the barge cargoes bound up the Hudson River and through the Erie or northern canals originated in the Port of New York. As cargo was transported in barges with notched sterns, it was natural that similar cargoes destined for ports on Long Island Sound and farther east were also loaded in notched-stern barges. Soon it became prudent to use the same barges for coastwise service towed on the hawser at sea and pushed when in smoother inland waters.

Pushing—Ocean and Coastwise. The increase in horsepower efficiency, speed, and fuel economy through the use of the push mode aroused interest in developing it for use offshore.

Allison C of Red Circle illustrates the medium-depth notch. Notice her push cables come from the barge skeg. When the tug disconnects, the cables are heaved back with the bitter end lying on the barge deck.

In the early 1950s, the Atlantic Cement Company built three ocean barges of over 400 feet and drawing 33 feet when loaded with about 20,000 tons of dry cement. Each had a moderate notch that allowed the tugs *M. Moran* and *Esther Moran* to slip in as deep as their forward side bitts. Use of the notch at sea was found to be impossible except in fine weather. These tows, usually bound for the Chesapeake, Savannah, Port Everglades, or Tampa were made on the hawser from sea buoy to sea buoy and in the notch from there to the dock. The success of this combined towing/pushing operation led others to follow, mainly those from Gulf ports to Florida and the lower East Coast. Each new operator has expanded on the technique of trying to hold the tug and barge together as one unit in the worst kind of weather. All manner of rigging has been applied to hold the tug in the notch. Because of the difference in draft

Sonat Marine's (now Maritrans) *Valour* illustrates deep-notched towing at sea. (Courtesy Maritrans Operating Partners)

and displacement between tug and loaded barge, it becomes necessary to get on the hawser in rough weather. As the main deck of these barges when empty is from 35 feet to 40 feet above the surface of the water, tug captains were unable to see ahead from the tug's regular pilothouse. Upper pilothouses were added and are now the vogue for this work.

The original pushing hookup consisted of a heavy wire pendant from each of the barge's outboard corners. The pendants were led to the tug's stern through fairleads in the bulwarks and again on deck. Here they were joined by shackles to the tug's towing hawser. There have been many variations of this.

Each of these variations has been claimed as the ultimate answer. Some of them are: hydraulic rams at the stern to take the sudden tension, a combination of heavy wire pendants and nylon straps, much deeper notches, heavy rubber fendering, hydraulic pads that come out to wedge tug and barge, and a bow stinger that fits into the barge.

Some of these systems have been patented. Some are highly guarded. Of these, a few have successfully operated between the Gulf and East Coast ports. One of these systems has an outstanding record of sailing from mid-Atlantic terminals to South America and staying in the notch all of the way. But for most, if not all, there is a time when the tug will have to get out of the notch and pull, or heavy damage will result.

Notched ocean barges are common on the U.S. East Coast and the Gulf of Mexico, and on Japan's inland sea. The waters of the U.S. West Coast, the Orient, and Europe are not receptive to this type of towing.

TUG-BARGE UNITS

There are two kinds of tug-barge units. The first, mentioned above, is a combination seagoing tug, designed and fitted out to push a specific single or group of notched barges one at a time. This type of tug may also get out of the notch and tow when at sea as well as carry out other unrelated towing assignments. Collectively they are known as integrated tug-barge units (ITBs).

The second and more controversial type of tug-barge unit has a misleading name. Because of this it must be included here. Actually, there is no tug per se. The unit consists of two parts, joined semipermanently as one: a cargo hull (the barge) for liquid or bulk products and the power unit (the tug); this is a TBU. Usually built by separate shipyards, they are joined together solidly to operate as one unit until such time as repairs or other unforeseen events cause them to be separated. Actually, together they are a ship. If, in an emergency or otherwise, the power unit becomes detached from the cargo barge hull, it cannot tow as there are no fittings, tow winch, or cable. The physical construction of most of these power units requires careful maneuvering just to maintain stability.

The original tug-barge unit was introduced on the U.S. Great Lakes in the early 1950s by the Cargill grain-shipping interests. A barge with a square void in the stern admitted a power unit that fit as tightly as a piece in a puzzle. It was solidly bolted by quickly removable studs. The result, named *Carport,* was introduced as a highly flexible and economical shipping solution that supposedly would be exempt from certain manning and other marine inspection regulations.

This announcement was a little premature; the U.S. Coast Guard at that time (1952) declared *Carport* was in reality a ship and should be so manned and regulated. Cargill promptly welded both parts solidly together and sold *Carport* as a ship, which for years ran around the Caribbean under various flags. Since then, some owners with long-term movements of bulk or liquid commodities have been intrigued by the name tug-barge unit and the financial possibilities they offer. The U.S. Coast Guard in taking a second look has apparently gone along with this feeling.

Realistically, a 199- to 999-gross-ton power unit designated as a tug and a 6,000- to 7,000-gross-ton barge unit, when loaded and combined with the tug, equals a 30,000-ton ship.

An opinion by the editor of a well-known shipping magazine states, "The Coast Guard allows the shipment of cargoes in barges with unlicensed (or minimally licensed) crew on the tug only. It makes no difference that the tug-barge combination is hydrodynamically inferior and has poor fuel consumption. The operations and debt service costs will be cheaper."*

OCEAN AND COASTWISE TOWING ON THE HAWSER

At sea, towing from a tow winch equipped with a steel tow cable is the accepted standard method. The types of tow winches vary. U.S. East Coast tugs and most large northern European ones use a large, heavy-duty, single-drum winch. West Coast and British Columbia tugs are normally equipped with a double-drum, side-by-side winch, or with one drum over the other in what is known as a "waterfall" configuration. Operations on the Puget Sound and the inside route to Alaska have had a great influence on the use of this type of winch. The barges are used for transporting general cargo and railcars. They are broad, averaging a 100-foot beam. The result is more buoyancy with a greater tonnage per inch immersion than with smaller high-sided yet deeper-draft East

*Marine Engineering/Log, February 1983, page 126, "Opinion."

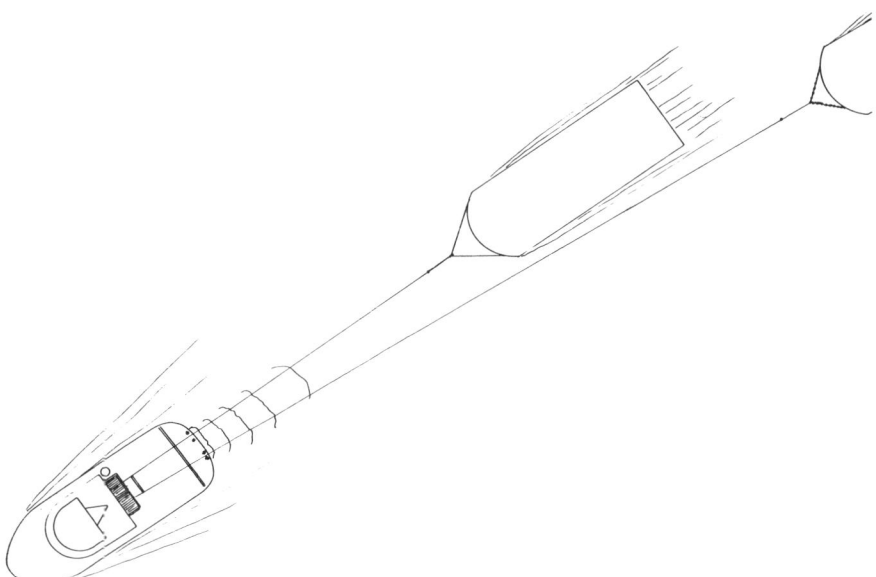

Tandem towing is a common method used on the Pacific Coast to Alaska and the Hawaiian Islands. Each tow cable comes from its separate winch drum and passes over the stern through vertical, hydraulically operated norman pins.

Coast barges. Added to this is the great depth of water, which quickly goes to 40 or 50 fathoms close to shore, both off the coast and through the straits and sounds. This allows tugs to stream out a tandem tow of two or more barges held on separate tow cables from separate winch drums.

One winch drum may hold up to 2,500 feet of cable, the other 3,000. The head barge, the first astern of the tug, may be held on 1,800 feet of wire, and the second barge on 2,500. Each cable is led over a wide stern roller in such a way that they are separated, often by hydraulically controlled vertical roller pins. In this manner each tow cable is clear of the other and may be adjusted as desired.

On single-drum tow winches used in ocean towing elsewhere, steel towing hawsers of from 1½ to 3 inches in diameter and from 2,100 to 3,000 feet in length are common. Passage of these hawsers over the tug's stern rail has always presented a chafing problem. Chafing or hawser boards clamped to the cable are common and troublesome in rough seas or on long ocean voyages. Some are made from a section of steel or aluminum H beam, or a rounded pipe, split lengthwise and filled with

Hawaiian Marine Lines, a Crowley Maritime company, provides cargo transportation between U.S. West Coast ports and Hawaii via tandem barge tows. This one is off Waikiki Beach outbound for Los Angeles. (Courtesy Crowley)

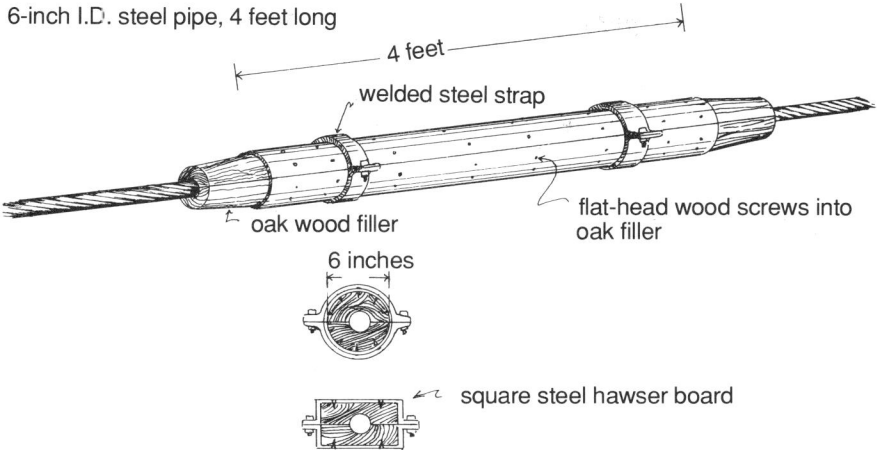

Several types of chafing gear used on U.S. seagoing tugs. They can be installed on the tow cable inboard and slacked to the stern taffrail. In stormy weather, they frequently give problems.

oak that has a groove the size of the tow cable hollowed through its length.

Another and most successful method of avoiding chafing at the stern is the installation of high towing arches, so that as the tow wire comes off the winch drum it is nearly horizontal at the same height until it reaches the aftermost tow arch, which should be close enough to the tug's stern rail so that under normal towing strain the tow wire will lead into the water astern without touching the rail. Many tugs have just one arch. If more than one is required, each should be of heavy-duty, 6-inch to 8-inch round pipe with a heavy bronze bushed sheave in which the tow cable rides. (See discussion of chafing gear in chapter 8.)

To safely and efficiently operate an ocean tug and her tow, much thought must be given to preparation. Once at sea, an hourly check on all conditions of the tug and tow must be maintained.

If the unit in tow belongs to the tug's owners, the tug master should be familiar with the company's guidelines. If the unit to be towed is owned or chartered by a customer, the tug is required, through her owners, master, and crew to furnish good outfitting and handling with efficient service and prudent navigation. It also must avoid any unnecessary or known hazards without deviating from the practice of good seamanship (see chapter 8).

5. Tug Design

IF possible, a tug or towboat should be designed for a specific type of work. Of course some consideration must be given to an alternate usage, which may be incorporated in the original design. Many huge tugs designed specifically to be stationed at critical locations for salvage operations have turned to long-distance ocean towage. Tugs originally built for offshore oil rig work have been called upon for coastwise, inland, and even harbor assist work, after some modifications. Unless guaranteed a long-term return on their investment, the owners who approach new building should carefully consider ancillary usages that may be possible for a proposed vessel.

Such considerations as hull form, type of propulsion, stability, maneuverability, speed, power requirements, and endurance are of utmost importance. Working and control areas must be compatible with the desired use of the vessel.

Many tugmen will argue that there is no perfect tug or towboat. A good tugboat must be designed to be completely functional. Such basic qualities as a strong hull, which is closely and ruggedly framed with adequate watertight bulkheads; heavy bulwarks; solid bitts, chocks, cleats, and other fastenings are required.

Beyond the strength of the vessel, safety is of equal importance. The first criterion is the exclusion of water from the vessel's interior. Watertight doors with freely working dogs and coaming of classification height, while a little rough on the shins, are standard requirements. Their hinges should be on the forward side. Those watertight doors on the main or lower deck should be fitted with the latest in quick-closing and -securing devices.

The continuing evolution of tug design is an interesting study. It still offers room for improvement. In the first quarter of the century, ships that tugs assisted were much smaller with straight sides and little flair forward. Tugs were steam powered, with tall stacks. The stacks were necessary in order to get a good draft for the fires in the boilers, as well as to support the escape pipe and ever-present steam whistle. Masts were primarily for the display of running and towing lights. In the

Peter Moran is a good example of the small U.S. coastwise tugs available during and after World War II. Most are diesel powered in the 1,200-HP range. Notice the handmade bow and side fenders, some with whiskers. (Courtesy Moran)

United States many had a tall mast aft. Some tugs also had a mast forward.

Abroad in northern Europe, the tug's mast was set forward and, with the smokestack, was close to the wheelhouse and open flying bridge. All were grouped about one-quarter of the way from bow to stern.

In the United States, cabin deckhouses were of wood. Some had elaborate paneling. On the main deck, the house ran from clear of the forward bitts to well aft. This gave the tug a very short open stern. Above, on the wooden canvas-covered boat deck, was a lifeboat in chocks plus a pilothouse and captain's cabin. On some tugs, the pilothouse/cabin top was all on one level. Later, following the pattern set by designers of coastwise passenger vessels, the bootheel pilothouse was introduced.

European tugs and most other non-U.S. tugs had only a small wheelhouse for use in inclement weather. Normally, the open bridge on top of the wheelhouse was used. There, open to the weather, all the

amenities of the wheelhouse were repeated. There was a steering wheel, an engine order telegraph, a whistle pull, and a tow hook release.

In New York Harbor, the classic railroad tug was developed. Whether steam- or diesel-propelled, this tug had lots of power. A high, enclosed

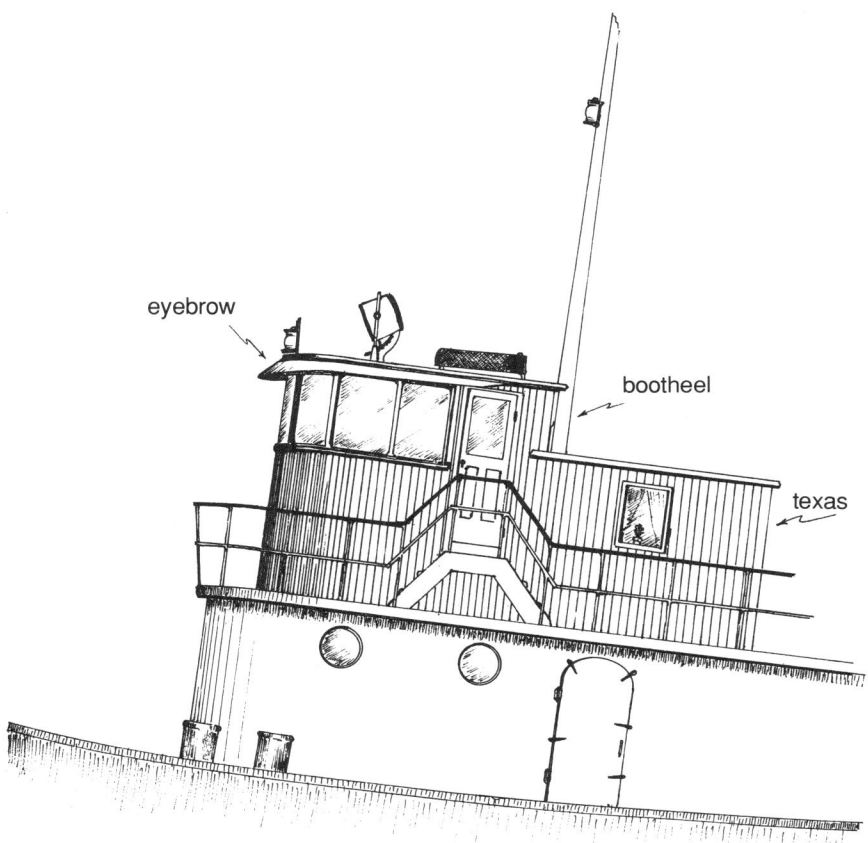

A bootheel pilothouse, a standard fixture on most older U.S.-built tugs

Opposite, above, United Towing's (British) tug *Englishman* is a good example of a World War II-class general purpose tug. Here she is acting as a ship assist in placing H.M.S. *Revenge* on a mooring buoy. (Courtesy United Towing, Ltd.) *Below,* SinMac's (now McAllister Bros., Ltd.) tug *Yvon Dupre* shows the open bridge used on many tugs. Her skipper, one hand on the steering wheel, peers around the stack at his towline and charge, which he is dragging into a Montreal slip. The tug holding the stern is believed to be *Graeme Stewart*. (Courtesy Canadian Pacific Railway)

Erie Railroad's diesel-electric *Cleveland,* flagship of the fleet, was a classic New York Harbor railroad tug. An estimate of the height of her wheelhouse can be made by comparing it with the deckhand descending the steps to the boat deck. (Courtesy S. Lang)

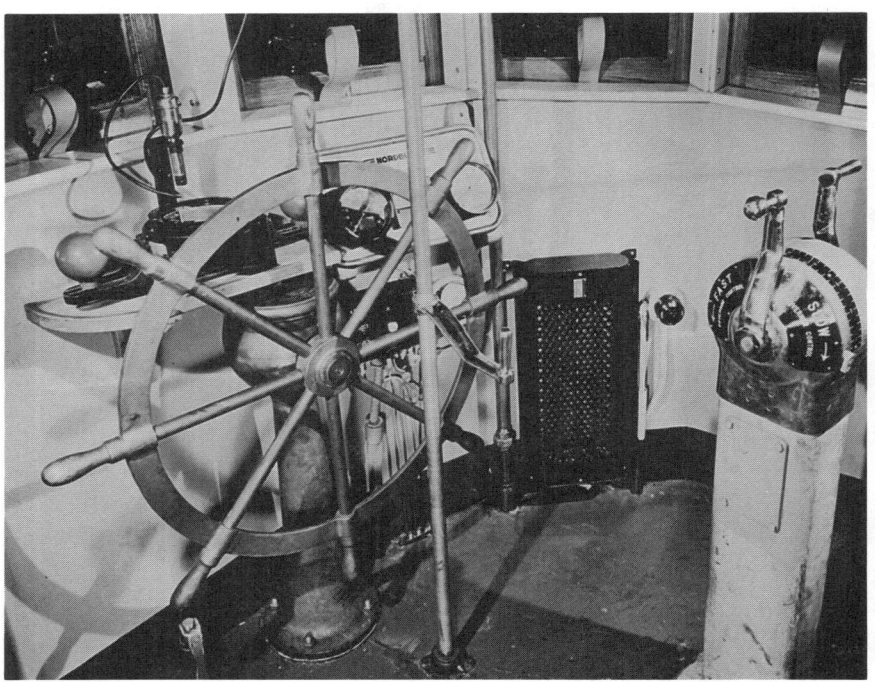

When the wooden-hulled *Brynn Foss* was re-engined in 1949, very little room was left to get around the engine controls and steering lever. (Courtesy Nordberg Engine Company)

The interior of a Schottel tractor tug showing all-joystick controls and automation. (Courtesy Schottel of America, Inc.)

pilothouse about 5 to 6 feet above the boat deck allowed the pilot to see over his tow.

These specially constructed railroad tugs were required to tow the various railroad car floats, stick lighters, and open and covered barges owned by the approximately fourteen rail systems. All of their freight car tracks ended either on the New Jersey shore of the North River or on the Brooklyn-Long Island shore of the East River. As these covered barges and car floats with railcars were so high, it was impossible to see over them from the wheelhouse of a normal tug. Each railroad had its own fleet of tugs, all built with a lone pilothouse some 11 to 12 feet above the boat deck. These old, well-built craft are gone from railroad service today. Fortunately, several U.S. companies have seen the possibilities of a new life for some of these tugs. So, in the ports of New Orleans, Tampa, and Port Manatee, Florida, they are at work, now modernized and converted to diesel power. Older tugs built in the hand-steering, bell-pull, or telegraph/Chadburn era, had their simple pilothouses updated. The

result was frequently an overhead and bulkhead festooned with new equipment squeezed in, making the chart table almost inaccessible.

PILOTHOUSES

When the switch from steam to diesel propulsion came, the wooden hull and wooden deckhouse disappeared. On newly built tugs, deckhouses,

A good example of a compact and efficient river towboat pilothouse arrangement. The engine controls are in the center of a stainless steel console between the steering sticks. The longer sticks are for the main rudders and the shorter ones for the flanking rudders. The gyro, radar, and electric searchlight controls are close to the place where the pilot would be seated or standing. (Courtesy Dravo Shipyards)

Opposite, above, I.O.T./Sonat Marine (now Maritrans) *Mariner* of 1965 had added in one decade two VHF radios, a Decca radar, autopilot, master gyro, and Loran. The master gyro is under the plywood cabinet to the left. The chart table and stowage were later revamped. (Courtesy S. Lang) *Below,* the wheelhouse of a Dublin Harbour tug is small, neat, and efficient. Equipment includes autopilot, RPM, and rudder angle indicator in the console at the right. Note the code flag locker behind the captain.

Note the difference in visibility for the tug skipper between London River tug *Sun XVIII* and the ship-assist tug below.

Rabigh 1 is a 2,600-HP ship-assist tug of the 1980s. Her visibility from the pilothouse is 360 degrees with overhead ports.

now of steel, have been shortened and pilothouses are reduced in size. They offer better visibility, both horizontally and vertically. The pilothouse or wheelhouse is the most important spot on a modern tug. It is the control and navigation center. Here on many tugs is a console that monitors the entire power and auxiliary plants: an electrical switchboard with back-up alarms. There are also start-up and stop switches for the main engine, steering, and other systems.

ELECTRONICS ON THE BRIDGE, IN THE PILOTHOUSE OR WHEELHOUSE

Dramatic progress in the navigational aids, communication, and automation of engine rooms has brought new equipment to the tug pilothouse. If it has been added piecemeal, it may not have been installed in the most accessible location, which would in part nullify its usefulness.

On newly built vessels a choice must be made by the designer and owner whether to incorporate a minimum of equipment in a traditional pilothouse layout, or to go to a completely new concept. In the latter case, much thought must be given to the personnel who will operate it. What may be an exciting idea to a designer may be less than practical to the tug captain. It is particularly important that an adequate chart table and chart stowage be provided.

Consoles must not block access to all the windows. On a tug, unlike a ship's bridge, some windows must be free so that watch standers when maneuvering may use them or open them to see the deck and fittings. The watch officer, even when seated in a pilot chair, should have clear, unobstructed vision over any consoles. Placement of electronic or other wheelhouse equipment should be such that there are no head-busters.

Modern tugs have bridge consoles and electronics that can offer such equipment as a gyrocompass. Master gyrocompasses are now quite small and can easily be stowed under or in lockers with a steering repeater located for the helmsman as well as an autopilot. Other items such as VHF and SSB radios, radar, and Loran are common. For deep-sea navigation, satellite receivers, weather chart displays, the latest state-of-the-art sounders, and electronic viewing equipment are now in use. Other very nice refinements are the automatic fog signal and electric whistle switch, internal and external voice communication systems, and strategically located loud-hailers. Switches for all deck, flood, and searchlights should be easily identifiable and located within the helmsman's reach.

Quite naturally not all tugs will have all or even a good part of the above equipment. Each should be designed, however, to have a minimum of equipment needed for efficient operation. Arrangement of the equipment should be worked out with the persons who will operate the

A Smit ocean tug's bridge console with navigation station is similar to those found on an ocean cruise liner. (Courtesy Smit Tak)

tug, not unilaterally by the designer or port captain. Many tug captains assigned to a tug in the last stages of building have had to revamp the pilothouse equipment layout, much to the owner's displeasure. Looks and efficiency of operation are often at odds with each other.

VISIBILITY

Because visibility is the paramount consideration at all times windows in the pilothouse should be of tinted glass to reduce glare, particularly at night and from aft. They should be high enough so that a tall man need not stoop, and low enough at the bottom so that a short watch officer does not have to stand on a raised platform to look out. (The requirement of keeping a good lookout is discussed in chapter 13.) Looking aft when at sea, the watch stander must have a clear view of the afterdeck and tow cable (see chapter 7).

The pilothouse deck should be level so that watch standers' feet are not forced to feel as if walking up a slight rise. Many designers have failed to realize the twice daily discomfort this gives the watch stander;

or they try to save construction costs by following the outside sheer and camber of the vessel. A few weeks' duty on one of these sloping interior decks would change their thinking.

AFTER STEERING AND CONTROL STATION

If an after steering station is provided, it should have all means for the control of the vessel's movement from that spot. The captain should be able to steer, adjust engine direction and speed, direct the operation of the towing winch, use a searchlight, sound the whistle, and give orders through a loud-hailer and intercom. The route from pilothouse to aft should be an easy passage, clear of all obstructions.

AFTERDECK ENGINE CONTROLS

Afterdeck controls should be so synchronized with those in the main pilothouse that no change in speed or steerage or stoppage of propeller would be necessary while going from one station to the other. Their movements should coincide. As a safety measure a latching security cover should be provided to prevent any tampering or movement of controls when under way.

TOW WINCH CONTROLS

On some towing vessels these winch controls are located just inside the after engine room watertight door. This is not a good arrangement. They should be located in a clear area on an upper deck. The captain or mate on watch when connecting up to a tow or letting go should have the full view of the towing hawser and the men working on deck. He should be able to feel the strain on the cable as the winch heaves in or pays out. He should be able to coordinate his actions in maneuvering while making such cable take-up, holding or paying out at the critical moment.

All switches controlling the tow winch should be independent, and when in use they should be synchronized so that the watch stander may leave a station and take control at another without losing speed or steerage.

DECKHOUSES

Deckhouses have always had a function. Once covering the cylinder heads and boilers, they are now much shorter, incorporating the galley and some crew accommodations. Often they do not follow the sheer of the hull. On the afterdeck, whether fitted with an H bitt, tow hook, or tow winch and cable, the position of these is now much farther forward than it was on the post-World War I U.S.-built tugs. These fittings to which the towline is secured are now often forward of the propeller, except in the tractor type tug—their propulsion is approximately one-third the distance from the bow or stern, and they are so built as to be able to tow from either end. The chances of this type of tug girding, or being "caught in irons," and capsizing is greatly reduced.

STABILITY

This is a matter of great concern in tug design. The center of gravity (G) must necessarily be low in order to insure an adequate GM (metacentric height), which is a measure of a vessel's stability. For, when the vessel rolls, its center of buoyancy (B) will shift as the vessel heels, and its new center of buoyancy (B1) will act as a force upwards toward the metacenter (M). If the GM is too small, the righting arm (GZ) will be insufficient to right the vessel, and it will capsize.

The metacentric height of river towboats will normally remain nearly stationary. In the construction of all other tugs, the naval architect must seriously consider the GM in relationship to the predicted G. Severe conditions are frequently imposed on tugs. These may be due to a sudden roll; extreme towline tension on one side, either from pulling or towing a heavy unit; or from the drag and sheer of a tow in a gale.

HULL FORM

In general, the river towboat hull will conform to that shape most pleasing to the operating owner, after careful consideration of the size of tows and projected depth of the waterway to be frequented. Several very successful river towboat hull forms have been produced and continually improved by such ship builders as the Dravo Corporation, St. Louis Shipbuilding, and Jeffboat.

Harbor, River, Coastwise, and Ocean Tug Hulls

These should reflect the depth and bottom conditions expected in their main work area. If it be a tidal area where skimming over mud flats or sand is required, a nearly smooth, watermelon-shaped "keelless" hull will allow movement and steerage while close to the bottom. In case of a grounding, the tug will be less liable to fall on her side.

Tugs that will work in shallow nontidal waters will respond better if given a wide V-bottom and moderate chine, giving a good upward run aft for propellers and rudders.

Tugs normally working in deep waterways will have finer seakeeping and towing qualities if given the traditional deep displacement hull used for years in wooden sailing vessels.

FENDERS AND GUARDS

A well-built shiphandling tug must of necessity be able to lie alongside almost all parts of a ship, with certain exceptions, without damaging either vessel. The exceptions are spots close to ship propellers, rudders, under the flare of bow, or under the overhang of stern. A good ship-assist tug will have a wide and heavy guardrail continuously from bows to and around the stern, with thick, heavy rubber facing, or continuous fendering.

Fendering Protection

River towboats are seldom fendered other than by heavy half-round guards built onto the hull. Most other tugs of all types have some sort of a fendering system. With the diminishing use of natural-fiber deck lines on tugs, the old rope side fenders and bow-and-stern whiskered puddings are only seen on privately owned rebuilt classic tugs.

The purpose of this fendering is to protect both the tug's broadside and that of the vessel she is lying against. The width of the tug's

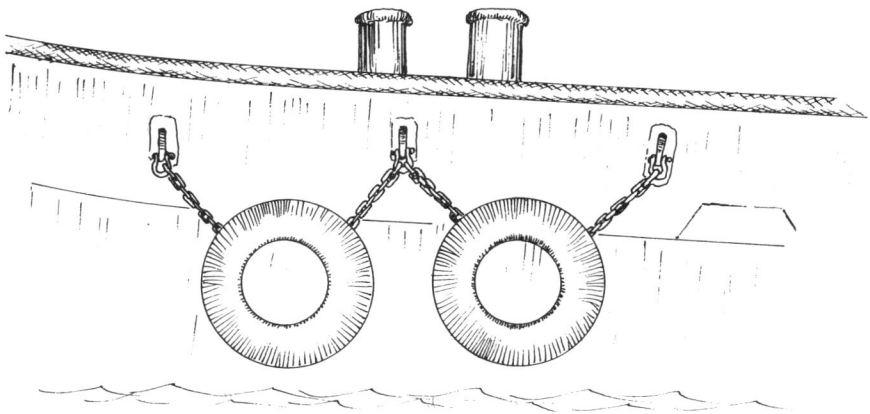

Tires as fenders held by chains are good on harbor and coastwise tugs. They do offer trouble in rough weather at sea, as the chains are noisy and frequently are washed on deck or torn off.

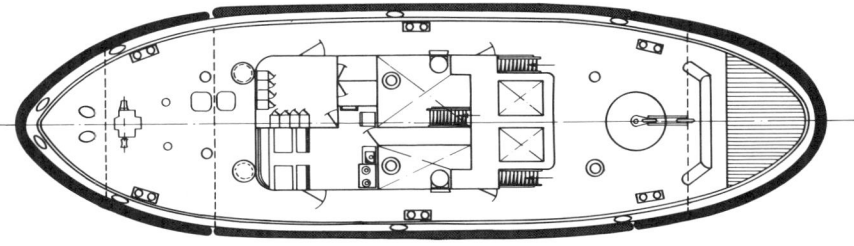

A view looking down on the Voith-Schneider water tractor *Vanguard* illustrating the rubber fender rail around the entire tug. (Courtesy J.M. Voith GmbH)

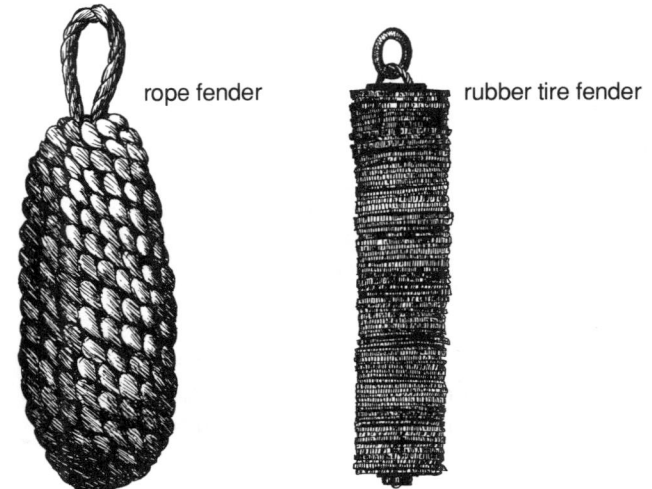

Handmade rope and manufactured rubber tire side fenders

guardrail should be such that when fendering is added, the tug's bulwarks will be well away from the towed unit. As an added protection, bulwarks should be sloped slightly inward. If rubber tires or hanging fenders are to be secured along the tug's side, a solid bar or other arrangement should be made for fastening. Before installing an elaborate fendering system, consideration should be given to the areas the tug will service and the type of work she will do. A continuous rubber rub rail will serve in almost all situations, backed up at the bow and stern by fitted rubber sections. Tires and hanging fenders are very troublesome in rough-water areas, as they fill with water as the tug rolls, get torn off or washed on deck in storms, become covered with marine growth if touching the water, retard the progress of the tug, and require frequent maintenance and replacement.

TUMBLE HOME AND DECKHOUSES

A tug should have some protective features built into her hull. In addition to the inward slope of bulwarks, a tug's broadside from the end of the sheer at the bow to the beginning of it aft should have a gentle curve. It should rise slightly outboard from the waterline to the guardrail then inward, similar to the side of an egg. This should continue up to the covering board strake and bulwarks. This is known as a tumble home, and it is most desirable. Some icebreaking tugs built for the U.S. govern-

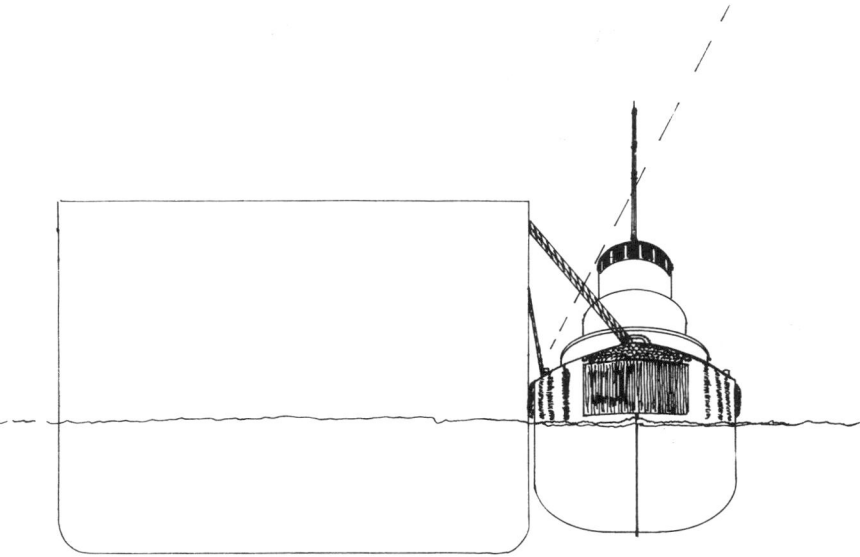

Superstructure of tug well clear of towed unit

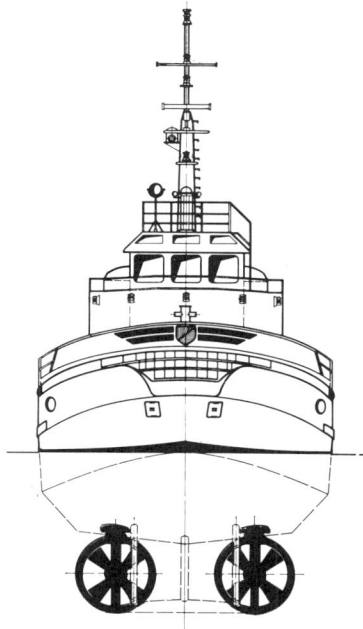

The Schottel tug *Karl* head-on illustrates the necessary tumble home. (Courtesy Schottel)

ment in the mid-1980s have deckhouses that are so high and close to the rail that if the tug is alongside a ship, and riding up on the ice, the tug's boat deck rail could easily be damaged. If the tumble home design, coupled with a wide deck and the pyramiding of superstructure, is incorporated in harbor and ship-assist tugs, the possibility of contact damage when the tug is working around other units will be greatly reduced. Tractor tugs are usually given these protective features, which older standard U.S.-built tugs have lacked. The possibility of damage resulting from the width of deckhouses and pilothouses is often visible in the form of rails on boat decks and bent eyebrows.

THE STERN—GENERAL LAYOUT

Very little towing is ever done over the stern of river towboats. Small inland push tugs do, on occasion, when forced by rough weather, tow from astern. A good set of bitts is required. A very handy design is to have a small single-drum tow winch with several hundred feet of appropriate cable. In lieu of that, a solidly secured set of H bitts could also be used. As on all tugs, there should be a quarter side bitt or cleat for mooring. Side bitts with lips should be fitted with a pin protruding sufficiently on each side to accommodate three turns of line as it is secured in a figure eight around the main bitt. If only a single bitt is installed, it should have a hook under the bulwark rail for the same purpose.

Over 50 percent of all towing work on hawser tugs involves the stern, or is done over the stern. Careful consideration must be given to a safe and workable layout. More personnel have been injured on the stern than in all other areas. The hazard is not limited to personnel injuries. Many times a catastrophe has been averted through quick action when a shackle or line has jammed around or under some protruding yet unprotected object.

Protective Measures for the Stern

To avoid these possibilities, protective measures include the following:

Installing hinging or removable sleeves on all cleats.

Having vertical fairlead rollers set on a rounded base.

Placing a crescent-shape pipe guard over any raised hatch or other obstruction if the tug is without towing arches.

Rounding the edges at the top of the bulwark rail, both inside and outboard.

Rounding off any angles on the forward edge of hawser gratings or lazarette covers, if raised over rudder quadrants or ram steering assembly.

Rounding all towing arches or bars on both sides.

Rounded edges on tow arch, bulwark rail, and fairleads will prevent shackles from catching.

All of these protective measures are designed to reduce the possibility of shackles hanging up either when paying out or heaving in tow cable connecting points. Fairleads installed for use with pushing gear should be protected by a towing arch. They should be laid out so that their sheaves are not exposed. There should be no sharp corners that might allow lines to be cut or snagged.

All lubrication fittings should be located in such a way that they cannot be struck or sheared off.

Towing Arches

Towing arches or bars are those convex beams on the stern running from rail to rail of most seagoing and some harbor tugs. Their purpose is to hold the tug's tow cable clear of the deck and all obstructions. They also give the cable a straight line from the tow winch out over the stern rail or roller. Most deep-sea tugs have three or more of these. They are made of heavy reinforced pipe on bent plate. Careful thought should be given to requiring the minimum number, as they are an obstruction for tug crews. They require constant greasing and are added points of friction for the hawser. When towing at sea, chafing gear or hawser boards are required. Vessels with shorter afterdecks frequently use a single arch of very heavy pipe on which a sliding sheave is fitted. There are heavy stops just above the rail at each end of the arch, which keep the towline from cutting the tug's rail. The arch itself must be placed in such a position that the tug hawser will clear the taffrail without resting or cutting on it except in the most violent of seas. Chafing gear should be applied at such a point prior to encountering heavy weather and the arch itself should be kept greased at all times. As grease is messy and frequently is washed off, a white silicon type such as Lubroplate has proven very satisfactory.

TOW WINCH FASTENING, LOCATION, AND CONTROLS

Tugs requiring a tow winch should be designed with adequate heavy underdeck beams and knees, framed and welded so as to give the absolute maximum strength required of the tug and tow in heavy seas—approximately twice the tug's bollard pull.

A tow winch should be located as far forward as practical to allow the tug to swing a full 90 degrees either way with a static tension, or with a towed unit under full control, without causing the tug to roll down more than 5 percent toward the object to which it is connected.

If the tow winch is located under cover of the boat deck, it should be adequately illuminated. Lighting fixtures should be placed so that any slack cable cannot strike them. Clearance over the tow winch drum(s) should be several feet when the full amount of cable is spooled in.

As indicated earlier, tow winch controls should be accessible both at the winch and at any after control station. A switching arrangement should allow power to be turned on or off independently at either location. In some cases, it may be desirable to locate winch controls in the pilothouse. Wherever they are located, they should be sealed in a watertight protective case with security against unauthorized operation.

TOW WINCH TYPES

Tow winch types include such well-known products as Intercontinental, Almon-Johnson, Markey, and Skagit in the United States. Some, such as Almon-Johnson, have automatic tension devices and huge springs that are constantly loaded and unloaded in a seaway. This allows a

Esteff S. DeFelice's double-drum, diesel-drive Markey TDSD-36 tow winch is visible behind the pair of white H bitts. The socketed ends of her two tow cables are shown ready for use. The one to starboard is $2\frac{1}{4}$ inches in diameter and the one to port is 2 inches. (Sockets are sometimes referred to as jewels or D's.) This is a U.S. Gulf-built rough-water petroleum service tug. (Courtesy Markey Machinery Company, Inc.)

slight seesaw movement of the hawser. Automatic tension devices require very careful adjustment and watching so that payout and retrieval are equal. Otherwise, the watch officer will find his tow about to approach his stern or the remaining cable on the winch drum dangerously short. The type of tow winch to select is subject to the owner's and operator's requirements. Tugs on the U.S. West Coast and many that have been built for the oil industry have double-drum tow winches. They carry up to 3,500 feet of cable on one drum and 2,000 to 2,500 feet of cable on the other. The cable size is regulated by the tug's bollard pull or horsepower and the type of units to be towed.

HAWSER TENSION DAMPER

To prevent the parting or breakage of the towline when shock loads do occur, it is either necessary to slow the tug or to have a tension damper installed. As this is a large piece of equipment, it is usually placed on the inside of the deckhouse of large ocean tugs. It may also be used on the deck of tug/supply vessels. One of the common types is the Murdock rope tension damper. It is made up to two sheaves mounted at opposite ends of a rotatable arm. The arm connects to elastomeric tension springs that

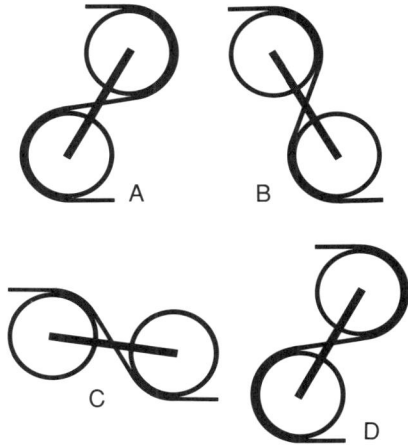

Diagrams illustrating how the Murdock tension damper works to minimize shock loads: *A*, in unloaded condition, sheaves are set to establish the proper force vs. length of travel condition; *B*, when load is applied, sheaves rotate toward a horizontal position, increasing the effective length of the line (as sheaves move, the unit's elastomeric torsion springs provide increasing resistance; *C*, at maximum force, sheaves are almost horizontal; *D*, as force is released, the elastomeric torsion springs begin to return sheaves to their original setting, shortening the effective length of the line. (Courtesy Murdock Engineering Company)

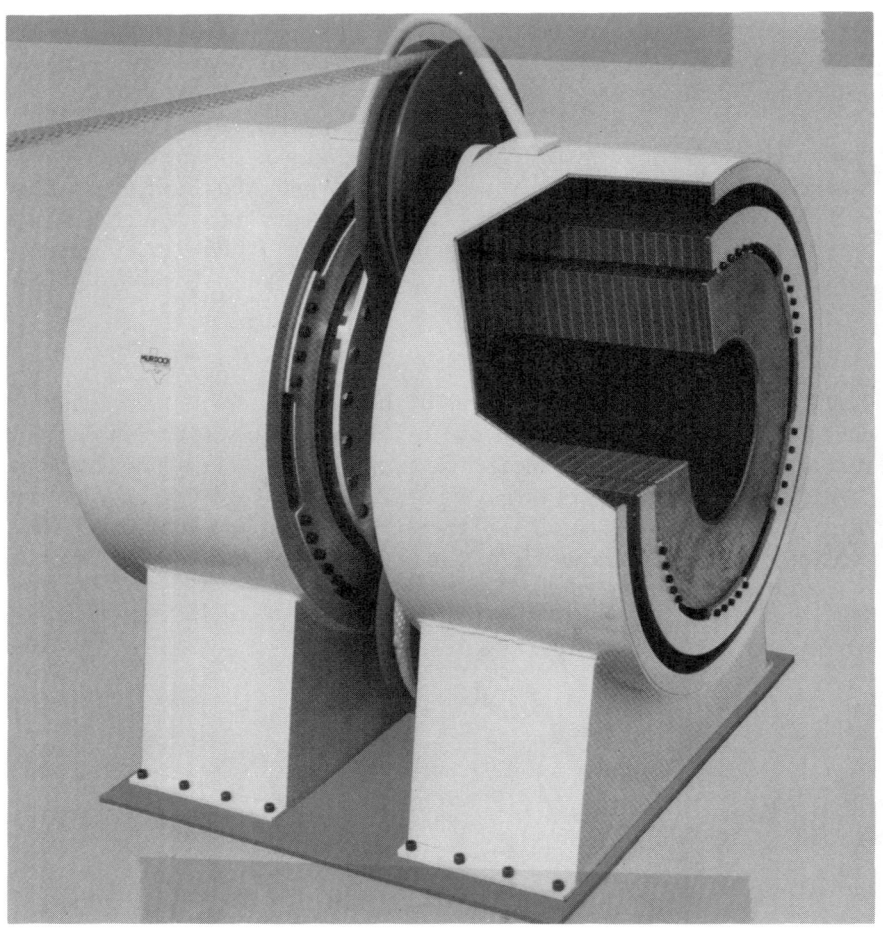

The Murdock tension damper (Courtesy Murdock)

are housed in a fixed frame. The Murdock tension damper works to minimize towline shock.

TOW WINCH POWER

This should be derived from an independent source that does not come as a take-off of the tug's main power plant. Most ocean tugs have a separate diesel generator set if the tow winch is electric or a hydraulic pump for that type of system.

TOW WINCH OPERATION

See chapter 7.

TOW WINCH CABLE GUIDE

In conjunction with the tow winch there usually is a set of vertical rollers, which are mounted on a diamond-cut round stainless steel bar, allowing the cable to be threaded back and forth as it leaves or is rewound on the winch drum. This latter action is often described in manuals as spooling.

TOW WINCH LINE RETRIEVAL AND WARPING DRUM

Most tow winches have on a horizontal axis a small drum used to heave in heavy fiber or synthetic deck lines. This drum, which has many names (in different places and different languages), can be operated independently from the main winch drum(s).

The lineup of the final run to this warping drum should be fixed through a fairlead so that the line being heaved in will easily go on the drum as the inside turns and will not cause a pileup or cross turning of one overriding another or jamming.

DECK CAPSTANS

Vertical deck capstans should have their controls arranged similarly to those of the tow winch. The drum should be fitted with whelps. These are wooden strips of from $1\frac{1}{4}$ to 2 inches in width and no less than 1 inch in thickness. They are spaced 10 inches apart and run from under the top lip of the drum to above the bottom lip. When sythetic ropes are being heaved in under tension, these whelps prevent slipping or burning. Nothing will prevent ice-covered lines or capstan drums from slippage. The height of the capstan's bottom lip should be such that a line leading from it to a set of bitts or deck fastenings should be horizontal or slightly downward. It is also most important that there be no open space between the bottom lip of the deck capstan and its foundation mounting. Otherwise, one might experience a repeat of what happened to a tug of a well-known New York operator.

One of their tugs was towing a large, empty oil barge down the Hudson River. The barge was being towed on two sternlines and was sheering back and forth. When meeting another tug and tow in the narrow channel above the East Kingston Bridge, the tug was slowed. The barge, not slowing as quickly, sheered to the right and one of the gate hawsers slacked to the deck on the inboard side of the capstan. When the tug

The forward deck capstan of the tug *Turmoil* is rigged with whelps, the vertical strips that aid in keeping synthetic or soft lines from slipping when they are being heaved in under a great strain.

speeded up, the line slipped under the lower edge of the capstan and became taut, catapulting the capstan drum and shaft overboard.

In mounting a capstan it must be kept in mind that the bottom of the capstan drum should be at the same height or slightly higher than the cavel of any bitt(s) or fairlead. This will keep turns on the capstan from overriding.

FAIRLEADS

Roller type fairleads are the only acceptable rope-saving device that allows a piece of line to be heaved around. If possible, they should be placed to give a lead angle of not more than 90 degrees. This is frequently impossible without cluttering the deck with a tripping hazard. The ideal situation is to arrange two trawler type open-roller fairleads near

TUG DESIGN   111

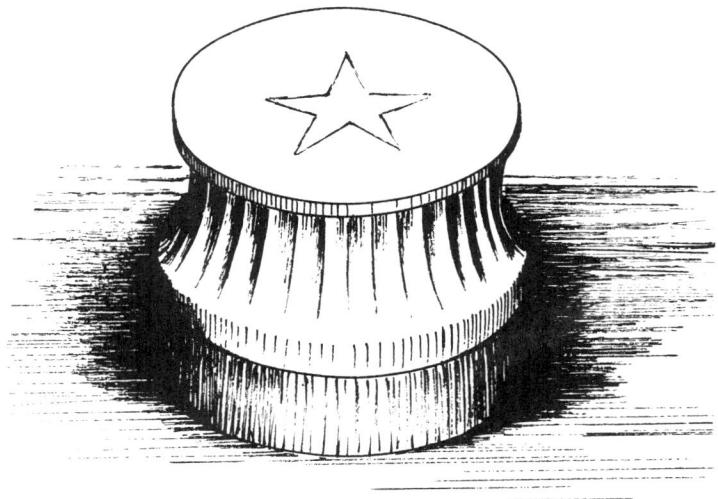

A properly installed capstan leaves no way that a line can get under its base.

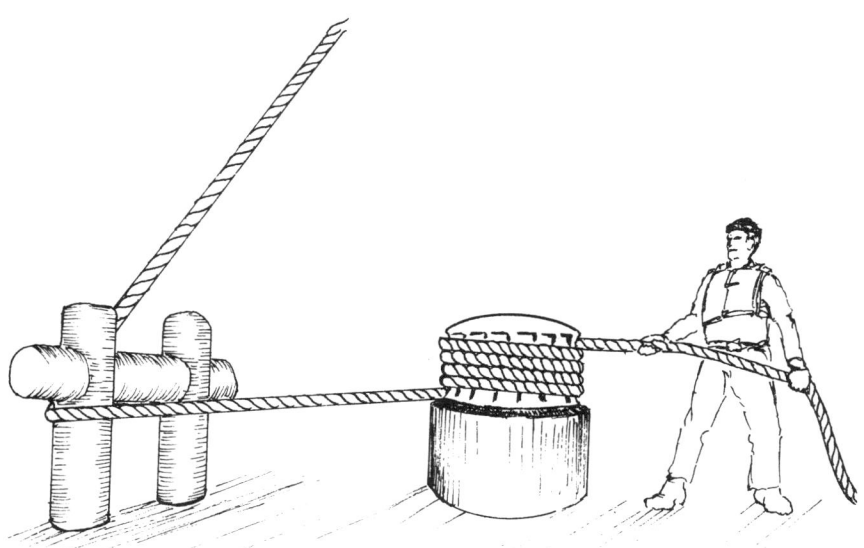

The lower turns of line on all capstans should be on the same level as any fitting the line will be led around.

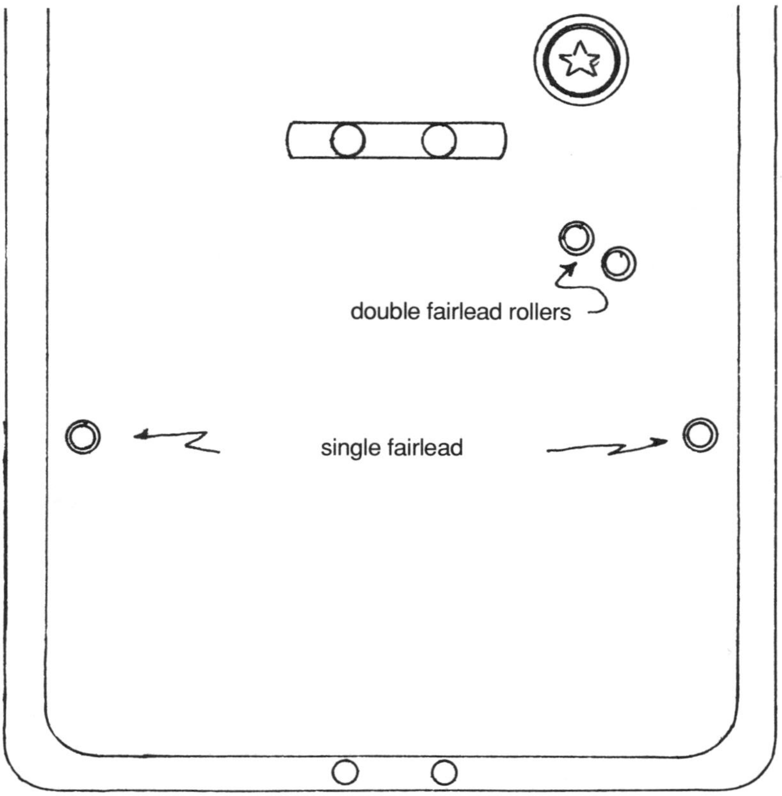

[diagram of tug stern deck showing capstan, double fairlead rollers, and single fairleads]

Any lines passed around single fairleads and through double fairleads should lead nicely to the tug's capstan if all are at the same level.

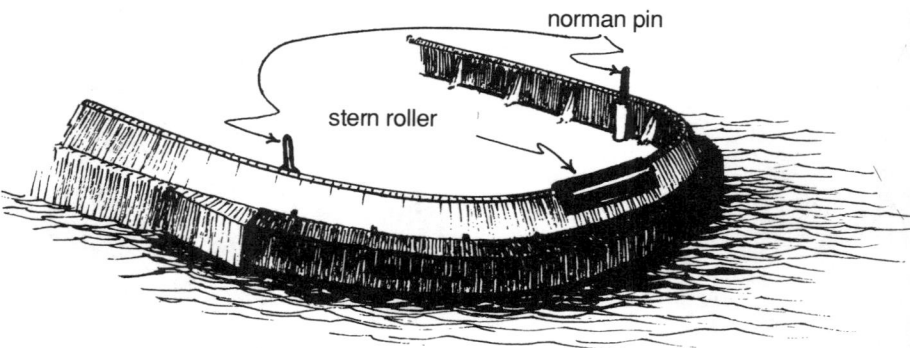

These norman pins are hand-removable. Others near the stern roller either hinge inboard or raise and lower hydraulically.

the bulwarks on each quarter, and one double-roller fairlead amidships. Roller fairleads are not usually necessary in any but the larger coastwise and oceangoing tugs.

VERTICAL AND HORIZONTAL STERN ROLLERS AND NORMAN PINS

Bay, sound, coastwise, and ocean tugs are all faced with the same problems: how to protect the towing hawser against excessive wear; how to keep it leading nearly directly over the stern; and how to have it lead directly either to the tow winch or to the capstan.

Protection of the tow hawser at the stern rail was accomplished by chafing gear of old canvas, fire hose, or various types of hawser boards. To keep the hawser from leading too far off each quarter, norman pins were installed. These were usually made of heavy solid round bar inserted into a pipe socket or hole in the bulwark. They protruded several feet above the rail. Because they were heavy and awkward to handle, great care was necessary to avoid injury. This type is still in use on some tugs.

THUMB CHOCKS

As an added precaution against allowing the towing hawser to lead forward around a deckhouse corner, a casting in the shape of a half-circle can be mounted on the bulwark rail on either side. Such castings are known as thumb chocks.

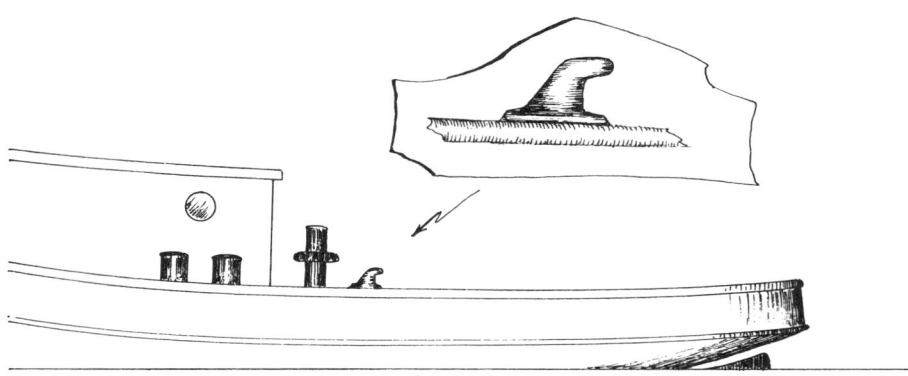

Thumb chocks are still included as a fixed fitting on some tugs. They offer a stopping point for tow hawsers and lines when being heaved in. However, if the tug should roll and the line jumps out, it may strike someone. It will also become slacked. Lines should not be run in or through thumb chocks to the H or other bitts when towing alongside.

HYDRAULIC PINS AND ROLLERS

Modern tugs now have norman pins and vertical stern rollers that are hydraulically controlled from either the after control station or the pilothouse. They are arranged amidships at the stern to keep the towline running straight from the tow winch overboard. They are raised when energized. They sink flush with the rail when not in use. Tugs with double-drum tow winches usually have two sets of norman pins—one for each winch drum, making a total of four hydraulically controlled vertical pins. However, there are some with three.

In deep-sea towing, it is a great advantage to have a single, hydraulically controlled norman pin on each quarter. Hinged vertical stern rollers have been used, but like the hand-installed norman pins, they require an alert, "heads-up" consciousness when being adjusted while the tug is under way with the hawser out.

HORIZONTAL ROLLERS

On some tugs, particularly those handling anchors for rigs, a horizontal roller of a sizable diameter has been installed. These, as well as vertical rollers and norman pins, should have a lubrication arrangement to keep them free-running.

In connection with these pins and rollers, the supporting bulwark area should be strengthened internally to prevent bending or denting, which would cause the rollers to jam, or the entrance of water, which might freeze and retard free movement.

All rollers should be made of hardened steel or a special tough alloy to prevent scoring or cable wear. There should be no space or crack between the base of vertical or horizontal rollers and the rail. Inevitably, a steel hawser will find such a tiny crevice and wear there.

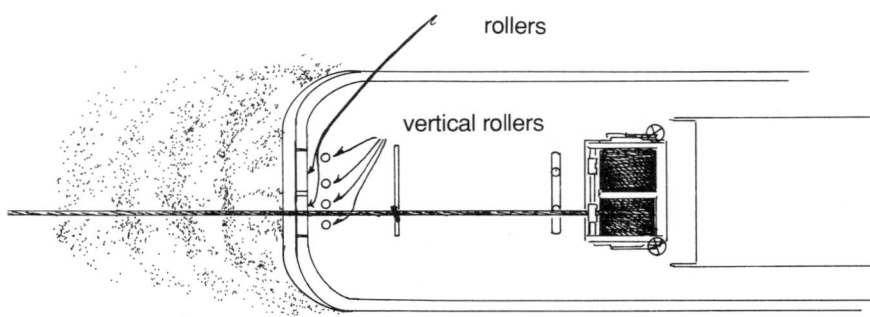

Vertical rollers are also called norman pins. They are hydraulically controlled from either the pilothouse, the after steering station, or the main deck.

CABLE HOLD-DOWNS

The tow cable and/or towing hawser will receive proper care if adequate chafing gear is provided. Chafing gear should be used wherever it fastens over towing arches, tow bars, or tow rails and the stern or taffrail. Nothing, however, will prevent it from remaining squarely and fairly between the norman pin rollers when the tug is pitching or rolling in heavy weather. It must be securely held down.

Various methods have been adapted throughout the world to achieve the above. One large fleet uses a solid steel 1-inch by 6-inch bar about 4 feet long. This is securely welded to a doubler plate amidships and about 2 feet inboard of the stern roller and norman pins. This 1-inch thick bar is pierced by holes for shackle pins. The holes should be able to accommodate the diameter of the largest shackle necessary to withstand the heaviest vertical stress that the tow cable will receive at that point when the weather and sea are extreme. To avoid damage from sharp edges, these pinholes should be smooth. To distribute the stress on the shackle pin, the holes may be built up on each side. When this method is used, the hold-down itself may be either a loop of heavy chain through a shackle riding on the tow cable or a heavy thimbled short length of wire also secured in a riding shackle.

SINGLE TOW ARCH AND SHEAVE

On tugs with a single tow arch and riding sheave, the tow cable is held into the sheave by a short piece of chain that runs under it. A shackle fitted over the hawser just forward of and just aft of the sheave will keep the hawser held in the sheave in almost any weather. Personal experience with this method was very satisfactory. There are, of course, many other ingenious methods of preventing the tow cable from flaying

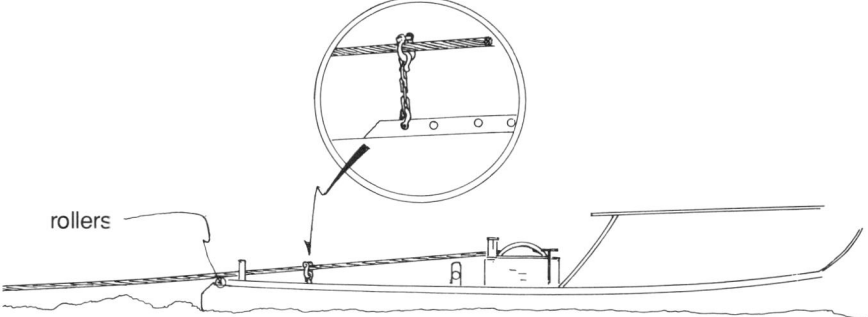

This is a popular type of hold-down gear. There are many others.

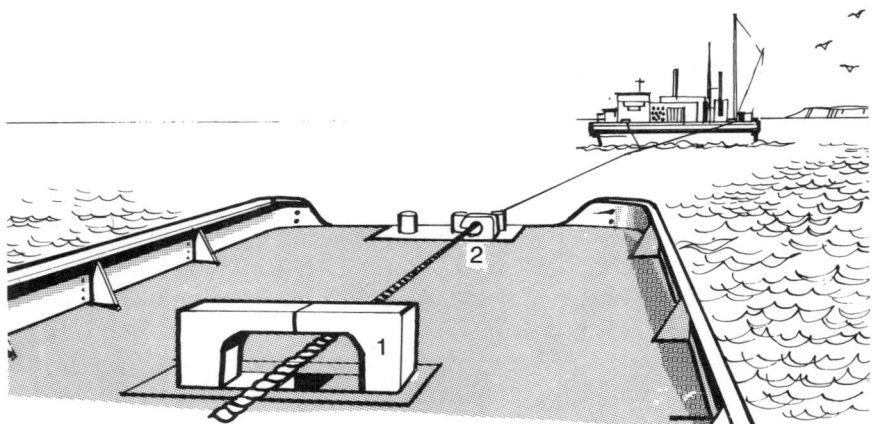

Retractable fairleads allow tug/supply vessels to lead a hawser to the stern. (Courtesy Western Machine Works)

1. retractable deck fairlead as on both vessels
2. retractable stern rollers and cable hold-down on both vessels

The retractable fairleads fit into recesses in the tug/supply vessel's deck.

or straying across the stern. British tugs have used a "gob rope" of cable-laid line or wire, one end of which was fastened to a cleat in line with the amidships run of the tow cable. A bight is run up over the cable and back through a snatch block and onto a capstan. This allows the cable to slack to one side or to be heaved down tightly.

HYDRAULIC CABLE CONTROLS FOR TUG/SUPPLY VESSELS

Some tug/supply vessels are rigged for towing and have long, clear afterdecks. Because of the distance from the tow winch to the stern, it is necessary to keep the tow cable from slipping over the side. This is accomplished by the use of a hydraulic set of double rollers, which recess into the deck about halfway between the tow winch and the stern. When they are in their upright position they may also be used in conjunction with two stern hydraulic pins.

FAIRLEADS AND MOLLY GOGGERS

British and some European tugs use a fairlead that folds down on a hinge to the deck when not in use. It is usually placed to lead a line through to a capstan when heaving in synthetic hawsers.

BITTS (STERN, SIDE, AND BOW AS MOORING FITTINGS)

Side bitts as previously described are necessary for mooring alongside, or working in docking or undocking ships, barges, or other units. Those known as side bitts, either forward or aft, should be built to follow the same tumble home as given the vessel. If possible, they should be placed so that their outboard side is at least 4 inches inside the outer edge of the bulwark.

The bulwark, in the area where lines will cross it when running to the bitt, should be curved to prevent line chafe.

STERN H BITTS

These are located amidships and normally just aft of the tow winch. They should have their cavels (the flat or rounded extension that runs horizontally through them) at a height above the deck equal to that of the bottom of the capstan drum. If the tug is without a vertical capstan then the cavel should be at a height of the waist of an average-sized man.

FORWARD BITTS

Located on the tug's bow, these bitts may vary in design from the single cruciform found on small tugs and the pawl post of U.S. Great Lakes harbor tugs. Some may have a fore and aft letter H configuration. On large modern tugs triangular-shaped bow bitts have been very successfully used. All bitts must have very strong cavels, with smooth surfaces to prevent damage to the lines that will be used around them.

A few older U.S. railroad tugs used for railcar float work may have a single bitt, called a button, halfway between the forward side bitt and the stemhead.

STEMHEAD

Most stemheads have been replaced by the heavy half-circle eyelet, closed chock. On new vessels as well as those already in service it is prudent to test them at regular intervals. Unless they have been carefully installed, they tend to crack at welds. One modern tug of about 9,000 HP, when backing down on a large barge, had a bowline through this type of closed bow chock. The line led upward to a cleat on the barge. While the tug was backing with the line under a strain, the chock let go and flew in the air as if fired from a crossbow. Luckily, no one was hurt. A poor welding job was blamed for this casualty, as the chock had been

The bow eyelet has replaced the round stemhead on many tugs. *Andy Head* also has been fitted with rubber fenders entirely around the vessel. (Courtesy J. Wilsky)

fastened to the top of the rail instead of being recessed into it with welds on both sides.

TOW HOOKS

Most European harbor and other tugs are equipped with a tow hook. One is placed close to the amidships tow winch; a second may be located farther aft. There are various types of tow hooks. Ocean tugs' tow hooks usually have a huge spring, while harbor tugs' hooks are rigid. All are fitted so that they may move in an arc while riding on a heavy half-circular bar with very heavy stops on each end. All should have a very easy method of being released either by a yank on a lanyard, a swipe with a maul, or by remote pneumatic tripping.

To prevent the possibility of girding, the tow hook on conventional tugs should be located well forward of the propeller(s). When a towline is in it, the tow hook's location will allow the tug to pivot on the hook. On Schottel Z-peller, rudderpropeller, and cycloidal tugs, the tow hook is usually installed in nearly an amidships fore-to-aft position. This gives greater safety from girding.

On tractor tugs and any with a limited crew of three or less, a release should be available in a safe spot on the lower deck. Tractor tugs

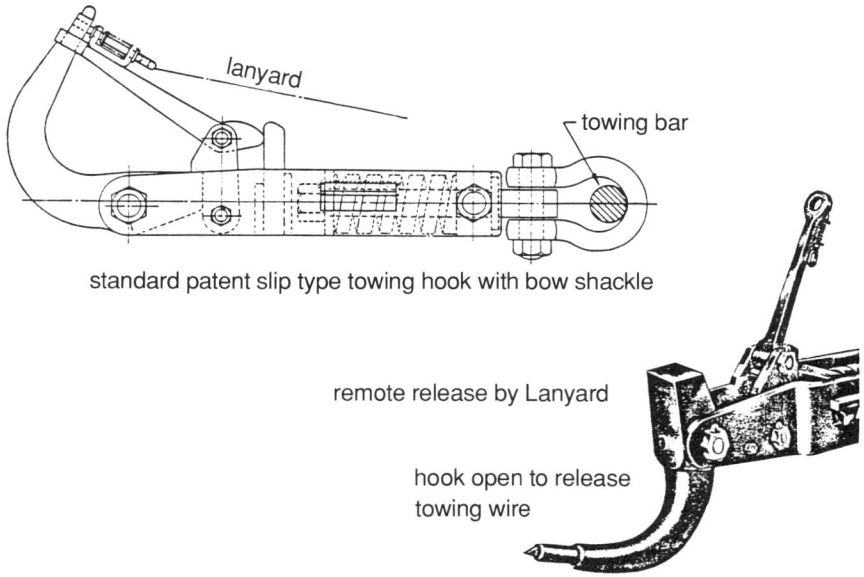

standard patent slip type towing hook with bow shackle

remote release by Lanyard

hook open to release towing wire

These towing hooks are frequently found on European and some Canadian harbor tugs.

Here are two different types of towing hooks. Both are used on ocean tugs. Notice the difference in release gear.

when operating with a three-man crew often have a tow hook release on the main deck, inside the engine room passageway, and the winch control stand.

DECK WINCHES AND SHEAVES

Deck push cable winches and sheaves that allow the cable to pass through the bulwarks should have easily located lubrication points. Power winches should have weatherproof protective covers in place when the winch is not in use.

VENTS

Vents from fuel, fresh water, and void space tanks should run up to a ball-check gooseneck. If at all possible, the gooseneck should be located on a deck above the main deck, and the vent itself should pass upward inside the deckhouse between the outer plating and inner finish. Vents previously installed on the exterior next to the deckhouse have been very susceptible to a buildup of rust on their inner sides. Rust in such a location is very difficult to remove.

STOWAGE

A study of the best stowage for deck gear should be incorporated in building plans. Tugs that tow heavy units for long distances require spare towing gear to be on board; hawsers, shackles, pendants, and line are among the most frequently needed sizable items.

Spare nylon or poly type towing hawsers of up to 12 to 14 inches circumference are frequently stowed on a grating on the boat deck. These should have a protective cover. The location should be close to the deck edge on whatever side the capstan or other retrieving device is located. As most decks have a flat bar facing with a lip of from 3 to 4 inches in height, a vertical pipe type roller should be installed in a convenient spot so that the hawser will run out freely as it is passed to the deck below or hauled back for stowage.

An added feature incorporated in many large tugs is a cherry picker type boom, operated by air or hydraulic pressure. It will have many uses and is most handy when extended over the side to lift drums of oil, coils of lines, chain, etc.

OUTSIDE ILLUMINATION

In addition to the normal watertight deck lights, the tug must have a good illumination of her deck working areas. Forward floodlights may be located under the pilothouse canopy eyebrow or boat deck coaming. Aft, they may be attached to the boat deck rail. These should be so arranged in order to light the entire stern. If the tug has an after control station, additional lighting should be located above it so that men working on deck at night can clearly see the person at the controls even when the other floodlights are on.

OUTSIDE LIGHT CONTROLS

All outside light controls should be from switches at the pilothouse console.

TELEPHONES AND INTERCOM SYSTEMS

A pilothouse-controlled two-way speaker system is recommended. Speakers covering the forward deck, aft control station, tow winch, and stern work area are usual. Each station should have a button to alert and energize the pilothouse.

Inside the tug, telephones are commonly used between pilothouse and engine room, captain's cabin, galley, steering gear area, and chief engineer's stateroom.

Due to the excessive noise, the engine room telephone buzzer should have a backup that actuates one or more flashing lights until the phone is answered.

ENGINE ROOMS

River towboat engine rooms are usually wide and off the main deck, with lots of air space. Tug engine rooms by comparison vary in size and location. Many on modern tugs are below the main deck and require an enormous amount of ventilation. The temperature of the high-speed internal combustion engine is extremely high. This heat not only rises but radiates from the engine to any metal close by. Ladders leading down to engine rooms should have their handrails protected or wrapped so that personnel will not get burned.

Cooling air may come down from blowers on the boat deck. Unless these are fitted with covers, they may admit seawater in violent weather. Auxiliary air should be introduced through a secondary system of louvers in the stack and mast. In addition to the heat of the diesels, noise is another consideration. At operating speed, it is almost impossible for personnel to be heard without shouting, while almost kissing the other person's ear. To avoid ear damage, it is imperative that engine room personnel wear special protective soundproof earmuffs.

On many modern tugs the engine room heat and noise have been overcome for operating personnel by providing an insulated, air-cooled, soundproof operations booth made of Plexiglas. These offer a full view of the engine room. All pertinent gauges, along with telephone, engine, and auxiliary monitoring alarms and a unit of the general alarm system, are mounted within.

ENGINES

They must be compatible with the hull size and strength. Requirements of hull plate and engine bed are worked out by classification societies. Choice of an engine and auxiliaries in a new tug depends on the estimated amount of daily use, the operating areas, and weather. Care of the engine must be practiced by fixed periods of maintenance procedures rather than by the previous method of human monitoring by sound and feel.

Some engines are high-speed and some are medium- and low-speed. All engine manufacturers stress fuel economy, and the new owner would do well to scan both the market and tugs that are operating with the type of engine in which he has interest. Some modern engines have more economical fuel systems, which burn blends of lower-grade products.

Auxiliaries

Auxiliary machinery such as pumps, air compressors, and generators should be carefully placed on either side of the engine room at a height for easy maintenance and mounted in such a way that essential service points are easily accessible.

Generators and electrical main switchboards should be placed in a protected position, keeping in mind the possibility of the entrance of seawater from a ventilator or other source.

Valves

Fuel and lube oil piping valves, manifolds, and any crossover piping should be run at deck level in the open for easy inspection. Sight gauges for monitoring all the operations and pressures as well as fuel, lube, and potable water tanks should be easily readable from the engine room log desk.

Automation

In automated engine rooms, all gauges are monitored by alarms and lights. Usually, there is one on each piece of equipment, one in the responsible crew member's quarters, and one in the pilothouse. These alarms should be checked frequently. Failure to do so on some tugs has been catastrophic. With the great increase in horsepower and resultant heat, a slight leak of diesel fuel in an unmanned engine room has been known to cause disastrous fires. Also, failure of back-up alarms in connection with either cooling water or lube pressure has resulted in bearings and cam shafts becoming overheated and distorted.

Popular Diesel Engines

Among the many diesel engines popular in the late 1980s, here are a few used in tugs in the United States: Alco; Caterpillar; General Motors, Electro-Motive type; Fairbanks Morse; and Nordberg. A popular group of non-U.S.-manufactured diesel engines, many of which burn heavy fuel economically, are: MAN-B & W, Deutz, Krupp MaK, Mirrlees, Nohab-Polar, Pielstick, Polar-Atlas, SACM, Stork-Werkspoor, Sulzer, and Wärtsilä.

Horsepower

This is what towing is all about—the best use of horsepower. This is how towing is marketed. The tug owner will always base his charge on his claim to horsepower provided.

The measurement used in rating horsepower or the pulling power of tugs varies. In the United States, horsepower is stated as "indicated" (IHP) or "brake" (BHP), while in other areas, the number of tons "bollard pull" is advertised as their strength. The rule of thumb is: 100 HP equals 1 ton of pull. The term *horsepower* came from the steam engine, whose power was found through a system known as taking and reading of indicator cards. Brake horsepower is the reading developed by the engine when new and on the manufacturer's test block.

Quite naturally, tug owners list their vessels' horsepower using the highest figure attainable. Often, it is a statistic furnished by the engine manufacturer from a full power-no load test. Results of a dynamometer test that gives "tons of bollard pull" are far more realistic. The term *bollard pull* has, however, been misused by some. Tug engines have been overloaded and a test propeller fitted in order to obtain the highest possible result.

At one time, tugs were connected stern to stern with several hundred feet of tow cable between them as in a tug-of-war. On signal, both tugs started to pull against each other until one definitely dragged the other. This was not a true test unless it was run for 15 minutes. In a shorter period, each tug would reach a peak or maximum bollard pull in between 10 and 15 seconds after the start. Then, turbulence and cavitation around the tug's stern would cause a decrease in tension until the water smoothed out. The first tug to reach the maximum bollard pull was usually called the winner. If this tug-of-war were to have continued for from 15 to 20 minutes, she might not have won.

To make sure that this expression of bollard pull is based on a uniform set of criteria, the Norwegian classification society, Det norske Veritas (D.n.V.), in consultation with some of the largest European tug owners, has formulated a test that will give the best possible indication of a tug's true towing power. This test stipulates that a continuous bollard pull be maintained for at least 10 to 15 minutes in order to arrive at an indication of a tug's true pulling power. Its results are stated in a bollard pull certificate issued by D.n.V. As an example, D.n.V.'s full bollard-pull-testing recommendations are shown in appendix II. The American Bureau of Shipping was the first classification society to issue a bollard pull certificate for tugs. Lloyds and other classification societies and bureaus have followed suit.

WASTEWATER AND SEWAGE

Provision for storage and disposal of waste, bilge, and sewage fluids must be considered. These provisions must conform to all government regulations for all the waters in which the tug may operate.

THRUSTERS

In order for tugs to be more maneuverable, as well as to maintain position without mooring, thrusters have been added at the bow of many large tugs and tug/supply vessels. As thrusters, they have also been introduced into the bow of tug-barge units, where their use in improving maneuverability in docking offers a great savings by avoiding a need for additional tug assistance.

TUG DESIGN 125

There are several applications of thrusters. Some are retractable into a well and some are fixed to the hull. AquaMaster, Inc., offers vertical, fixed, or outboard units with 360-degree azimuthing as well as retractable and draft-adjusting models up to 4,000 HP. Schottel offers transverse thrusters, bow-jets, cone-jets, and pump-jets for the various requirements of draft and maneuverability.

STEERING SYSTEMS

Whatever type of steering system is chosen or is in use, it should have an independent backup. This should not be an extension of the primary system. There should be a hand steering gear, if feasible, or an emergency tiller, if possible, to cover these safety requirements.

If a hydraulic system moves rudder rams, a separate hydraulic pump, line, and emergency motor should be an integral part of the system. The method of switching from one to the other from the pilothouse should give immediate action. If separate stations are installed, each should be in conjunction with the other, so that the person handling the

The AquaMaster bow thruster unit. (Courtesy AquaMaster Inc.)

tug can shift from one station to the other with the rudder remaining at the last command and no switching necessary.

Some large tugs have up to three separate steering systems moving twin rudders and one for flanking rudders. Those tugs with an upper wheelhouse should have available at least one back-up steering system in addition to the primary.

Steering systems on most tugs that venture out of harbor are linked with an autopilot. There are many autopilots on the world market. Some of these have been indirect spin-offs of the old Sperry (U.S.) "iron mike." The latest are the Sperry Universal Gyro-Pilot and Sperry A.S.M.-Adaptive Steering System, which compete against such systems as Anshutz, Decca, and others.

"JOYSTICK" STEERING

This is the unique system used on rudderpropellers (Z-drive), or cycloidal propulsion drives. As they are related to propulsion they are discussed under "Propulsion and Controls" in this chapter and in chapter 6.

RUDDERS

Other than steerable propulsion units, individual rudders remain as the directional control unit of the towing vessel. To attain the highest precision, many factors should be considered at the drawing board or computer.

The standard single rudder, placed slightly behind the propeller, has been modified in shape and size. Its bearing surface should be designed to get the greatest effect from the propeller at various angles of turn. This has been accomplished by trimming unnecessary corners at the rudder's top and bottom.

Fine tuning of the trailing edge does not always result in fine steering. In considering a rudder shape, size, and location, the designer should include the possible fractional speed reduction due to drag versus the fuel savings from steering a better course.

It has been found that some rudders that are tapered to a thin trailing edge tend, unless given considerable counterbalance, to allow the tug to wander and require continual turning of the steering wheel and system. This has been corrected on some vessels by building out a "salmon tail," which is a vertical wedge on each side of the trailing edge of the rudder. On single-screw vessels, the edge on the starboard side (if the propeller turns to the right when going ahead) is made slightly larger. If the tug has a left- or port-turning propeller, that side would be the larger. The result of these applications gives steadier steering and a better-turning tug with little loss of speed.

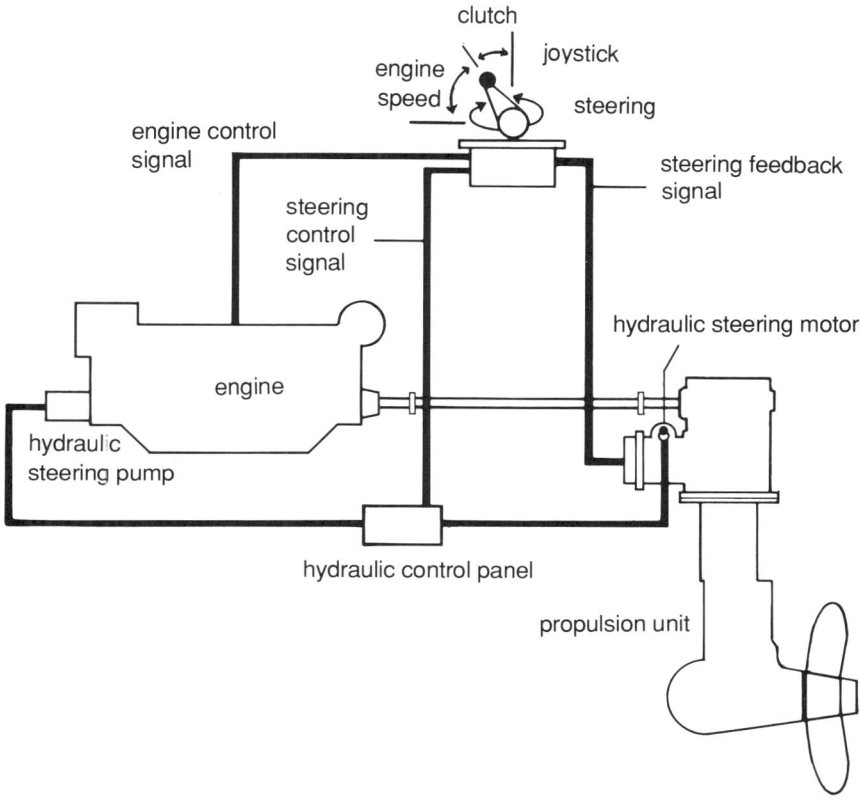

One of several models of Schottel rudderpropeller systems. (Courtesy Schottel)

Rudder size and location are also criteria affected by the work area and its weather. In shallow waters the rudder should not be at the lowest point of the tug's draft.

River towboats usually have several propellers, with a rudder behind each one and flanking rudders placed ahead and to one side of the propeller. All are above any extension of the vessel's bottom or keel.

Small workboats and tugs with single engines may have a rudder hung by pintles into gudgeons on the stern frame or stepped into a socket on the end of the skeg. Some small tugs and workboats that are dredge tenders have a protective cage around the rudder and propeller.

Construction of all rudders should be rugged, with interior web framing. Holes should be provided for the filling and passage of anticorrosive fluid. A filling plug on the top and a drainage plug on the bottom are necessary.

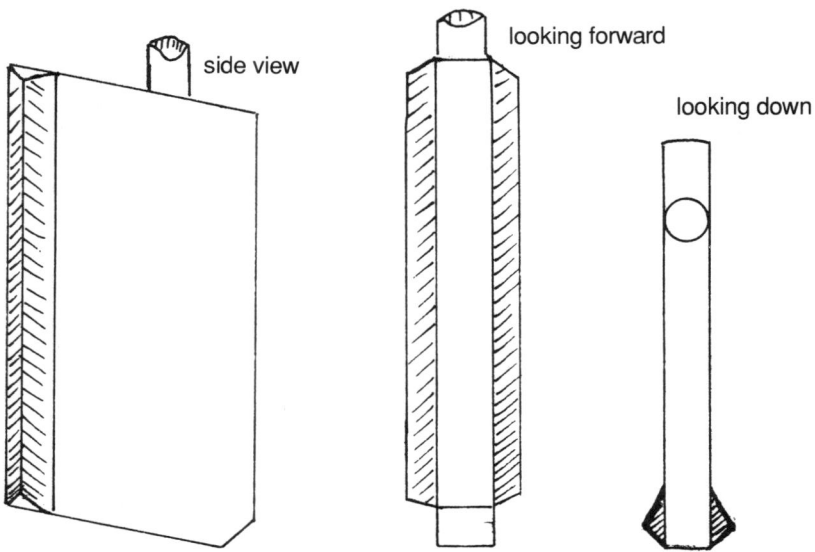

Cheekpieces or salmon tails as shown may be fitted to almost any rudder on a single-screw tug. Steering should be greatly improved.

Rudders on twin-screw tugs, which are hung in the open from a flange on the rudder stock, are more susceptible to damage than those fitted to the nozzled propeller. Working around log rafts, breaking heavy ice, or maneuvering near hard gravel or rocky sloping beaches are risks that with one touch can bend or break off the open unprotected rudder.

Degree of Rudder Angle

The degrees of turn to "hard over" (full rudder) left or right is never less than 35 degrees from amidships. Any additional angle on a hung rudder, i.e., one swinging on the sternpost, places too much strain on the steering gear. Counterbalanced rudders, those with from one-eighth to one-fourth of their surface forward of the rudderpost, could accept a much greater swing and resulting small turning circles.

One of the most successful rudder steering improvements is the Tow Master Rudder System.* This is a series of three fins or shutters, small counterbalanced rudders fitted side by side on the after side of any

*Designer: Burness, Corlett & Partners Limited in Britain, Michigan & Wheel, U.S. licensee.

The Tow Master steering system of triple rudder vanes fitted behind a nozzled propeller greatly reduces the tug's turning circle. (Courtesy Michigan Wheel Corporation)

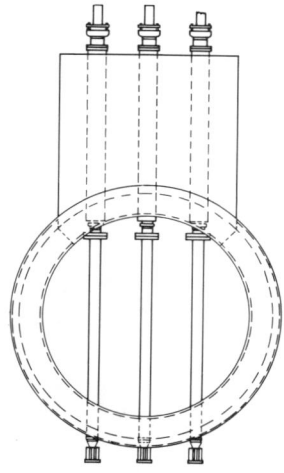 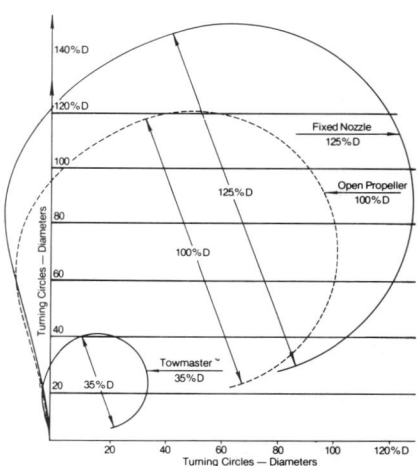

This turning circle diagram illustrates Tow Master's advantages to tugs' steering. (Courtesy Michigan Wheel)

nozzle or direct propulsion system. They have been installed on many tugs worldwide. The claimed results are astounding: they permit a larger turning angle of 60 degrees, which slows the vessel and creates a tighter turning circle. The rudder angle is the same or somewhat larger than that of a conventional single rudder with less torque.

FIXED PROPELLERS

Propeller technology has changed gradually. In the early 1800s Napier introduced cast iron blades on propellers that were tested by some of the British navy's "wooden walls" at Gibraltar. Due to breakage of blades, bronze propellers were introduced. Propellers with three blades were standard on tugs and most ships until the mid-1900s.

Up to that time, a suitable diameter and pitch had been arrived at through naval architects' formulae coupled with expected hydrodynamic flow. Once in service, if vibration and cavitation were very noticeable, a trial-and-error adjustment was made. Through technological experience, several expert propeller companies evolved. The small work tug and pleasure craft gave craftsmen the opportunity to experiment in fine tuning propellers for fast-turning engines. From this, four- and five-bladed propellers were introduced, all with fixed blades.

With introduction of the Kort nozzle and added blade area (achieved by filling out and squaring the blade tip), tugs were able to derive 40 to 50 percent more power thrust with such ducted and squared propellers.

An AquaMaster unit ready for installation in the hull of a tractor tug. Notice the close fit of the propeller blades. (Courtesy AquaMaster)

THE NEW TUG GENERATION

Twin-Screw Steerable Rudderpropellers

The next advance used in tugs was the introduction of the steerable propeller. There are several adaptations of this method. In the Aqua-Master system the propeller is housed in a duct or nozzle, which may be turned 360 degrees as directed. This removes the necessity for rudder(s). The Niigata Z-peller is similar. In the Schottel system, the propeller is offered open, on the base of a rudderpost type shaft, also turnable in a full circle. All of these propeller systems are common in tractor tugs and

A Niigata Z-drive ready for installation. The real water tractor shows how a Schottel or any other Z-drive can be protected. This location allows the tug to operate from either end. (Courtesy Niigata Engineering Ltd.)

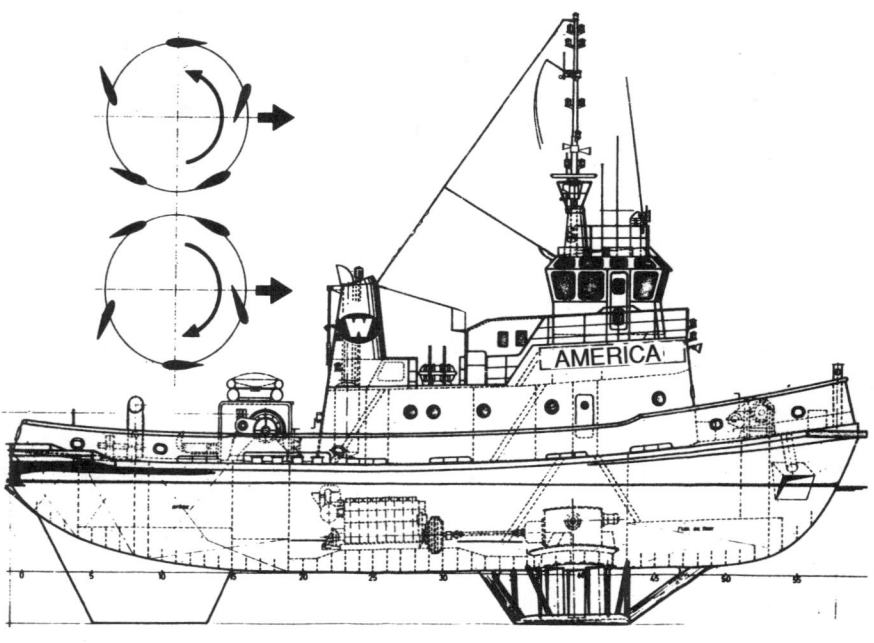

Here is a cutaway of Willamette Towing's cycloidal propulsion system on their harbor and seagoing tug *America*.

have been adapted to conventional tugs even in new buildings.* Voith-Schneider's cycloidal system is older and very popular in many areas.

Efficiency of Steerable Props. Comparisons in Europe apparently have shown that a Z-peller tug produces more bollard pull than a cycloidal tug of equal size and horsepower. It is also claimed that Z-peller propulsion is more economical than cycloidal.

Controllable-Pitch Propellers (CPPs)

These were previously too complex or expensive but have, through design progress, become much lower in cost; and out of very limited experimentation in the 1930s, they have become commercially viable in towing vessels. For the tugboat, CPPs offer some interesting advantages over the fixed-blade propeller.

By various changes of pitch, maneuverability is improved. By increasing the pitch over what would be normal, the tug acquires a greater thrust in towing and higher speed when running free. By the reduction of pitch the tug can reduce speed or pull or remain almost motionless. At such a slow speed with a fixed propeller, a direct-drive engine would stall or the reduction gear transmission would be endangered through the lack of engagement.

When Z-drive steerable propellers are to be installed, designers usually try to accommodate the largest propeller diameters the tug can handle. This offers one obstacle: The tug's draft may be deeper than that of a cycloidal system.

CPPs do not require that the engine, reduction gear, or propeller shaft stop turning, or that it be reversed. Reversing response through blade adjustment is immediate. Idling by feathering the blades is easily accomplished using the same simple bridge control system. With such a range of pitch versus speed, an economical operating procedure can be attained. Maintenance requirements of the main engine should be reduced due to the smooth continuous operation. Maintenance of CPPs and individual blades is claimed to be easily accomplished, and individual blade removal may be done without shaft disturbance.

In summation, CPPs, whether installed on a rigid propeller shaft or a steerable system, offer the ability to immediately vary the pitch and enable the tug to run at the most efficient speed at considerable fuel savings.

Notable manufacturers of CPPs are Escher Wyss of Ravensburg, West Germany; Lips Propellers of Chesapeake, Virginia; Bird-Johnson

---

*Manufacturers are AquaMaster system, Hollming Ltd., Rauma, Finland; Niigata, Tokyo, Japan; Schottel-Werft, Spay, West Germany.

A five-blade controllable-pitch propeller as installed on some tugs.

Systems of Walpole, Massachusetts; and Liaaen Compass Thrusters of Aalesund, Norway.

Voith-Schneider Propulsion

For years, the vertical blades of the Voith-Schneider cycloidal propeller system have been used on harbor tugs in Europe. Presently, they are very often found in tractor tugs throughout the world. The principal difference between the steerable or rudderpropeller installation and the cycloidal system is that the vertical blades of the cycloidal propeller turn. The unit itself remains protected. Because of their size, rudderpropellers either in the open or in a nozzle require that the tug have a much greater draft.

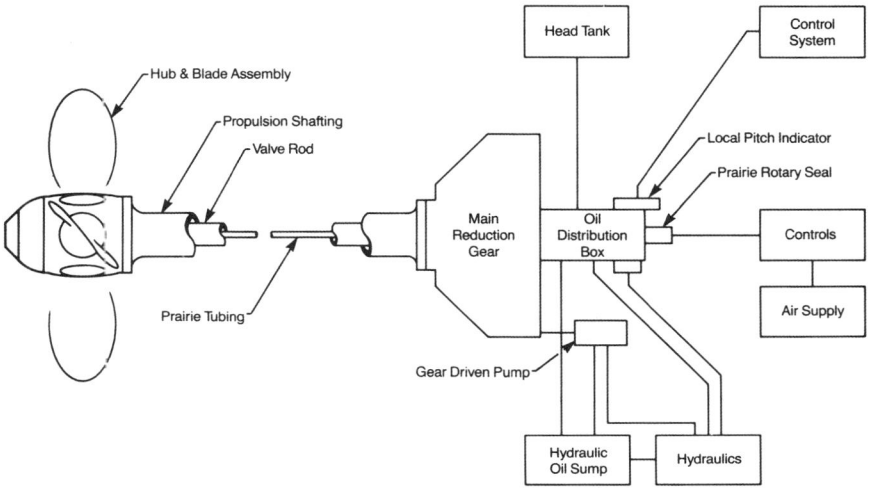

Controllable-pitch propeller system diagram (Courtesy Sulzer—Escher Wyss—GmbH.)

Location of either type is forward, about halfway aft from the stem. A rigid fin under the stern acts as a keel. Its depth is the same and it is in a horizontal line with the bottom of the propeller forward.

Engine Controls of Standard Tugs

When the control of the direction and speed of the tug's engines began passing from an engineer who was answering bells or telegraph signals up to the hands in the pilothouse, it might have been compared to going from the automatic transmission in an auto to the stick shift. The transition was not always smooth. No longer could the poor engineer be blamed for not answering the bells fast enough. Now, when the captain or pilot handles the throttle, no one else can be blamed for the results.

The revolutions of the diesel engines were so much faster that the tug would jump ahead or astern if the propeller was engaged for more than several seconds. Many engine controls were a combination of a direct air injection and fuel follow-up. One had to be careful not to use up too much of the compressed starting air until under way with the tow. It was most embarrassing to run out of air while maneuvering or making up. It was equally disconcerting for a captain who was shifted within the fleet from a steam tug to one with diesel-electric drive. On the latter, the pilothouse control handle had to be brought back slowly from full speed to zero or stop and rest there for 10 or 15 awful seconds before pushing into astern. Any cheating resulted in a continuation of going ahead—often with disastrous results.

Pilothouse controls for diesel engines improved gradually. Some air types still engage and lock the clutch in any direction if care is not exercised. To overcome this, it is necessary to watch gauges in the console, cutting off the air before the clutch is fully engaged. Otherwise, there is a 10- to 12-second delay before the clutch can be disengaged. This is most distracting when making up the tug to a unit and watching the men working on deck with lines.

The proper clutch for the modern tug is one that smoothly and quickly shifts the propeller and/or clutch and reduction gear from ahead to astern to attain any speed. Many of these types of controls are now available and in use.

In the large pilothouse, controls should be placed at convenient points so that the tug captain can shift from one side to the other and pick up control of the vessel without stopping. However, it is not prudent on tugs with an upper wheelhouse for the operator or pilot to shift from lower to upper or vice versa while the tug is in motion and the propeller engaged unless a knowledgeable and responsible person is standing by in the other wheelhouse ready to take over if necessary (see chapter 6).

Engine Controls of Tractor Tugs—Z-Peller and Cycloidal

After 15 years of improvement in the tractor tug, the latest "joystick" controls are simple to handle and fast in response, both in steering and in the application of power. There are two types of joysticks. One is the single upright lever shown as Type A. Type B is a circular horizontal ring follow-up with a rudder angle indicator built in close to the joystick. There is an open area that always indicates the direction of thrust. The inner bar regulates power. Each propulsion unit has its own control. Type C is a computerized version.

These joysticks or similar controls can operate from a single steering stand, so positioned that the pilot can see to maneuver or work his tug from either end. By turning the stick in an arc, the thrust can be directed in a full circle—360 degrees. By raising or lowering the angle of the joystick, the power and direction are controlled.

The tractor tug's turning circle and distance to stop from full power are usually about the length of the tug. To use an old Yankee cliche, "They will turn on a dime and give you nine cents change."

The owners of Zidell's Portland, Oregon, cycloidal-powered tractor tug *America* claim she can shift from pulling a ship to pushing a ship in 15 seconds. Most nontractor tugs in the Portland area require at least 3 minutes for the same maneuver.

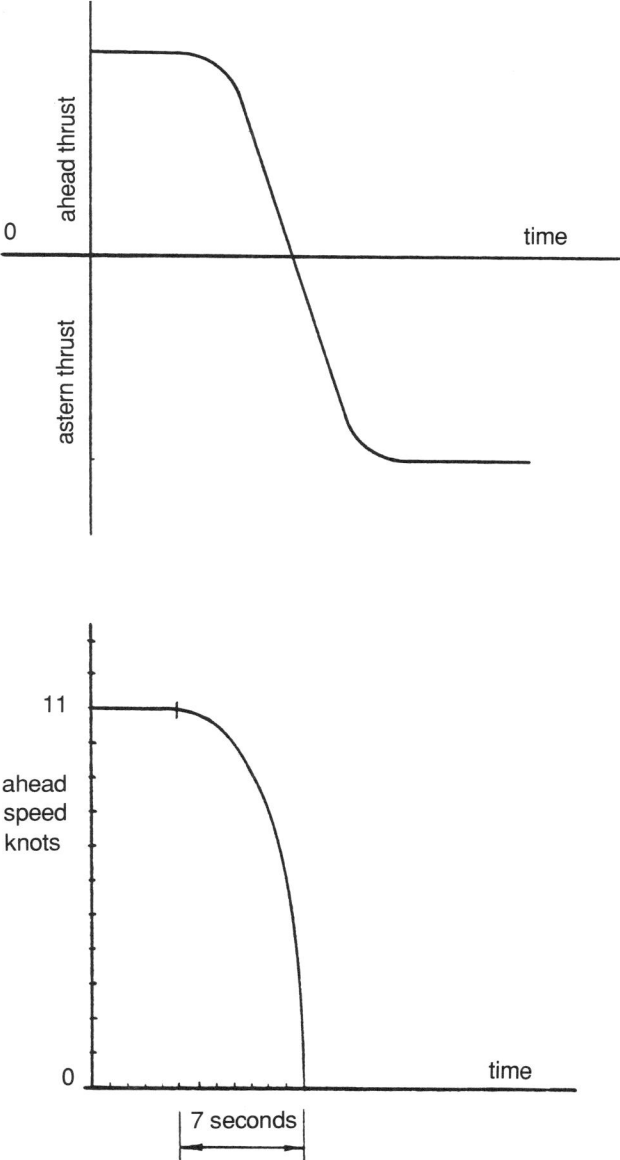

Full ahead/full astern thrust/time diagram and speed/time diagram for reverse tractor Z-drive tug *Portland*. (Courtesy Shaver Transportation Company)

ANCHOR AND CABLE

A well-equipped tug should have anchoring gear that can be released or heaved in from controls in the pilothouse. Most North American tugs are given an anchor as an afterthought, but are not fitted with anchoring gear. Usually it is just an anchor that by someone's eyeballing is thought to be large enough to hold the tug if she ever had to use it. It can be found stowed under bulwarks or in the shelter of the overhang of a deckhouse, where it will collect rust unless given an annual chipping.

Thus the tug becomes a wandering gypsy, hanging on to other people's wharves or moored equipment. This may be fine for ship-assisting tugs, but for those involved in any long-distance towing, either offshore to oil rigs or to distant ports, consideration should be given to installation of an anchor and chain with a proper windlass.

Anchors fitted with chain through hawsepipes to a power windlass are common accessories in most European tugs. This gives tugs freedom to stop and anchor instead of running distances to tie up between tows and, in some cases, offers all kinds of operational flexibility and fuel savings. When an anchor is installed in a hawsepipe, the hawsehole should be recessed so that the anchor when stowed will not protrude and

Dublin harbor tug *Clontarf Monarch* has anchors recessed in hawsepipes.

hang up on the deck of a loaded barge or scrape on a wharf or other object.

PAINTING AND RUST PREVENTION

There are a thousand theories on preventive hull maintenance. Most reputable paint manufacturers will offer a complete protective plan that may or may not be suitable for the waters the tugs will operate in.

Primarily, steel used in construction will show much less deterioration if subjected to a pickling solution prior to fabrication. Light blasting of welds for slag removal should follow with a touch-up of the same pickling over new work. After seasoning, the entire surface should be treated with the latest state-of-the-art sealer and rust inhibitor. This will add years to the life of the tug's hull and house and greatly reduce maintenance.

Bottom paints have advanced so greatly that there is a wide range of choices, the best antifouling paint to smooth, cold, and very expensive plastics. Tugs that will operate on rivers or in waters that contain silt, flowing sand granules, or drift of any kind should not be coated with expensive long-lasting plastic bottom paint, which can be chipped and reduced to the efficiency of cheaper coverings.

DRY-DOCKING

Every new building should be provided with an adequate set of docking plans. Particular attention should be given to showing the location of all through-hull fittings and suspended structures such as rudders, bilge keels, keel coolers, open propeller shafts, and struts.

Tractor tugs that have underwater Z-peller or cycloidal units require the most careful protection in dry-docking. A copy of the tug's docking plan should be kept on board for presentation to repair yards. It is equally important to see that the plan is returned to the tug.

In some tidal areas, particularly where no shipyard is available, small and medium-sized tugs are grounded out for repairs on a grid next to a seawall or heavy pier. This is accomplished by giving the vessel a slight list toward the pier and securely holding her in that position. To improve upon this method, Nils Lucander, a Seattle builder and designer of tugs and small craft, has introduced what he calls the "three-point design" for shallow-draft twin-screw tugs. The normal keel runs only a short way from the forefoot and ends. Then on either side, under each propeller shaft, a keel runs out from the hull, acting as a skeg into which is fitted a rudder. Not only does this arrangement allow the tug to be set aground upright, but proponents also claim that it provides about a 25 percent increase in propeller thrust.

ZINCS AND ELECTROLYSIS

Before placing zinc pads on new hulls, some reference should be made to studies regarding electrolysis and its prevention.

TAIL SHAFT AND OUTBOARD BEARINGS

These should be checked, protected, and maintained at the level required by whichever of the various classification societies will issue the vessel's load line and class certificate.

FIRE EQUIPMENT

In addition to an independently operated battery-powered general alarm system to alert all hands, tugs should have at least an adequate supply of strategically placed fire extinguishers. As many fires have started in galleys and engine rooms due to the extreme temperature, these areas should receive special attention. A long hose from several bottles of dry smothering foam should be installed where it will reach the galley and the engine room.

For an engine room that may be subject to a leaking fuel line, a Halon or similar type of automatic or manually operated system is highly recommended. This is particularly true if the engine room is automated and unmanned.

Tugs that handle petroleum barges or other flammable cargo carriers have been fitted with the Fire Boss system manufactured by Chemetron Fire Systems, Inc., of Illinois. This is a 50-gallon dry chemical tank into which a charge of nitrogen is introduced by releasing a pin. A long hose on a reel offers efficient use. Its reach makes it possible to cover any part of the tug or other vessel. This results in sweeping and quickly smothering all types of fire in a remarkably short time. Many tugs that handle tankers in oil loading and discharging ports are fitted with water cannons of the movable type, which will also telescope. Stationary fire monitors are also installed on many older harbor tugs. The Swedish SKUM fire-fighting equipment made by Svenska Skumslachning A.B., Stockholm, is a popular installation on European and Asian tugs.

LIFESAVING EQUIPMENT—EPIRB AND SURVIVAL SUITS

Tugs do sink and tugmen do occasionally fall overboard. Equipment provided to keep them afloat until rescued has not always proven successful. It takes a series of steps as described in chapter 7 to make survival possible.

Belfast tug *Coleraine* is equipped with a telescoping fire monitor.

One of the first steps is the installation of an Electronic Position-Indicating Radio Beacon (EPIRB). This is a floatable wandlike tube with a tiny transmitter-antenna that floats clear of a sinking vessel. As soon as this EPIRB is released, it transmits signals that aircraft and U.S. Coast Guard receivers can pick up and home in on.

These emergency position-indicating radio beacons come in two classes. Class A is IMO and USCG approved. Its lithium batteries are good for 80 hours and normally do not require replacement for 10 years. They broadcast on frequencies received by satellite aircraft and the USCG.

The small Class B EPIRB is for use with a life raft or survival suit. It only broadcasts on channels monitored by the Coast Guard (VHF channel 16), although aircraft have picked up its signals.

As to equipment, the latest proven flotation device for individuals is the survival suit. It protects against hypothermia, the greatest cause of quick death from exposure.

Life Jackets and Work Vests

While these are required, they are difficult to work in around decks. However, they should be provided for all hands. Life jackets offer fair flotation in warmer water if properly fastened.

Both survival suit and life jacket should be fitted with a mouth whistle and a hand-activated chemical lightstick.

Life Rafts

A life raft should be properly stowed so as to float clear, and it should be large enough to accommodate all hands. Unfortunately, when tugs have foundered it is usually very rough and windy. Life rafts tend to blow away more rapidly than the men in the water can reach them.

Large seagoing salvage tugs seldom sink but do have their own powerboats, some of which are rigid and some inflatable. The latter offer many possibilities as auxiliary lifesavers.

Life Rings

These are normally required and must be rigged with a long retrieving line and a light that activates when the ring is deployed. They are best for assisting in the retrieval of someone who is already in the water.

Personnel overboard buoys, free-floating or attached to life rings equipped with a strobe or overboard light, may also be a very prudent addition to the vessel's safety equipment.

NAVIGATION LIGHTS AND SEARCHLIGHTS

Not only should all navigation and towing lights required by law be mounted for clear visibility, they should also be pilothouse-controlled. A buzzer alarm should be provided to indicate individual failures.

One of the greatest aids in night tug operation is a good searchlight. Every tug and towboat should have one that will send a clean beam for several miles. A seagoing tug should have one aft that will clearly light up the tow. A river towboat as well as a tug that pushes in the notch should also have a small powerful floodlight on each after corner of the deckhouse for use when entering locks, turning in restricted areas, or leaving docks.

TUG INTERIORS

Today, tug owners realize that to generate interest in sailing on tugs, they must provide their seagoing team with the finest seagoing fabrics and appointments. The galley must be modern, and laundry and entertainment equipment must be provided. The tug should be a seagoing habitat second only to home ashore.

Designers, owners, and operators should understand, when considering building or rebuilding tugs, that female crew members are no longer uncommon. Both sexes have a right to privacy and protection regarding quarters, bathing, and toilet facilities. River towboat operators met this requirement years ago.

Interiors of large ocean tugs on which crews live for months as they wander around the world are usually much roomier than those on tugs that operate in certain areas where regular crew changes are possible.

On these huge tugs, masters, mates, and chief engineers all have bedroom cabins separate from their combination offices/lounges. The latter are usually furnished with a desk, bookcase, lamps, file cabinets, a small coffee table, and a cushioned settee. Tug/supply vessels' interiors are very similar in seagoing style to the comforts of river towboats. They have all of the latest amenities for their personnel. An example is Halter Marine company, whose tug/supply and tug/support vessels are noted for the company's in-house design staff, who create pleasant, color-coordinated, and highly livable accommodations for crew members.

The interiors of fixed tug-barge units resemble accommodations of the modern ship that they replaced. Personnel are assigned separate rooms and may share bathing and toilet facilities with the crew member in the next cabin. These facilities are frequently built so that they can be entered from either cabin. Consideration of privacy should be given in such arrangements.

Passageways

Tugs should have passageways so that it is not necessary to go out on deck to reach the pilothouse, galley, quarters, or engine room. Watertight doors on inside passages should be located at strategic points so that various portions of the vessel can be sealed off in case of fire, flooding, or casualty.

All passageways and ladders should be of sufficient width so that crew members going in opposite directions may pass. The decks of all passageways, galley, staterooms, quarters, pilothouse, and lounge should be as near level with the surface of the sea (when smooth) as possible.

It has been difficult to get some tug designers to accept the design characteristics detailed above. Some otherwise finely crafted tugs have had inside passageways and cabin decks that followed the outside deck sheer. This is most uncomfortable for watch standers: The pilothouse deck slopes upward toward the windows, and cabin decks slope so that a chair wants to slide downhill. Other designers seem to have been able to overcome this structural torment. Tractor tug design, for instance, tends to be more level on the interior.

Galley

The galley should have good counter work space, deep double sinks and/or dishwashers, and garbage compactors. Equipment such as electric ranges, ovens, and large refrigerators are normal appliances. The galley should be positioned close to, but separate from, the mess area.

Lounge

So that crew members off watch may have a place to spend leisure time watching television, reading, or just relaxing, a lounge away from the galley should be provided. Unfortunately, designers of U.S. tugs are limited by an archaic admeasurement system in giving the tug crews roomy quarters.

Quarters

The captain and chief or other engineer, as assigned, should be given spacious quarters with an adequate desk, file, and ship's safe, plus clothes dresser and hanging locker. If the tug is large enough, the master's stateroom should be separate from his office and have additional toilet and shower facilities.

Other officers' rooms should be like those of the other crew members and, if possible, single occupancy. Inclusion of an individual washstand is normal. Toilets and showers may be located to accommodate each deck level. However, it should not be necessary for a mate or engineer to have to come down to a lower deck or for a cook or deck member to climb to a higher one to use a head.

The rooms for unlicensed crew members should, if possible, be on the main deck. If below this level, they should be well insulated against condensation and cold.

Heating and Air Conditioning

These are best provided by individual units that allow the occupants to set their own temperature range.

Soundproofing

The noise level of diesel towing vessels has been given considerable study by Canadian tug operators. An agreement has been made and an allowable decibel limit has been set. This study should be universally accepted.

OUTSIDE LADDERS

These ladders to upper decks should be located out of the weather where they cannot become iced in the winter, and, like outside deck work areas, they should be treated with nonskid paint.

TUG SIZE AND TONNAGE

A comparison of the ratio between tug lengths, horsepower, and tonnage of U.S.-built tugs and those of other countries can be very misleading. In the United States, due to an antiquated admeasuring system, a close relationship exists between gross tonnage and the manning and safety requirements of the Coast Guard.

Since about 1968, the gross tonnage of most tugs built in the United States has been arranged to come out in the magical "under 200 gross tons" bracket. This is often accomplished through the most ludicrous deduction of certain areas. Although the use of designated cubic cargo space has been given over for crew quarters, this is cheating.

As of 1988, regulating bodies in the United States had taken no action to adjust the relationship between a tug's length and its displacement tonnage. Much to the detriment of the U.S. towing industry, this has led to cramming the greatest horsepower into the smallest hull in order to avoid safety and other requirements that are necessary for U.S. vessels of over 200 and 300 gross tons. Excluding tractor type and other strictly sheltered-water, ship-assist tugs, there are many of these quasi-ocean-coastwise towing vessels with short squat overpowered hulls and undersized crew quarters.

Some U.S. companies, particularly in the Gulf, have with some compassion built larger hulls with fair-sized crew quarters. The key to a reduced gross tonnage under U.S. admeasurement laws is, among other items, to be able to deduct from measurement any area that may be used for cargo stowage. Thus, staterooms are laid out on either side of a main deck level, with all bulkheads bolted and removable. These are claimed as cargo spaces. Belowdecks are many empty sealed voids. This all adds up to tugs of 199.9 or fewer gross tons which, if they had been built in 1930, would measure with steam propulsion as nearly 500 gross tons. For owners, this new lower tonnage avoids a certificate of inspection and other regulations that many consider irritants involving the U.S. Coast Guard and a manning scale that, while safer, is not to their liking financially.

Such skirting of size, seaworthiness, and safe manning is not considered by most European tug operators. Other technological and operation economies have produced reduced manning in short towing operations. Offshore, a system of three full watches is standard.

COMPARATIVE SIZE OF TUGS AND THEIR CREWS

The size and use of tugs in relation to the number of crew members aboard is detailed in the following table.

Type of vessel	Length, in feet	Horsepower	Crew size
A. Harbor, United States	83–100	1,000– 4,000	3–5
Harbor, United States tractor	83–100	1,000– 4,000	2–3
Harbor, European conventional	83–100	1,000– 4,000	7–9
Harbor, European tractor	83–100	1,000– 4,000	2–3
B. Coastwise, United States	110–145	2,500– 5,000	5–8
C. Ocean	150–210	3,500–26,000	12+
D. Anchor-handling, European	175–210	2,000– 6,000	7–8
E. Tug supply, European & United States	180–220	2,000– 6,000	7–8

CLASSIFICATION

Most commercial tugs built for hire will for various reasons require a survey by a reputable classification society, such as Lloyds, Det norske Veritas, or the American Bureau of Shipping. If properly built, the vessel may receive an A-1-A-, Maltese cross designation for ocean, coastwise, or whatever type of towing operation she was designed for. Beyond this seal of approval indicating that this is a vessel built to standards acceptable throughout the maritime world, the classification society gives the vessel a Plimsoll mark and load line certificate, items very necessary in international towage, insurance, and customer confidence and approval.

Owners should make every effort to keep the vessel up to classification standards or better through regular maintenance and dry-docking.

SUMMATION

There are two schools of thought on what should be the ultimate results in tug design. The first and oldest is that except as justified by exceptionally clearly defined circumstances, a tug should not be used out of the sphere of its original designed service. The other is to build for diversification. Some New York Harbor towing companies are fine examples of diversification. Turecamo tugs with tow winches and an upper pilothouse can dock and undock ships, then let go and grab a loaded oil barge to be pushed to a port on Long Island Sound or the Hudson River. They also may take one on the hawser for a voyage up the coast to the Great Lakes. Such a concept leaves many options for the owner. This practice is not limited to the East Coast. Foss and Willamette tractor tugs in the Pacific Northwest go to sea as needed when not involved in ship-assist work. Most tugs built for outside work in other parts of the world do not become harbor tugs.

Opposite and following pages, profiles of tugs and towboats of the 1990s

TUG DESIGN 147

East Coast harbor and coastwise tug

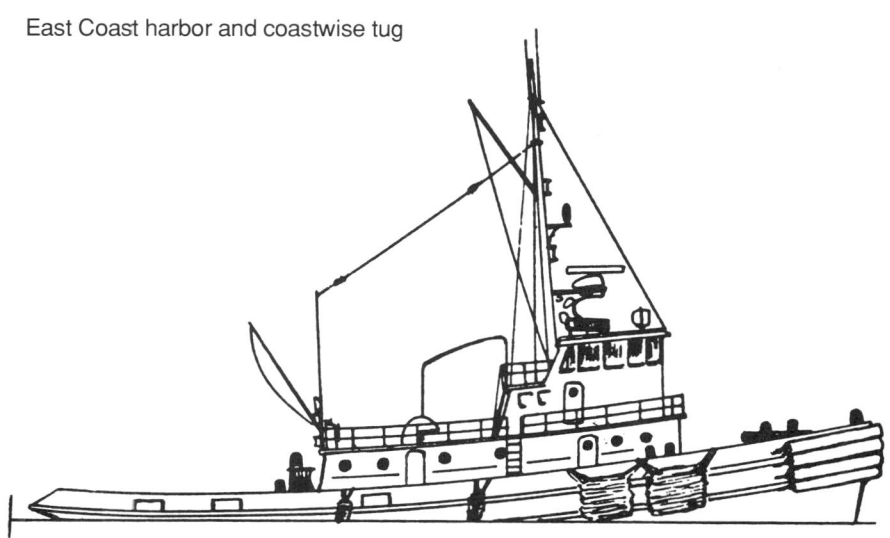

YTB harbor tug, built to meet navy commitments worldwide

Gulf Coast harbor tug

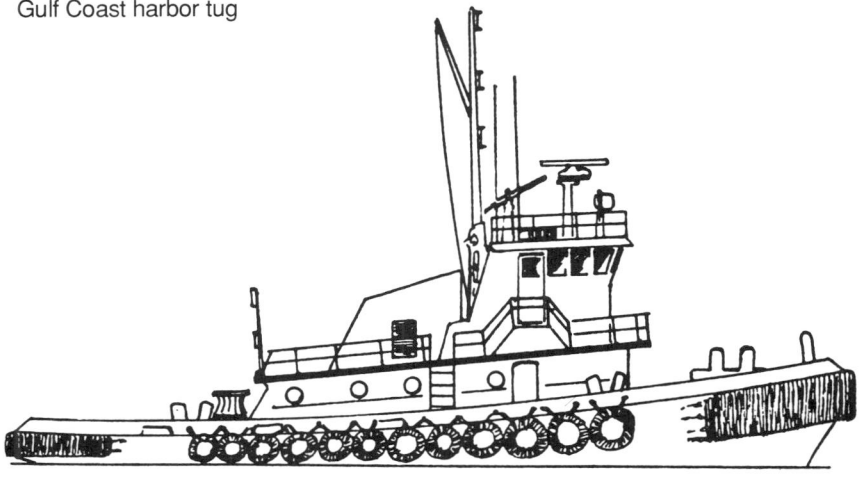

Artubar tug—these 150-foot, 7,200-HP tugs are designed to connect with five-deck articulated barges

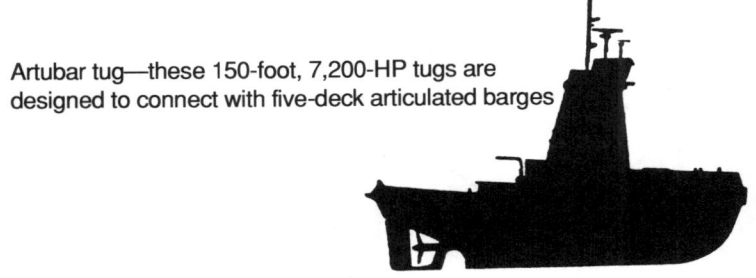

Atlantic and Gulf Intracoastal Waterway type towboat

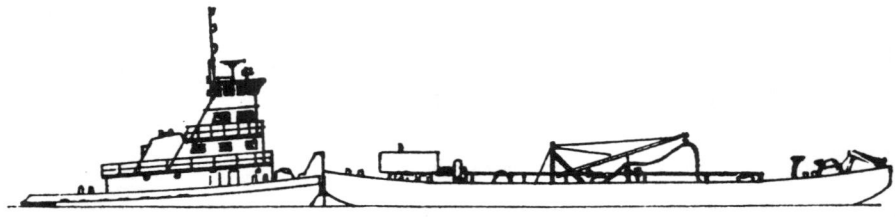

ATF fleet tug, primarily designed for navy salvage work

typical modern U.S. West Coast harbor tug

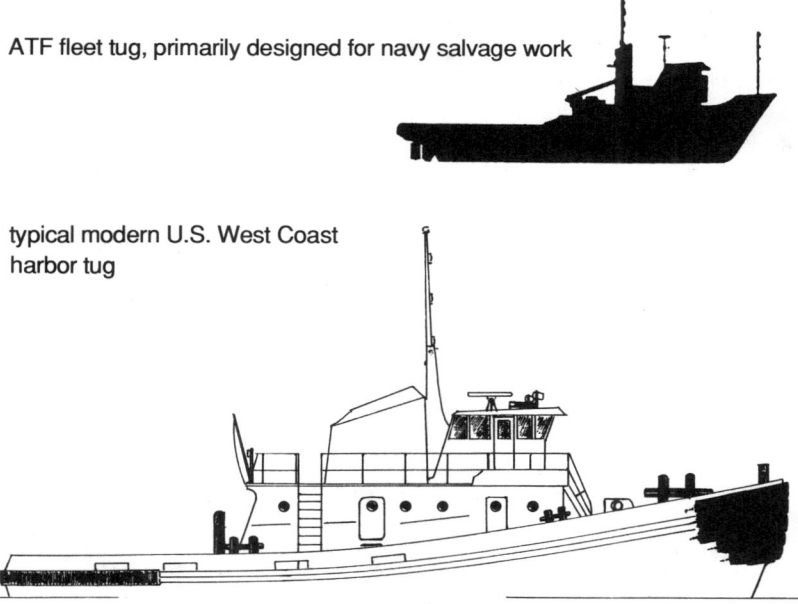

TUG DESIGN   149

the new generation of ocean salvage
and rescue tug

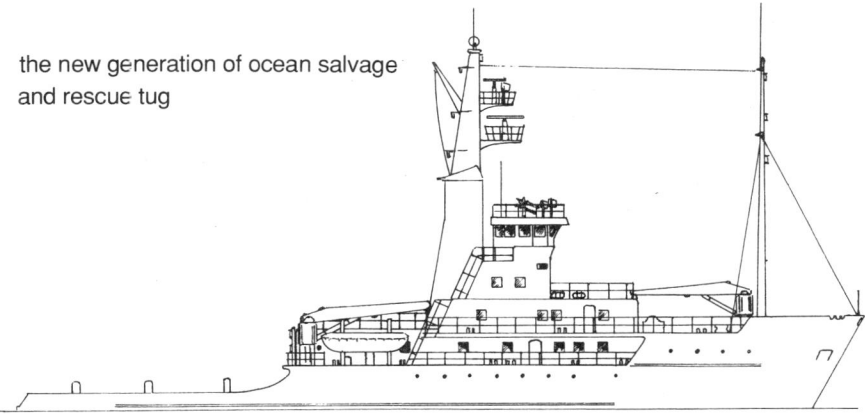

a standard, moderately notched barge and tug

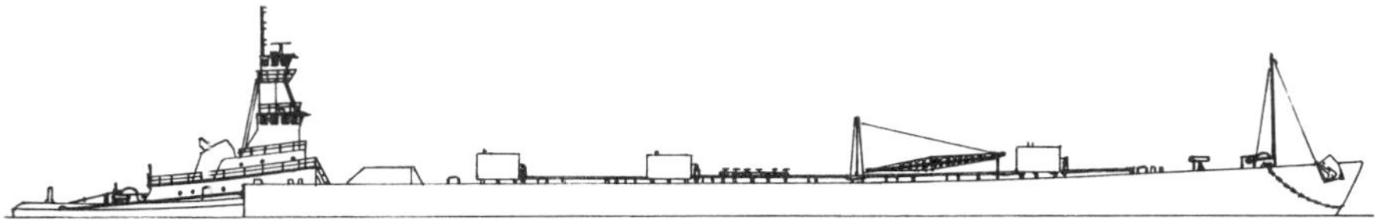

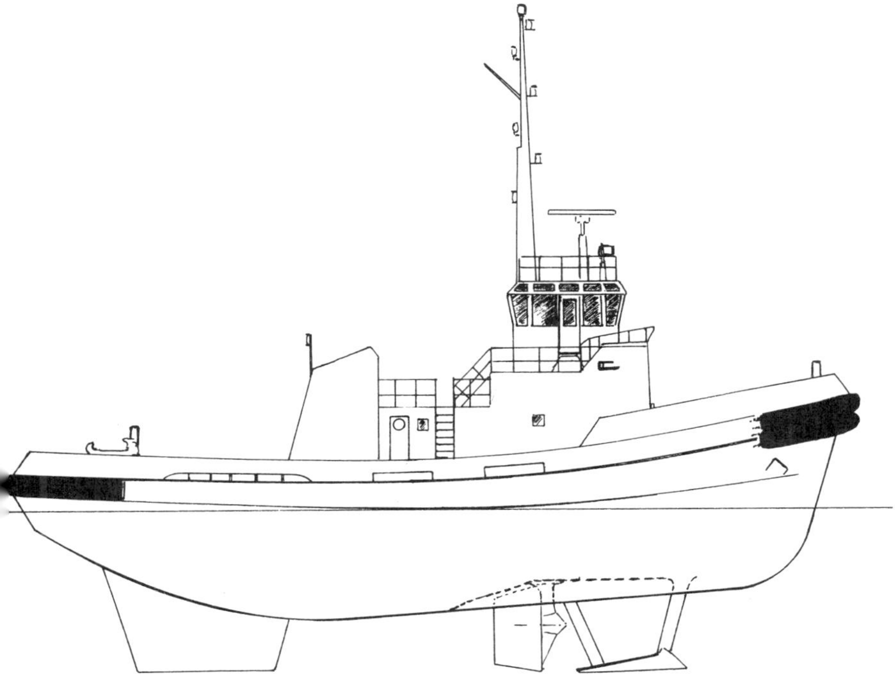

Schottel propelled water tractor

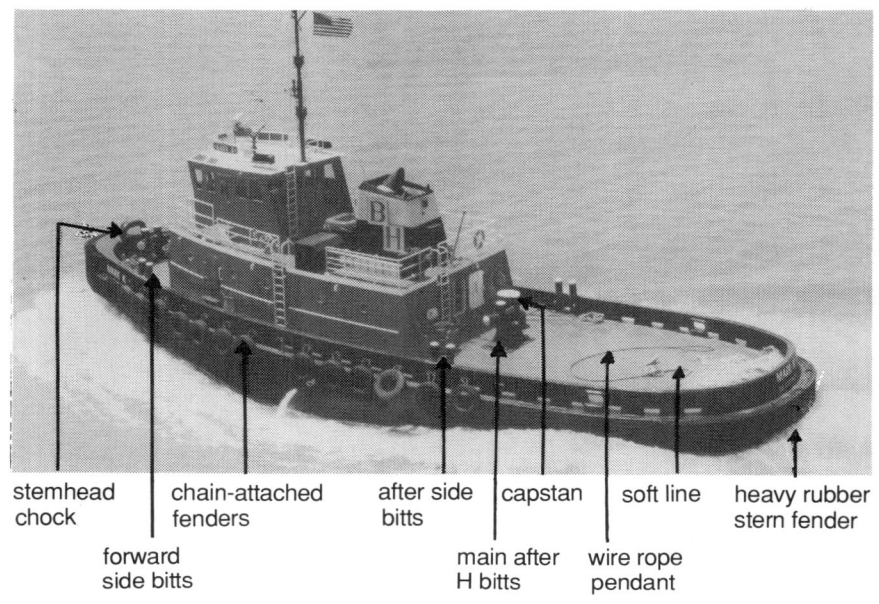

Mark K is a typical Texas harbor tug. (Courtesy B & H Towing)

The new U.S. Navy SeaLift Command tug *Mohawk* (Courtesy Marinette)

PART II
Operations

6. Getting the Tug Under Way

REPORTING ABOARD

As an example, assume that a U.S. towing vessel has been laid up and a whole new crew consisting of a captain, a mate, two seamen, one cook, and one engineer are reporting. They are to take the craft on an assigned towage. They may or may not arrive en masse. If any of them had worked on the vessel before, their experience will be of great help in getting her ready to sail. This is particularly true if the captain or the engineer had served aboard previously.

While the engineer is opening up, finding the generator, and bringing the boat back to life, the captain should locate the vessel's document, enrollment, and license to see that they are up to date. He should also check all other required papers and their location, such as an ABS, load line certificate, radio station license, and the date of the last compass deviation table, etc. On non-U.S.-flag tugs, the same type of document review should take place, following the regulations of the nation where the vessel is registered. The captain, regardless of nationality or registry of the tug, should enter all of this information into the tug's log with the date and time.

While waiting for the lights and power to come on throughout the vessel, the mate should inspect the decks, checking lines, shackles, and all necessary fire and lifesaving equipment with the deck crew. As soon as the engineer has the generator running and has checked sea valves, bilges, etc., he will want to start the sanitary and freshwater systems and cut in power to the galley, so that the cook can check his range and refrigerator.

When power energizes the deck machinery, the capstan, winch, or hydraulic boom hoist should be tested in various directions at slow speed. Tow winch controls must be studied so that the tow cable on the drum is not slackened or heaved in. The winch drum must be disengaged. The anchor winch wildcat must also be unlocked and held by brake, if a spool capstan is attached and being tested.

With power in the wheelhouse or bridge—and hopefully after a short break for coffee and readying of quarters—the captain should gather the crew and check their papers, licenses, and shipping cards to make sure all possess those documents their positions require (including I.D. and alien registration if not nationals) and to commence the payroll.

Further progression in readying the vessel will include testing communication with the engine room, the steering gear, whistle, and navigation lights. If the vessel is equipped with radar, autopilot, depthsounder, Loran, and other computerized units, they should be checked out before sailing.

It will take from one-half to several hours for the master gyro to settle on the correct heading. This should be checked on the chart for the true heading of the tug's berth.

It is the master's duty to see that the vessel has a set of up-to-date charts for the areas he is to sail in, with notices to mariners, through whatever the latest chart system may be. All radio equipment (VHF-SSB), weatherfax, Loran, radar, and sounders should also be tested.

The final presailing checkoff would, for the deck crew, include greasing movable rollers and sheaves, winch and capstan; seeing that the outside dogs on watertight doors are free and the hinges are oiled; and topping off potable water tanks. The engineer will have sounded his fuel and lube tanks. If he has identified any necessary spare parts and equipage, he should, with the cook and mate, inform the captain of the stores required prior to sailing. The cook should have knowledge of the number of days of grub he will need, and if it will be necessary to carry a reserve.

While awaiting this delivery it would be a good time for the captain to go over, with all hands, the towing project about to be undertaken and his usual method of connecting up, etc. The crew should also be familiarized, at this time, with the location of fire equipment and emergency gear (the condition of these items should have been checked already). Location of life jackets and survival suits, if provided, should be made known. If possible, a drill in the use of fire and possible abandon-ship procedures should be held.

MASTER'S RESPONSIBILITY

It may seem academic and frivolous to include a checkoff list. It has been done for two reasons. Every conscientious tug master must realize his major responsibility for everything that occurs on or with his tug, and to its personnel. He will, for the protection of his crew, himself, his master's certificate, and his company, want to see that all of these items and pieces of equipment are checked and in good working order.

The secondary reason is to point out to shore operational personnel and dispatchers that a tug that has been laid up cannot, like an automobile, be started up by having a key put in the ignition. It takes time and careful, thorough checking before the tug is properly seaworthy. If any problems should arise, a failure to have done so may bring back the word "unseaworthiness" to haunt the owner.

Of course if the tug has been operating and this is only a change or relief of personnel, such as the captain, an engineer, or a deck person, such persons will want to check on things prior to getting under way. Of particular interest would be unfamiliar equipage. Engineers in particular will want to ensure before starting that all controls are in a neutral and completely disengaged position (see chapter 10).

As soon as practical, the master of a U.S. tug or towboat should take care of the following: The Department of Transportation, U.S. Coast Guard, requires a complete and current "Master's Report of Seamen Shipped or Discharged." This is Form CG-735 (T). The form is available at local Coast Guard Merchant Marine Inspection Offices in most ports. Particular attention is directed to item 5 on the reverse, and to citizenship. This form should be accompanied by CG-718A, a book of Certificates of Discharge, and CG-718E, which includes forms for Record of Entry in Continuance Discharge Book, if any crew member possesses one. Forms 735 (T) with copies of 718A and E should be mailed to the Coast Guard Headquarters monthly or at the end of voyages when there is a crew change.

MAIN ENGINE WARM-UP

While the preliminary steps have been taken, the engineer has checked the main engine, air compressor, and fire and bilge pumps and is prepared to start up. Care should be taken to see that the stack's exhaust is clear and that all remote controls to the engine are disengaged. By using the engine room controls alone, the engine may be safely started and run at idle speed until lube oil pressure and cooling water temperatures are up to operating level.

If a dock trial is desired, the mate or captain should be notified and the mooring lines should be checked. Also, it would be advisable to see if the propeller area is clear of any lines, floating objects, or other craft that might be disturbed by the prop wash.

GETTING UNDER WAY

This is a good time to check out the intercom from the pilothouse to the deck. As this will be the first time of letting go, orders should be clear. If the tug has no intercom system, the signal to let go can be given on the whistle. Around New York Harbor and East Coast ports, this is signaled by one blast followed by two shorter blasts. In Gulf ports, one blast followed by one short has been used. Different areas have their own customs, but captain and crew must understand each other. This is teamwork. Shouting is unnecessary except in an emergency.

HAND SIGNALS

A series of hand signals between deck and pilothouse personnel is usually worked out and used in signaling which line to let go, "all clear," and in docking barges. Signals are also used to indicate the distance off at the stern when backing up to units, and when entering locks. Standard signals by crew members to pilots, and on rivers, are shown in chapter 7.

LETTING GO—DEPARTURE

Prior to letting go, the captain or pilot will have to figure out his maneuver and the order in which he wants his lines to be taken in. If there is no one on the dock to let go his lines, they must be free enough so that the deck person can flip them off bollards while remaining aboard the tug. If the distance to the shore bollard or cleat from the tug makes it necessary to put a man ashore, it may require putting a slip line around a piling or running the bight of a springline out so that the tug can work ahead slowly while the other lines are let go and taken in.

If the tug is on a pierhead in the current, the final departure should be made when traffic is clear so that the tug can use the current to breast her off. If the tug is single-screw and the current is from ahead, hard over rudder and a slight touch ahead on the engine will allow the current to pull the bow off. Then with the rudder amidships, she can proceed without the stern striking pilings or other objects. If the current is running against the stern, it will be safer, if there is room, to hold a head line and let the tug swing nearly around before backing away.

On twin-screw tugs, various combinations of one engine ahead on the inboard side, and the other astern, with the rudders opposite to the direction you wish to go are used to walk the tug sideways away from her berth. It takes a little practice and adjustment of speed on either engine to keep the tug straight as she works off. The direction of the propellers' turn will also affect the results, which may differ on each vessel. Tractor tugs with Z-pellers or cycloidals can move off sideways from a berth with no difficulty.

Departure from a Slip

If the tug is in a slip and can be turned, this is the safer procedure in departing. Whether you are able to turn and go out, or must back out, a good lookout for traffic must be kept. If view of the tug is obscured from the waterway outside the slip, pier, or quay, be sure to blow one long blast and, if backing, three blasts. Have a lookout on the stern to signal

GETTING THE TUG UNDER WAY   157

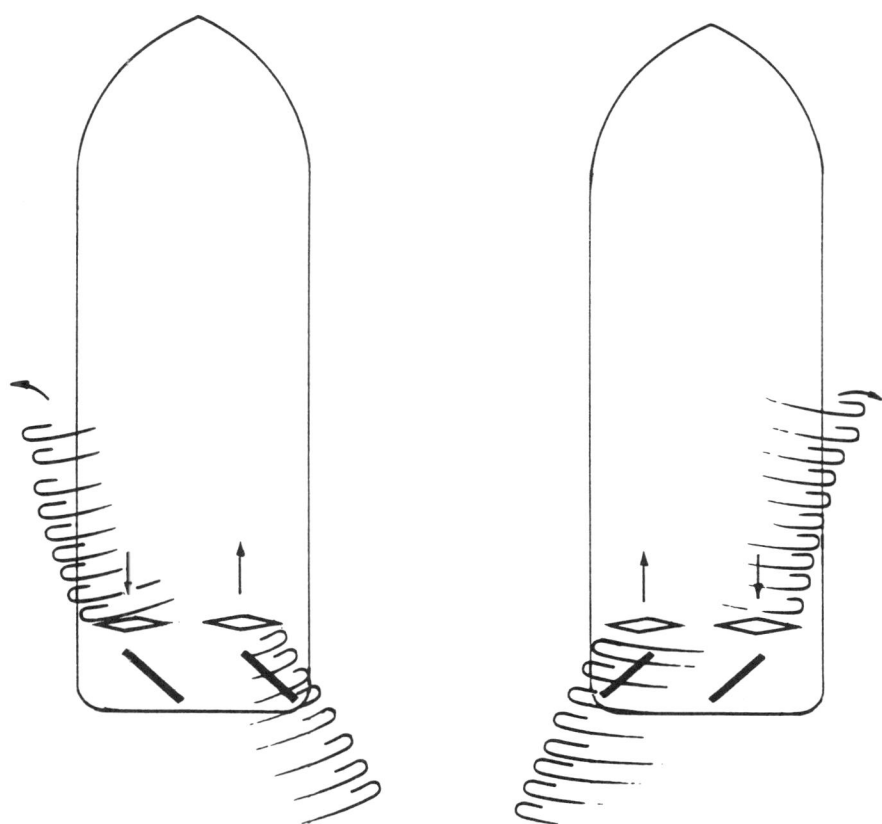

A tug, and sometimes her tow, may be walked sideways with the rudders in the position shown.

if the area is or is not clear of traffic. A few fine tugs have been run over and sunk with some loss of life several minutes after letting go, and all because of failure to see an oncoming ship or a loaded tow while backing too rapidly out of a slip.

FEELING OUT THE TUG

Almost anyone assigned as captain or mate of a tug will have had some experience in boat handling; however, in the instance of a new or strange vessel, the captain or mate may wish to feel how she handles. Arrangements with shore personnel should have been made and time allowed for this.

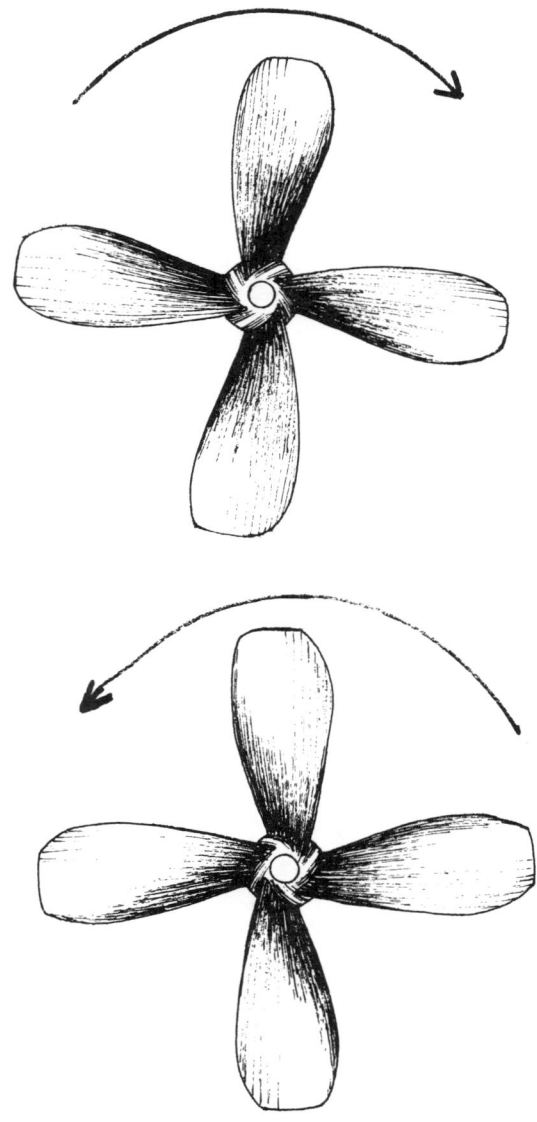

Above, right-hand turning propeller and *below*, left-hand turning propeller. Right tends to go to port—left to starboard.

GETTING THE TUG UNDER WAY

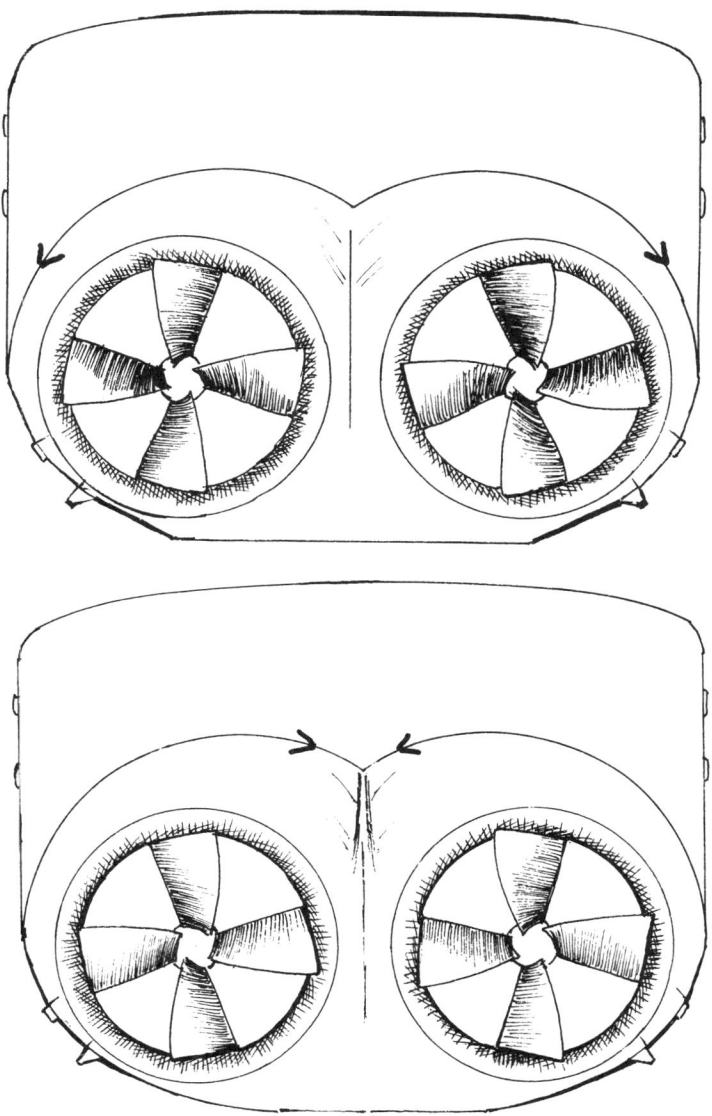

Above, outboard turning propellers and *below*, inboard turning propellers. Right hand backs to port.

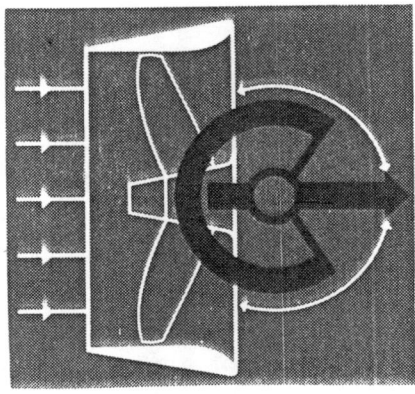

The control wheel or lever of the single hand copilot control unit for the Schottel rudderpropeller combines the steering control (thrust direction), engine speed control, and clutch control. The position of the propeller is shown on a separate indicator-unit and corresponds with the angle of the control wheel/lever on completion of a steering command. (Courtesy Schottel)

The first thing is to get the tug to a roomy area clear of traffic with sufficient depth for maneuvering. En route, the heading of magnetic, gyro-, and steering compasses should be checked against the chart course. If any great error is noted between the gyro- and magnetic compasses, arrangements should be made for a compass adjustor.

If it is some distance to the proposed trial area, it may be desirable to try out the autopilot. This equipment will usually steer a steadier course than is possible by hand steering.

STEERING A TUG WITHOUT A TOW

Like any large vessel moving through the water, the single- or twin-screw tug is affected greatly by the design of its hull, trim, and speed.

The short, round-bilged, powerful tug will tend to root and sheer even at half-speed. The long, deep, peapod-shape hull of coastwise and ocean towing vessels is much steadier. In between these two extremes of types are a million variations of tug hulls and horsepower. Each will react differently when running light. On river towboats, this means running "with nothing on the head." Any tug will hold a reasonable course in still waters at slow speed. It is when running against heavy current and through narrow channels and in shoal water that excessive rudder action is required.

GETTING THE TUG UNDER WAY

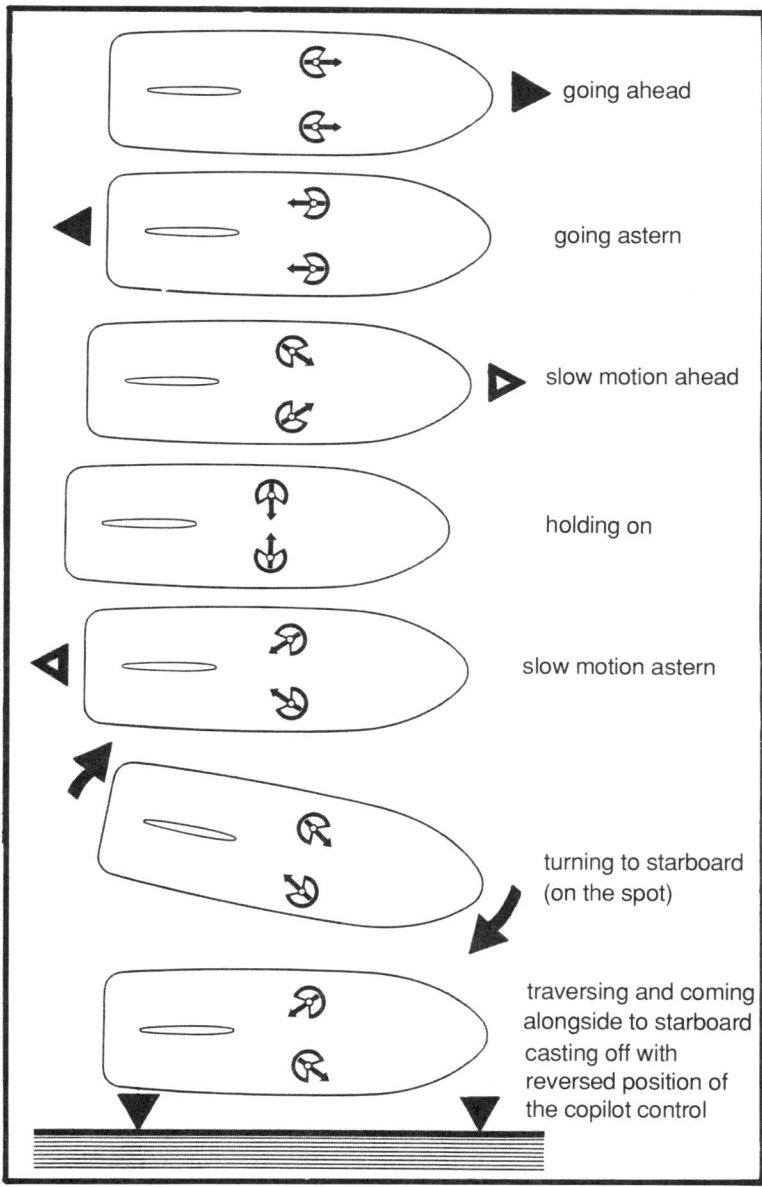

Diagram by an experienced tug master showing the handling of a twin-screw Schottel tractor tug with the aid of a separate single-lever control unit for each rudderpropeller combining the operation of the disengageable clutch, engine speed control, and steering. (Courtesy Schottel)

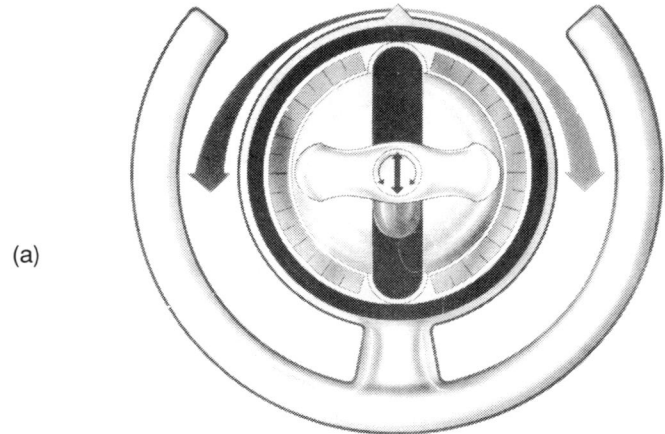

(a)

The word "joystick," a term used in aviation, has its origin in the simple tiller and whipstaff for steering a boat or ship. The Schottel joystick steering system is a microcomputer-controlled device for vessels equipped with more than one rudderpropeller. Through an input gear (a combination of a lever and horizontal wheel), the desired sidestepping direction and rotary motion of the vessel, as well as the force of the movement, are fed in continuously at a set value (see figure a). At the same time, the lever adjusts the sidestepping direction as well as the sidestepping force. The direction of motion is indicated by the direction of the lever, the force of motion through the deflection of the lever.

The horizontal wheel allows superimposing of a rotary motion. The force of the turning motion is set through the angle of the horizontal wheel. A superimposed motion of the vessel can be achieved by adjusting the set values at the same time (see figures b & c, *opposite*).

The set values are fed into the microcomputer, which makes the necessary calculations for carrying out the order for the required rudderpropeller steering angle and the thrust (engine RPM or CP pitch). These values are then transmitted to the electronic remote control (see figure d, *page 164*). However, when the Schottel joystick steering system is installed, the old Copilot steering system required by the classification societies is also installed, This gives the helmsman instant control of the tug should there be any failure or breakdown of the black box. (Courtesy Schottel)

GETTING THE TUG UNDER WAY 163

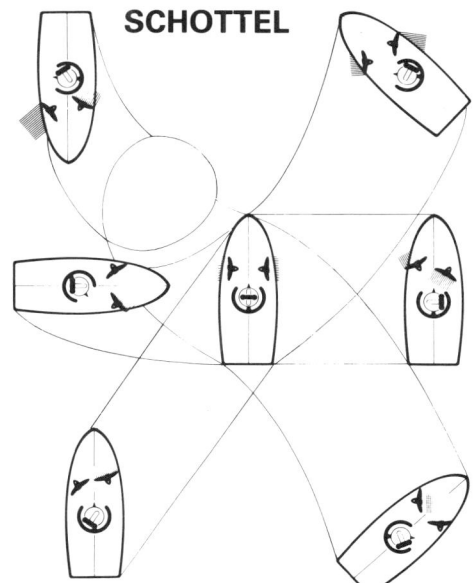

(b)

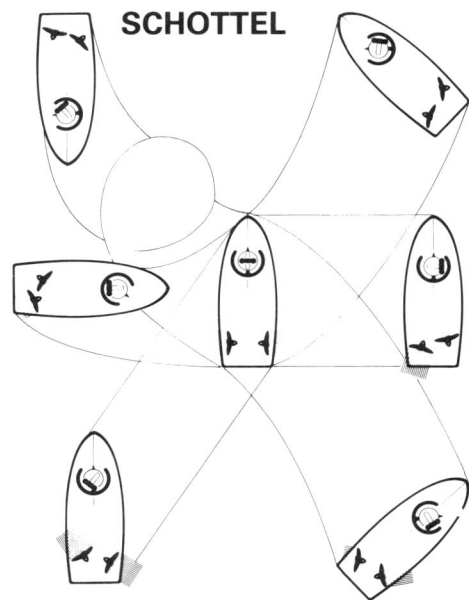

(c)

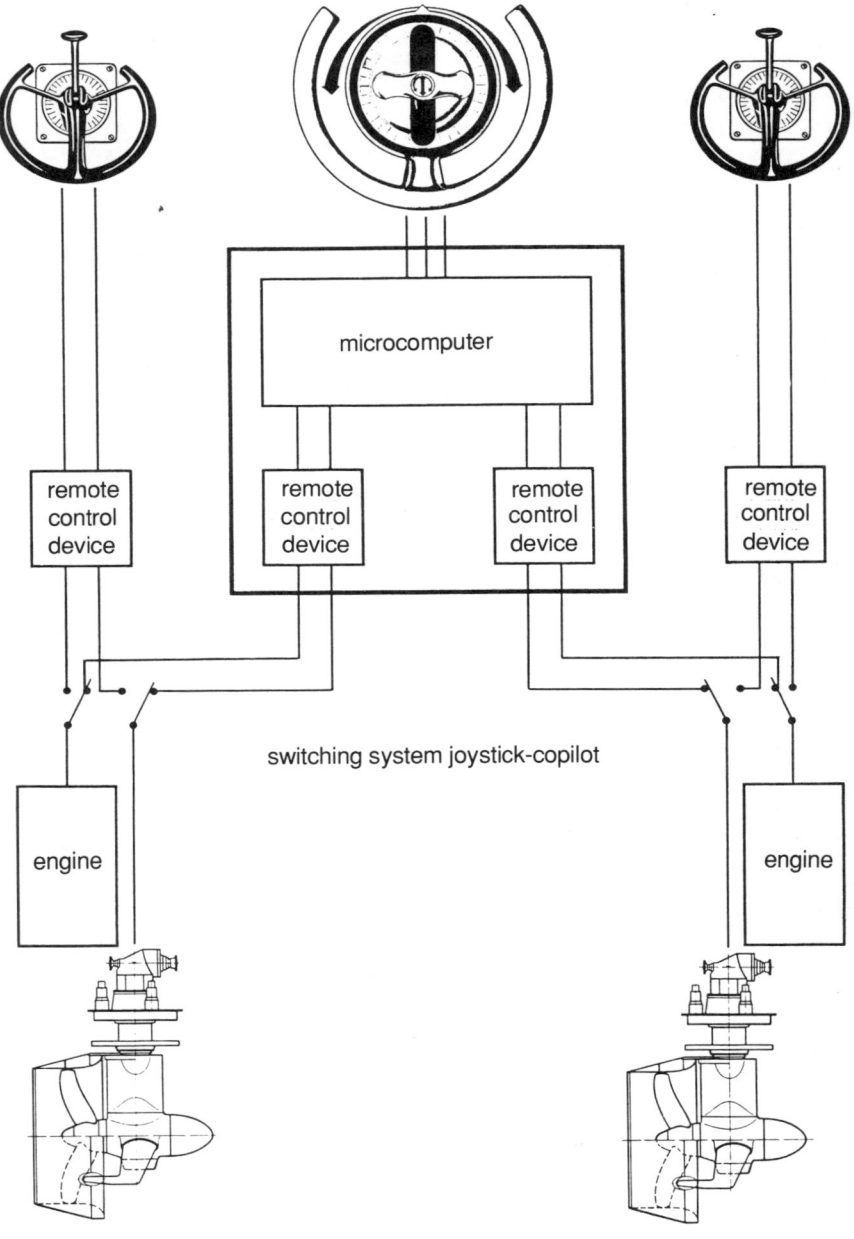

(d)

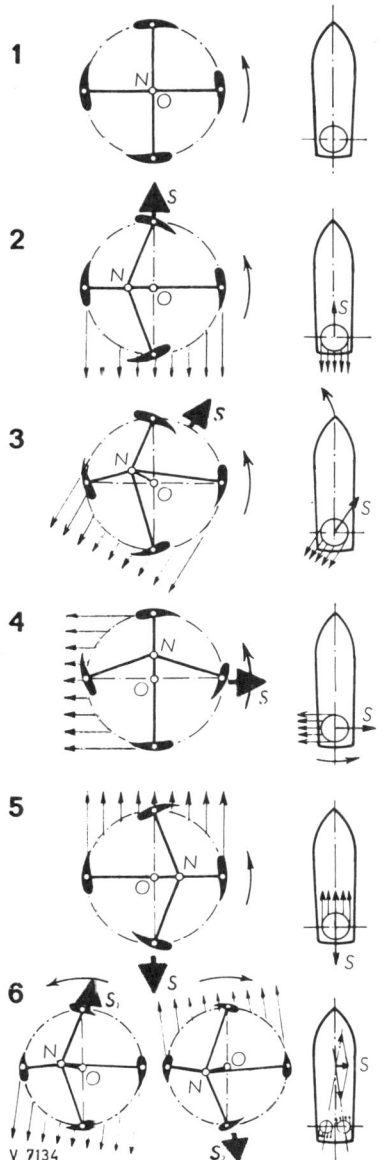

Operation of the VSP

The pivoted propeller blades project from a circular rotor casing which is flush with the ship's bottom and which rotates at constant rpm about an axis which is generally vertical. Links which lead to the steering centre N are fitted on the blade shafts inside the rotor casing. When this steering centre is moved from the centre O of the circle, the blades perform an oscillating movement about their axes and a propeller jet is produced with a thrust acting in the opposite direction (see sketches).

Assume that the propeller rotates in the direction of the arrow. In sketch **1**, the steering centre N is located in the centre O of the circle. While rotating, the blades remain tangential and no thrust is developed (**'idling' of the propeller**).

In sketch **2**, the steering centre N is shifted to port, and the blades are so controlled that the verticals to the chords of the blade profile intersect in the steering centre N. Each blade then performs an oscillating movement about its axis. The leading edge of the blade is directed outwards in the forward half circle and inwards in the rear half circle. Thus, in the forward half, water is thrown into the blade orbit, and, in the rear half, away from the orbit, though in the same direction. In this way, a water jet astern is produced and, as reaction, thrust S, which provides forward propulsion: **the ship moves forward.**

The thrust is at right angles to the line O-N, and its magnitude is proportional to the distance O-N. Because of the rotational symmetry about the axis of the propeller, similar considerations apply to any other location of the steering centre N.

In sketch **3**, the steering centre is shifted to port and simultaneously forward. Propeller jet and thrust S are again at right angles to the line O-N; in addition to a longitudinal component, the thrust includes an athwartship component, or in other words, the propeller provides a steering force: **the ship turns to port.**

When the steering centre, as shown in sketch **4**, is shifted forward, a pure thrust athwartships to starboard is produced: **the ship turns on the spot.**

If the steering centre N is shifted to the right, as shown in sketch **5**, the resulting conditions are opposed to those shown in sketch 2; the thrust is directed astern: **the ship moves astern.**

If, in a ship equipped with 2 Voith-Schneider Propellers, one propeller is given a forward oblique thrust and the other an astern oblique thrust with both thrusts directed to the same side (sketch 6), then the resultant of the thrusts S1 and S2 is a transverse thrust S which acts about midships. **The ship moves transversely.**

For clearness' sake, the schematic links inside the Voith-Schneider Propeller are simplified in the sketches. Actually, the movement of the blades is controlled by linkage systems (kinematics).

Operation of the Voith-Schneider propeller. (Courtesy Voith-Schneider)

TURNING THE TUG—ADVANCE VS. TRANSFER OF FIXED-PROPELLER TUGS

In tug handling, one of the first things to find out is how small a circle it can make and how much room is required to turn the tug. This is called advance and transfer.

If the tug were an automobile, you would turn hard right or left on a narrow street and the car would make a complete half-circle within its length without the rear end sliding sideways. The car's track would be a perfect half-circle with no difference in radius.

A power vessel on the water is not on a solid surface. In turning it has various forces affecting it, such as the pressure and resistance of water on the outside of its turning circle vs. the water momentarily displaced on the inner side of the turn. This sliding sideways in a turning circle is called the advance. The distance required from start to completion of the turn with the tug headed in the opposite direction is called the transfer.

You do not have to deal with these names. Through careful practice, find out how much room is required to turn your tug under certain conditions. Try this in the slip using slow speed. If the tug is single-screw, you can back and fill if the slip is narrow. This is done by placing the rudder amidships, or leaving it hard right when backing and alternately coming ahead and going astern. By carefully watching the swing of the stern, the pivot point of the tug will be noted. This is of great help in small spaces. If the stern swings well and the bow much less, the choice may be to place the tug's bow where there is less room, leaving the more open area in which to swing the stern.

There are many variations of this backing and filling to turn the tug and, after some practice, the best method for the particular draft and trim, weather, wind, and current will be found. Frequently, backing at dead slow will allow some steerage if the stern has been given a kick or quick propeller thrust ahead in the desired direction. Another method to try is going astern at slow speed, and ahead dead slow with the rudder placed hard over in the direction desired. A heavy tug with sternway will often swing in that direction with the propeller turning ahead dead slow while the tug still has sternway. This is a very safe method for use in small areas.

Twin-screw tugs will usually stand still in one spot and turn completely around by placing one engine ahead and one astern. It may require more power on the propeller going astern to match the thrust of the one going ahead to keep the tug's pivot point stationary.

A single-screw tug when under way will turn more easily to the right using less room than to the left, because the pressure of water from

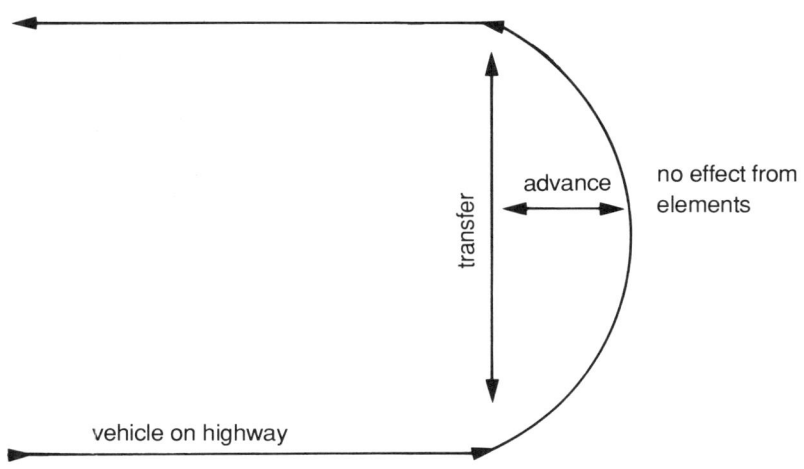
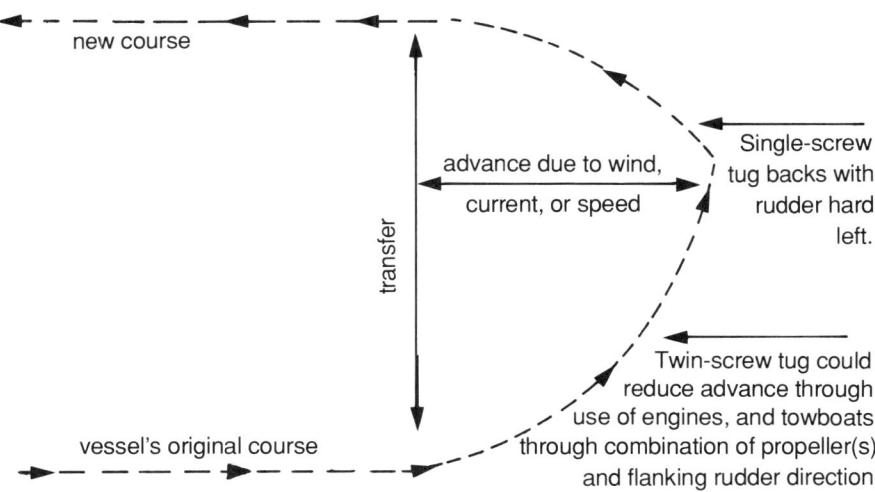

A comparison of turning circles for a single-screw tug and an automobile. The automobile will not slide unless turning on ice. The amount of advance or radius beyond a tight semicircle is small. The distance from the beginning of the turn to the point where the car is heading in the opposite direction is called the transfer. Note the difference between the advance and the transfer of a tug or vessel in the water. Of course, those fitted by AquaMaster can greatly reduce their turning circle.

the right or pushing side of the propeller will strike the flat area of her hard-right rudder.

Tugs with flanking rudders, also known as backing rudders, which are placed ahead and to one side of the propeller(s), can turn much more easily and can be steered when going astern.

Tractor tugs can turn in their own length, pivoting as if a spud or pole went through the hull to the bottom.

TRIAL RUNS—NEW TUGS

If a tug is on a builder's or owner's acceptance trial, all of these maneuvers should be undertaken slowly and carefully. In this manner, the captain will feel out and gain confidence in his ability to handle the new vessel. Parts of this trial require the tug to go through some rugged tests. There should be full-power turns with the rudder placed hard over from one side to the other. Speed runs over a measured mile and crash stops are usually accomplished. The time and distance from full ahead to full astern should be noted, as well as how efficiently the decks free themselves of water. Sometimes, such a test at full ahead with rudder hard over will reveal how far the vessel will lie over and if she will dip the bulwarks under.

USE OF TUG'S POWER

The person at the tug's controls should realize the power held in his hand and how to use it. The first impulse usually is "let's see how fast she will go." This may be fine in an open bay away from traffic, but in harbors and restricted waterways it is not prudent or acceptable to run a light tug as if responding to a fire—unless you are. To throw a wake and waves over someone else's tow, washing over floats and causing small craft in slips to roll five minutes after you have passed will eventually result in damage claims against the tug's owners.

COMING ALONGSIDE A STATIONARY OBJECT

Under normal conditions, the most successful way of coming alongside any fixed object is by using a combination of nature's forces and as little of the tug's power as possible. Nature's forces are the current and the wind. They should be used in the order of their relative strength. If you are approaching a dock, barge, ship, or other stationary object, it should be noted whether you have a fair or head current or if it is angling onto or off your point of landing. The same applies to the wind. Approach slowly. If in doubt, stop and drift. Let the tug show you which force is stronger—wind or current. Then proceed with care. It takes a while to feel the effect of such forces and to gain confidence in the best approach to make a safe, gentle, seamanlike landing contact.

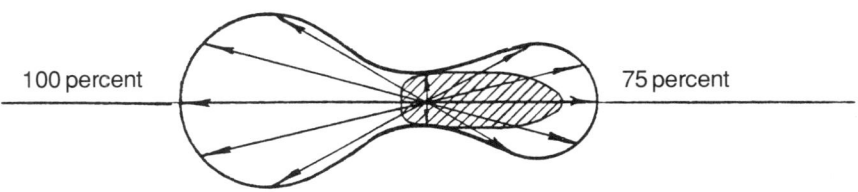
conventional twin-screw tug

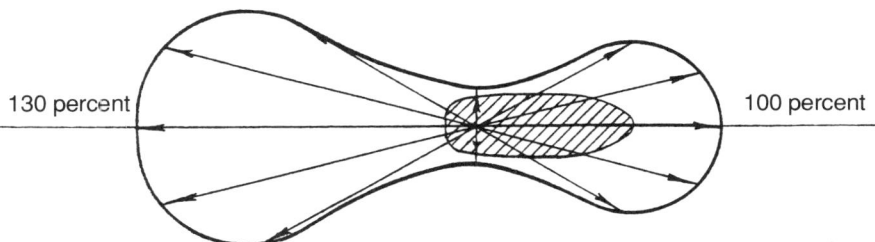
conventional twin-screw tug with nozzles and flanking rudders

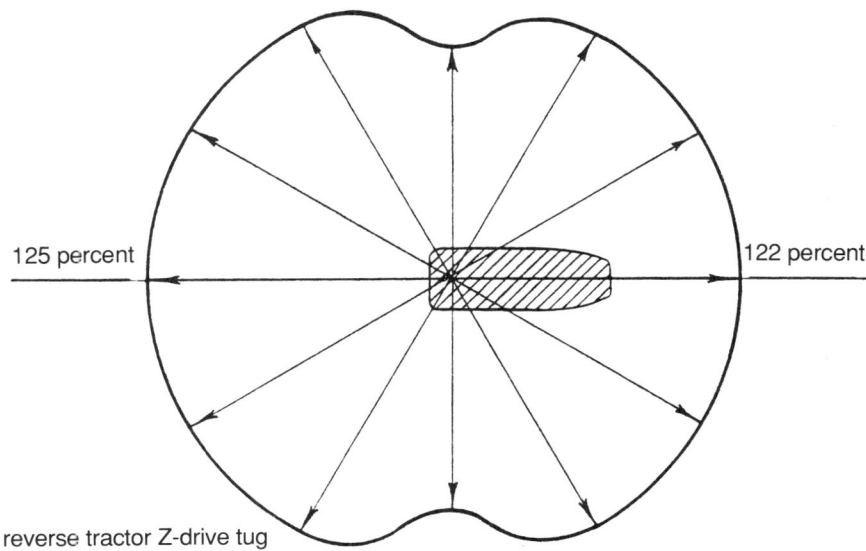
reverse tractor Z-drive tug

Comparative thrust diagrams (Courtesy Shaver)

Some tug captains love to be showmen. This is fine if the maneuver is successful. It requires teamwork between deck and wheelhouse and is best reserved for emergencies. It is not prudent, even if successful, to come barreling in to make a landing that requires full astern power at the last moment. As water is not compressible, it will soon be noted that the faster the tug is traveling, the faster the water it displaced returns to the hole or space the tug has just vacated. In coming alongside a slip or another vessel or bulkheads with solid sides, this return of displaced water will, if the tug was moving at a good clip, force it away from the intended landing spot unless a good deck person is able to quickly lasso a bitt on the object the tug is coming alongside.

TOWING ORDERS

At some point prior to leaving the dock or while under way, it is assumed that the tug master receives towing orders either verbally or via VHF radio. If ordered to a ship-assist job, the tug master will need to know the point of meeting, the type of assistance required, and the destination berth if he is to go aboard the ship as a docking master. This information is relayed to the tug crew with any instructions as to specific mooring lines to be used or any particular way in which the job is to be done. Teamwork is always the key to good tug work.

COMING ALONGSIDE A MOVING OBJECT

This task requires cool-headed confidence on the part of the tug's skipper. Now in addition to his own tug's hole in the water, there is another, possibly bigger, hole moving as fast or faster than his vessel. Realize that the hole in the water the unit being approached has made will have two effects on the tug. First, the displaced water will be reentering behind the stern of the ship or barge, then it will be running parallel to it alongside, but being pushed away near the unit's bow. This will tend to suck the tug into the stern of the moving unit if you approach too closely. It is better to parallel the moving unit at about the tug's width away and at the same speed as the unit. By careful rudder and propeller adjustments, the tug can be worked into a gentle landing. You will have made the cook and engineer happy: No spilled soup or broken dishes, no wrenches in the bilges.

If the object you are approaching is another tug with a hawser tow, be particularly careful not to approach at a sharp angle. Stay off and clear of her hawser and her propeller suction until parallel and abeam and then work in slowly. Once alongside, a slight speed and rudder setting toward the tug or other unit should hold that position without pushing the other unit off course. It takes practice and a few quick but fine speed adjustments to ensure a perfect landing. The frequency of such approaches helps greatly in establishing confidence and pride in one's ability to land gently and smoothly alongside another moving vessel.

COMMUNICATIONS

Communications between tugs and tug captains and ships or pilots are usually carried out on VHF radio frequencies. These vary from port to port. In some places, a general channel such as 13 may be used. In others, each company may have its own in-house one, such as channels 10, 9A, 11, 12, or 18A. In ship-assist work, the tug should acknowledge all orders from the ship's pilot either by voice or whistle, whichever is the local usage. If more than one tug is assisting, the pilot may indicate the individual vessel by vessel name or by the captain's name.

THE WELL-EQUIPPED TUG

What equipment should a tug carry in addition to required safety devices and engine spares? This depends of course on the tug's size, power, and normal working assignments. A space should be made available for the stowage of hawsers and the extra gear that tugs usually acquire. Rope deteriorates more lying on a deck where dirt and fresh water get into the coils. Most tug owners have a shed, barge, or workshop at their tie-up yard. Extra towage equipment should be stored ashore if the work schedule of the tug does not indicate its use in the near future. This does not mean that a well-run tug should not have many spare pieces of towage gear aboard; however, valuable equipment, if not stowed correctly aboard the tug, becomes questionable in value.

Fenders

For side fenders, many tugs use tires, varying in size from those of a small airplane for small tugs to huge truck tires for heavy tugs. The most satisfactory way of securing tires is by chain run to pad eyes welded on the hull. The chains may be passed through holes cut in the tire or to a bent plate inside the tire. Rope also may be used, but it does not wear well in rough usage. Other side fenders are manufactured from old tires cut in squares and strung on wire.

Many new tugs have a solid rubber insert in the guardrail and strips of rubber around the bow. Large salvage tugs may have huge Yokohama type fenders that can be lowered if it is necessary to moor the ship alongside a casualty.

Side Fenders

River line towboats, shifting boats, and other boats with towing knees rarely have bow or stern fenders, but may in some work require a few small side fenders.

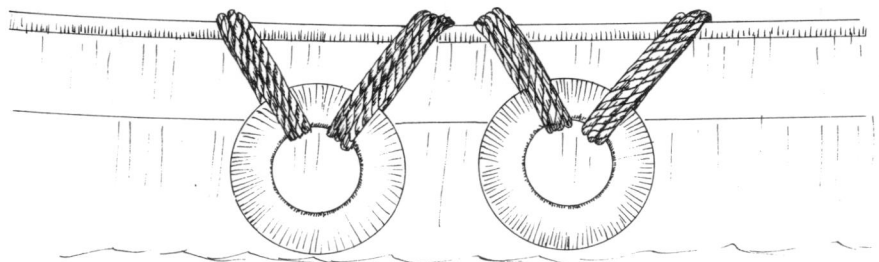

Fenders hung by rope frequently move and the rope chafes on the tug's rail. They require frequent attention as they are often washed inboard in any seaway unless holes are cut in their bottoms to drain water.

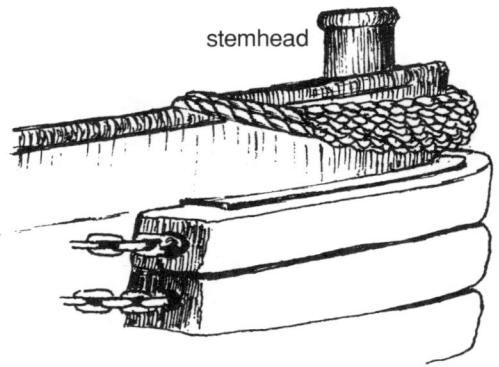

Rubber bow fenders are common on most tugs today. They should start higher than shown. The stemhead has been replaced on most tugs by a U-bolt or eyelet. These horizontal bow fenders can be washed aboard in extremely heavy weather.

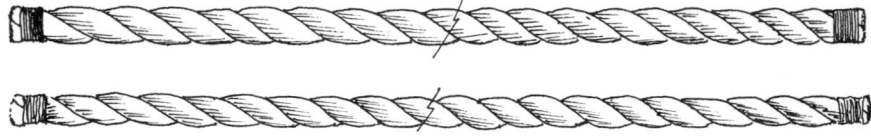

Long line for fenders

Bow puddings and stern fenders were once made using old manila deck lines. This was done by tug deckhands, the shore gang, or riggers. This is an art form of seamanship not seen today on commercial tugs. The required material is not often available. A few towing lines and private owners of classic tugs still make these rope fenders.

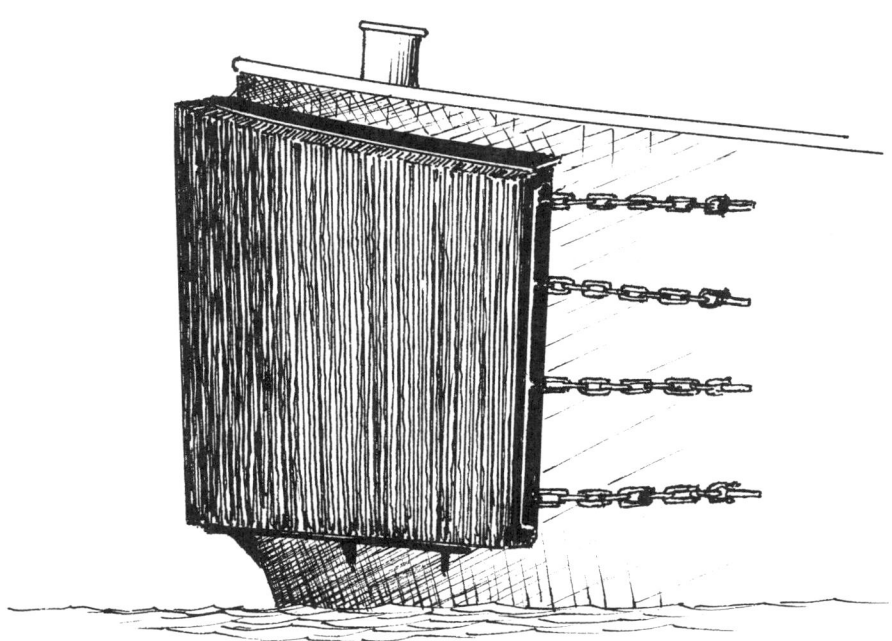

The vertical single bow fender made of strips of rubber tire is nearly immovable and most satisfactory.

Briefly, to make a rope side fender, several long sections of manila or poly type line of from 5 to 8 inches in circumference are required. Measuring off the length they are to hang from the tug's rail, two good pieces are cut and their ends whipped with rope yarn. At this point, these two pieces should be formed by doubling back in a loop lashed together and hung from a hook or davit.

Two similar pieces a little shorter are now added by passing them through a loop and lashing around the other body of the fender. By adding shorter pieces to the outside the fender should be built up to the desired size. The latter pieces need not go through the open loop at the top.

The next step is to select very long pieces of good three-strand line and whip the ends of each strand. Now unlay to get three separate strands as if about to make an eye splice.

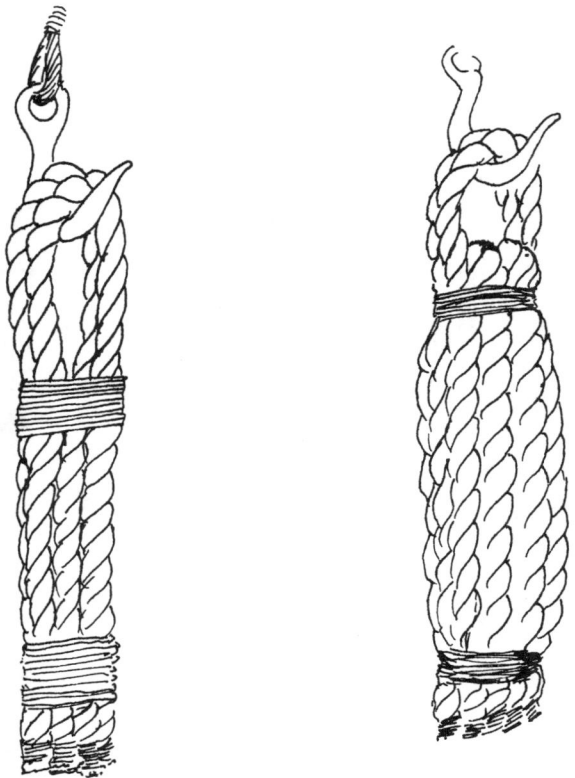

Rope cut and hung to start side fender   Fender filled and shaped

The whipping and opening of strands to be used in hitching

These are known as the hitching. This is the simplest method of making tug side fenders of rope. Purists will use all new rope and for hitching use smaller unstranded line. This, of course, is for looks on yachts and naval officers' gigs and is not financially practical for commercial vessels.

GETTING THE TUG UNDER WAY

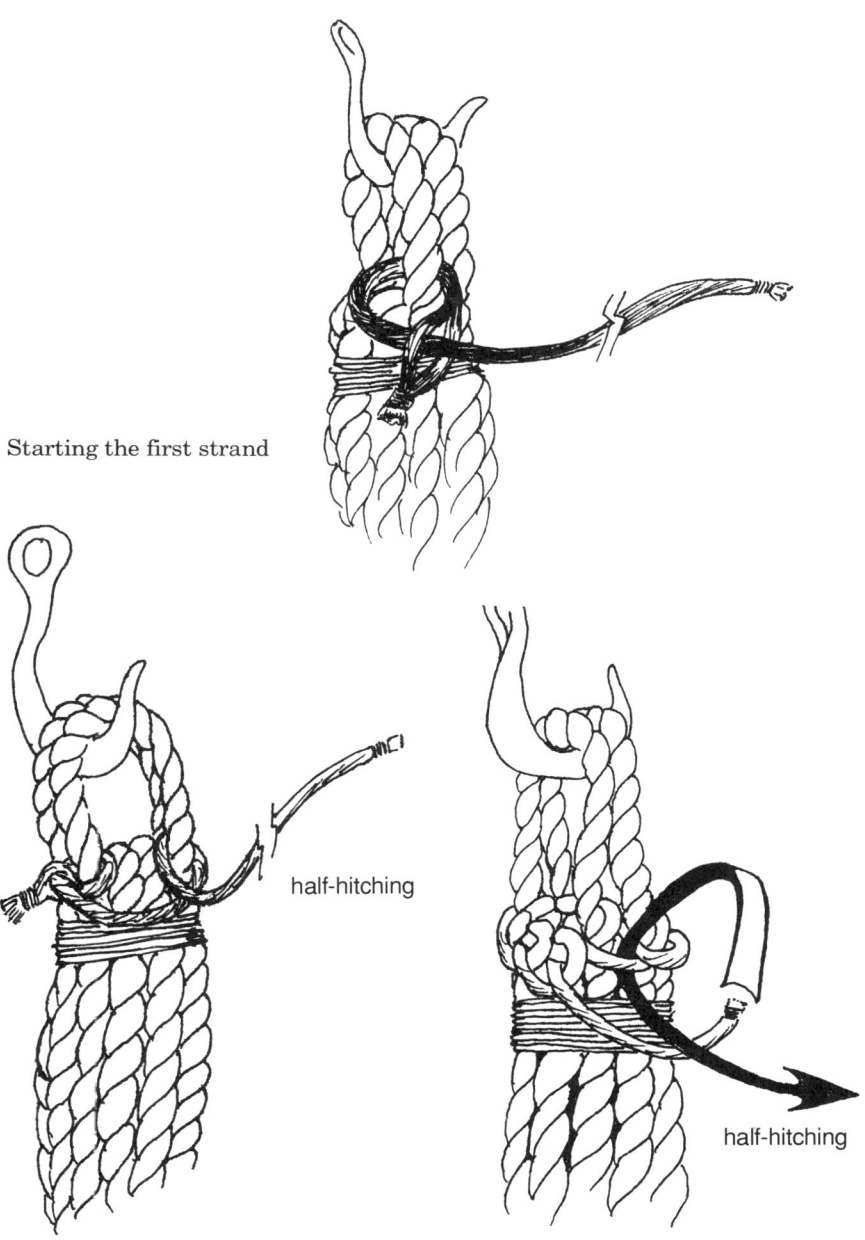

Starting the first strand

The second tuck—a half-hitch

Starting the half-hitching continuously around the fender

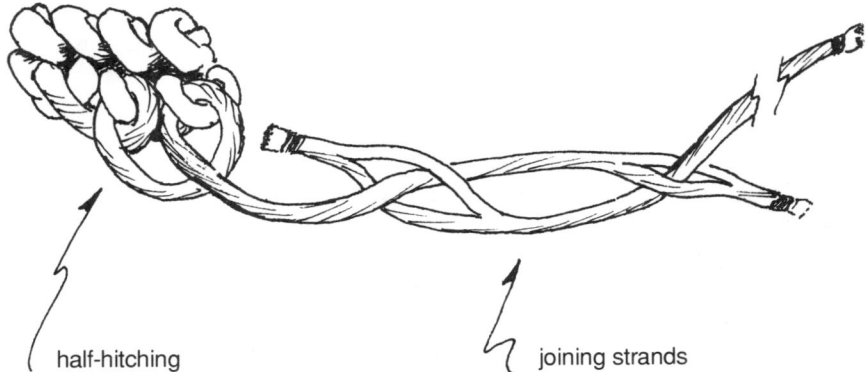

half-hitching   joining strands

This is how the hitching strands are joined.

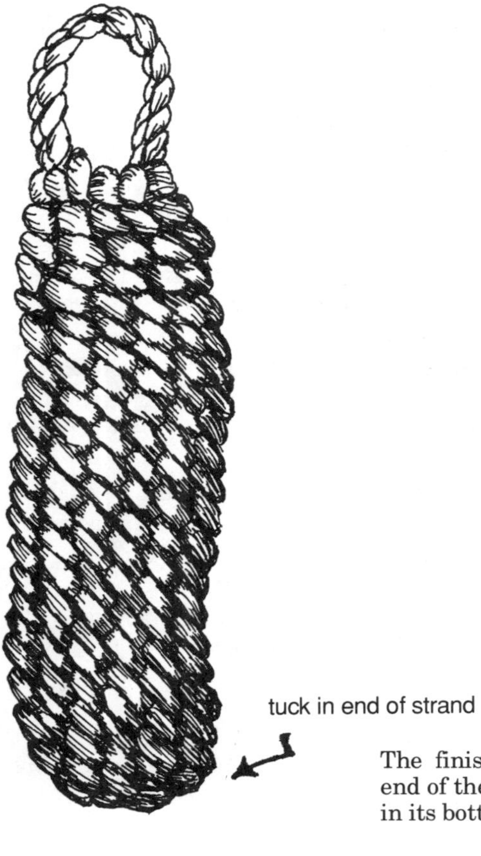

tuck in end of strand

The finished fender with the seized end of the hitching tucked out of sight in its bottom

Once a pile of strand has been made, the final step is the hitching. To start, pass the end of a strand through the eye of the large loop, then, opening this strand just inside where it is whipped, pass the other end through.

Now with the other end, commence making half-hitches around and around to the right, tightening them as you go. To add strands as each is used up, pass one end through the other; the new end through the finished one and the finished one through the new, decreasing the size of the fender by keeping the half-hitches tight as you progress to the bottom. This requires skipping one or more hitches in each layer. The final end must be seized inside to another strand.

Bow and Stern Fenders

The bow fender or pudding and stern fender are started by using similar material with a piece of three-fourths- or seven-eighths-inch galvanized chain as a core. Around it are laid and lashed the heavy pieces of line to shape the desired form.

In each case, the hitching starts in the center. A single strand should circle the entire fender and be held in place by a length of rope yarn. Hanging the fender may be difficult. It can be worked on horizontally in the beginning by hitching from the center with a series of half-hitches toward each end, reducing at every other row to keep the fender tightly knitted.

When half of the fender has been completed, it may then be hung upside down, and hitching the opposite end will go much easier.

If whiskers are desired, they may be added by opening strands across the front with a fid and inserting short pieces of line, which are seized together and then combed out with one's fingers. This saves the face of the fender from wear. Nothing saves your fingers from wear. Broken nails and bleeding cuticles are a common occurrence in making fenders by hand.

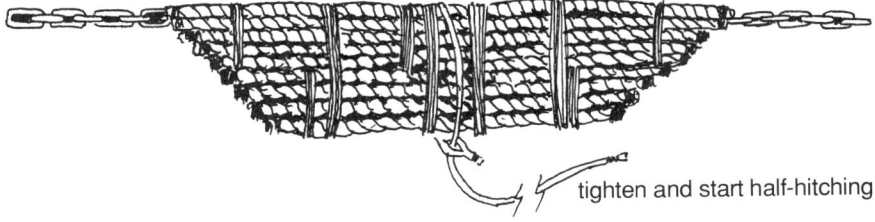

tighten and start half-hitching

The start of a bow pudding or fender. The upper lines are secured around a piece of good chain. Additional lines are reduced in length and lashed until the proper shape has been attained. Hitching starts in the middle and works to one end.

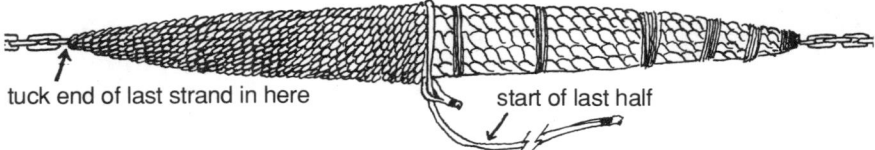

tuck end of last strand in here   start of last half

When half of the fender has been hitched, it is usually hung with the hitched end up so that, in hitching the remaining half, the hitching is pulled down tightly over the exposed line.

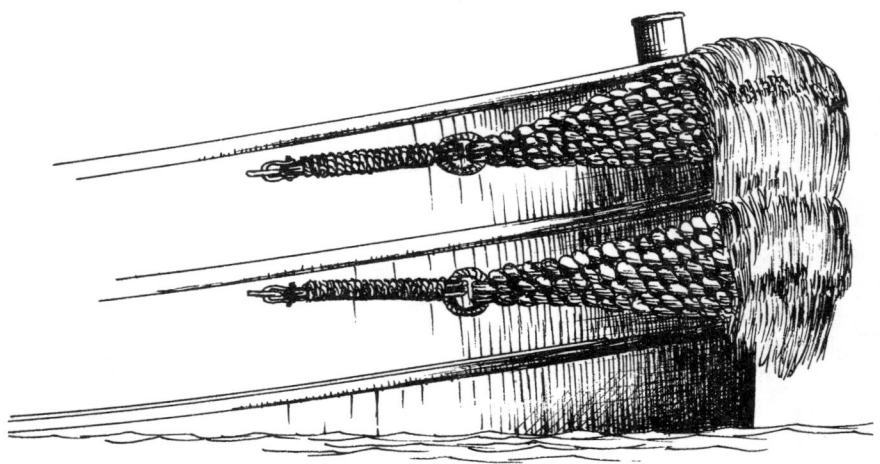

Bow rope fenders with whiskers. Strands are inserted through bottom rows of hitching and worked upward.

Deck Lines

Every tug requires a full set of deck lines, hawser, heaving lines, messengers, and stoppers as necessary. Onboard spares must be considered both to enhance the efficiency of service and to avoid possible charges of unseaworthiness if the vessel becomes unable to service its charge during a towing operation because of a lack of spare towing gear.

The type of cordage available to tug owners is being continually upgraded. From the once-basic manila the industry has gone through pure nylon and polypropylene to many combinations of poly and other fibers. In addition to the normal three- and four-strand lay, there are now multistrands and braided ropes available.

Where once cordage manufacturers produced rope mainly for ships' use in mooring or running through block and tackle, they now present cordage for every conceivable job requirement. The towing industry alone has available to it specific state-of-the-art cordage for deck usage in various work areas. Among the newer ones are Araflex, an Aramid rope; 2-in-1 nylon; and 2-in-1 Power Braid.* According to their manufacturers, these ropes have proven to be excellent for tug hawsers. Kextran-B, 2-in-1 Power Braid, and 2-in-1 Stable Braid are fine for mooring and alongside towing and push cable facing ropes.

Dupont, producer of the ultimate in synthetic fibers has introduced both Estalon and Kevlar, which are the basic fibers woven into these braided and Wirelay ropes. Tubbs Cordage† produces both three-strand and braided under the brand name KARAT. Tubbs claims that, in addition to being "about half the weight and diameter of polypropylene of equal strength," KARAT rope "is the lightest, strongest, and easiest-to-handle mooring line ever manufactured."

Araflex is one of the Wirelay ropes produced by Greening Donald Co. Ltd.‡ It is based on an Aramid wire which allows a steel wire rope configuration of totally synthetic rope. Kextran-B has been used successfully to replace facing wires on some towboats on U.S. inland waters and ship-assist tugs at Roberts Bank coal piers in British Columbia. Kextran-L has recently been added to these Greening Donald Wirelay ropes. It is designed for many seagoing uses including mechanical termination, which would include reeling onto a tow winch. The manufacturer claims that all of these Wirelay ropes are one-fifth the weight of the steel cables they replace, are easy to splice, leave no fish hooks, are very flexible, and have a urethane protective coating.

Samson Ropes claim their braid 2-in-1 ropes offer greater strength, more controlled stretch, and are nonkinking and -hockling. Here are a few comparisons of various type deck lines and towing hawsers made from synthetic fibers as used on various towing assignments. The table is not offered as an approval of the product over any other available.

Manufacturer	Type	Circumference, in inches	Breaking strength, in pounds
Samson	2-in-1 Power Braid	7	142,000
Tubbs	KARAT	7	78,000
Samson	2-in-1 Power Braid	11	396,000
Tubbs	KARAT	11	162,000
Greening Donald	Wirelay Kevlar	2$7/8$ (diameter)	558,000

A complete table of ropes appears in appendix II.

*2-in-1 nylon and 2-in-1 Power Braid are products of Samson Cordage Works, Boston and Seattle.
†Tubbs Cordage Company, Orange, California.
‡Greening Donald Co. Ltd., 55 Queen Street, Hamilton, Ontario, Canada L8N 3J3 or 550 Hadley Road, South Plainfield, New Jersey 07080

Small Lines

Most tugs require small lines of various sizes for use as heaving lines, messengers, stoppers, and for securing loose items and hawser covers. Most frequently used and lost or stolen are heaving lines. These should be made up of the softest yet strongest pliable line of from 50 to 75 feet in length. On ship-assist tugs they should be long enough to reach the deck of a large, high-riding ship. It has always been popular to make a separate monkey fist that can be attached or taken off the heaving line. If a nut or piece of lead is used as a core in the monkey fist, it should travel straight and true when heaved.

Long messengers of from 2 to 4 inches in circumference or larger should be made up so that they will reach to whatever may need to receive the tug's tow bridles, hawser, or push cables. Stoppers, 6-foot

Monkey fist as used on many heaving lines

lines with an eye splice and about 4 inches in circumference, are necessary to hold the lines being taken from or to the capstan, or in connecting push cables or the main hawser.

Tugs that do any long-distance hawser towing should carry a spare or "insurance" hawser of nylon or other synthetic fiber. This is for use in an emergency.

U.S. coastwise tugs usually stow these emergency hawsers under cover on a rack on the boat deck. This is not a very satisfactory location when suddenly the crew is called upon to break it out. The belowdecks hatch, served by a deck plate directly below a horizontal spool capstan of a tow winch, as incorporated into many European tugs, is a more feasible and convenient stowage arrangement. These vessels also carry a spare wire rope towing hawser, on a reel down in the same hatch.

All large synthetic rope hawsers 10 inches and larger in circumference should be fitted with metal rope thimbles. Those made of nonrusting, nonsparking, metal alloy of nickel, aluminum, and bronze will outlast the line itself and can be used again.

Shock Lines

The entrance and exit of many U.S. Gulf ports are less than 50 feet deep with swift-running tides. Sea conditions are often confused and rough. Yet the depth will not allow for any great length of tow cable to be used. To shorten cable when transiting these entrances could result in parting

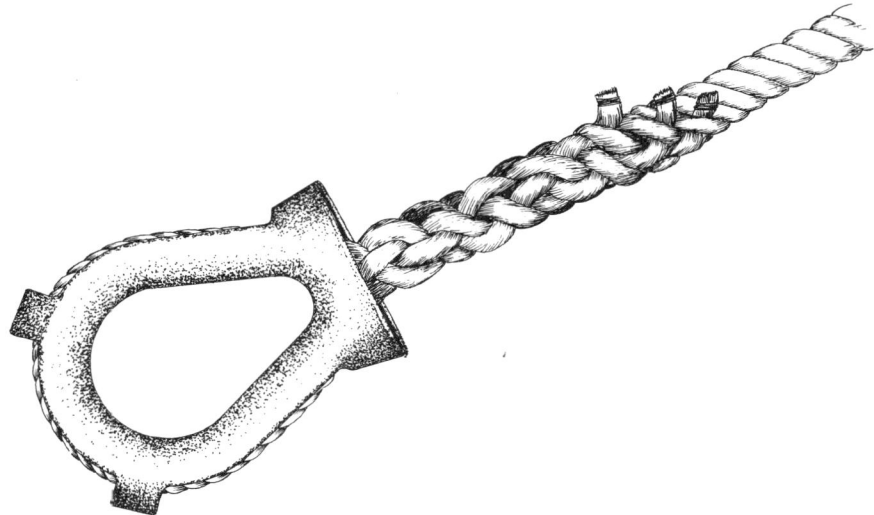

Heavy-duty bronze thimble. (Courtesy Newco Manufacturing Company)

it. To overcome this surge problem in pulling a large, loaded barge in very choppy seas, a 200- to 250-foot piece of 10- to 12-inch nylon, with a thimble in each end, is connected between the barge bridles and the tug tow cable. This is called a shock line.

Tugs that regularly tow barges in these areas usually carry a spare shock line, stowed in the same area as the spare hawser. Both should be kept covered. Direct rays of the sun are most damaging to nylon rope.

Pendants and Bridles

In some areas pendants are referred to as pennants. The well-equipped tug will usually have spare wire rope pendants of approximately 90 feet. These should be in various sizes up to at least one equaling her main tow cable. Spare pairs of towing bridles of various lengths and sizes should be carried. All of these should be protected against deterioration by treatment as required with a preservative that will penetrate to the wire core and into splices. Bridles that are frequently used may be stowed and lashed inside bulwarks or in other convenient deck locations.

Push Cables

Tugs and all inland towboats that are rigged for pushing should carry at least two spare sets of push cables for each power winch, and for tugs, two for the main tow winch. As these are often very heavy, they should be stowed either in a locker off the main deck or, if on the boat deck, near a power hoist or davit.

Shackles

One item of top priority on any towing tug is the shackle. There must be shackles available for all types of connections, from 3/4 inch and 2 inches on tag lines to fit over bridles on small or large push cables, to the 50-75-100-ton towing shackles of deep-sea operations. Spares of all sizes are a necessity.

All shackles that will be used in overboard connections between bridles and pendants and pendant and tow cable should be of the safety type.

Shackles that are used to connect the tug's hawser to a pendant of a tow bridle should be of nickel steel or similar extra-strength metal with a breaking strain no less than that of the tug's hawser.

To ensure that the nut of the shackle will not come off, the pin should be given a quick tightening when connecting and a cotter key or welding rod should be placed through a hole in the end that protrudes beyond the nut.

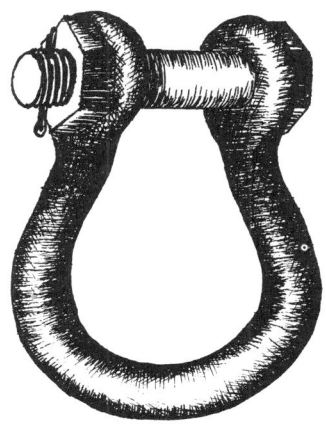

The safety shackle should be used in all overboard connections, such as between the tug hawser and pendant, the pendant and flounder plate, and the flounder plate and chain or wire bridles with thimbles.

Lubrication—Grease

Shackle pins should be greased when the shackle is stowed and the threads should again be lubricated before being connected to the hawser. A good grade of water-resistant grease is recommended. Lubroplate has been used with great success. It is white in color and resists salt water while adhering to such areas as shackle pin threads, towing arches, winch bearings, fairleads, and roller chock pins.

River and Inland Towboats—Spares

These powerful towboats have a gear locker full of hard rigging, short wires, steamboat ratchets, and the necessary bars, mauls, and shackles. This is usually located forward on the main deck close to the towing knees. Spare cable for all deck winches may be kept there or in a locker aft of the engine room.

Deck Equipment—Ocean and Coastwise

Tugs working offshore or for hire should have a good supply of spare wire pendants, shackles, short lengths of chain, chafing gear, hawser boards or covers, intermediate hawsers, and splicing equipment for cordage or wire rope. For connecting up hawsers, tow cables, etc., several small mauls or 10- to 15-pound hammers, cold chisels, cotter keys, short pry bars to turn shackles when connected, and large mauls to knock out tow hook latches are essential. Short pieces of heavy pipe to be used as auxiliary norman pins and full burning and cutting equipment are commonly carried. A portable salvage pump, gas- or electric-driven, with full suction and discharge hoses and auxiliary generators, is often seen on ocean tugs. Those working in areas where fires may occur on tankers or oil terminals should have fire-fighting equipment and fire monitors. This basic list is not intended to be complete—much more can be added.

Instead of carrying diving equipment, many salvage tug owners now fly divers with their equipment to the location of a casualty when required.

7. Under Way—Towing on Inland Waters

TUG admirers, like the writer, get the same urge to get a good-looking tug under way, to pick up a tow or assist a ship, as a sports car enthusiast gets when looking at a new model. Towboating is contagious. Participation in a big project, depending upon where you stand, can be exciting, exasperating, boring, or a battle with the elements.

For the tug master, success is due not only to himself but also to his crew as a team. His leadership and their response through mutual respect give the tug a good name, and both owner and crew can be proud of that well-earned reputation. Training and preparedness mold a good working tug. What follows are some of the methods commonly used in various types of towing.

SHIP ASSISTING

This is the most common use of tugs in world ports. The first movement is to meet the ship. In approaching a moving ship, *keep off!* Wait until the ship has slowed to 6 or 7 knots. Parallel the ship's course at least half a ship's width off, until abreast the midpoint of the ship. By then, the tug's speed should match that of the ship. If the ship is going too fast and the tug is needed alongside at that moment, request the ship's pilot to slow her down. Then work in to lie alongside. Usually, a twin-screw tug will hold her position there by working her outboard propeller dead slow with the rudder 5 degrees toward the ship. Rudderpropeller and cycloidal tugs can maneuver and hold themselves alongside a ship much more easily than a fixed-propeller tug, as their simultaneous engines and steering controls give immediate response. Through fine-tuning of the throttle, a gentle landing is possible. If, after the tug is alongside, the pilot on the ship wants a line to be passed up and secured, be sure to let him know when it is fast. Do not fail to advise him if you feel he has increased speed, causing the tug to heel and take water on deck.

Working the Ship

When all fast alongside ready to work the ship in response to a docking pilot's orders, the tug captain should not be a robot. He should remem-

This 1986-built rudderpropeller can gently ease alongside the ship she assists at Jacksonville. (Courtesy McAllister Brothers Inc.)

ber that while the person giving orders on the ship is the ship's servant with overall responsibility, the tug captain is still responsible for his tug and the safety of his crew.

What Tugs to Order

In many harbors throughout the world tug companies maintain a fleet of various horsepower tugs to cover their local towing requirements. Not all are equal to the maximum required for certain jobs. At times, a tug of less than adequate horsepower is dispatched and may manage to accomplish her mission satisfactorily.

Ship pilots, agents, and customers, however, always want the best for their ship in appearance, reliability, and power, with some reserve.

To fulfill clients' requirements, port authorities and others have worked out a ratio of a single tug's horsepower vs. the deadweight tonnage of a ship. Tractor tugs' horsepower must equal 5 percent of the ship's deadweight, a standard propeller tug, 10 percent. For a 72,000-deadweight-ton ship, a 3,600-HP tractor, or two 3,600-HP propeller tugs

Cates Towing of Vancouver, British Columbia, is proud of its new tractor tugs. The company's Z-pellers like *Charles H. Cates II* have a unique telescoping boom by which they can pass their lines to a ship. (Courtesy C. H. Cates & Sons Limited)

would be required. A 144,000-deadweight-ton ship would require two tractor tugs or four standard tugs. Certain piloting authorities not only accept this ratio but demand its use in ship assist or dockings.

Ship Assisting—Containerships

Unlike a tanker fully loaded with liquid cargo to the outer shell, a containership has air spaces inside her hull between the frames and the container cells. Due to their size and deadweight, containerships require powerful tugs to assist them in docking. So that the tug's captain and/or docking pilot will not pick a spot at random, these ships frequently will have areas marked on their hulls where tugs may place their bows to push. This prevents dents in the ship's hull plating between frames. Other hull areas, such as bulbous bows and the location of bow thrusters, are also marked to assist the tug captain.

Many containerships have a square stern and a bow thruster. In turning one of these ships, an assist tug often places her bow at the center of the ship's stern and runs two lines off at 45-degree angles to the ship's chocks. With the tug secure, as a movable rudder and power unit aft, and the bow thruster forward, the ship may be turned off a pier in her own length. She may also move herself ahead or astern and fit into close quarters between other vessels by using her own engines.

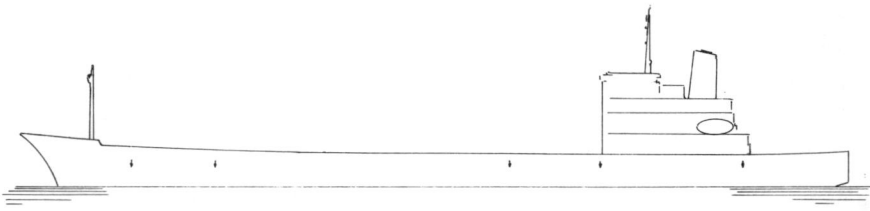

Most modern ship hulls are lighter in structure than their predecessors. To prevent indentations from powerful tugs during dockings, marks are placed on the ship where tugs should push.

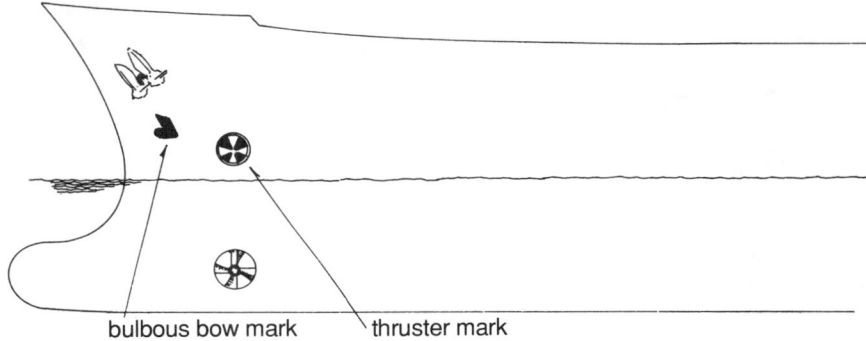

Ships with bow thrusters and a bulbous bow indicate this by marks near the stem.

Ship Assisting in General

A tug assisting a ship in the Manchester Ship Canal passes or takes two tow wires or lines from her bow to the vessel's stern. This appears as a Y-shape or bridle arrangement. In San Francisco this is known as the Lurline method.

If, when in the basin, the ship must be turned, the stern tug crew can let go one wire and carry the other one aft and place it in the tow hook, if this method is to be used in turning and docking.

Turning a ship, putting a ship into a slip, a lock, or a basin, and pulling one out are among the most dangerous of tug-assist assignments. The size of the modern ship often prevents the docking pilot from seeing a tug working at or near the bow. This is particularly true of a ship whose deck is piled high with containers.

The general pilot/tug master relationship will vary from port to port. In those ports where the towing company offers the services of docking pilots, the relationship is very personal. This type of docking pilot is actually a tug captain who has temporarily changed hats. As a

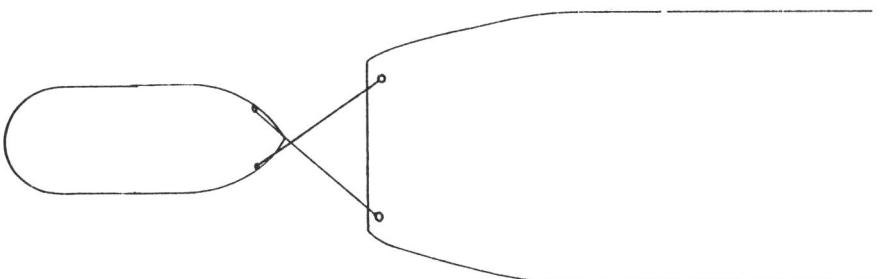

Tugs frequently make up to work across the stern of container and other flat-stern ships.

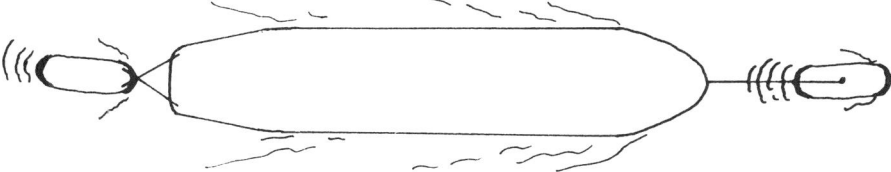

Tug makeup in ship assist on Manchester, England, ship canal

pilot, he is on his own. He may be a member of a self-employed docking pilot group.

It is the bar, river, or harbor pilot with little or no towing experience and a rough hand on the telegraph who can cause trouble to the docking tug. This may occur when, in turning a ship in a tight basin, a tug may have to shift positions. The tug may have to change from holding back on a ship to pulling her stern around. She may have to let go altogether and take a line from a different spot. If the ship has too much headway on during these moments, there is the possibility of a line parting or of the tug being girded.

THE TRACTOR TUG

Tractor tugs have proven to have excellent maneuverability and outstanding operational safety in assisting seagoing vessels. They are equipped with two propulsion units, which offer combined movement and steering. The propellers may be fixed or controllable-pitch. They are mounted in nozzles that offer great hydrodynamic efficiency. The entire unit, which extends below the hull of the tug, is rotatable through 360 degrees. This omnidirectional property enables the tug to maneuver equally well going ahead, astern, or sideways. The tug can stop or turn in its own length.

Shaver's Z-peller *Portland* shows careful design in tumble home, placement of side bitts, and closed chocks around her bow. (Courtesy Ackroyd)

The function of each propeller is entirely independent and controlled separately. These two electrohydraulically controlled units can be programmed to operate through a single lever. They may be mounted under the tug either forward or at the stern depending on the tug's operational area and permissible draft. If necessary, forward units may be recessed into the hull to reduce draft.

Tractor tugs have also been proven in deepwater operation with the forward drive and are extremely stable. The forward-mounted unit offers maneuverability in heavy seas and has a protection plate below it. These tugs, when fitted with a tow hook near the stern and forward-mounted propellers, have proven to possess positive stability when working with a hawser. The danger of capsizing due to girding is nearly nil.

Schottel tugs with stern propulsion are ideal for ports in areas with limited water depth as they combine minimum draft with excellent maneuverability. The control of the Schottel tug is claimed to be far superior to any other so-called joystick control available at the time of writing. The Schottel control, trade named Masterpilot, is in fact not a stick. It consists of an outer wheel that turns 180 degrees and an inner lever that also turns 180 degrees, much like a conventional engine throttle control which moves from one extreme to another from forward to neutral or reverse. The outer wheel controls the turning of the vessel. The inner lever controls the sidestepping or sideways movement of the vessel as well as forward and aft. The deflection of the same lever con-

The Schottel Masterpilot control is simpler to operate than some joysticks. (Courtesy Schottel)

trols the clutch and speed. After a little practice, most tugmen become proficient in the handling of the Schottel-propelled tractor tug.

PILOTS AND DOCKING METHODS

Most tug captains doing ship-assist work are familiar with local docking pilots and their operating habits and personalities. Regardless of the docking pilot's instructions, however, to protect himself, the tug, and the tug owners, the tug captain sometimes must make an immediate decision whether to hang on or let go. If the ship is in a ticklish position, he must be sure that by continuing on he can succeed in getting her and himself safely out of the tight spot.

Complete understanding and cooperation between tug captain and pilot are necessary. Methods of operation differ in each port. In European ports with tidal basins, tugs usually handle ships on stern lines. On large ships, up to four or five tugs will each have a line from a certain position until off the berth. A hawser tug may hold each end of the ship while others may push her into place at the berth.

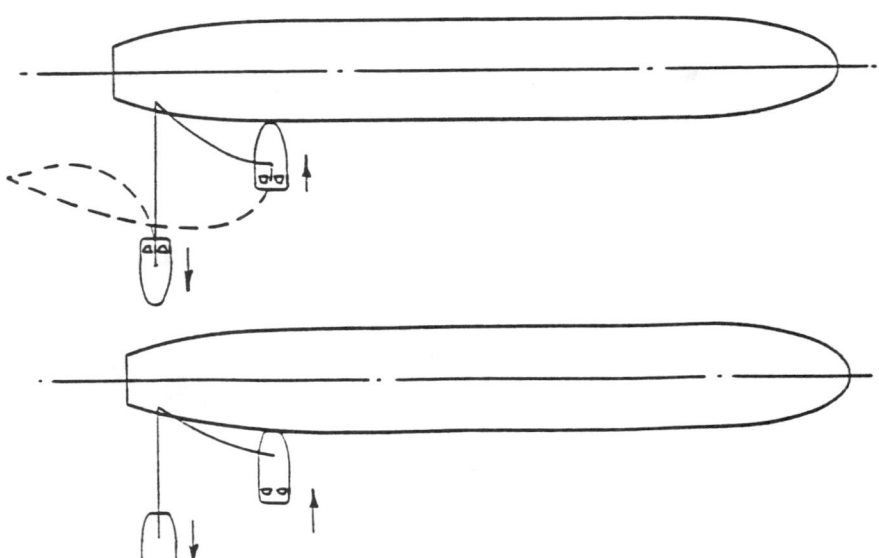

Comparison in change of maneuvers from pulling to pushing in a ship assist. In the upper diagram, a conventional tug takes 3½ minutes, while the Schottel or Z-drive rudderpropeller type takes 25 seconds.

Culver is specially equipped with a high fire-fighting nozzle that can reach some decks on ships similar to the *QE 2*. *Albert* is a very modern harbor tug. Her trusting crew stands on deck as both tugs turn this famous ship. (Courtesy Ajax)

UNDER WAY—TOWING ON INLAND WATERS 193

Standard U.S. East Coast type harbor tug. (Courtesy S. Lang)

Modern ship-assist tug used in many non-U.S. ports. (Courtesy Damen)

Port of Dublin Authority harbor tugs and line-handling vessels

Quebec Tugs Limited *Capt. Ioannis S.* was built for hawser work in the St. Lawrence River, Gulf, and ocean. (Courtesy Quebec Tugs Limited)

In London, Thames River ship docking and handling is frequently done on the hawser. In New York and other U.S. East Coast ports, tug assistance is accomplished by working alongside.

At Canadian Maritime and St. Lawrence River ports, a combination of both methods is used in docking and undocking.

Tug *Challenger* and another tug assist *Port Chalmers* as the ship approaches a dock in the Thames River. (Courtesy Port of London)

A different method of ship docking is used in New York harbor than that used in the British Isles. Here, Moran tugs are docking the liner *United States*. (Courtesy Moran)

At U.S. Gulf ports, various combinations are used. Passenger liners at Port Everglades, Florida, are docked via the alongside method. Miami and Nassau use both alongside and hawser tugs. Most U.S. West Coast ports use alongside tugs or a combination of both if the maneuvering area or the passage through drawbridges requires it.

Above, in Saint John, New Brunswick, the tug *Ocean Hawk* tows *Beaverford* toward her berth. (Courtesy Canadian Pacific Railway) *Below,* docking a ship in a Saint John pier requires holding the ship against the river current. Here it is being done as a hawser job with *Ocean Hawk* pulling the bow into the slip while a tug whose stern is barely visible, holds it from closing on the pierhead. Another has a hawser holding the ship's stern against the current. Today, Irving Tugs in the same port may handle such a docking in a different manner. (Courtesy Canadian Pacific Railway)

Opposite, top, Quebec's tugs tow a ship to an inner berth. (Courtesy Canadian Pacific Railway) *Middle,* APL's new *President Lincoln* being turned in the Mississippi River. (Courtesy Crescent Towing Company) *Bottom,* APL's *President Lincoln* arrives on her maiden voyage off her berth in Seattle. Foss's tractors *Wedell Foss* and *Arthur Foss* assist. (Courtesy Foss)

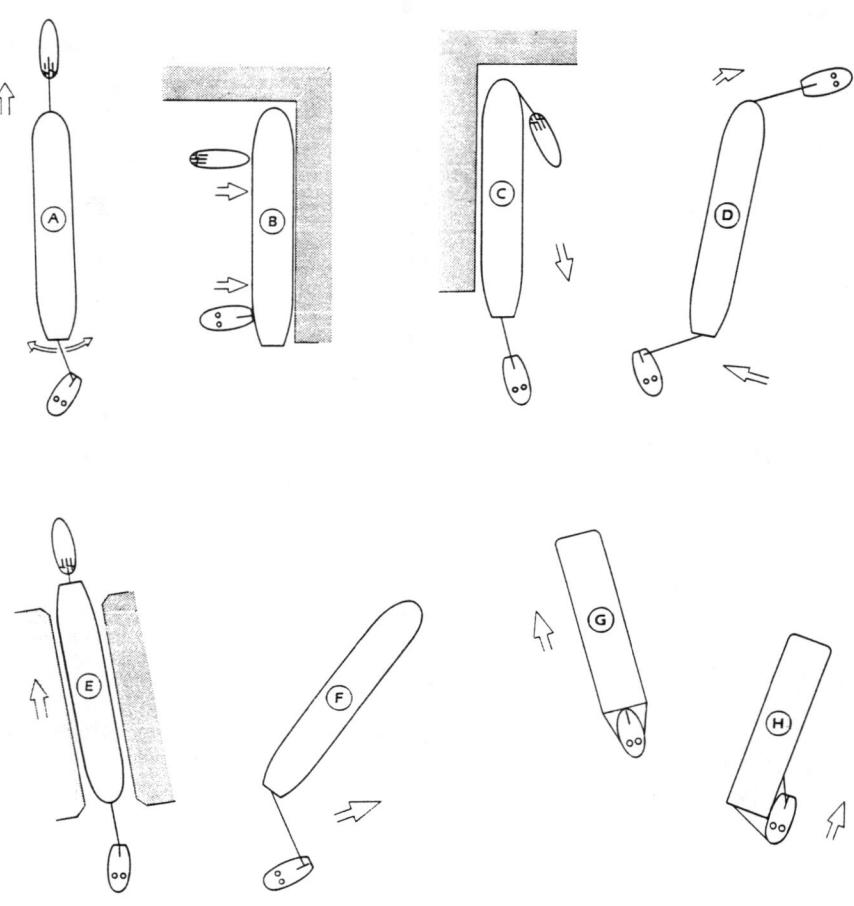

A, working with other harbor tugs the Foss water tractor retards the ship's forward motion and exerts steering control side to side; *B*, berthing is accomplished by pushing with the tractor's stern while maintaining the towline connected; *C*, maneuvering a vessel from her berth is accomplished with single lines—if close quarter handling is required, the tractors can pull/push on command; *D*, controlling the rotation of a ship is a single line maneuver with the water tractor's 360-degree power and thrust; *E*, transiting bridges or channels with minimum clearance and maximum control and safety is accomplished by "tight lining" the maneuvered vessel forward and aft; *F*, with ship power on, one tractor astern, on a single line, can retard the vessel's forward travel and exert steering control side to side; *G*, pushing a barge is accomplished by the tractor placing her stern against the barge with propulsion units aft for maneuvering in tight quarters; *H*, pushing a barge alongside is much like pushing ahead with the tractor's propulsion units aft to ensure course stability. (Courtesy Foss)

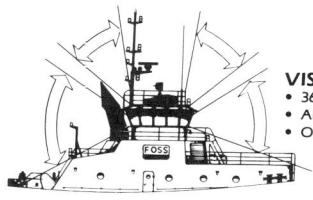

VISIBILITY
- 360° on the horizontal plane
- All work areas viewed from pilot house
- Overhead visability 360°

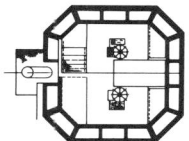

PILOT HOUSE
- Second generation design based on successful SHELLEY FOSS
- Window slope minimize internal reflections to maximize visibility at night
- All controls centralized
- Fire monitor on top for increased fire protection in any area of operation

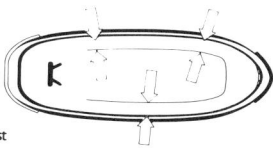

HULL FORM/WORK AREAS
- Unique hull configuration to assure safety and maneuverability
- Special fender areas and arrangement
- Seakeeping both fore/aft
- Wide side decks
- Generous area around deck machinery
- The propeller guard increases propeller thrust, provides effective protection against grounding and debris

The wheelhouse arrangement of a tug showing the U-shaped control console with radar and compass, giving the tug captain ready-to-hand control of the vessel.
The two steering levers or wheels of the SCHOTTEL STEERING SYSTEM S 600 combined with the rudder angle-indicators are clearly visible.
The large windows on all sides of the wheelhouse allow the helmsman to have an excellent all round view and with the U-shaped console, this means that the tug can go ahead, astern, move sideways or turn completely without him having to change his position between the two control panels.

Design characteristics unique to the Foss water tractor tugs. (Courtesy Foss)

SHIP AND BARGE WORK ON THE HAWSER

When on the hawser working a ship the tug captain must be aware of the tug's capacity to list and right herself without capsizing. There are definite limits beyond which the tug will be in extremis. Getting "in irons" or girding is an ever-present danger when working on the hawser. The conditions that lead to it are:

A tug's tow bitts located too far aft between the tug's pivot point and propeller and rudder.

The tug pulling at an angle causing her to list toward the object.

The object or ship not responding to the tug's pull and actually pulling the tug toward it.

A heavy unit attempting to pass the tug so that the towline leads off the tug's beam or forward of it.

The tug's rail going under, either from this pull or from the ship's propeller wash.

The tug being unable to free herself and as a result capsizing (girding).

Many bays, coves, and inlets are fully water-covered to various depths at all stages of tides. They often have narrow dredged channels, which may or may not be marked. When towing a barge or similar unit in one of these narrow dredged channels, great care must be taken. If there is a junction or cut-off that requires a sharp turn, the tow should be slowed as much as possible. Assuming the towed unit is on two short stern lines, the tug shuld be turned so that a strain is taken on one line, to determine if the tow will swing in the desired direction. If it will, fine. If the captain can feel the tow will turn but needs a little more power to start the swing, then he should give it.

If, on the other hand, due to size, current, available turning space, etc., the tow will not begin to respond, the tug captain must take immediate action to protect the tug and his crew. He should try to get ahead of the tow and back against it to stop it, or let go one line and swing around alongside head-and- tail. He should not get into a position that might cause the tug to become girded and to capsize.

Extreme situations can occur in seconds. A tug pulling a tanker stern first out of a slip on the Neches River in Texas was laid over so that the edge of her boat deck was in the water. The cause was a "telegraph

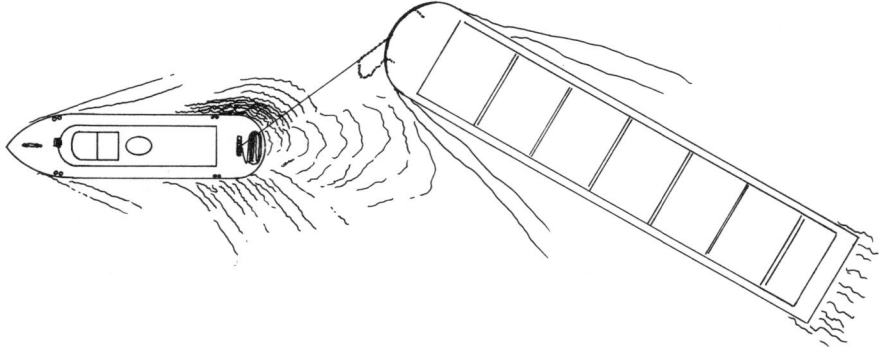

Tug's tow bitts are too far aft to control her tow when it sheers.

happy" ship pilot who became nervous and suddenly gave the ship a shot astern at half speed. Luckily, the tug captain quickly shifted the tug's rudder to right his craft and went full astern, slackening his towline. This tug was an older conventional U.S. harbor tug with tow bitts about 16 feet from her stern.

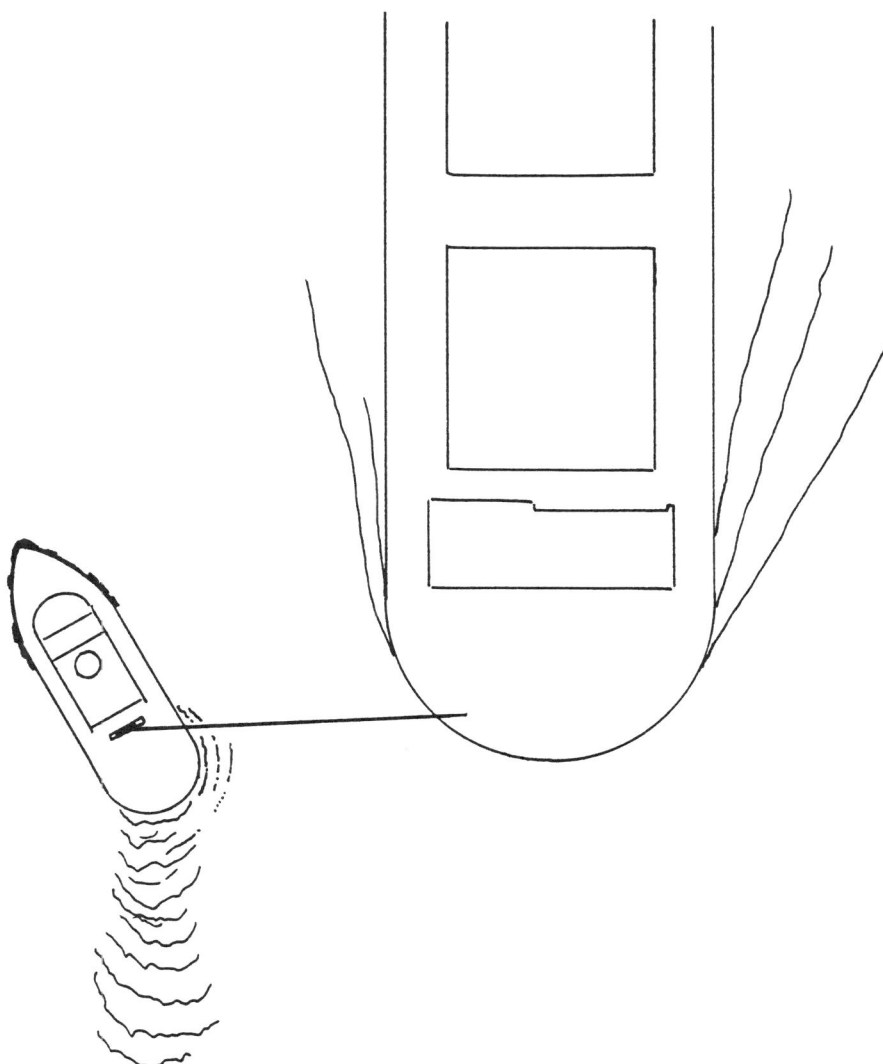

Tug is in very poor position if the vessel is moving as indicated or if it is a ship with sternway.

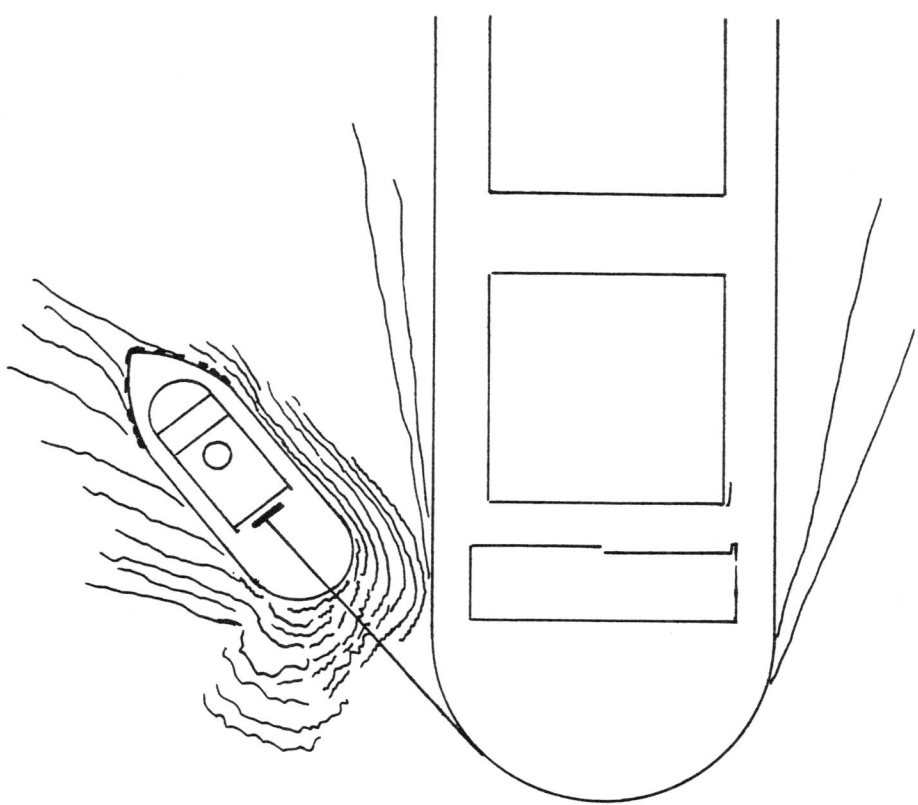

As the tug pulls, the unit does not respond. The best way to save the tug is to work her alongside of the unit, backing if necessary to keep from taking water on deck.

The same type of maneuver with a tug having a low-slung, short deckhouse and a long stern has been seen. The tug was pulling a ship stern-first from a side basin into the River Seine at Rouen. She had the ship's line in her tow hook, which was nearly amidships. The ship hit the Seine's current with some sternway on and the pilot started her engine ahead. With the line from her tow hook running off her port rail at about an 80-degree angle, the tug did not list and held the ship's stern in position as the ship swung and headed downstream in the current.

TOWING A DEAD SHIP

In the United States, the terms "dead ship" and "flat tow" refer to one without power, whether she is laid up or her engines are broken down.

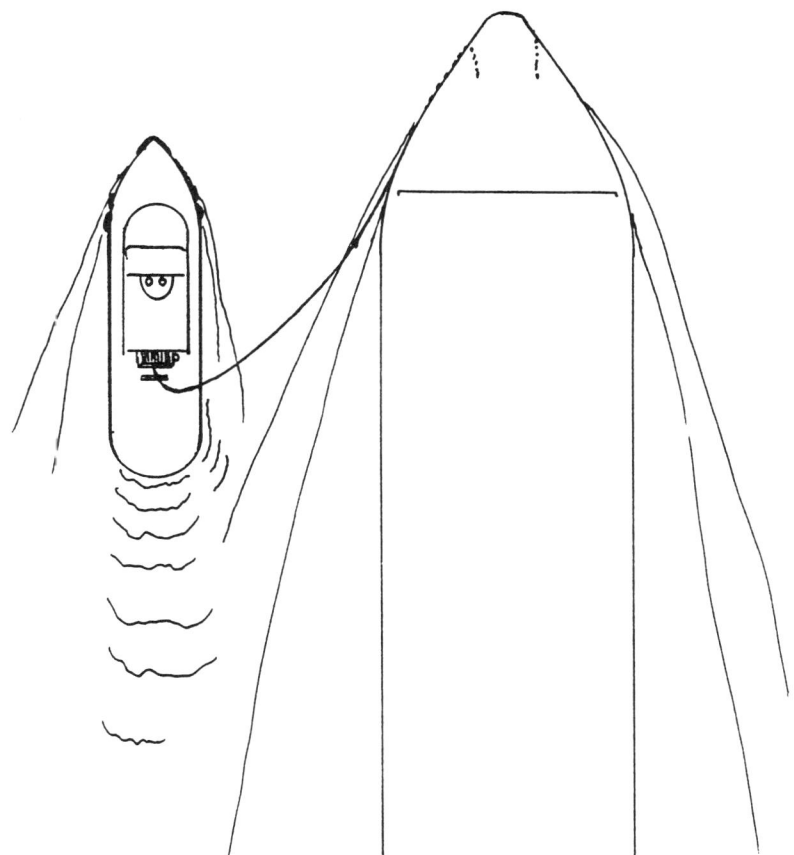

A very dangerous situation. A heavy ship is passing the tug. To get out of this predicament, the tug must quickly slack her tow cable and run ahead as far as possible in order to check and take charge of her tow.

Moving such a vessel from one spot to another within the port or confines of its area or tributaries, requires just as much care and planning as any other tow. Operational people ashore should, unless an emergency exists, plan for the movement to occur under the best visibility, tide, and current conditions. Line handlers and a riding crew may be necessary if the ship is unmanned and out of service. An assisting tug is normally required.

If you are on an assist tug or if you use one in assisting you with a tow, there must be a very careful and clear understanding of the general

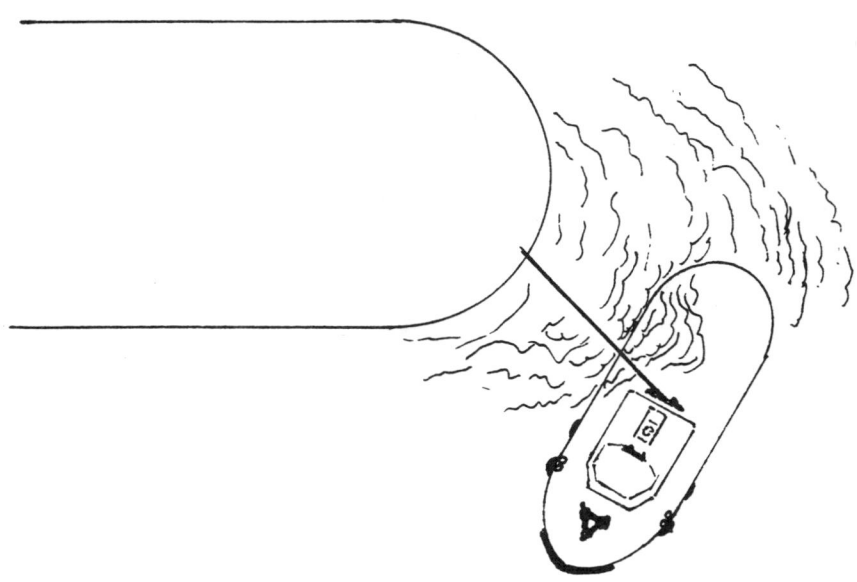

This is what can happen on a standard tug if she is caught in such a position close to the propeller wash of a ship. A tractor tug would have been able to keep herself at a better angle and still give the desired towing results. The tug shown is listing toward the ship and already has water on deck.

method to be used in the movement; the voice radio channel to be used; and any alternate sound or emergency signals to be used. Both parties also must be in agreement and understand each other.

The captain of the towing tug should make sure the ship's rudder is amidships and that all appears safe aboard the ship for this particular movement, including the hoisting of the proper day shapes or navigation lights. He should also know which side of the vessel will be moored at her new location and be aware of any and all possible maneuvering and/or problems to be overcome en route. These should be discussed with the assist tug captain.

If the port is strange to the tug master and a pilot is provided, the tug captain should go over all of the details of the movement with him. Complete understanding between tug captain and pilot should be established. The lead tug captain must realize that ultimately he will be held responsible for all that happens en route, particularly if the pilot rides on his tug. An open and immediate line of oral communication between

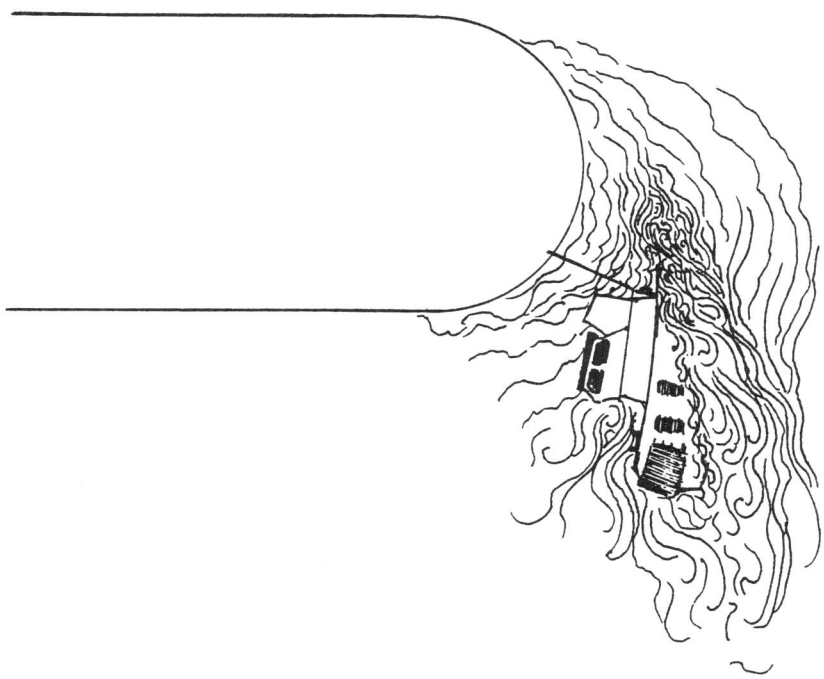

This is the horrible fate of a tug that is caught at a 90-degree angle and pulled over. It is called girding.

the tug captain and the pilot and between them and the assist tug and any other pilot on the flotilla must be maintained. All orders by voice radio must be acknowledged and carried out at once.

When ready to get the ship under way, if at anchor, both the lead or hawser tug and the assist tug should be alongside the ship. The assist tug would have made up on the ship's after quarter and the lead tug would be hanging onto the opposite side in a manner and position necessary to work with the other tug to keep the ship straight and to work the ship ahead to gain slack in the anchor chain as it is hove in.

In such a movement the ship would have only auxiliary power. Once the anchor is secured the headway of the ship should be nil. Then the lead tug should get into position to pass up her towline or hawser to men on the ship's forecastle head. Once this is secured the tow is ready to get under way. Of course, the ship's position would have been maintained during this connecting maneuver by the assist tug.

If the ship was at a dock, the assist tug would work head-on near amidships to hold the ship alongside while the mooring lines were let go. When only a single line remains fore and aft, the hawser tug should pass her line to the bow. When the ship's two remaining lines are taken in, the assist tug will either back the ship off the berth as the hawser tug starts to pull at an angle or she may put up a line to the ship's stern to back on and later steer as directed.

There are many combinations used in such maneuvering until clear of docks, locks, slips, and drawbridges. The main thing to remember is speed. Keep it as low as possible while still maintaining control of the ship. Remember to stop pulling, and have the assist tug backing, if there is any doubt of clearance ahead. This might occur due to other traffic, a drawbridge failure, or a lock tender's signal to hold off.

While most captains and personnel in charge of other facilities you pass through will respect your difficulties, there is no law that gives your tow a priority or right of way over other traffic unless a specific movement permit has been granted and made public by authorities of the port who are authorized to do so.

If the ship to be towed was in a laid-up fleet she would undoubtedly be broken out and handed over to the towing tugs while held in the stream until the hawser was connected and the assist tug was fast. The time of turnover should be acknowledged only when both of the towing tugs are fast and have taken charge with the tow commencing to move under control. If, for any reason, the ship is unwieldy, unmanageable, or unacceptable, the movement should be stopped at once and the tow refused. However, it should be held and returned to the tugs which broke it out of the fleet using all possible care. Then a decision on whether to proceed after further tug assistance arrives or to refuse the tow completely is in order. Each of these actions, like all movements, should be carefully entered in the tug's log (see the section on the tug log, this chapter).

Mooring at the tow's destination should be approached slowly, as with any large barge, using the current and wind as advantages if possible. Added tug assistance may be required. The added charge for assistance is always cheaper than the cost of damage claims should they be incurred. If the tow is to proceed to sea in charge of a larger tug, a safe area to transfer the tow must be chosen.

WHISTLE SIGNALS BETWEEN PILOT AND TUG

If for any reason the radio contact between tugs fails or if the pilot is riding on the ship and his radio fails, there are recognized signals used by most U.S. docking masters and tug captains. These are as follows:

When tug's engines are:	Whistle signal	Means:
stopped	one blast	slow ahead
stopped	two blasts	slow astern
moving	two short followed by two short	dead slow in the direction propeller is turning
moving	five or six short rapid blasts	full speed in that direction
at full speed ahead or astern	one blast	slow
at slow speed ahead or astern	one blast	stop

Usually these signals are blown on a police type mouth whistle. If possible, the pilot in blowing them usually looks toward and points at the tug he is signaling to. If the ship has power on her whistle it is often used and this signals action by the lead hawser tug only. The pilot's mouth whistle then is directed only to the assist tugs alongside or astern. These signals are also used in U.S. harbors in ship-assist work. The following also apply: one long blast and a slight pause followed by two shorter blasts, the whole repeated three times, signals "all right—go ahead—pull." If pulling a ship out of a slip or basin be sure to blow one long blast (known as a bend signal) in accordance with Rule 34 (e), International and Inland Rules. Above all else, when coming out, proceed at dead slow speed and if entering a river or tidal current, have sufficient hawser out so that your tug will not be run over or girded. The assist tug should be placed on the bow opposite the direction of the current in order to hold the ship up and aid in the turn before going back alongside to make up.

DEFENSIVE ACTION BY THE TUG CAPTAIN

In every maneuver the tug captain plans or undertakes, he should act defensively. He should be aware of any possibility that may endanger his crew, his tug, or the tow. During ship-assist work, he should if at all possible keep his tug in position for the next order the pilot may give. He should see that slack is taken out of his lines and advise the pilot of anything he sees that may endanger the ship, the tug, other vessels, or the shore structure. This can be done either on the voice radio or by several short rapid toots on the tug's whistle. If any contact is made or damage incurred, the tug captain should note this in his log and give particular attention to any warning he gave to the pilot or action that he took or could not take to prevent such damage. (For further discussion of a tug log see the tug log section of this chapter.)

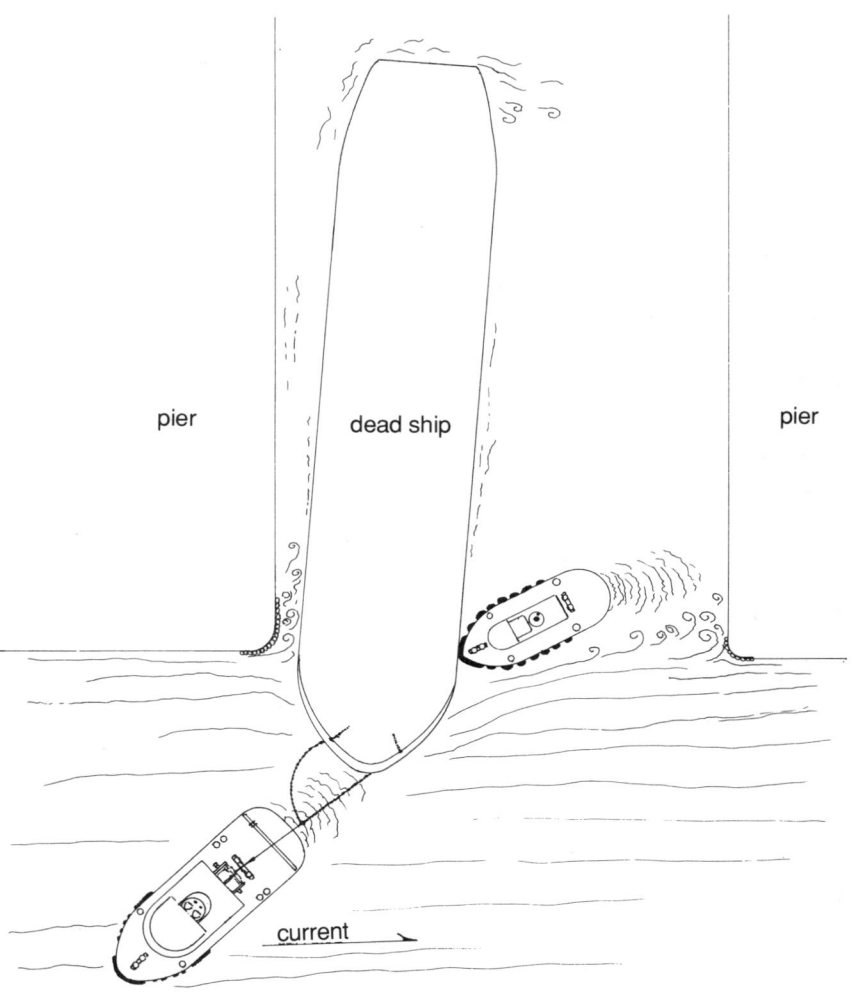

When pulling a ship or any large unit out of a slip, basin, or lock where there is a strong current across the entrance, an assist tug may be required to hold the unit from dropping against any abutments on the opposite side. The assist tug may have to slide toward the unit's stern as it enters the channel.

SHIP ASSISTING, GREAT LAKES PORTS

Tug assistance in U.S. Great Lakes ports is done in a unique manner. Most Great Lakes ore and bulk carriers and other ships trading in the Great Lakes give assisting tugs their line(s). The forward towline is a

Tug *Wilfred M. Cohen* towing AlgoCentral Marine's S.S. *AlgoLake* on the busy St. Clair River between Michigan and Ontario, Canada. The *Cohen*'s tow hawser is connected to the ship's towing pendant which leads from a pad eye located on her stem between her anchors. Such an arrangement is normal on all ships entering into the Great Lakes system requiring tug assistance. (Courtesy Rev. Raymond M. Donahue)

short wire rope pendant shackled into the ship's stem several feet above her loaded draft. The outboard end of this wire has a small thimble spliced into it. It is shackled into a similar thimble in a piece of soft synthetic line of suitable size. While the ship is under way, the end of this towline is kept secured on the ship's bow.*

A heaving line from tug to ship or vice versa allows the line to be passed to the tug. In some berths or waterways the ship may enter, such as the Niagara or Buffalo River, the Cuyahoga in Cleveland, and south Chicago Harbor, the tug will alternately hold this line on her stern bitt and shift to put it on her forward pawl post (bow cruciform bitt). If two tugs are required due to sharp bends, the one aft will hold a short line from the ship's stern chock.

Great care must be taken when pushing on or under the stern to avoid damage either to the ship's rudder or propeller or from them to the tug's superstructure. It was to avoid such damage and still be able to work closely around ships in restricted areas that the Great Lakes Towing Company, known as "the big G," designed their tugs with very low profiles.

A tug's pivot point when on the hawser is nearly amidships. It is when pushing on the bow of a light draft ship that a tug may become endangered as its movements are governed by the orders of the ship's captain. On occasion, a tug has swung a ship around smartly in a basin and the ship captain, fearing the ship was turning too fast and would go beyond his desired heading, has rung full ahead on the engine telegraph

*See Great Lakes Towing Company's towline arrangement in appendix VII.

Notice the graceful lines of these "Big G" tugs moored at the Detroit Marine Terminal. *Delaware* and *Indiana* retain the low lines they were given when stean powered in the 1920. *Wisconsin* has a raised pilothouse added. Bow fenders are small as the tugs work either from their pawl post or after bitts. They carry one spare ship line and possibly two mooring lines. All other lines are furnished by the unit in tow. (Courtesy Rev. Raymond M. Donahue)

(Chadburn) and dragged the tug, rolling her rail under. Such a situation usually results from a bit of carelessness on the part of both captains. The tug captain should know the waterway and the ship's normal heading for the exit well enough to realize how hard to push and when to ease off. The ship's captain, high above, could see what was about to happen and stop the tug in time or, if fighting a heavy wind, he should have ordered more than one tug.

TUGS AS ESCORTS

Oil spills are disasters, especially if the spill involves crude oil from a ruptured tanker. The names *Amoco Cadiz, Arrow,* and *Torrey Canyon* bring shudders to every maritime environmentalist. Tests have revealed that a loaded 190,000-deadweight-ton tanker proceeding at full speed will, if her engines are placed at full astern, require 1 mile of travel before coming to a complete stop. With this knowledge various groups in port areas have sought injunctions against the admission of supertankers. One such area is Puget Sound. No tanker of greater than

125,000 gross tons may enter. Any tankers of a little lesser size must have two tugs as escorts from Port Angeles pilot station to the terminals at Marsh Point, Cherry Point, and Ferndale, all in Washington.

Tugs assigned to escort are usually of 2,500 HP or greater. They run with the ship, one off each beam, at a distance of approximately 100 yards. They constantly monitor the ship's pilot on a specified VHF channel and are ready to take immediate action if required, for instance, by providing power on either bow to hold the ship in deep water until her officers have her under control. The tugs should be ready to pass a hawser or hold her until she has her problems solved and is ready to proceed again under her own power. These same tugs may, with or without added ones, dock the tanker on arrival at the terminal.

BIG SHIP DOCKING AT EXPOSED OIL TERMINALS

With the advent of very large and ultra-large crude carriers (VLCC, ULCC), terminals to load or discharge their products have had to be built in deep water so as to match their loaded draft of from 60 to 65 feet. In many places, such mooring piers are in exposed areas. Seas of 5 to 7 feet are common. Placing the ship alongside so gently that it would not crack an egg takes great professional judgment by the docking pilot and teamwork from the tugs.

A 200,000-ton ship, 1,050 feet in length, moving ever so slightly sideways against a wharf with fendered cluster pilings can do a great deal of structural damage. This was learned at Milford Haven, Wales, and other oil piers even though ships came alongside great rubber fenders with almost no visible movement. The ships must be stopped by the tugs when at least a foot away and then heaved in slowly by mooring lines.

To handle and stop these huge seagoing behemoths requires heavy tugs with great horsepower. Tractor tugs are fine. It is recommended that the ship lines from the tugs be of the braided type equal in strength to 10-inch nylon. There are various methods used in these dockings. Up to four 2,500-HP tugs are required.

Avoiding Damage When Docking ULCCs and VLCCs

On occasion, these huge tankers also receive damage on the docking. This cannot be blamed on either the tug master or the docking pilot alone. It is the result of a combination of things. Often a ship is delayed by weather causing reshuffling of an already tight berthing schedule. The terminal may have another vessel due immediately upon the discharge of the vessel now delayed.

The owner of the first and delayed carrier may plead for immediate docking even though the tide or current is not favorable and some of the

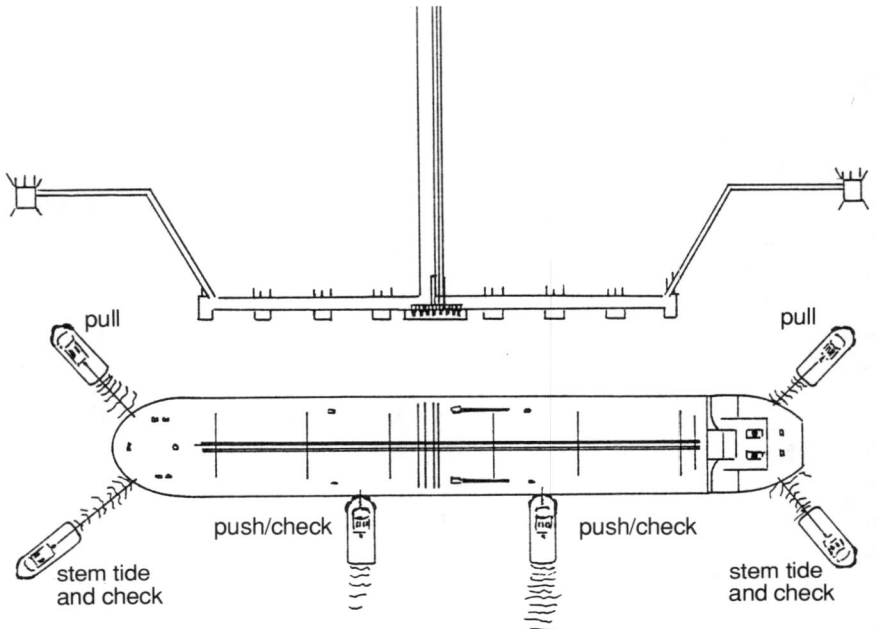

Placing a VLCC loaded or empty alongside a pier requires very gentle handling with almost a complete avoidance of pressure when it comes alongside.

tugs originally assigned to help this ship are busy elsewhere. Those substituted may have insufficient power and a difficult time with the particular vessel.

Tug owners should, therefore, be wary of falling into a trap by trying to accommodate customers who disregard nature and the facts of life. This was summarized at the Second International Tug Conference. Paper Number 2 read in part, "It has been stated that in docking of every large crude carrier of from 200,000 to 300,000 dwt, there results an average of over $15,000.00 of minor hull damage."

BARGE WORK AND THE TUG

Of course, there are all types of barges throughout the world. In the Americas the word "barge" covers a variety of floating units, some of which hold cargo on deck and are also referred to as deck scows. There are small barges with cargo holds such as LASHs and SeaBees. On inland rivers in the United States there are both open- and covered-hatch cargo barges of standardized size, which are pushed by huge towboats and made up into tiers by small fleeting tugs. The crew of the large linehaul tugs carry their own wires, steamboat ratchets, and other gear so

Seaspan tugs holding a huge OBO alongside as she tops off a cargo of coal at Roberts Bank, British Columbia. (Courtesy Seaspan)

as to make up the barges as one unit. Usually, the tow grows up to five tiers across and five in length with the towboat pushing at the rear of the center tier.

Making Up Alongside

Nearly every tug captain whose vessel is used in alongside towing has his own ideas as to the correct way to make up his tug to various units. It would be simplistic to offer any one method as the correct way. In general, most tugs will handle making up alongside to a barge better if the tug's stem is pointing toward the forward corner of the unit being towed. This is easy to accomplish after a towline has been put out by standing in the center of the pilothouse and looking over the stem. The forward corner of the object about to be towed should be in line of sight.

Certain principles will be found to apply when making up alongside. These will depend on the tug's deck fittings, capstan, tow winch, and hull shape. If the tug has a straight-sided hull amidships, she will probably not require as much care in selecting the point to tow from as the tug whose hull curves from stem to stern.

If possible, the towline should lead in such a way that, with the tug working dead slow ahead, the tug will be pulled toward the barge. The bowline normally should be tightened by working the tug ahead easily, with the rudder over toward the barge until the satisfactory angle mentioned above is reached. Then with the slack of the bowline (whether single part or double part) taken in and secured, the rudder can be altered away from the tow. This will cause the bowline to be tightened.

Thirty loaded barges downbound on the Mississippi. (Courtesy American Waterway Operators)

The stern of the tug will now come in. At this point, the stern line can be heaved in. Some tugs use the after capstan with the line first led under a cleat. Others use a pendant from the tow cable, which is heaved taut using the tow winch. The lead of the stern line should, if possible, be from an amidships chock, bitt, or cleat on the unit to be towed. This offers better leverage when the tug is turning the tow. Once the tow is made up and under way, deck personnel should check both the bow and stern lines to see if there is any slack to be taken up.

Tugs picking up small, light-draft units may be able to handle them with a towline or tow strap run from the tug's stem or bow plus a backing line and a stern line.

If the unit to be towed is low and loaded, it is good practice to double all lines, and possibly to put out three parts on the bow if this is the primary point of strain in backing. If the towed unit is high and light (empty), it may be necessary to run towlines from the tug's bow bitts in order to get a lead that will hold the tug in toward the unit as well as to clear the tug's deckhouse.

Tractor tugs, because of their configuration, make up slightly differently. As they have power windlasses, they frequently only require a

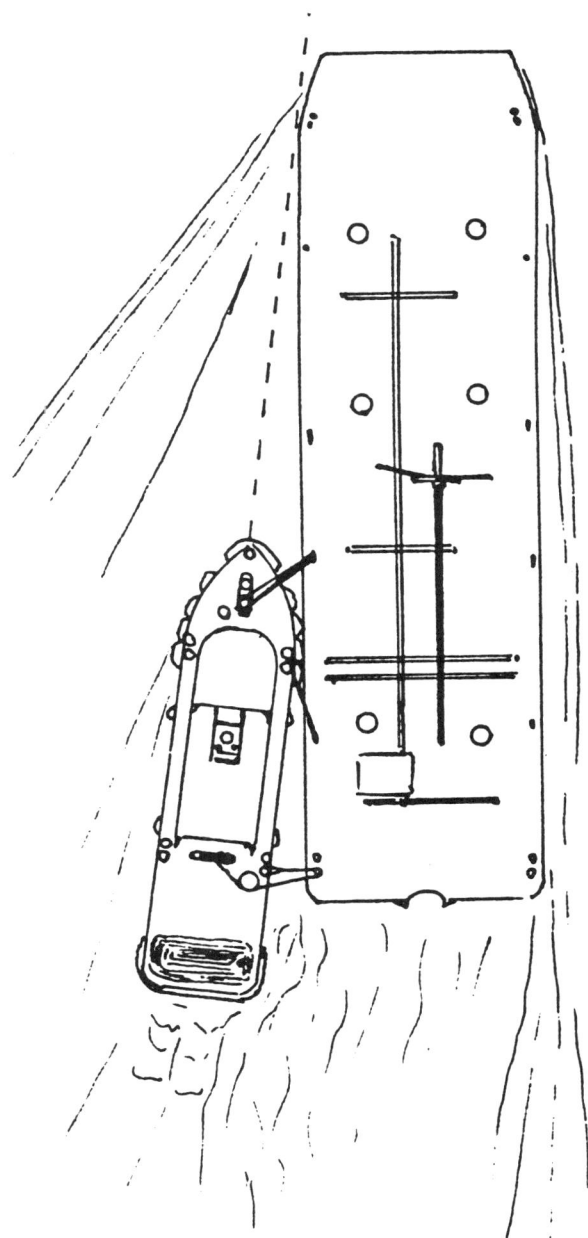

A tug towing on the hip with her centerline pointing toward the bow of the oil barge. This has proved to be the best way to make up whether towing the barge bow-first or skeg-end-first.

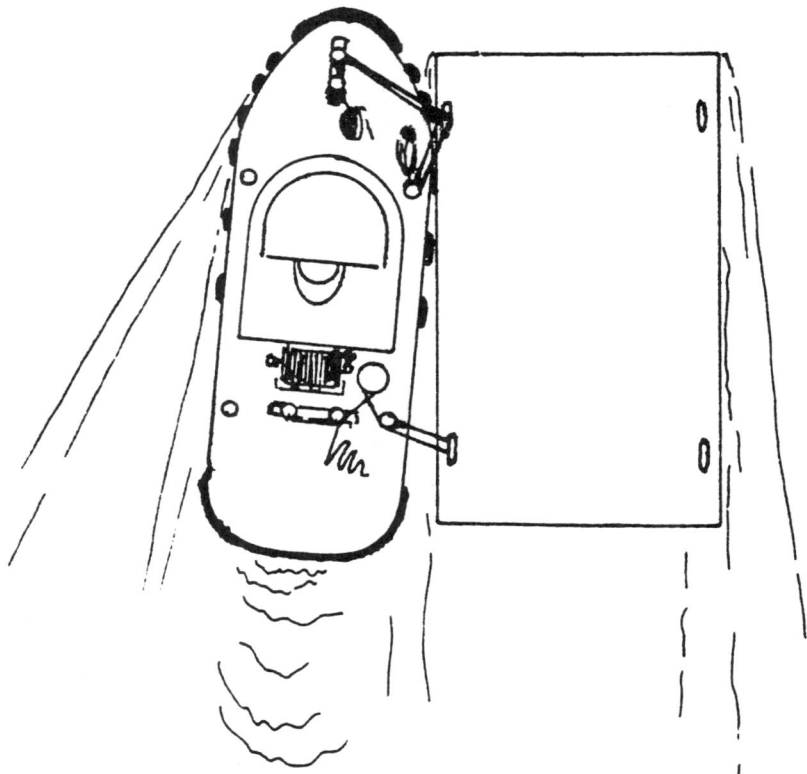

Tugs with power can make up and handle small units in this manner.

bowline leading forward and a stern line in the opposite direction. These will hold the tractor tug in a position from which she will be able to apply her power and steering in any direction.

GETTING A BARGE TOW UNDER WAY

For the tug captain about to tow any barge, there will be several important facts he must consider. First, is the barge loaded or light? His tow orders and the barge's location should indicate this to him. He must decide and so inform his deck personnel as to how he plans to pick up and tow the barge. Will it be on the hawser; on short stern lines (gate lines); alongside (on the hip); or pushing in the notch if the barge has one? This decision will be based on where the barge is lying, how easy or difficult it will be to get it away from the dock, and how the wind and current may help or force a definite manner of departure. Should a request for an assisting tug be made?

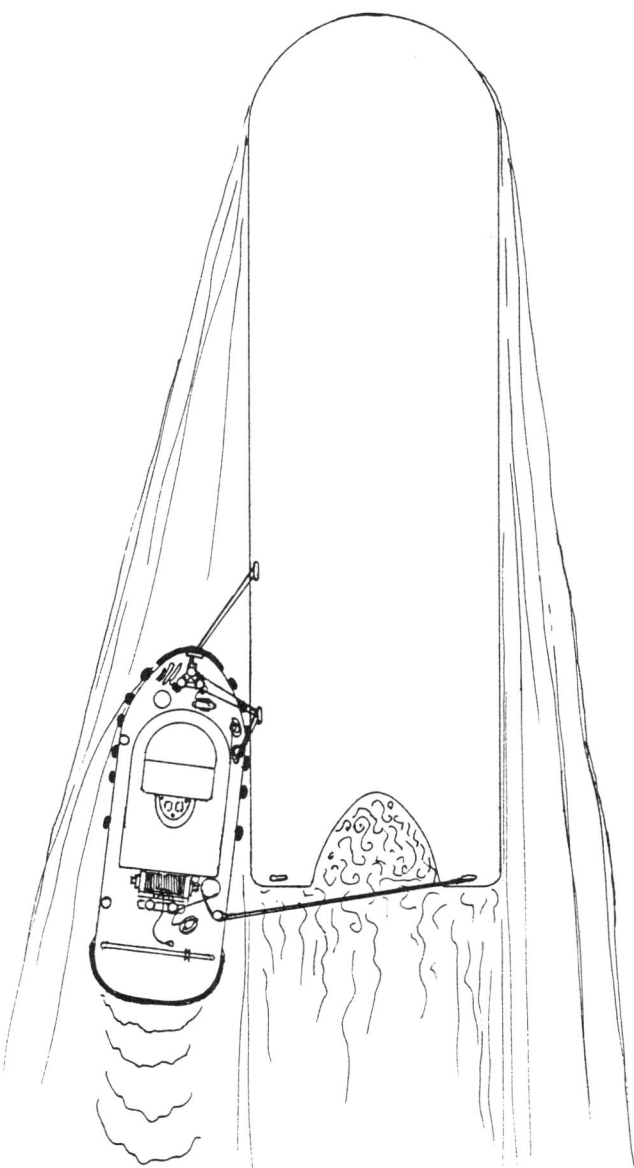

When towing a heavily loaded barge, all lines should be doubled up. Placing the stern line on the unit's outboard corner gives the tug a greater pivot point in turning.

Answers to these questions should point to the proper approach to the destination and the effect wind and current will have at the expected time of arrival. It is always prudent to consider and use nature's forces—the tide, the current, and the wind—either in getting away from the berth or landing.

After the choice of how the tug will pick up the barge and make up to it, the method of nicely clearing the wharf or pier is next. This is contingent upon the barge's size, whether it is loaded or light, and if it is manned with its own crew. If the barge is manned and empty and the tug will take it away on the hawser, the tug captain must indicate any particular way he wants the barge lines to be let go. Good bargemen usually know the best order of letting go their lines and will signal the tug visually and vocally when all are gone.

In such a procedure, the tug will have backed in under the barge's towing end, normally the bow, and the tag line on bridle wire pendants will be thrown to the bargeman. When he has pulled them up and dropped the eye over the bitt or cleat, the tug should ease ahead, slacking her cable just enough so as to be able to lie with her head about 45 degrees off the line of the wharf. This should allow the connecting point of the bridles to clear the stern. The tug should lie still until all the lines on the barge are in and the signal to proceed is given. Leaving the dock with a barge on the hawser, particularly an empty one, and awaiting the release of the barge lines, it should be realized that even the slightest movement of the tug will cause the barge to start following. If the water is deep and a few fathoms of wire have been slacked, its weight alone will cause an empty barge to creep toward the tug. This should be avoided if possible because sometimes it is necessary for bargemen to get on the wharf to free a line which may have been jammed. To return aboard they may have to climb a ladder and any movement would endanger them.

If the barge is 400 feet or more long and has its own bridles of wire or chain, it may be prudent and necessary to have an assist tug hold it to the dock, especially if the barge is unmanned. The towing tug's crew may have to let the barge lines go. When all the lines are in and the crew back on the hawser tug, the assist tug can back the barge off as the lead tug works slowly ahead.

With smaller barges, a slight touch ahead and then a stop with the tug's engines will often be just enough to swing the barge bow off the dock. As the outboard bridle starts to come taut, a second touch ahead should check this swing and, with the barge fairly astern, slow ahead with the tug's engines should take the barge cleanly away from the wharf. Whoever is handling the tug should be at the after controls during such a movement. Like all other maneuvers, it takes practice, timing, and teamwork on the part of deck and barge personnel. They should have con-

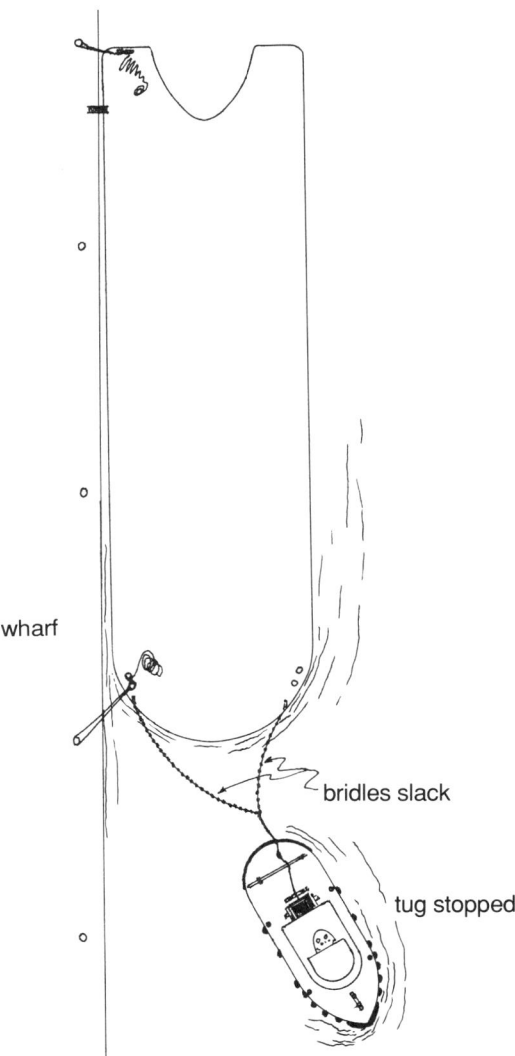

On leaving a wharf with an empty barge on the hawser, the tug's bridles and tow cable should be left as slack as possible until all lines are in. The slightest strain will cause the barge to move, particularly if the bridles are of chain.

The tug captain should work from the after controls when the hawser is being connected and streamed unless perfect visibility and communication are available from the pilothouse.

fidence that the man handling the tug is watching out for them and should also be aware and familiar with what he is trying to do.

If the barge is to be towed away with the tug made up alongside, it may be necessary to hold a line and back the forward end off the dock. Oil barges when empty and towed alongside will handle better if the skeg end is ahead. This actually means that the barge is being towed backward or stern first. The reason is that this puts the barge pivot point nearer the bow. By being towed stern first it does not tend to slide sideways away from the tug as much as it would if made up with the bow ahead.

Another method of getting an empty barge away from the dock is to place the tug head-on near the barge's middle. Place one line, preferably the towline or tow strap, from the tug's stem to a barge cleat. It having been decided on which side of the tug the barge should be towed when made up, this towline or strap should run to a barge cleat on that side.

There must be sufficient room to maneuver off the dock. With the tug working ahead, the barge lines can be let go while the barge is held against the wharf. Then, by backing the tug slowly, the barge will come off with the end that the tug's towline was directed to, swinging slowly toward the tug.

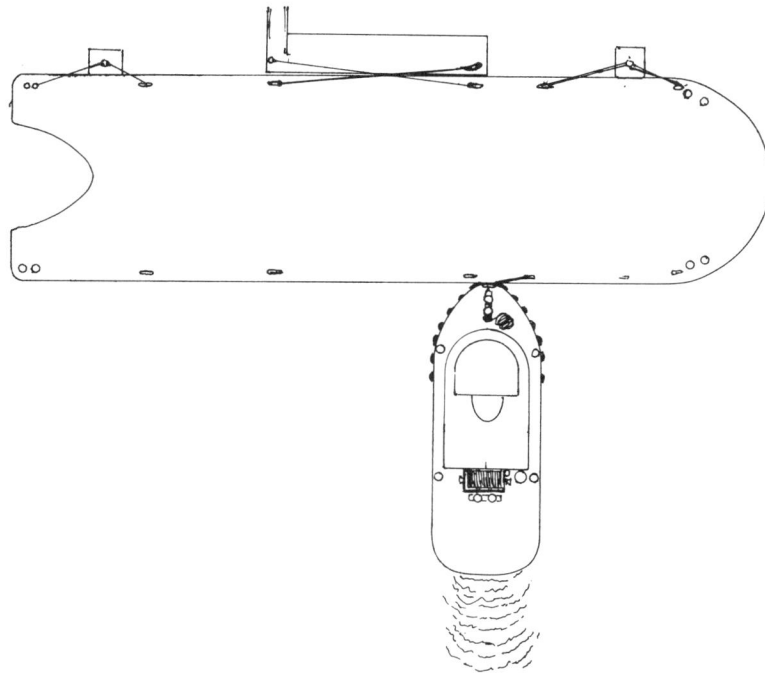

Position 1: Head-on, holding barge against pier while lines are being let go. Note lead of tug's bowline.

With the barge clear of the wharf and some way on, a quick application of power astern should force it to swing rapidly toward the tug on the side for making up. If the tug is a single-screw with a flanking rudder, she can help this by backing her stern toward the barge. If twin-screw, power applied in opposite directions will give the same results. As the barge approaches parallel to the tug, the tug comes ahead with the rudder hard over to swing her own stern in. A backing line and stern line are now quickly made fast and the tow is under way. This type of maneuver requires coordination and speed by deck and barge personnel. It should not be attempted in traffic lanes or in small waterways.

Harbor Oil Barge Towing

The most frequently moved single barge unit in the United States is the oil barge. Oil barges run in size from 15,000 to over 250,000 barrels

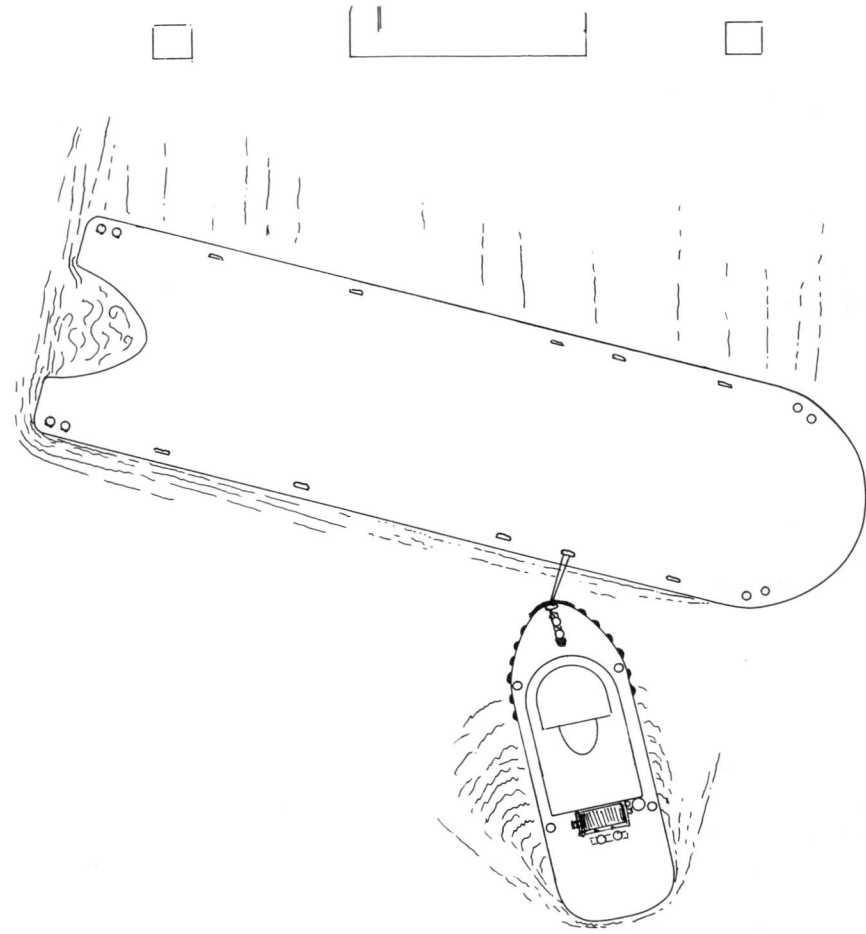

Position 2: Tug backing slowly using rudders or propellers to maintain position as she draws the barge away from the pier.

Opposite, above, Position 3: The tug has now swung alongside and the other end of the barge is swinging toward it which will allow the tug to come ahead and make up. *Below,* Position 4: Barge is alongside; tug made fast and under way.

UNDER WAY—TOWING ON INLAND WATERS 223

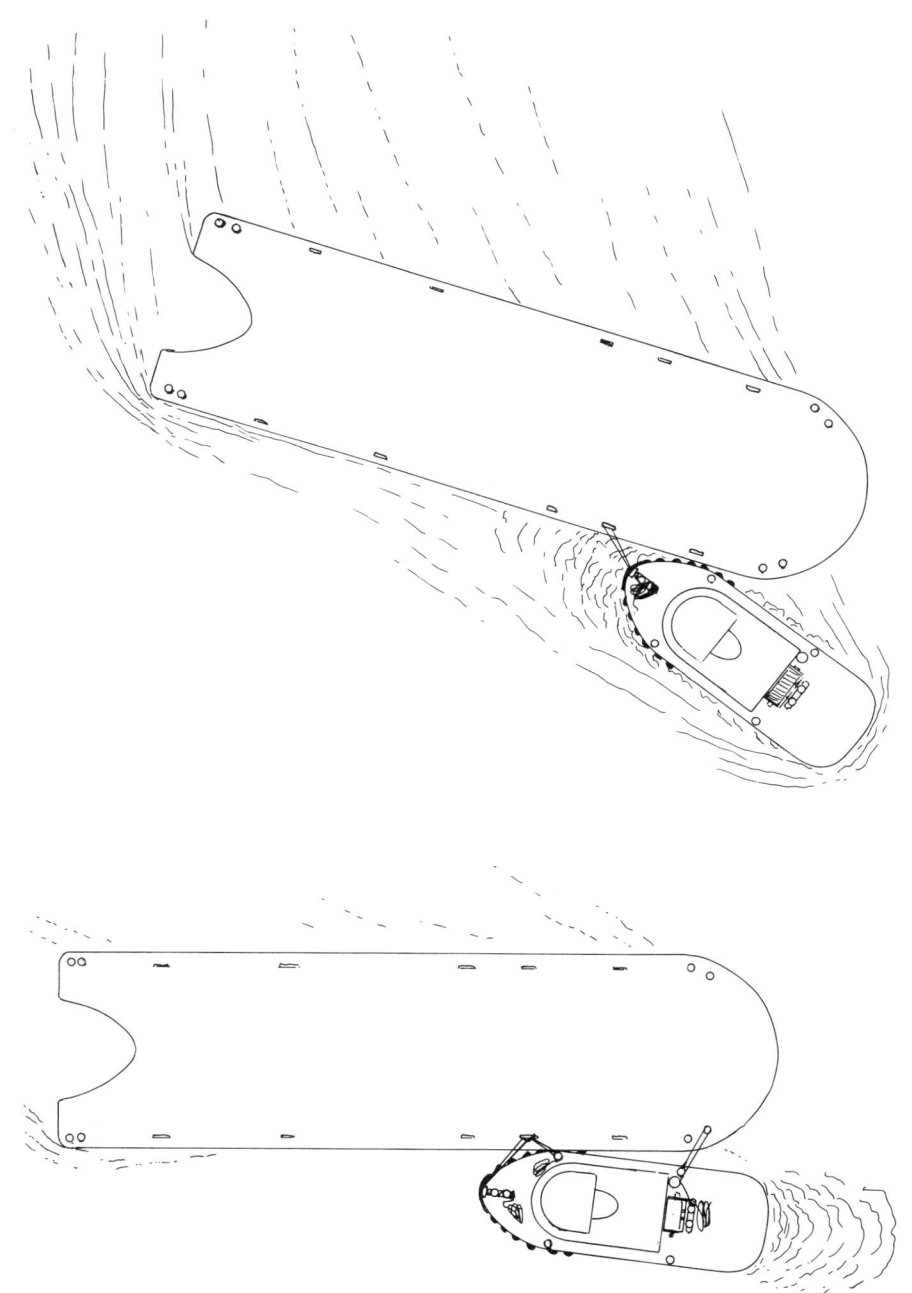

capacity. On the East Coast, they are from 150 to 550 feet long, with beams of from 35 to 80 feet and equivalent drafts of 35 feet. On the West Coast many are 400 by 100 by 25.

The larger barges are mainly used in lightering products from ships to shore storage and refinery facilities. Smaller oil barges are also used for this purpose but more often to carry loads between two or more shore facilities or to powerhouses and generating stations or distributing terminals.

The tug captain about to tow an oil barge must go over the same important considerations as mentioned earlier in towing other barges. On landing or leaving the dock with large petroleum barges with either the tug alongside or in the notch, it is common practice for an experienced tugman or mate to direct the tug's pilot via hand-held radio regarding distances off, mooring lines out, or let go and taken in, and clearances. If an assisting tug is used, the person acting as a docking pilot may give orders to both tugs or the tug master may wish to give orders to the assist tug from his own vessel.

ADVANTAGES OF PUSHING

The advantages of pushing are speed and maneuverability. The evolution of a tug or towboat getting behind a barge to push it may have started on a river somewhere. This method was evident early on in steamboating on the Mississippi-Ohio river system in the United States.

It was found that by placing a single barge or scow of about the same size, beam, and draft as the steamboat, there was very little change in overall speed against the current. Soon, a second and then a third barge was added. Today, they are tiered out to a width of five and occasionally ten in length making fifty units into one.

Integrated barge tows of petroleum, liquid sulfur, and other cargoes move daily up the Mississippi-Ohio river system. The units are held together as one by hard rigging of wire, turnbuckles, and ratchets. The pusher towboat with triple propellers, rudders, and flanking rudders can maneuver such a tow around tight bends and into locks, or hold it either in midstream or "to the bank."

None of this could be accomplished with either the same speed or control if the tow was on the hawser or held alongside.

Control over these huge tows is almost unbelievable. Towboat pilots can take them up- or downriver.

Push towing is not relegated just to the Mississippi River system. On the Gulf Intracoastal Waterway, there are hundreds of small push boat tows. On the Columbia River between the states of Washington and

UNDER WAY—TOWING ON INLAND WATERS 225

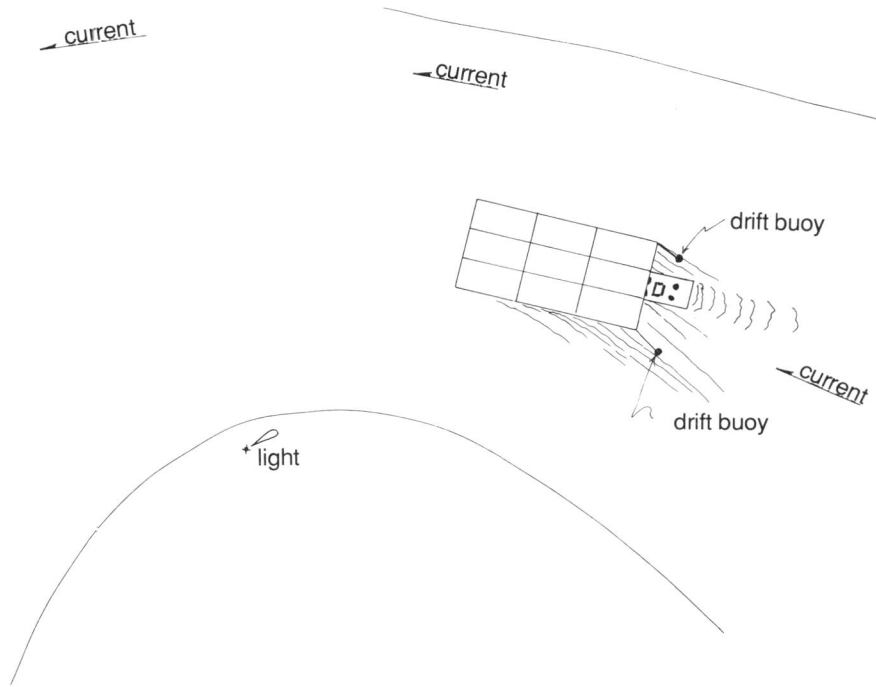

Downbound river tow using drift buoys. These are particularly helpful to the pilot when rounding a bend in low visibility. The angle at which they trail indicates how much the tow may be sliding sideways.

Oregon, towboats may push three or four high-sided barges ahead, or a five-barge flotilla with two barges ahead and three on the side.

These advantages of speed and maneuverability in pushing when handling a single loaded liquid cargo barge led to the development first of a pad in the middle of the barge's stern to hold the tug's nose. Then, barges were built with a V-shape bracket attachment, and finally a recessed notch to receive a portion of the tug's bow.

On the New York State barge canal system it was found that by increasing the depth of the notch through allowing an increase in the barge's length, more cargo could be carried with about one-third of the tug now buried in the barge. Economically, this was a great feature, as the tug and barge completely filled the lock, leaving no open water areas except close to the tug's stern. Also, the tug could enter and transit locks quickly without letting go of the tow.

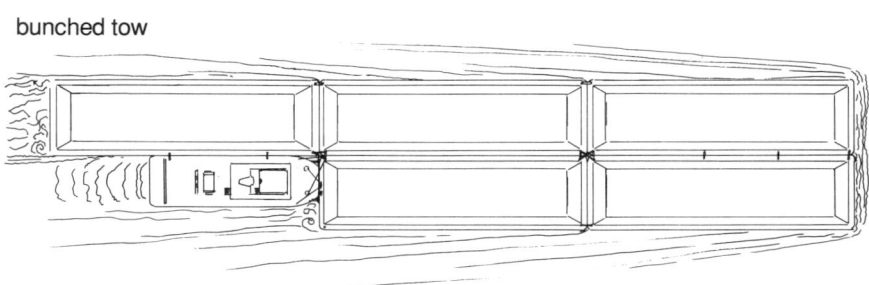

Above, Columbia River tows are frequently made up in a single tier, as opposed to the bunched tow *below.*

Tidewater's *Captain Bob* pushes her tow into Ice Harbor Lock on the Snake River in eastern Washington. (Courtesy Harbor Images, Matt Lyon)

Approaching the Barge Notch

Before getting a barge under way with the tug in the notch the tug must be positioned. Like a ferryboat entering a ferry slip or rack, some careful and fully controlled maneuvering is required. Almost at once, the following problems must be solved mentally by the tug captain. Are we entering with a fair current, no current, or a crosswind? Are there any mooring lines in the way? Is the notch free? Will I need someone to take the tug's lines or can I hold the tug in the notch? Will holding the tug in the notch by working slow ahead disturb the barge's mooring lines, dock bollards, and particularly any loading apparatus or hoses through which product may be flowing? All of these questions and their answers will normally flash through the mind of a good tug captain. He will *never* attempt to get his tug into the notch of a petroleum or chemical barge when hoses are out to a dock or ship, unless requested to do so by the personnel attending them on the barge.

Getting into the Notch

Assuming that the barge is ready and it is now safe to enter the notch, positioning for this maneuver requires as much care as that of a pool player lining up his cue and sight on the next shot. Entrance into the notch should be slow, straight, and smooth—no bouncing off one side or the other. Using good judgment as to what the wind, current, and tug's response will be if all is correct will give the desired result.

If the barge is moored at a dock or alongside a ship, going into the notch is usually quite easy, particularly if the barge is loaded. If the barge is empty and the rake is out of water, great care is required so as not to get the tug's gunwale or bulwarks under the rake. It may be necessary to hold the tug's nose against the inner center of the notch by working dead slow ahead until lines are secured or push cables are taut.

Making Up in the Notch

There are many ways to secure a tug in a barge notch. Each company (and often individual captains) has its own tried-and-true method. The size of the tug and the barge and the depth of the notch are the criteria around which barge connections for pushing have been worked out. The size of main push cables and/or any connecting pendants from the barge should be equal in diameter and strength to that of the tug's main tow cable.

Here are a few well-known methods of making up a tug in the notch:

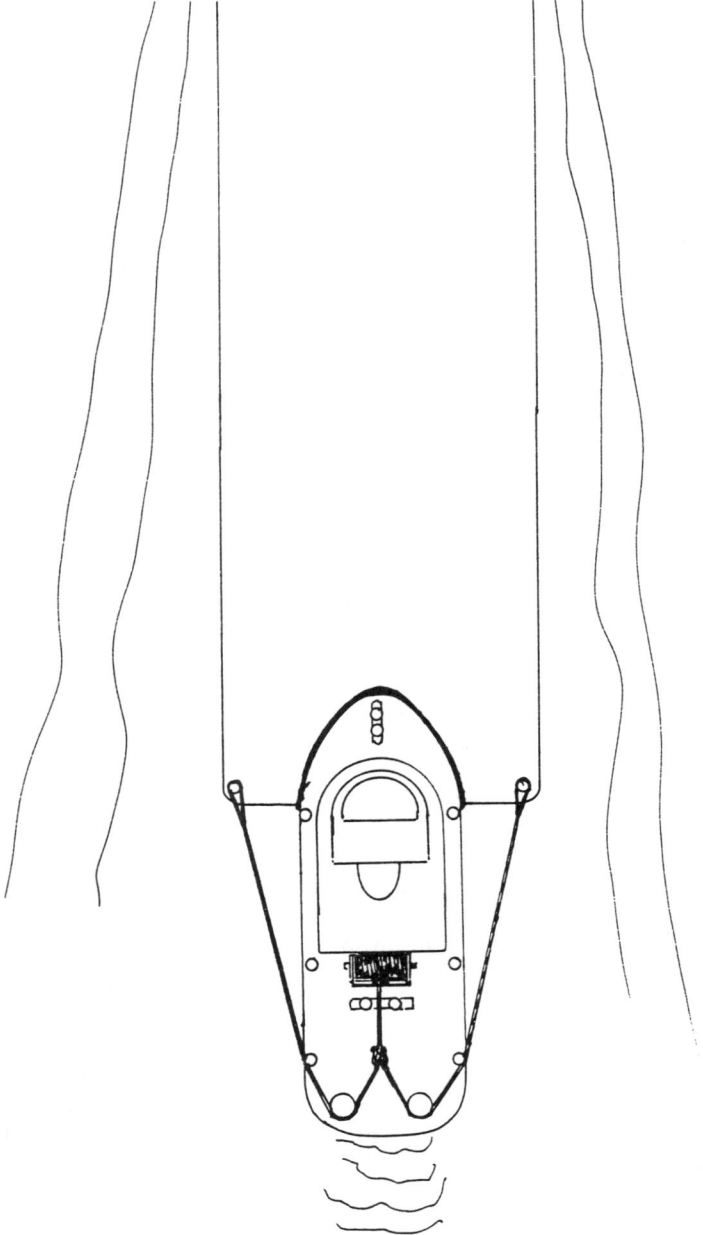

Pushing in a tight notch using single cables, which remain on the tug when disconnected

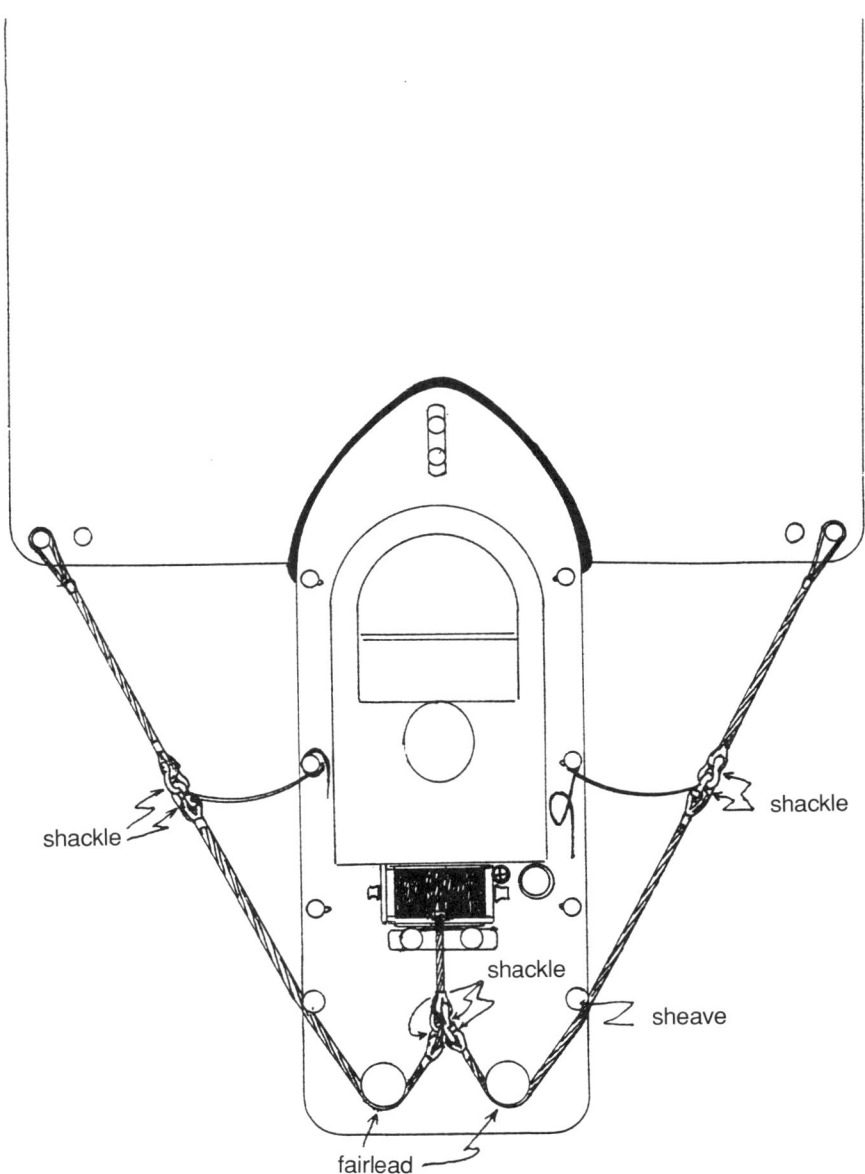

In this rig, the shorter cables remain on the barge. It is a rather awkward way to connect as the cables from the tug must be hauled aboard and released from the hawser. This is a popular method used by several towing companies in the Gulf.

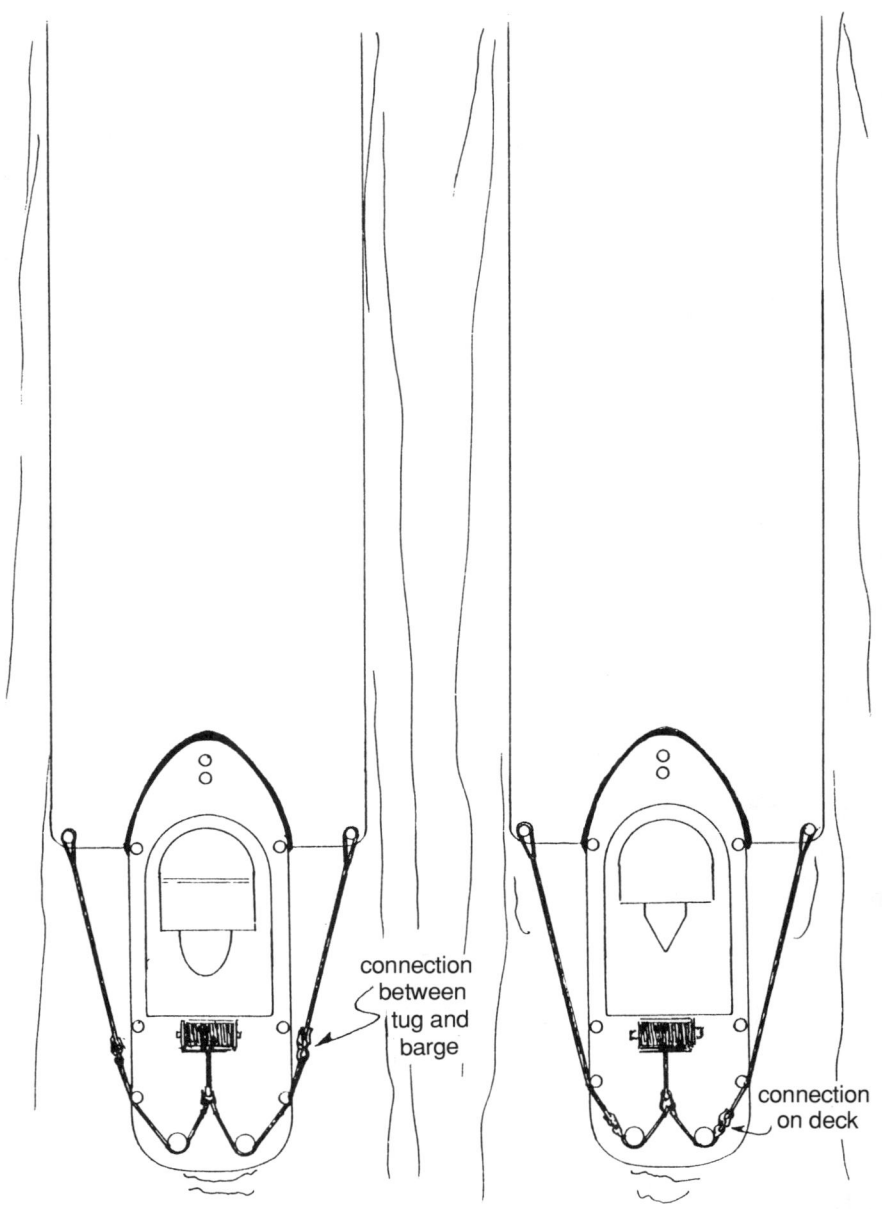

Left, in another method where the push cables are kept on the barge, they are connected to each of the tugs at the bulwark rail by slacking the main tow cable. *Right,* these push cables are heaved aboard from the barge and the connection is made on deck.

The 3,500-HP *Sarah Hays* is in a shallow notch. Careful examination will reveal that her main push cables run through roller chocks on the barge to bitts. In addition, there are double winch wires operated by hand from either side of the barge's stern. In such a makeup, the tug cannot endure much rough weather without parting some of these cables and must do most of her outside towing on the hawser. (Courtesy Watkins)

The Big Notched Barge

These notched barges are usually built and rigged so that one or more of the company's tugs with about the same characteristics may serve them. Some are built in conjunction with a tug to which they are constantly assigned. Their main push cables may be fastened on the barge either on a deck pad eye or on the hull near the barge's loaded draft. On some large barges of from 400 to 500 feet in length, each main push cable is shackled into a pad eye on the after edge of the skeg.

Because of the great difference in draft, which would be about 3 to 4 feet when the barge is empty and 35 feet when loaded, the lead of the main push cable at a median angle offers the most satisfactory results.

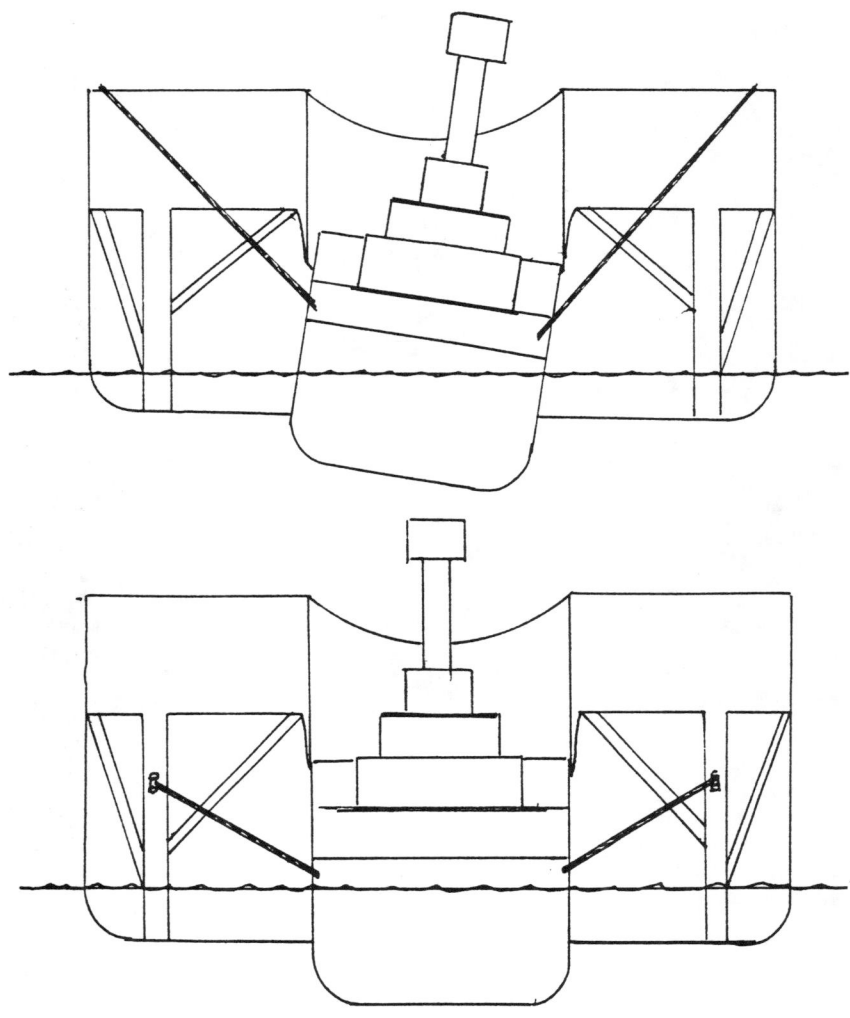

Above, when push cables run from a tug's stern to the deck of an empty barge, the tug will list if rudder beyond 15 degrees is suddenly applied. *Below,* barge push cables leading from skegs give better control.

It has been found that a tug of from 3,500 to 9,000 HP when in the notch of an empty barge (that is, if the push cables lead up to a deck 25 feet or more above the water) will tend to lie over dangerously if much rudder is applied.

It is most important, then, to arrange to have the main push cables from the tug's stern to the barge lead as nearly horizontal as possible when the barge is empty. When the barge is loaded, the lead is usually only a few degrees—10 to 20 percent from horizontal, depending on the size and freeboard of the barge.

Some very deep-notched barges have been built that admit up to two-thirds of the tug in a snug fit. Arrangements to hold the tug in a deep notch vary from using a heavy set of lines from the tug's bow bitts to the barge to hydraulic pins coming out of the sides of the bow and fitting into a vertical spot in the barge. With such deep notches, tugs are normally rigged with a vertical rubber-faced pad extending as high as the boat deck. These are located slightly aft of amidships. They would match a similar set on the barge.

Safety Lines in the Notch

Although the tug may be held tightly, there must be some sort of heavy cables to restrain the tug from backing out of the notch if she were forced suddenly to go full astern. These are referred to as safety lines. Their use should always be considered.

Rapid Connecting and Disconnecting at the Notch

As most changes of the tug from being on the hawser to getting into the notch are accomplished near harbor, bay, and river entrances, they must be accomplished quickly with the barge barely drifting so that the tug can maintain some control of her tow while getting the main push cables out and tightened. Therefore, a careful study must be made of the simplest and most direct, yet foolproof, method of hauling push cables aboard or passing them to the barge.

The first lines out from the tug should be from the bow through the stemhead chock followed by a safety line on each side. With these, the tug should be able to maintain control, work ahead dead slow, and steer as necessary while the balance of the connections are made.

Big tugs with big barges require much more space and time to shift to and from the notch than small harbor tugs with medium-sized barges. Small tugs without a tow winch usually get a bowline first. This is doubled off the stem. In the New York area a wire pendant is passed to the barge's outside corner bitts. Each pendant has a large eye splice on one end and a good-sized thimble in the other. Into one of these thimbles a good deck line is run with its eye on the tug's aft quarter bitt. The choice of the quarter bitt to use for this line depends on the side on which the tug's afterdeck capstan is located. This fixed line should go on the opposite side. The tug's stern is then swung slightly in that direction as a deck person picks up slack of the soft line and runs it around the bitt. As

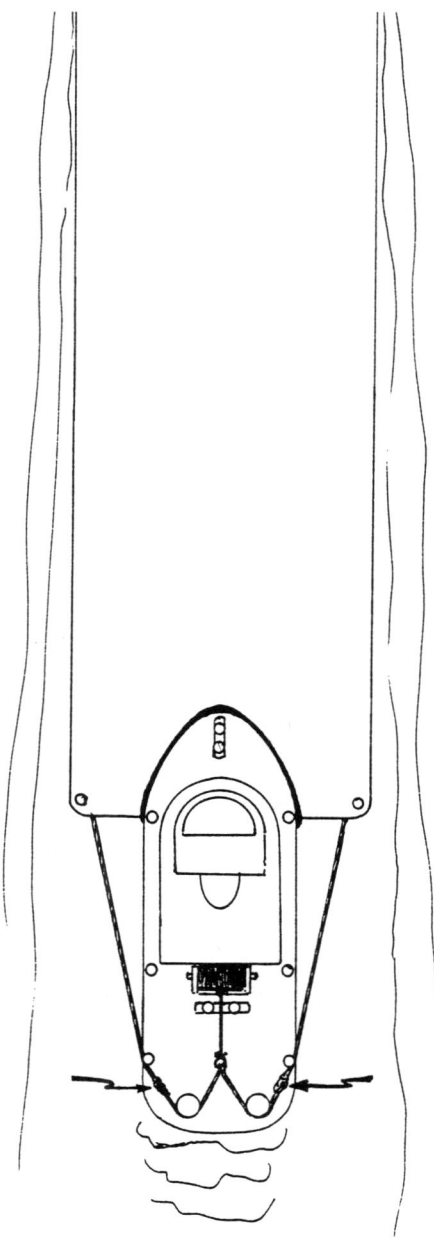

Push cables from barge skegs are connected on the tug stern.

Bulk Marine Corporation's *Victory* illustrates deep-notch towing in open waters without cables. (Courtesy Bulkfleet)

the rudder is placed in the opposite direction, the other pendant, rigged in the same manner and presumably already passed to the barge, will have the eye of its deck line also on the quarter bitt on its respective side. The other end will be run through from the thimble and under the quarter bitt horn to the capstan. With four or five turns on the capstan drum, it will be heaved in and the line on the other side slacked a little if necessary until the tug is absolutely straight behind the barge. The line on the capstan remains there with the bitter end secured back on the tug's main tow bitt.

SHORT HAWSER TOWING—DEEP-DRAFT UNITS

Deeply loaded ships and barges towed on a short hawser may, under some conditions, sheer greatly. To correct this, one or more assist tugs should be added and the tow's speed adjusted to minimize its wandering, possibly through adding more scope of cable if circumstances will permit.

TOWING FROM A HOOK

Very few tugs in the United States are fitted with a tow hook. In other areas, tugs frequently put the eye of another vessel's line in their tow hook when handling them in sheltered waters. It is not uncommon in Ireland and some other countries for the tug to pass a combination fiber and wire docking line from her tow hook to a ship. As recently as 1985, Red Funnel tugs in Southampton used this method.

Due to their size and the reduction in tug manning scales, those fitted with tow hooks must have a quick and easy method of heaving the ship's line to their hook. More important is the ability to release any line from the tow hook immediately. The most recent development is the pneumatic release. This is an adaptation of the hand-pulled lanyard method of tripping the hook.

When towing a vessel whose height above the water would cause a towline from the hook to lead upward at an angle, care should be taken so that, when the tug turns, the tow hook does not pass beyond the end stops of its radius. To avoid this, sufficient line should be paid out, particularly if the tow will come under a very heavy strain, or if the towed unit sheers radically.

DREDGE TOWING

There are a few dipper and clamshell dredges still in operation. A dredge, including those with suction cutter heads, is usually a simple unit to tow. A powerful tug can normally tow one alongside, with a dump scow on the tug's opposite side, if necessary. Dredges follow well when towed on the hawser. If they have spuds set on the bottom, they are easy to pick up or pass stern lines to. As they arrive at a new landing or position, they will assist by dropping their spuds. Care must be taken to avoid damaging these spuds by not having too much way on the tow when approaching the new location. Dropping a dredge's spud when it is moving over 3 knots is very dangerous to the spud and its spud well.

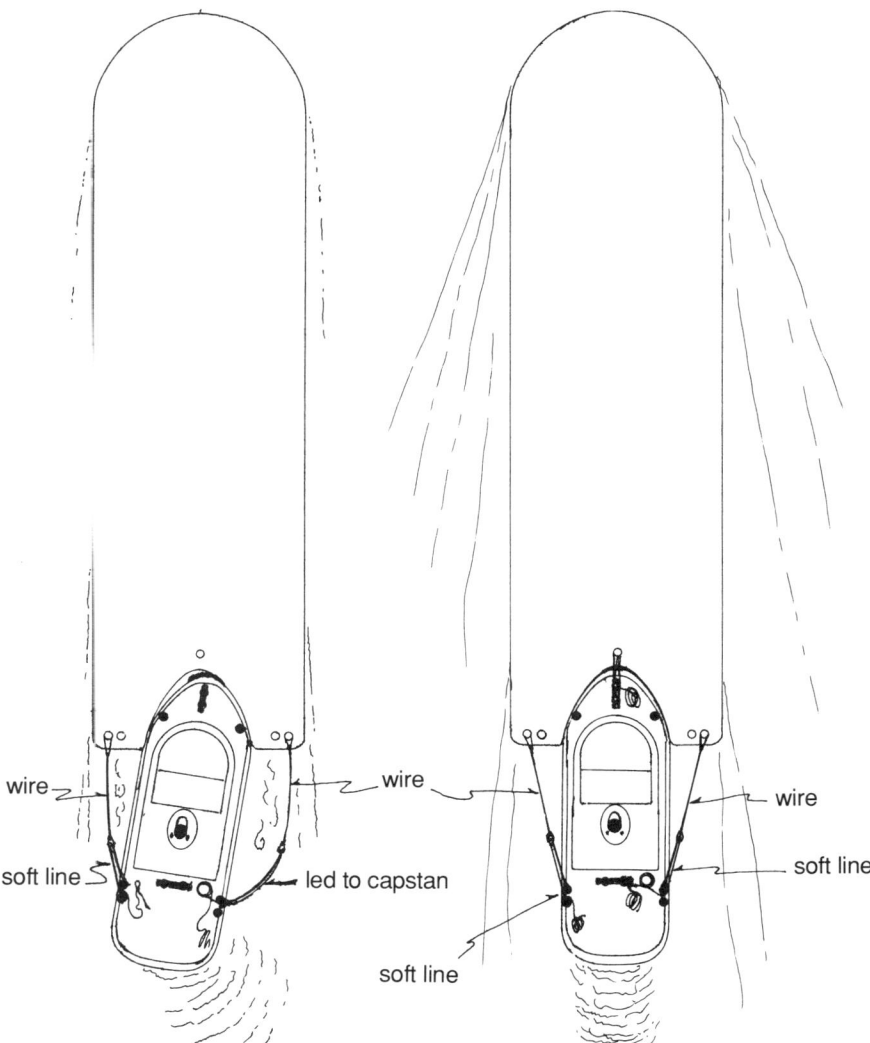

Left, harbor tugs without a tow winch use a combination of wire and line. By applying right rudder as shown, all slack is taken up on the port side. The starboard wire and line are led to the tug's capstan. *Right,* when the harbor tug has slacked the port line, the starboard line is heaved in on the capstan until the tug is square in the notch. Both of these soft lines from the wires are doubled. With the tug in the notch, bow safety lines are usually put out.

When handling dump scows, dredging tugs usually have a heavy tow strap measured specifically to fit from the tug's forward alongside tow point (bow side bitt or stem), to a cleat on the dump scow. Bow- and stern lines are run out to the dump scow as they would be to any normal alongside tow.

If the dump scow's load is released by hydraulically opened or chain unwinding pockets, it may be necessary to slack the tug's bow- and stern lines. The scow in dumping will assume different positions as each pocket is dumped. To avoid getting the tug splattered with mud, it is best to have the tug positioned on the windward side. The tow should be completely stopped when dumping, with the tug made up alongside. Soft watery mud, gravel, or rock does not tend to fly up. Heavy clay or material that will stick in large chunks to the sides of pockets frequently goes with a huge splash, causing a spout of muddy water to fly high enough to land on the barge and on the tug's deckhouse.

Many contractors have tugs with towing knees that push their dump scows. In pushing, all of this worry is unnecessary. Slack the facing wires while dumping and tighten again when the scow is empty.

It is usual practice when picking up a loaded dump scow from a dredge and returning it to the dredge to have the side on which the pocket-releasing shaft, rams, etc., are located away from the dredge.

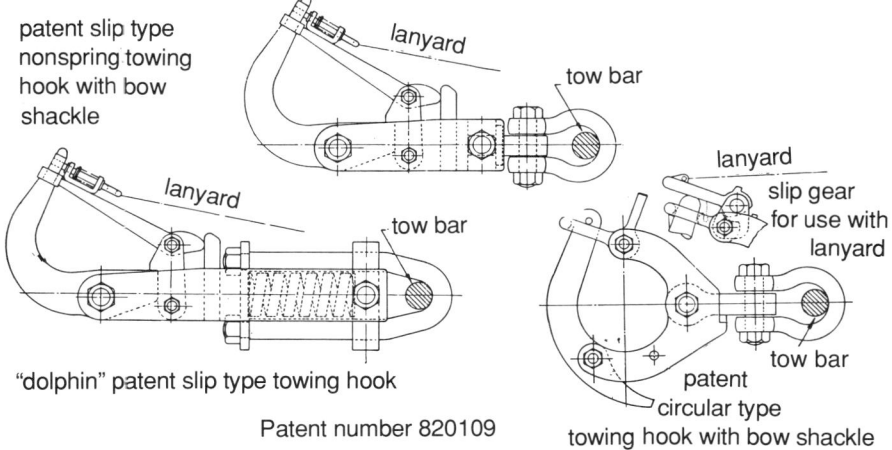

The tow bar to which each of these tow hooks is fitted has a heavy stop at each end of its semicircular track. (Courtesy Loveridge Ltd.)

Opposite, above, Clontarf Monarch of Dublin shows how her shiphandling line is held in the hook. Notice how guards have been placed over hatch covers and deck winch. *Below, Coleraine* has tow arches to keep her towline from catching obstructions on deck.

This is so that the dredge's bucket does not pass over the dump scow's shaft during loading.

Towing dump scows and other disposal barges to offshore sites is covered in chapter 9.

RAILROAD CAR FLOAT TOWING

This is also known as car-barge towing and refers to those long, sometimes narrow, shallow-draft floats upon which are two to four sets of parallel railroad tracks. They are used to shift freight cars from waterside railroad trackage either to a waterfront terminal or its own internal railway that is inaccessible by land linkage. This form of towage was common in most U.S. East Coast ports from Boston to Norfolk. It has nearly disappeared, but some form of it remains on runs between Puget Sound ports and to British Columbia and Alaska.

Handling these older floats in harbors is not difficult. Most are designed with spoon-shape ends. The one upon which the cars are admitted is the bridge end. The end with the track stops or bumpers is the stern. In making up alongside these particular car floats, the tug is usually placed so that her stern cleat or after bitt is even with the first set of cleats forward of the barge's stern.

In handling these long, narrow, harbor type car floats, it should be remembered that, loaded or light, they hold their headway for some time. They are designed for speed and ease of handling. They do not tow well on the hawser. Those that are hawser towed between Cape Charles and Little Creek, Virginia, are rigged with rudders and steered from a house that straddles the tracks at their middle.

In Puget Sound, the rail connection between Seattle and Port Townsend is a tandem tow of two car-barges. An intermediate hawser is strung between these car floats from a winch on the stern of the head barge. This is necessary as the sound can become too choppy to allow the barges to be made up close together.

In Port Townsend Harbor, the hawser tug alone rounds up, lets go, shortens up the tow, and puts the floats one on each side before shoving them into the float bridge slip.

As the depth of water in Puget Sound and the various straits of British Columbia is so great, Seaspan tugs tow single Canadian car-barges on a long scope of hawser from Vancouver and the Fraser River to Seattle. On arrival in Elliott Bay, Seattle, they pick up and dock the float.

Other car floats are handled in Seattle by Crowley-Red Stack tugs. These floats are especially built for Hydratrain's rough route across the Gulf of Alaska to Whittier. They depart and arrive as a seagoing single or tandem barge tow. An assist tug will take one of these barges away,

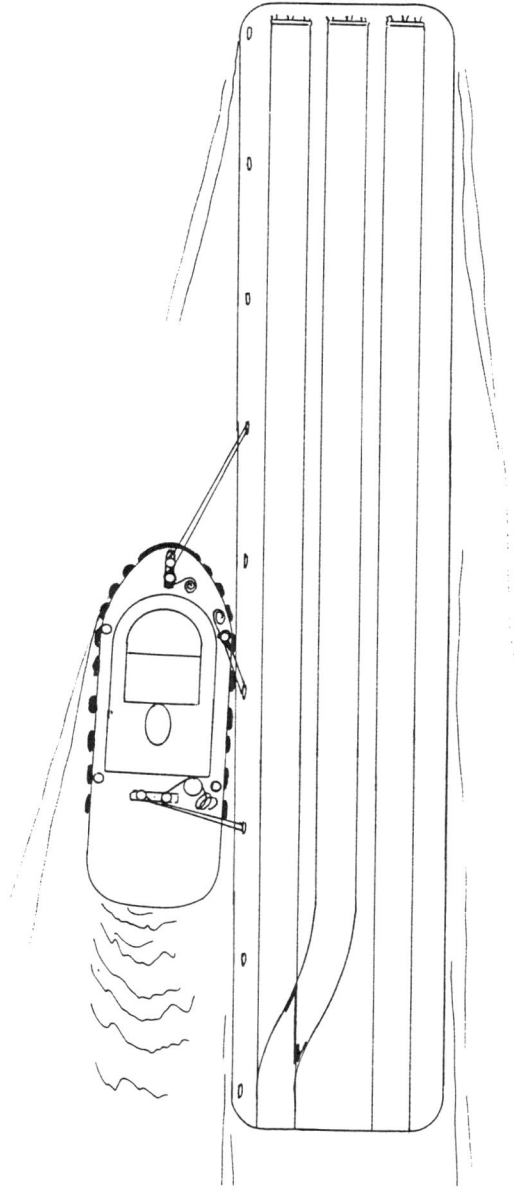

Tugs make up to railroad car floats farther forward than they would on ordinary barges. The reason is the tug's draft is usually deeper than that of the car float and the car float's length makes it more difficult to handle when made up close to either end.

as the hawser tug has a double-drum winch. The hawser tug then heaves in the second barge and takes it alongside to the dock. All of these docking movements are done under the direction of a mate using a handheld radio as he stands on the car float bow offering clearances to the person on watch handling the tug.

HANDLING TOWS IN LOCKS AND CANALS

Ascending

It would not be practical to cover methods used in towing in all the canals and locks of the world. Each is covered by certain local regulations and customs. All have a few things in common. In most areas, the tug captain must possess local knowledge of the waterway or take a pilot. Often, both are required. In some cases, if towing a dead ship or a large unit, additional tug assistance and pilots also may be required.

The locks in the Netherlands at IJmuiden and the lock gates to basins in the harbors of Le Havre, Cherbourg, London, Southampton, and Liverpool are good examples of the necessity of additional tugs. In such places, a tug on the bow and a steering tug astern of the tow is a common method. Radio orders from the lead tug direct the operation.

Locks must be entered slowly and lines given to lock attendants as ordered. Once the tug has approached the lock-edge area, movement orders should be quickly obeyed as they are received from the lock tender. It is most important to know ahead of time what the condition and status of your tug and tow are, well before the final approach. Is the red light on? Is the tie-up wall clear? Is another vessel there? Is there room enough for your tow or must you wait? Is the vessel, if one is in the lock, being lifted or lowered? Is the lock about to be dumped?

The tug captain will want to know all of the above so that he can decide whether to lie back or creep ahead to the tie-up wall. He can usually find out these answers by voice radio to the lock master or from other vessels closer to the lock. Some locks are equipped with two-way loud-hailer systems.

Many locks on rivers, especially in Canada and the United States, have a dam running from one bank to the lock, which is located close to the opposite bank. A few in the New York State canal system have dams on both sides. Wherever one of these dams is next to a lock, the water over it is tumbling into a wide area known as the forebay. Depending on its shape and how high the river's pool above is, this swirling water can cause problems in trying to get the tow past it and into the guide wall of the lock. Sometimes, an unexpected eddy will run in toward the lock gate. It is best, if possible, to learn beforehand any such peculiar conditions that may exist at each lock the tow will pass through.

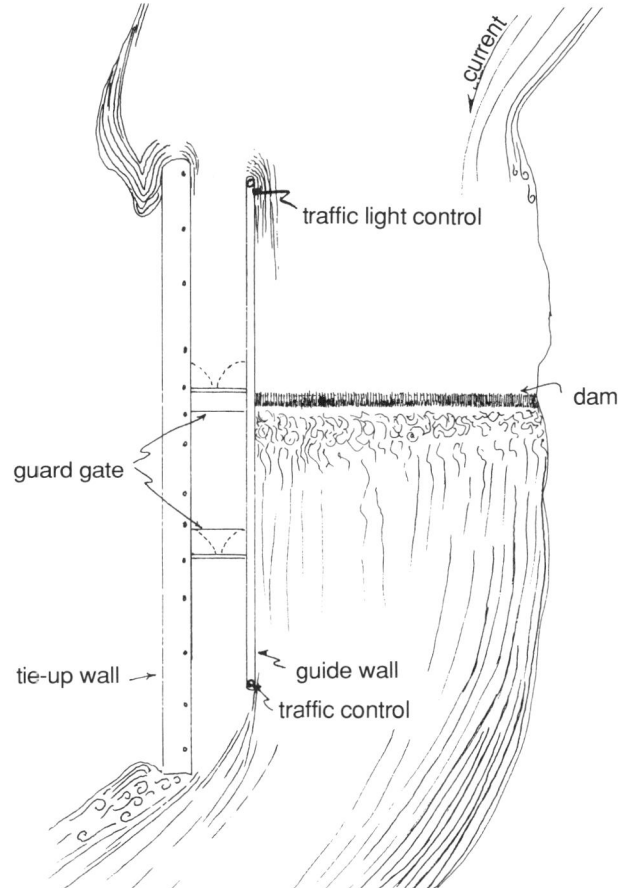

A typical river lock and dam

Lock Sizes*

Size also has a bearing. If the lock is small, such as in the New York State canal system (300 by 45 by 9 feet), you and your tow will undoubtedly fill it. Most petroleum barges using the New York State canals are between 230 and 240 feet overall with a 40- to 43-foot beam.

Pushing is the best operating method. If a tug is pushing a large loaded oil barge and is entering from the lower level, the barge's draft will just clear the lock sill. With just a foot of water under and around each side of the barge, the tug will have to push at about three-quarters to full speed for a portion of the barge's entry. This might be compared

*See table in appendix VI.

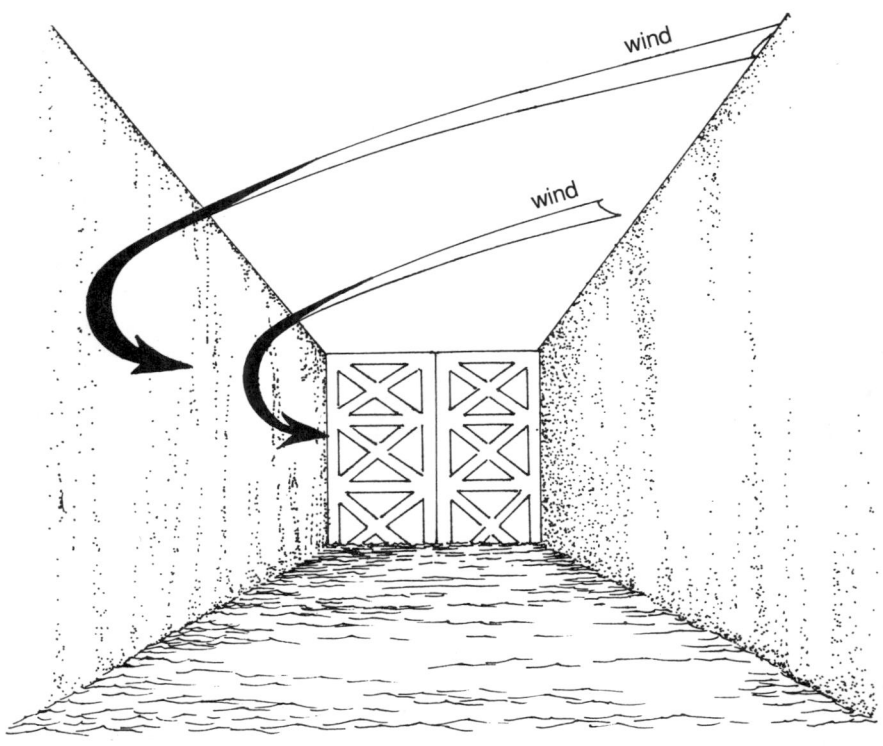

Wind will bounce off or deflect from lock walls when the chamber is empty.

to shoving a finger into a test tube full of water. As the water will not compress, it will slowly run under and fill the area astern. As this happens, it will begin to push the tow and the tug must be ready to reverse engines and check the momentum.

The above action also applies to large, deep-draft tows filling large locks such as are used by big river towboats on the upper Mississippi, Ohio, Tennessee rivers, and the Tenn-Tom Canal.

The longer the lock, the more water will fill up astern, so progress should always be very slow even though full power may be required at times. Lock tenders are quick to warn tug masters if speed is excessive or too slow. Frequently, the entrance of a lock that is empty will appear as a dark, yawning cavern, and voices from the lock wall above are difficult to understand. Tug masters should realize, whether running free or entering with a tow that will not fill the lock chamber, that the wind that may be blowing across the lock above often hits the lee wall, and its downdraft will be reflected in the opposite direction in the bottom of the lock.

The St. Lawrence Seaway, Welland Canal, and Great Lakes
The wind effect mentioned above is particularly noticeable in the 800- by 80-foot locks of the St. Lawrence Seaway and the Welland Canal. If the tug and tow are small, orders may be directed to lock through with other vessels or a ship. Some of the U.S. locks have floating mooring bitts that ride up or down as the water level changes. Whether these are used or mooring lines are run to the top of the lock, one of the requirements spelled out in the *Seaway Handbook* is that all transiting vessels including tows must handle their own lines at tie-up walls and in the locks. Another requirement is that each vessel shall have closed chocks so that mooring lines will not slip off due to the upward angle when the lock is being emptied. These requirements must be complied with before a seaway transit permit is issued. If the tug and tow are newcomers, a tie-up for inspection may be required.

Entering the St. Lawrence Gulf and River with a Tow. If proceeding from the Atlantic or Nova Scotia coast into the St. Lawrence Gulf and River, arm yourself with the latest annual Canadian Coast Guard *Notice to Mariners*. It will be noted therein that pilotage and routing in Canadian waters are compulsory in accordance with the International Maritime Organization's adopted routing system.

According to the weather and the size of the tow a choice may be made whether to enter via the Strait of Canso between Cape Breton Island and mainland Nova Scotia or through Cabot Strait between Cape Breton Island and Newfoundland. A Canadian or U.S. Coast Guard *Pilot Manual* gives all communication information and the required voice radio reporting stations and point. A 24-hour notice must be provided to pilot stations. Movement will be via VTS (vessel traffic service). If the choice is via Canso, ECTZ (Eastern Canada Traffic Zone) will control movement as you approach and advise the condition at Canso Lock. This lock is located on the Cape Breton side of the Canso Causeway. The approach walls are 651 feet on the south or seaward side and 700 feet on the Gulf of St. Lawrence side. The lock is 820 by 80 by 30 feet (see latest Transport Canada *Notice to Mariners* for maximum vessel size).

This area can be very windy at times. The lift of the lock is governed by the tide on either side (see *Tide Tables*). There is little current when westbound from Port Hawksbury, but there are several large tanker terminals in the Strait of Canso. Be prepared to meet large ocean vessels. They do not often pass west or transit the Canso Lock.

The Cabot Strait route and traffic system on the north of Cape Breton Island leads across the Gulf to a point off Percé Rock on the Gaspé coast, where tows through Canso would rejoin the main route to the river.

At this point, a tug master should call Sept Iles (Seven Isles) radio and report in, giving his position, dimensions of his tow, average speed, and destination. Canada is bilingual, and in the Province of Quebec, the French language often is used first.

Sept Iles will advise you of traffic and of where to give position reports. They will request a new ETA for Les Escoumins pilot station. Pilotage is compulsory and the tug and tow will be required to have one pilot for the 700 miles to Montréal and through portions of the St. Lawrence Seaway, if it is bound for the Great Lakes. This is another 182 miles.

Your best speed will determine whether more than one pilot will board. Most tows require two. The St. Lawrence River pilots are excellent. Most speak good English, but will converse bridge-to-bridge, and bridge-to-shore, in French. The pilots who board at Les Escoumins will be relieved at Quebec City. They will have made all of the required reports at passing points. If the tow cannot make better than 6 or 7 knots in still water, the Quebec City-Trois Rivières pilots may ask that the tow be anchored below Richelieu Rapids if a swift ebb is running. With the flood, a tow will proceed to Trois Rivières and another change of pilots will take place. On arrival at Montréal, if the tow is to proceed into the seaway, it may be docked for inspection. A seaway preclearance should have been arranged and a current copy of the *Seaway Handbook* should be aboard the tug. At this point, the first St. Lawrence Seaway pilot will be provided. If the tow has been accepted, and the pilot has permission to proceed into the seaway, the tow will be leaving the river at an oblique angle. On approaching the seaway entrance, there is at times a very swift crosscurrent. The tug master should realize his full responsibility in handling the tow and be able quickly to accept and apply the advice of the pilot while acknowledging that the pilot may not be an experienced tugboater (see table of locks in appendix VI).

A short distance inside the seaway entrance are several bridges and St. Lambert Lock. A curving 14-mile canal cut leads to Côte Ste. Catherine Lock and then back into the river. Twenty-four miles farther on is Beauharnois Lock whose entrance is again at an angle. Close by, water pours over a dam on the upstream side. All lockage in this area is controlled by the St. Lawrence Seaway Authority at Cornwall. On arrival at Snell Lock in Massena, New York, 46 miles upriver, control passes to the St. Lawrence Seaway Development Corporation. Here tows enter the $3^{1}/_{2}$-mile Wiley-Dondero Canal, which leads to Eisenhower Lock. Approaches to Snell Lock are a bit difficult. Current coming from the Grass River to port and from around the bend of Massena Point and the Moses-Saunders Power Dam to starboard will affect the tow until about even with the lock's lower tie-up wall. On leaving the canal cut above Eisenhower, it will be noticed that the aids to navigation provided by the U.S. Coast Guard and maintained by the St. Lawrence Seaway Development Corporation are difficult to observe and to judge the distance off, in comparison with those much superior ones in Canadian waters. From Eisenhower Lock, the 22 miles to the lower wall of Iroquois Lock is through the pooled waters of Lake St. Lawrence. Iroquois Lock is on a bend on the west or Ontario shore. It has a dam to port. The approach to the wall should be made carefully. The current eddy will assist in landing. The lift is slight, often only several feet.

It is upon leaving the upstream end of this lock wall that tugs with tows have to take particular care. A small man-made island to starboard divides the river's flow so that a current runs directly across the outer end of the tie-up wall. Tows that have exited this lock wall without meeting this crosscurrent have been swept across to the tip of the island to port, above the dam.

Downbound and descending Iroquois Lock, a tow should be headed as if to strike the outer end of this tie-up wall. The current will set the tow clear.

Also when downbound and entering the Wiley-Dondero Canal, there is no difficulty. But at Beauharnois, just as on river locks on the Ohio, one must be careful to allow for current and wind. Steer to close the tie-up wall if the lock is not ready and the lock gate is not open.

APPROACHES TO OTHER LOCKS AND CANALS

Each lock has its own peculiarities when approached from downstream, seaward, or the lower water level. Approaching from the upstream side of many locks requires much care. Here are a few to approach warily: The Flights above Waterford, New York, when leaving the Mohawk River; Mechanicville on the upper Hudson; Harvey Lock and Industrial Canal Lock in New Orleans; and Bonneville Lock on the Columbia River. Locks on the Ohio, upper Mississippi, Snake, and Columbia rivers should be approached with *great* care when these rivers are in freshet stages. In flash flood conditions, it is best to try to hold a tow against the bank, aground, or made fast to something solid until the current velocity drops to a safe-to-maneuver level.

Tugs and tows have been caught in such maelstroms and kept from making the lock tie-up walls. Some have been swept against dams or bridge abutments and totally destroyed.

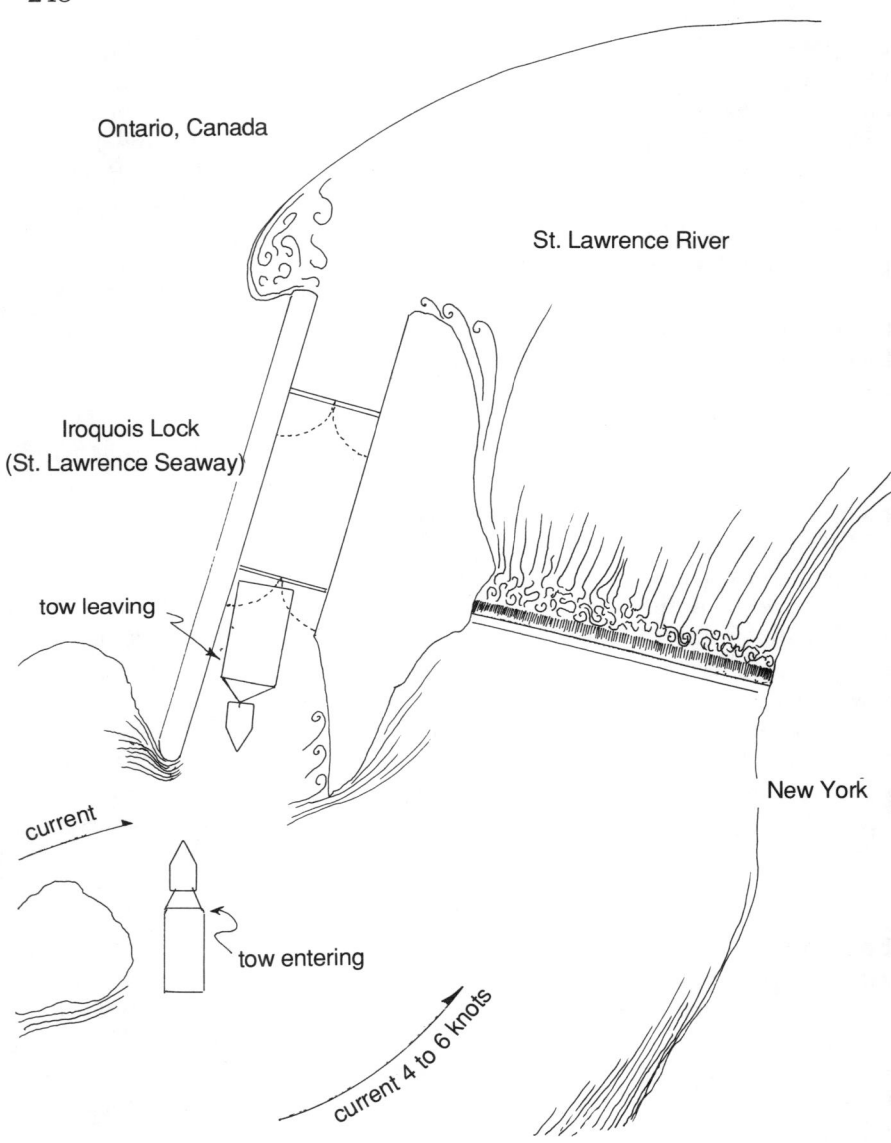

Iroquois Lock can be difficult to enter with a tow due to the crosscurrent.

Tugs on the Seine and other European waterways push single barges. The tugs must be low to pass under the many bridges.

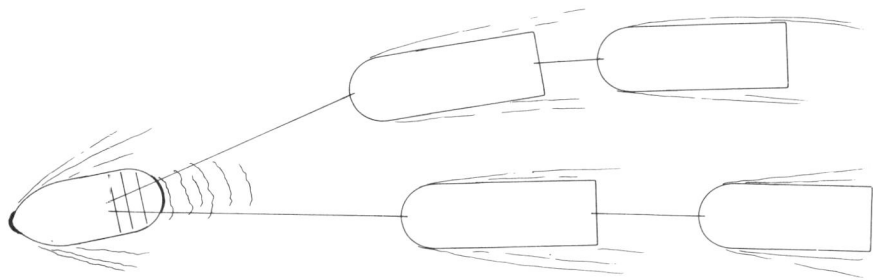

Typical Rhine River barge tow

There are both large and small canals of some length that are used primarily by ships. Tows and pleasure craft frequently are accepted. The method of towing on these waterways varies greatly. On the Rhine-Rhone and other waterways of Europe, those steel canal barges which are without power are towed astern in fleets by small, shallow-draft tugs. Towlines are either run from the barge to the tug's tow hook or from a winch on the tug. Each barge has a rudder, and landings are accomplished by rounding up in the current. Pushing has been introduced in some areas, particularly on the Seine around Paris, where there are many bridges without fenders.

Due to the uniqueness of the small canals of England, the Göta in Sweden, and others throughout Canada, there is a limit to the size, draft, and overhead clearance, which has resulted in localized towing methods. It is doubtful that outside tugs or their tows will enter these waterways. LASH or SeaBee barges would be the exception.

The Panama Canal

Of the big ship canals, Panama leads in the number of locks, and the use of assist tugs with Panamax-size* shipping. All movements are under the direction of the Panama Canal Authority and Pilots.

If canal tugs are available when an outside tow arrives at either end, at least one tug will be assigned to assist, if indeed tugs do not com-

*Panamax—The maximum-sized ship allowed to pass through the Panama Canal is length 948 feet, beam 106 feet, draft 34.8 feet.

pletely take over the tow's transit. This, of course, would be after the tow's mooring chocks and other requirements have been approved. Any towed unit arriving without approved Panama Canal type closed chocks will be refused transit until they are provided.

Because of this possibility and any other problems that might arise, it is good to have had an agent appointed prior to arrival. There are many excellent ones, all familiar with towage problems.

The Panama Canal Authority has berths at Cristóbal and Balboa where repairs and installations can be made.

Occasionally, tugs are allowed to take their own tows through the canal if made up alongside, pushing, or on the hawser. Transiting the canal with a tow alongside is not difficult. Usually, due to its slower speed, such a tow follows the last vessel in any convoy or may proceed well ahead of it. Pilots will direct line handling at the approach to each lock. The tow will be pulled both into and out of lock chambers by use of wire cables to small locomotives on the wall. Canal personnel ride the tow and handle all lines.

Afternoon winds are often gusty from Gaillard Cut to Miraflores Locks. Meeting traffic in this area when towing is not recommended. Most meetings are timed to occur in Gatun Lake. On exiting the canal locks and approaches on either end, there is plenty of room to string out a tow for sea.

The Suez Canal

Here there are no locks, but movements are regulated and piloted by the Suez Canal Authority of Egypt. It is a long, slow, hot passage. Assist tugs are required for large tows. An agent should act for all towing companies planning to transit the Suez.

The Nord-Ostsee Kanal

This canal at Kiel, West Germany, joins the Baltic Sea with the North Sea through the Elbe River. It is frequently used by tows. Its locks are

A closed Panama chock as required on all vessels passing through the Big Ditch

large and transit requires prior arrangements regarding the size of the tow and the horsepower of the tug.

The Noordzee Kanaal

Tows bound into this canal, which runs from the North Sea at IJmuiden to Amsterdam, must watch the weather and sea conditions when approaching the jetties. Tug assistance may be necessary. Pilotage is mandatory. ETA at the pilot station at IJmuiden should be reported at least four hours in advance by radiotelephone addressed to Scheveningen Radio (PCH). When within range, call Pilotage IJmuiden on Channel 12 VHF. Tugs assist ships and tows in lockage as necessary. Lock dimensions are given in appendix VI.

The Manchester Ship Canal

This canal has its own tugs, whose expert skippers handle all traffic, large and small.

The Locks in the Welland Canal

These locks are all in the Province of Ontario, Canada, and come under the authority of that country's Department of Transport. They are large and deep and some are doubled. Tows are frequently locked through with a ship. The entrance from Lake Ontario is wide and between breakwater jetties, which lead to a long tie-up wall at Port Weller.

Your ETA should have been previously established in accordance with the Canadian Coast Guard *Notice to Mariners* and directed to the Traffic Control Section as shown in the *Seaway Handbook*. When several miles east of Port Weller entrance or west of Port Colborne on Lake Erie, your presence should be announced by VHF radio. All movement thereafter will be upon the orders given by either Weller Traffic Control at the Lake Ontario end or Long Point Traffic Control near Lock 7. The speed limits imposed are 7 to 8 statute miles.

It is seldom difficult to enter from Lake Erie with a tow unless there is an easterly gale. Then, Toronto Harbor or Oshawa on the north shore may be the better place for shelter. At the Lake Erie end of the canal, berths at the Port Colborne tie-up walls become quite crowded when there is rough weather outside. Lake Erie, due to its shallowness, can become nasty in a matter of minutes. If bound into the Welland, which is a very difficult entrance in rough weather, tows frequently are anchored in the long cove in the lee of Long Point. If westbound and caught in bad weather near the middle of the lake, Presqu'ile Harbor is the best shelter.

The Locks at Sault Ste. Marie

These locks are approached from the upbound channel at De Tour. All movements and approaches from Lake Huron or Lake Superior are controlled by the lock authorities, that is, the U.S. Army Corps of Engineers in the United States, and the Department of Transport in Canada. As on most lock approaches, there are traffic lights and a pilot is required.

Cape Cod Canal*

This is situated between Cape Cod Bay and Buzzards Bay. Operated by the U.S. Army Corps of Engineers, it is transited by many tows and ships. Movements are controlled by both VHF (channel 13) and closed-circuit TV through the canal dispatcher at Buzzards Bay. There are also traffic lights at Sandwich and Wings Neck. Due to heavy tidal currents of 5 knots or more, tugs with tows are started through the canal in either direction about half an hour before the current turns fair in the direction they are proceeding. Dispatchers rarely allow ships and tows to meet in the canal. A request for permission to enter should be made half an hour before arriving at either end of the canal. Tows frequently wait at anchorages in Buzzards Bay near the Cleveland Ledge Light because of heavy weather off the other end of the canal at Sandwich.

The Chesapeake and Delaware Canal

This is also a Corps of Engineers-operated waterway. It runs between the Delaware River in the vicinity of Pea Patch Island to the head of the Chesapeake Bay. Permission to enter should be requested on VHF (channel 13) half an hour before approaching either entrance. The Corps of Engineers traffic dispatcher is located at Chesapeake City. The canal entrance and channels are monitored by closed-circuit TV. Tows and ship traffic frequently meet and pass in this easy-to-navigate waterway. All ships have pilots who monitor channel 13. During certain times of the year the canal is closed because of fog when it may be partially clear at either end, and during cold snaps in the winter it may be closed because of ice. Watch the crosscurrent on flood and ebb at jetties off Bulkhead Shoal Channel at Delaware City.

Sacramento Ship Canal

This has one lock at its head just above the berths of the Port of Sacramento. The lock is operated by the port and it opens only at certain times. Tugs and tows anticipating use of this lock should inquire ahead

*See complete regulations in appendix VI.

of time. After passing out of the lock toward the Sacramento River, vessels must avoid a middle ground shoal by holding to the south or downriver shore. A traffic light and holding area for river traffic bound into the lock and canal is located 500 yards downstream from the lock forebay on the south riverbank. River current varies due to runoff, which can reach 6 miles per hour in the springtime.

Columbia-Snake River Locks

These, too, are operated by the U.S. Army Corps of Engineers. And again, traffic control is by signal lights and voice radio. Bonneville Lock and Dam has a forebay that empties with the powerhouse outfall and makes a rapid current below the lock. When downbound and above this lock the river runs swiftly from Cascade Lock and the Bridge of the Gods. There are tie-up piers for tows both above and below this lock.

All of these river locks have upper and lower tie-up walls with floating mooring bitts in the walls of their chambers. The lifts are from 80 to 105 feet. Single-tier tows may have other craft in the lock with them. Double-tier tows fill the locks. Ninety percent of all tows on these rivers are pushed. The exceptions are contractors' equipment and boom stick and log raft tows.

After Bonneville there are eight more locks through which tows must pass to reach the head of barge navigation at Lewiston, Idaho. All have dams, and pool water levels are regulated.

Washington Ship Canal

This canal has one set of locks at its entrance. The pair, known as the Hiram Chittenden Locks, is operated like many others by the Corps of Engineers, with VHF and traffic light controls. The larger of the two locks will handle ships that enter at Shilshole Bay, and through the canal they may exit at Lake Washington. These locks and canals have many small pleasure craft and much commercial traffic consisting of tugs, tows, and medium-sized ships. The two locks are side by side. The smaller one will take a light tug or fishing vessel. The larger one will take a tow. At the discretion of the lock master, other vessels may be placed in the lock to transit with a tow.

MOVING DRILL RIGS

Drill rigs present myriad considerations. When they are to be moved or towed out to locations or over long ocean towage to another part of the world, a great deal of preparation is necessary.

Tow-out differs from a short shift of site in that the rig is usually in a sheltered area when the movement commences. Rig draft, condition of the waterway, tide, current, and traffic problems will all have been worked out prior to the arrival of the towing tugs. Most rig movements require two and some up to six powerful tugs. In rivers or in sheltered waters, jack-up rigs may use several tugs behind and pushing, with a lead tug on the hawser.

When tugs report to a drill rig they are about to tow, they should approach the side that has a walkway with wires hanging from it unless otherwise instructed. Whether the rig is a jack-up or a semisubmersible, it will be held in position either by its legs on the bottom or by anchors.

The tug will be ordered by the tool pusher in charge to stand by or to come in to connect up her tow wire. Before making such an approach, both wind and current should be studied. The tug should then approach at an angle that will allow her to be lying with her stern close under the rig's platform walkway and in such a position that the effect of the tidal current and wind will set her on a heading from which she will be ready to pull. This may take more than one attempt. If the tug starts to get out of shape, cast off lines even if connected, and start over again.

In connecting, it is important to keep a tug's stern as close under the rig's walkway as possible. A messenger line may be dropped from the walkway to the tug. A strain should be kept on it so that when the towing pendant and its retrieving cable are dropped, they will land partially on deck and cannot be picked up by the tug's propeller(s). Once connected, pay out enough cable so that the tug can maneuver one way or the other while the remaining tugs connect their cables. It is in this type of work that tractor tugs are most efficient. Tow pendants are a part of the rig's towing gear and are shackled into pad eyes usually located at a low point on the rig's body or legs.

When all tugs are hooked up, one will assume command, giving orders as to when to start pulling, the course, and any other instructions.

Smit Rotterdam and *Smit New York* begin tow-out of Frigg Control Tower from Stavanger, Norway. (Courtesy Smit Tak)

On arrival at the new site, or spotting and anchoring, tugs are usually released in the order most convenient to the tool pusher or whoever else might be in charge.

Anchor Handling

Tugs specially rigged to handle the heavy anchors used by drill rigs normally have a sloping low bulwark at the stern fitted with heavy rollers. Anchors are laid on the open deck by the rig's crane. As the tug runs off in the desired direction, pulling chain or cable as paid out from the rig, the anchor is held by a trip line. Upon signal, it is dropped at once. The chain and anchor will run out over the tug's stern rollers.

The anchor may also be slung over the vessel's stem with a heavy quick-release stopper around the chain. This is a very safe method of spotting anchors. Every anchor should have a marker buoy attached, and a cable heavy enough to break out the anchor and hoist it back on deck. Smaller rigs and dredges with lighter anchors on wire cables will have the tug pick the anchor up by running a short line around it and holding it on its stem. By having the tug back away, the anchor is run out and easily released. Care must be taken in securing this short anchor line. It should not be secured in any manner that will allow it to jam. Deck personnel should stand well clear when letting go or releasing anchor gear.

To retrieve anchors, heave them up on the tug by retrieving cables and tow them back to the rig as the anchor cable itself is heaved in. Smaller dredge tenders use the same method by towing a small derrick scow alongside. Some tugs will underrun the cable by placing a hook on a short line at the dredge and backing out until over the anchor. When the anchor is broken out from the bottom, the dredge heaves in her cable, pulling the tug to the dredge. The tug should continue to have her engines in reverse so as to hold a slight strain, which will keep the anchor almost "two-blocked" at her stem.

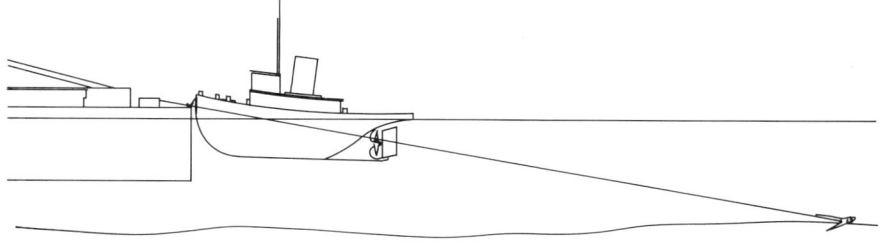

Tug ready to back out on dredge anchor cable

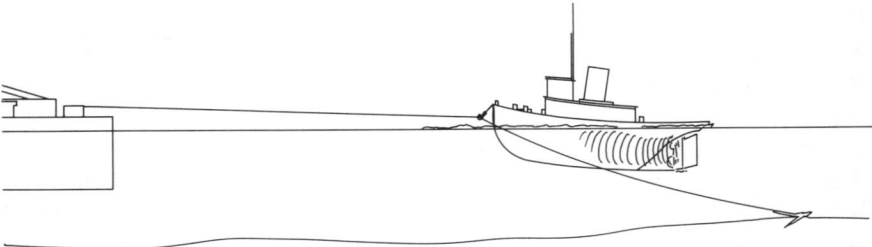

Tug backing with dredge anchor cable through a hook suspended from her stem

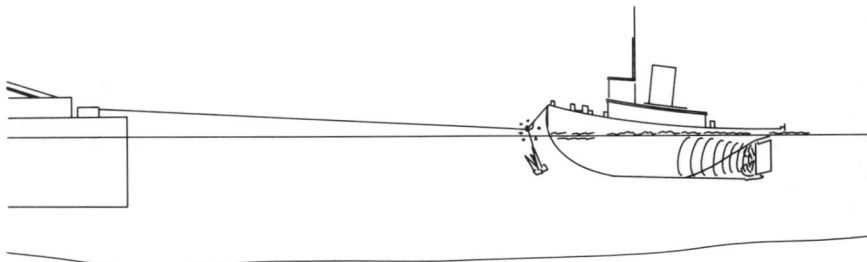

With the anchor off bottom, the tug backs until its shackle appears close to her hook. As a result of the tug backing slowly and the dredge heaving in the anchor cable, the anchor will arrive at the dredge's stern.

DRAWBRIDGES

These are structures that have never been admired by tug captains and pilots. They are obstructions to navigation created for the convenience of land travel. In their planning, little thought was given to the hazard they would present to water traffic. Yet we have to live with them. Supposedly, they are properly fendered to protect their own abutments and the craft that will pass through.

In the United States, most operable bridges are properly maintained, but there are those whose structures are old and from which the response to a vessel's whistle signal or VHF radio is annoyingly slow. The time from acknowledgment to complete opening may also extend the delay.

Bridges over canals, waterways, and locks that are controlled by a government authority seem to respond promptly and open quickly. In any event, the navigator of a tug or tow should approach any bridge, particularly a drawbridge, defensively. He should make his presence known early and promptly. He should also sound his whistle signal, which in the United States with few exceptions is one long and one short blast. If no voice radio reply or other answer is forthcoming, he should be prepared to stop and hold his tow. This can be difficult and hectic if he has a large barge or ship astern.

Bridge regulations and their operating hours are listed in most *Coast Pilots* of various countries. Any person in charge of a tow who has not transited a drawbridge before should have complete knowledge of these regulations and the direction of the tidal current as it flows onto or through the bridge. The best teacher is a local tug operator with a "no damage" record. For information on passage through drawbridges, refer to chapter 10.

Bridge dimensions are printed on navigational charts. It is most important for the tug operator and pilot to be aware of their tug's highest point. Can she pass under any particular fixed span other than the one which opens? Passing through narrow draw spans is often an art. It requires complete control and knowledge of the best way to handle the tow, whether it is astern, ahead, or on the hip. Many such spans have a cross- or diagonal flow striking them. Here are a few that require extra caution:

Fore River, Portland to South Portland, Maine, outbound

Piscataqua River, Maine to New Hampshire, outbound above Portsmouth.

Chelsea River, East Boston to Chelsea, Massachusetts, McArdle and Central Avenue bridges, either way

Taunton River, Fall River to Somerset, Massachusetts, outbound on ebb current

Connecticut River, Old Saybrook, Connecticut, railroad bridge

Westchester Creek, New York

Bronx River, New York, bridges, heavy traffic

Raritan River, Perth Amboy, New Jersey, railroad bridge

James River, Hampton Roads Bridge Tunnel, Virginia

Intracoastal Waterway at Morehead City harbor, Morehead City to Beaufort, North Carolina, bridge

St. Johns River, Jacksonville, Florida, area bridges

Old Tampa Bay, Tampa to St. Petersburg, Florida, Gandy Bridge and Courtney Campbell Causeway

Towboat masters have had difficulty piloting tows through several drawbridges that span rivers on the West Coast of the United States. These bridges are at Coos Bay, Oregon; Reedsport, Oregon; Willamette River, Portland, Oregon; Seattle and Tacoma, Washington.

Most bridges in Canada, the United Kingdom, and Europe are of sufficient height to allow vessel traffic free passage. In the United

States, many railroad bridges remain open except when a train is approaching.

PILOTING THE TUG AND TOW

Piloting the tug with or without a tow is a very serious matter. Many towboaters operate almost daily in a limited area, the waters of which they may think they have learned by heart. But there always comes a time when on a dark night, or in a thick fog, heavy rain, or snow, visibility is completely lost. This is when some tugs also seem to become lost.

It is always prudent to have a chart of the working area out where you can scan it. Frequently used courses should be noted. Average running times with both flood and ebb currents between buoys and other turning points should be noted. Any cross- or diagonal set of current at particular points in bends, on buoys in channels, and pierheads should be studied and remembered.

A good way to organize this is to build a notebook of facts so that as a knowledgeable watch stander you will have a journal of pertinent statistics to assist you as needed.

Charts should be corrected through use of *Notice to Mariners*. This is a service that responsible tug owners should provide. A great aid in harbor or river piloting is the comparison of the radar presentation with a good small-scale chart. It is important to remember that riverbanks and low banks do not give a good presentation as to their exact meeting with the water's edge. A resolution appearing as shoreline may be solid land on a higher point some yards in from the high-water mark. The possible presence of this "unnavigable" area should be investigated to ascertain how far off the beach the tug and tow are passing.

Some very modern tugs involved in sophisticated operations have installed the Viewnav Electronic Chart Navigation System. * This is a new generation of visual devices using digitized charts and a computer display of the vessel's position on an electronic chart. More than one person is required on watch to operate this system, as it is not possible to maintain a good lookout while steering and feeding information into the Viewmaster. Its single display shows it all—ship's speed, heading, course, distance to turning (way) point. It is unique in that its electronic chart presentation appears as an actual NOS chart with aids to navigation in place. If an aid has been moved, its true location is shown on the Viewnav display. Radar targets appear in different colors.

Viewnav electronic charts are available for various areas and all are kept updated by the maker.

---

*Viewnav Electronic Chart Navigation System by Navigational Sciences Company, Bethesda, Maryland.

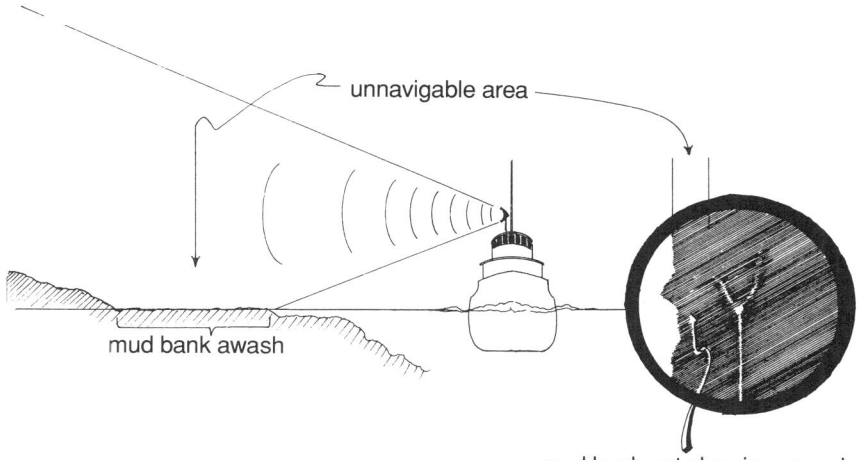

mud bank not showing on radar

Pilots can be fooled by radar reception that does not show land that is awash. Grounding can be avoided by checking the distance between dry land on a chart and that shown by radar. Chart depths and soundings by the tug's equipment should be compared.

Some of the basics in tug piloting are: adhering to all of the '72 COLREGS (Rules of the Road); knowing when to anchor in restricted to zero visibility; and being aware of the tide stage and current and how to use them to one's advantage. There are many places where passage should be made only at or near slack water, with the change of current about to run in your favor. Good examples are the Cape Cod Canal, Hell Gate in New York's East River, from Cape Mudge through Seymour Narrows in Discovery Passage, and Johnstone Strait in British Columbia.

There are many river entrances with bars and inlets such as the Columbia River and Pacific Northwest ports where safe passage in or out must be accomplished in good weather. This is particularly true along the Pacific Coast from Point Conception in California to Cape Flattery, Washington. On the East Coast, many movements of large tows or deeply loaded barges are regulated by the stage of the tide. Arrival at a berth must be planned for high-water slack and departure time set with enough of the flood remaining to allow the tow to get clear of the narrow tidal entrance. Portsmouth, New Hampshire, is a good example of these tidal conditions, which apply in many places between the Connecticut River and the Bay of Fundy ports. In Europe, this condition is applicable in the Bay of Biscay, Bay of the Seine, and the Bristol Channel. There are many other extreme tidal areas throughout the world where timing of vessel and tow movement is necessary.

When running free with a light tug, if large rollers or sets of waves appear on one quarter of your wake, slow down, check the depth, and alter course toward the other side. If these large following rollers appear on both quarters, you are in shallow water.

One of the finer points of piloting is knowing how to approach entrances to rivers, basins, or wharves with a tow pushing or towing alongside or astern. In approaching a wharf, how much headway can be allowed in order to stop? How does one shape up the tow to pass through certain drawbridges? The correct solutions to these problems come from practice and experience. A few near-misses are common among towboaters. Just make sure they are misses—situations you can get out of with no damage and from which a lesson is learned.

THE TUG LOG

One of the best ways to protect oneself, the tug's owners, and the company's customers is to keep a good log. On harbor tugs, this can be accomplished by making short notes and entering them in a log at the end of the watch. On coastwise or line-haul towboats, entry should be made at regular intervals, passing points, or course changes. If a course is steered, enter it, and the type of compass, magnetic or gyro. Weather and visibility are also normal entries.

Most important are notations covering any damage or injury and/or action taken to avoid same. Care should be taken in writing the log to list who was on watch, who maintained a lookout, the speed, course changes, and conditions of the tow at regular intervals. This should be done regardless of whether towing on the hawser, alongside, or pushing.

Some companies provide so-called log sheets that tend primarily to reflect interest in engine operation and time. While this does reflect economy of operation or lack thereof, such reports fail to realize the loss that may occur in a litigation over alleged damage wherein proof of prudent action is lacking due to poor log-keeping. Ocean and coastwise tug logs are discussed in chapter 9.

CARE OF THE TUG

Proper care of a tug is a many-faceted program. Here is a brief outline. The care and maintenance of a tug should be divided into that which is covered daily, weekly or during other regular operating periods, and annually. Daily care applies to four levels—deck, engine room, galley, and sanitary. Daily deck care would refer to the knowledge of the condition of deck lines, heaving lines, lifesaving equipment, survival suits, fire extinguishers, towing equipment, and fenders, and condition of stores and replacement requirements. Basic daily duties include: cleaning of quarters, galley, and decks; removal of garbage; wash down and cleaning of pilothouse windows. Watertight integrity must always be maintained and requires frequent inspection of the vessel by checking such items as outside vents, deck plates, hatch covers, and watertight doors. Interior problems arise through water in bilges, rudderpost leakage, or other piping and fuel line leaks, or any possible cause of fire. Weekly checks should be made of machinery, towing winch, electronics that may not be in frequent use, oil filters, potable water, lube oil, and fuel remaining. On the outside of the tug, deck and navigation lights should be checked for water in the glass globe, salt-covered lenses, and loose or burned-out bulbs. Lubricants should be applied to hinges on freeing ports in bulwarks, tow winch fittings, watertight door dogs, fairleads, outside boom hoists, davits, and any threaded deck sounding caps.

Regular checks of machinery should follow a manufacturer's warranty or planned schedule. Annual maintenance will differ according to the towing company's policy as to assignment to a shore-maintenance or tug crew. Usually, the chief engineer will oversee all mechanical overhauls and, if done in an outside yard, the port captain may check on all other work. Engines may be opened up, generators and compressors checked, tanks inspected and treated. The vessel may or may not be hauled out for bottom inspection until 18 to 24 months have passed, at which time propeller(s), rudder(s), nozzle(s), shaft bearings, zincs, and all underwater fittings must be checked.

On deck, it is important to fill in any wear spots or grooves on rollers, fairleads, stern rails, or tow arches that have been made by the tow cable.

Any classification society to which the vessel is certified, as well as the national regulatory authority, must be notified. Their individual inspection and approval must be met so that the tug may continue to be certified for operation. Additional care of seagoing tugs is covered in chapter 9.

SAFETY

This is a daily requirement on board all vessels. The backbone of safety is a constant awareness by every crew member of the ever-present possibility of accidents and how to prevent them. Prevention comes through training. Instruction and training must come through leadership and good communication. This leadership and communication must come from the top—the tug master. Most tug accidents involve personnel and are caused by failure to recognize elements leading to quick injury—specifically, working on the stern and not watching the lead of a tow cable, which may jump a few feet; standing between the bulwark rail and a line or cable leading over the side; standing in the bight of a line, wire, chain, bridle, or hawser that is being slacked out; holding onto a line that is being heaved in around a capstan at a point too close to the winch head; slipping on deck or placing hands on the casing of a swinging watertight door. Many personal-injury suits have been presented to courts as claims against the tug's seaworthiness. Frequently it is the crew member who is acting in an unseamanlike manner.

To make tug crews safety-conscious, some tug companies take the time to use visual aids. All efforts should be backed up aboard the tug by a sort of buddy system in which any crew member will watch and warn another of a dangerous or unsafe situation he is about to become involved in. Personnel who continually indulge in unsafe practices should be warned. If no change or effort toward safety is noted, dismissal as a protective method for the person, his shipmates, the tug, and her owners should be considered. Many companies require personnel working on deck to wear hard hats and work vest type life jackets when the seas are rough.

Besides personnel safety, safety of the tug itself must be considered of equal priority. Grounding or girding are constant operational hazards that must be guarded against through precise navigation and piloting and ship handling. Fire and sinking are high on the list of tug casualties. Protection and prevention require constant monitoring. The author's personal practice has been to go through or look into every part of the tug daily, correcting any unsafe condition immediately. Defects found included flammable material, interior leakage, wear on towing gear, slight leaks (particularly of diesel fuel, which under pressure leads to fire and explosions), and loss of lube oil or water.

All of this can be done aboard an assist tug while en route to her first daily assignment. On tugs that operate around the clock, monitoring should be done hourly when under way. Watertight doors should be kept closed and dogged in heavy weather and during any operating condition when the tug may be subject to listing or rolling and the possibility exists of water coming on deck. Tugs have sunk very quickly because an engine room door was left open, while all others were closed and dogged. Some engine room watertight doors are located under the boat deck overhang behind a tow winch. One particular tug took a sudden list which pulled her stern under and the engine room quickly filled. Seawater hit the main switchboard, killing all electric power. This entrance of water during a storm at sea has had a domino effect on the seaworthiness of tugs. As the unchecked volume increases, generators or switchboards short out and the steering gear goes dead from a lack of power. One after another, systems fail and the tug becomes a dead animal lying still and rolling at the mercy of the sea, to be swamped, capsized, and sunk.

With the reduction of manning, safety, watertight integrity, and fire prevention should become constant challenges. Automated engine rooms do not ensure that all fail-safe systems are covering every contingency. Regular personal inspection should be made in the engine room and other spaces.

Fire, Boat, and Abandon-Ship Drills

These should be held weekly if possible and should not be a repetitious muster. Certain possible problems and locations both in fighting fire and abandoning the vessel should be a part of weekly drills. Location of all emergency equipment should be checked. All tug personnel should be familiar with where fire monitors are turned on, where fire hose connections are, and how far each hose reaches. In addition, the location of hose nozzles and spanners should be checked. If a Fire Boss is installed, personnel should be instructed in its use.

For abandoning the vessel, practice donning survival suits and life jackets and find the best area to do so. Also study the release of the life raft, EPIRB, and ring buoys with a view that they might be blown away from the vessel.

Safety consciousness saves lives. The ocean has no compassion.

8. Under Way with the Tow and at Sea

CONDITION OF THE UNIT TO BE TOWED

EVERY tug master has a right and a duty to know the condition of any unit he is about to tow. If it is manned, he has the right to ask about it if he has any doubt of its seaworthiness. If unmanned, he should either inspect it himself or delegate some knowledgeable and trustworthy person to do so. Particular attention should be given to water standing in bilges and the possible free surface effect, and to all watertight closures and vents. If cargo is stowed on deck, its manner and lashing should be studied, with thought given to the possibility of any loosening and loss in heavy seas. If the movement is to be done on inland or sheltered waters, shifting is also a possibility. Unlashed cargo can be disturbed by the wake of passing vessels.

To be involved in litigation that may result from failure to make such inspections is most unpleasant. If an unseaworthy condition has been noted, the tug master should take remedial action by advising the tug owners of his refusal to depart until seaworthiness is restored.

INLAND WATER TOWING

On the major U.S. river systems, nearly uniform barge dimensions have been developed. Most of these barges are owned and operated by regular barge lines and pushed by line-haul towboats. On the Mississippi, Ohio, and Missouri systems, there are three barge designs. These are the hopper barge with an open hold for dry bulk cargo such as coal that can withstand weather; the covered barge used for grain and any cargo that requires protection; and the tank barge for all liquid cargoes.

There are additional types and sizes of petroleum and chemical barges designed for special operations by their owners, who are usually petroleum or chemical companies.

When first designed and fleeted in multiple tiers, these barges were known as "standards" and were 175 feet in length. Next came the 195-foot "jumbo" barge found in most Mississippi and Ohio river tows.* For ease of towing, standard and jumbo barges have both of their ends raked.

*See barge dimensions in appendix I.

Mississippi/Ohio hopper barge (Courtesy A.W.O.)

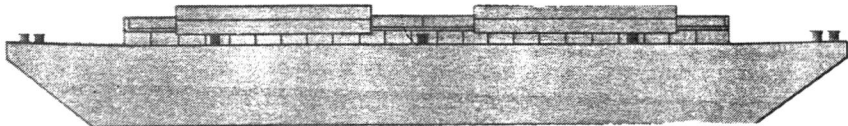

Covered barge on the Mississippi/Ohio system (Courtesy A.W.O.)

Tank barge on the Mississippi/Ohio system (Courtesy A.W.O.)

THE INTEGRATED RIVER TOW

These are tows of several specially built units that fit together as one. The bow unit may have a detachable spoon bow piece and be fitted with a thruster to assist on bends, on docking, and in turning, or the forward unit may be raked. The two or three other intermediate barge units will normally have flat ends so that when butted and secured in a single line tier, all resistance will be at the bow or forward end. Such unit tows are employed by oil and chemical companies, which require rapid transport of large liquid and hazardous cargoes.

It is understood that barges that are towed in various tiers are not necessarily all loaded with the same cargo or at the same terminal or transfer point. Nor are they all in a loaded or empty condition, nor are they all delivered to the same terminal. Some may be picked up or dropped off en route from the towboat's and tow's original departure point until arrival at the tow's final destination.

When tows are broken up, and single barges are awaiting delivery to a consignee's wharf or, if empty, are waiting for a tow, they are usually moored at designated spots along the riverbank where a fleeting service exists. Here, several shifting tugs bring or remove barges to and from tows or deliver individual barges to local wharves. Such facilities and services exist at nearly every large port on the Mississippi-Ohio river systems.

Making Up a Mississippi River Tow

In making up an upbound Mississippi River tow, the towboat will wait until eight or more barges in several tiers are lined up ahead of it and moored to the bank. At this point, the towboat will ease up until its knees are against the barge's stern. Which barge this is depends on the number and size of barges that will be in the tow. If the tow is to have one, three, or five tiers, the towboat will face up to the center tier. This is called "facing square." If the tow will have two, four, or six tiers, the towboat will straddle the two center tiers. This is called "splitting on heads."

With the towboat in place, her deck crew will commence carrying the various pieces of rigging required out onto the barges to make a solid connection. River terminology for this towboat rigging is a bit different from that used aboard ocean and coastwise tugs. Most of these river barges and towboats are secured by wire rope and pieces of chain and shackles. They are all tightened by steamboat ratchets between barges, and by capstans and winches on the towboat. This is known as "hard rigging."

Rope or soft lines of various sizes and lengths are used in handling and swinging barges and in locking. Other tools of the trade are the pike

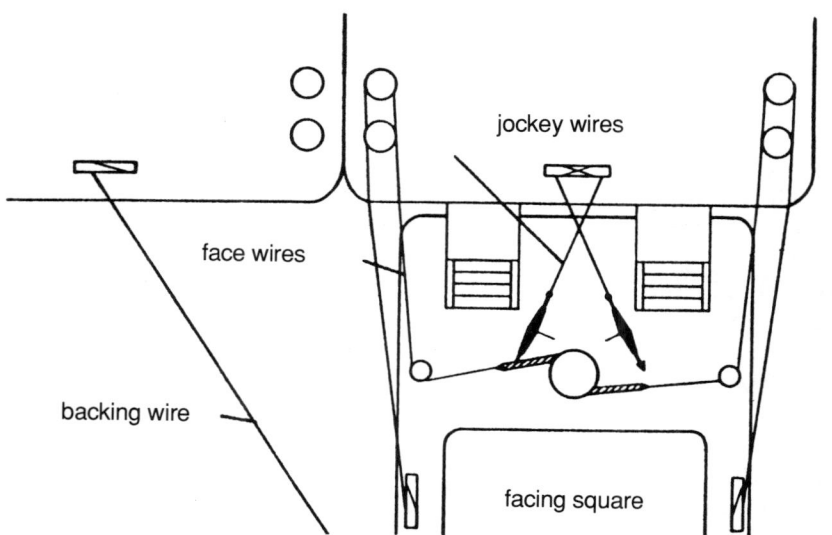

Facing square in the center of a single barge; method commonly used on the Missouri, Columbia, and Tenn-Tom. (Courtesy A.W.O.)

pole, used to reach and pull over lines and wires; the cheater bar, a long, hollow pipe that slips over a ratchet handle and offers more leverage for tightening. There are also the sledge and the toothpick, a long marlinspike that when thrust into chain links keeps them from turning or wires from twisting.

The deck fittings on river towboats used to secure lines and wires running to barges very nearly resemble their counterparts used on other craft. Only the names differ. A pair of bitts become timber heads, cleats become cavels, and stationary fairleads are called spools.

Coupling the Barges

With the towboat in place, coupling of the barges begins. This requires care by deck personnel, who must carry these fittings and wires out onto the tow. Work vests should be worn.

Fore and aft wires are a basic method used by most barge lines to couple tiers of barges. There are always newer ways using better rigging. Here are the proven methods.

Some companies use barges with stationary hand winches or turnbuckles. When using a hand winch, care should be taken. See that the dog is not stuck and runs free and latches so that the winch will not un-

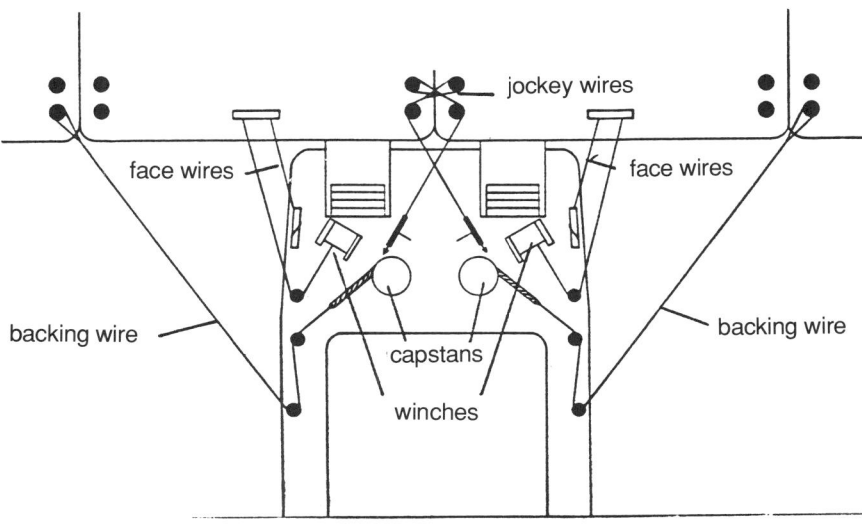

Facing up the towboat—with four strings or tiers, the towboat is made up between two barges. This is known as splitting on heads. Note that the towboat's facing wires run from her deck winches, while the jockey wires run from fixed ratchets on the deck of the towboat. (Courtesy A.W.O.)

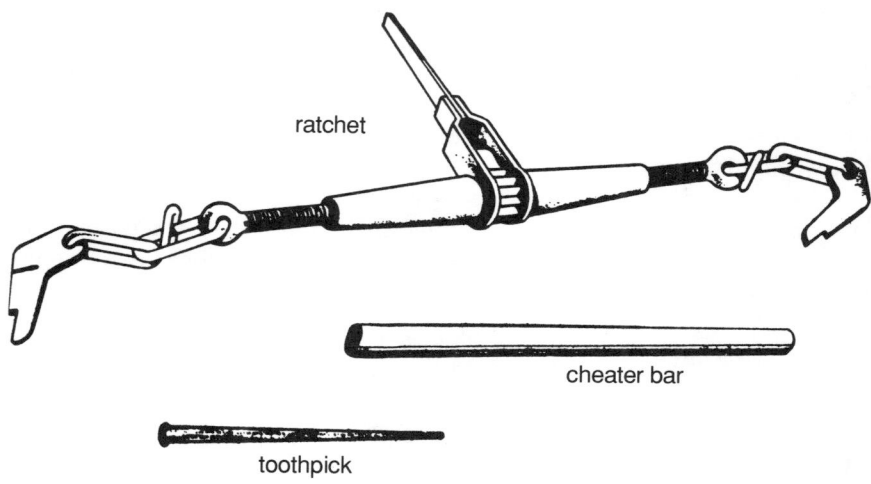

Top, the steamboat ratchet with pelican hooks on each end. *Middle,* the cheater bar, used to get added leverage on ratchet handle. *Bottom,* the toothpick, used to shove through thimble or link to keep it from turning when tightening (Courtesy A.W.O.)

wind wire. When releasing, do not use a bar on the handle or stand where the handle will strike your body as the tension is relieved.

The string of barges directly ahead of the towboat is known as the "push string." Any barges secured on either side are known as the "drag" barges. When a complete tier is made on either side, it becomes a "drag string." In order to keep all barges even with others, snake lines and check lines are used by deck personnel as they signal the pilot, or the pilot signals them. When all units in the tow are even, towlines are placed on the head ends of barges and backing lines on the rear ends of barges. The tiers are then held together by cross wires and jockey wires.

Frequently, in making up a river tow, the line-haul towboat or shifting boat will have to pick up one or more barges by easing alongside while made up to one or more barges. Occasionally, this requires the barge that has been picked up to be swung or topped around as its bow is facing the tow and is opposite to the direction of the others. To get this barge swung and topped around requires perfect coordination and timing between the towboat pilot and deck crew. Check and swing lines that are about to be used should be in the best of condition and crew members should wear work vests.

There are many other maneuvers and tricks of the river trade used in accomplishing them when picking up or dropping out barges from the tow.

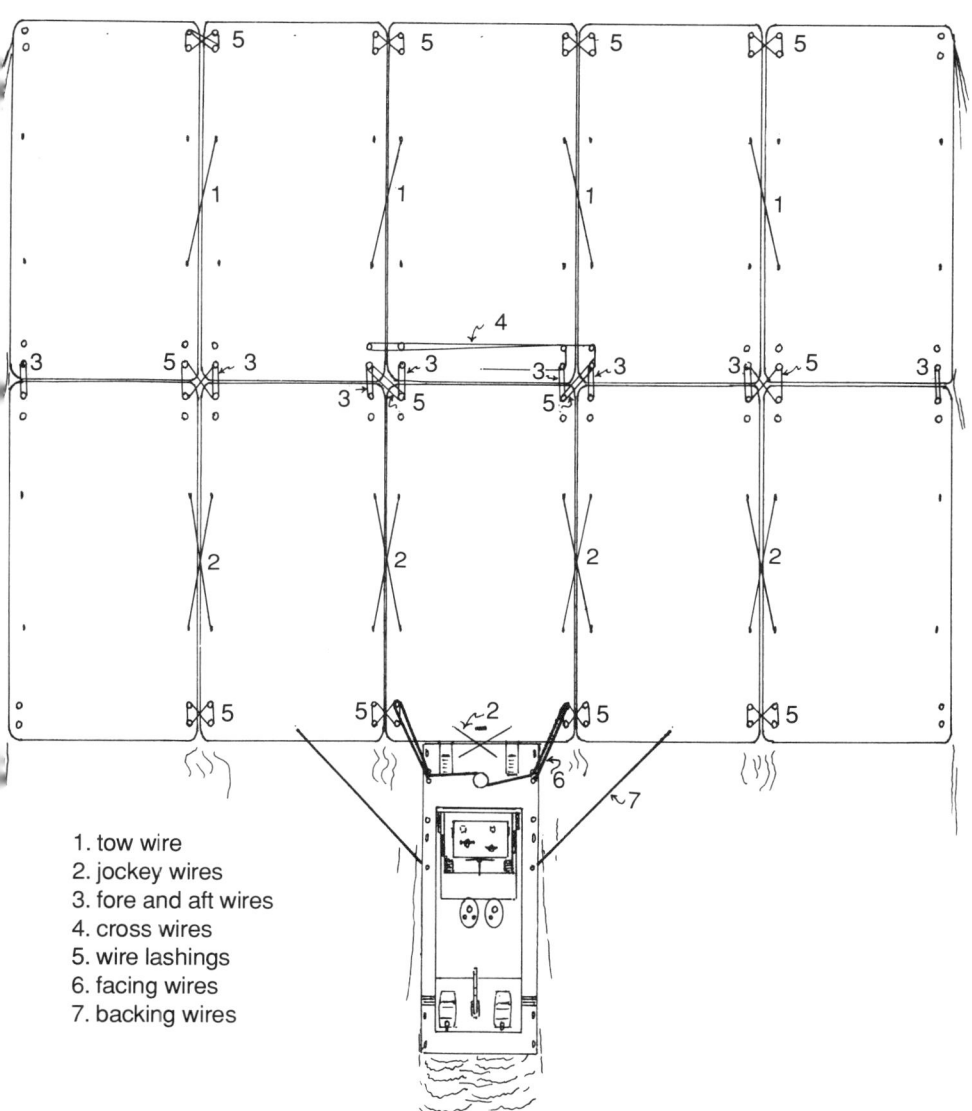

1. tow wire
2. jockey wires
3. fore and aft wires
4. cross wires
5. wire lashings
6. facing wires
7. backing wires

Makeup of a Mississippi River tow

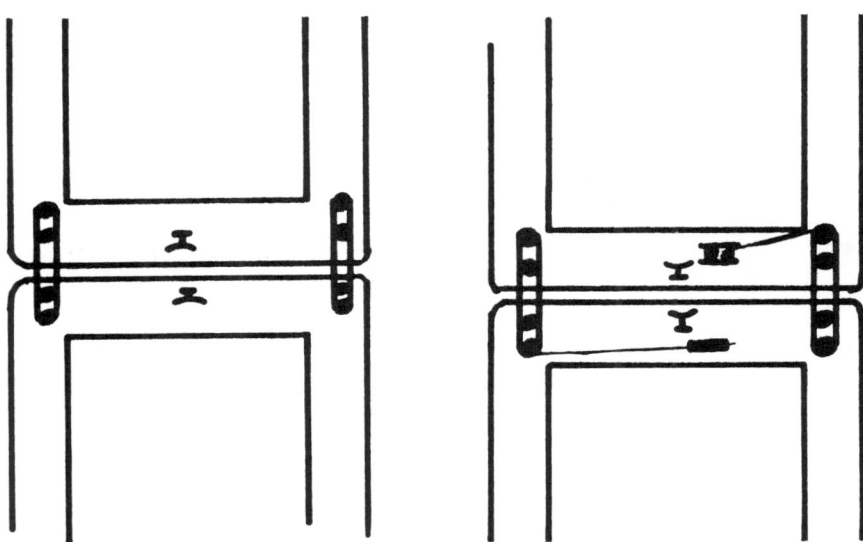

Left, portable rigging—wire and ratchet brought from the towboat. *Right*, stationary rigging—winch or ratchet fastened to the barge

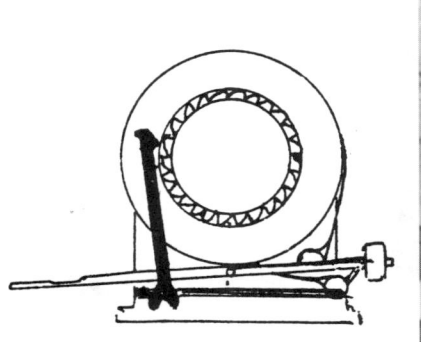

Hand winch as used on barges

Four barges lashed together

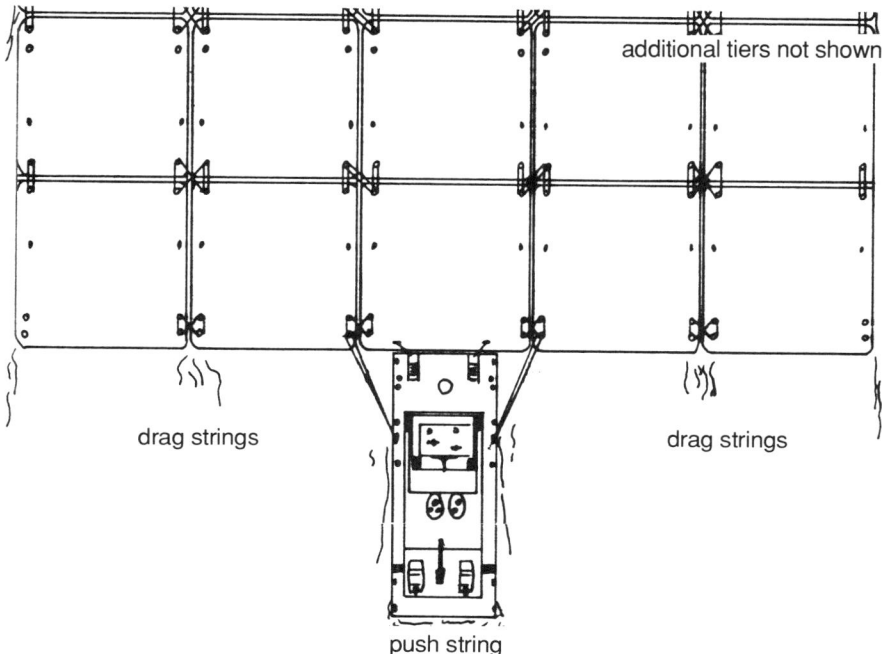

Tiers in river tows have names: the outboard ones when facing square are called drag strings; the tier ahead of the tug is the push string.

Once the tow is under way, that intimate knowledge of current, crossovers, bars, and landings stored in the head of the pilot comes into play. To aid him in steering there is the jack staff, a long pole with a pendant or windsock and usually painted white. It is set in a socket or stand at the center or the head of the tow. This allows the pilot to observe the speed of the swing of the tow and to use it as an object to line up with the channel or range light. For night use, a small mirror or reflector may be placed on the backside of the jack staff. There also may be a gyrocompass repeater, called a swing gauge, which gives the same information.

MAKING AND ENTERING A LOCK

Entrance into most locks, whether on the Mississippi-Ohio river systems, Tennessee-Tombigbee (Tenn-Tom) Canal, St. Lawrence Seaway and Great Lakes, Panama Canal, Columbia River, or waterways in the United Kingdom or on the Continent is controlled by traffic signal lights, a loud-hailer system, and voice radio on VHF channel.

With a tow, the approach or immediate entrance to the lock chamber must be done with great care and at a speed under which complete control and stoppage can be safely maintained. Remember when working around lock areas that the lock master is always the boss.

The tug or towboat captain or pilot must know well in advance if his tow as it is presently made up will fit into the lock chamber, or if it can be jackknifed to fit, or if it will have to be broken up to pass through in two sections.

There are several locks in U.S. waterways whose dimensions are smaller than others in their particular system. Bonneville Lock on the Columbia River is a good example. Here, a six-barge tow may have to be broken up. The tug will go through with three units and then return for the balance. Several hours may elapse before the tow is ready to proceed again.

Usually, there are tie-up walls or racks of pilings above and below locks. These are located so that vessels may be tied up clear of the lock approach and exit while waiting for other craft to clear. Traffic signal lights and loud-hailer speakers are usually arranged so that pilots can see and hear orders if not already in contact with the lock master by VHF.

DOWNBOUND TOWS—U.S. GREAT LAKES AND WESTERN RIVERS

It is inferred that Rule 9 (a) (ii) of the Inland Rules of 1980 applies to tows controlled by a "power-driven vessel." Thus, a tow proceeding with the current downbound shall have the right of way over upbound traffic and shall initiate passing signals "as appropriate."

A vessel proceeding upbound against the current should hold back "as necessary to permit a safe passage." Upbound tows about to exit a lock should hold back if a downbound tow is approaching until it is safely moored or its pilot states that it is safe to pass.

Many times, upbound Mississippi River tows consist of empty barges, the exception being those loaded with phosphate, petroleum, or other products. Downbound tows will have up to 30 loads of grain or coal. These seemingly unwieldy tows are floated around sharp bends by a sort of sideways approach, sliding into the turn at reduced speed while flanking and twisting and under full control. When the tow is about headed for the straightaway, full power is again applied. To note the amount of side drift, many pilots stream a plastic balloon-shape drift buoy on a lanyard off the outer corners of the last barge in the outer strings.

Upbound tows should lie below the bend until the downbound unit clears. This same downbound procedure is recognized for tows on the Detroit and St. Lawrence rivers and the Seaway to Montreal, and by prudence alone is by no means limited to these waterways.

PUSH TOWS ON OTHER INLAND WATERS

In a spin-off from the Mississippi River system, towboats with towing knees for pushing barges ahead have been developed for use on the inland waterways of the world. They have been strengthened for ice in the Beaufort Sea and on the great lakes of Canada's Northwest Territories. During the summer season, towboats both cross Great Bear Lake to Port Uranium City and descend the Mackenzie River. The barges are pushed unless it is too rough; then they are placed on the hawser. These areas are difficult to navigate due to quick changes in weather patterns and swift-running waters. Pilots must be able to read the current. Tows are usually one or two barges.

Tows on the New York State barge canal system from Waterford to Oswego or North Tonawanda are normally one barge with a notch pushed ahead of the tug. Bound north or west with a loaded oil barge is a much easier transit than south- or eastbound with the same barge in ballast.

Winds and currents never seem to be favorable when one is approaching lock walls. The combined efforts of a good deckhand with his hand signals and an experienced pilot are required to enter some locks. Sufficient power for good control is needed to keep the barge from sliding sideways. A touch on a wall that is hard enough to cause the barge to bounce off will throw its bow across into the opposite side and lock gate with resulting damage. Too slow an approach may allow the barge to land or brush the outer end of the approach wall with equally embarrassing effect. Barges are not to be bounced off anything. Their hulls are normally subject to denting if a blow concentrated in a small area is experienced.

As in playing pool or billiards, a smooth entrance with a fair current and wind of any kind requires the precise lining up and deft stroke of the pilot to get the tow to smoothly enter the pocket—the lock chamber. Once entered, good backing power may be required to stop. Such conditions and procedures as above would apply to similar situations at locks throughout the world.

DESCRIPTIONS OF TOWING PRACTICES AND MANEUVERS

Any descriptions of maneuvering with tows are from actual practice and may be accepted for use in other areas where conditions are similar. Likewise, nothing is implied or should be construed that these or any methods or maneuvers described herein are the only or absolutely correct methods. They are offered as methods that have been proven safe and efficient. Towboating itself has been continuously updated by the ingenuity of towboat men themselves.

Landing Barges in River Currents

In rivers where there is a steady outgoing flow with little or no tidal current, barges are usually landed with the bow heading into the current (head-upstream).

With a loaded or empty tow, it is relatively easy for the tug to check down its speed until approximately equal to the outflow. The tow will not be making any headway although the tug's propeller will be turning just enough to maintain control. By use of the tug's rudder, the tow when so held off a landing can very slowly be worked sideways to gently come alongside.

Whoever is on the barge guiding the pilot should make very clear hand signals. He should be sure that there are no obstructions between himself and the man in the wheelhouse. He should not turn his body in such a way that his arm cannot be clearly observed. He should stand facing or partially facing the bow and make all signals with his hands at shoulder level or higher.

Leaving River Terminals with, and Turning, Tows

At river terminals, the normal practice is to have ocean or coastwise barges landed head-upstream. Departure after they are loaded or discharged can be accomplished with either some thrills or ease.

Before completely letting go all lines a check on river traffic must be made both visually and by VHF. If a VTS is in force, contact, advice, and acknowledgment should be requested from that facility. If in an area where none exists, a security call should be made giving the tug's or towboat's name and what is being towed; the method—pushing, alongside, or on the hawser; and whether bound upstream or downstream.

If the barge is light or loaded and bound upstream, departure from the river wharf or terminal should be fairly easy whether the tug is behind or alongside. By letting go the bowlines and holding a stern line, the head of the tow will swing out into the river. When the bow is at least 30 degrees to 45 degrees off, the stern line can be taken in and the tow should proceed carefully so as not to have its stern strike a wharf piling or the corner of any barge or unit that may be moored ahead. To clear the latter, a short push at slow speed and then a stop will allow the river current to shove the barge sideways downstream so that good clearance is available.

The towboat's rudder(s) should be adjusted just enough to meet the current and to keep the tow from turning downstream. Most river wharves and terminals have very high decks to allow for rises of the river up to flood stage. It may be necessary, if the barge is low and if there are no line handlers available, to run a line around a stout piling both at the bow and at the stern of the tow so that all the regular lines may be taken in. These easily released slip lines will hold the tow.

If the tow is to head downstream, nature again will help the tug or towboat. After checking for traffic and reporting either to VTS or by a security call, the tug should, if possible, allow herself and the tow to drop with the current until the bow of the barge(s) clears the downstream end of the wharf or terminal. Most river terminal wharves are set out from the bank to where there is adequate depth for loaded barges to come alongside. This provides a good depth for the tug to hold herself in when downstream of the wharf while commencing to turn the tow.

The trick in turning the tow, whether it consists of several strings of river barges or a single 500-foot ocean barge, is for the tug, whether single- or twin-screw, to stop the downstream movement. Then, while holding the tow stopped in the current through adjustment of engine RPMs, the rudder should be placed hard over toward the river's mid-channel.

The tug is then held as the pivot point if possible. On a twin-screw vessel this is easier by adjustment of ahead and astern power. Ahead power should be just enough to keep the tow from moving upstream or out into the river. Astern power on the opposite engine will hold back the tug's advance. The tow's bow will be out in the current. It should be allowed to fall with the current until almost crossways to the river channel. At this point, more astern power should be applied. This will cause the tow to swing downstream rapidly without drifting sideways. A single-screw tug will have a bit more difficulty, as she must use slow speed ahead with hard over rudder and then half speed or stronger astern to get the tow swinging without getting in a dropping sideways position.

In no event should an attempt be made to leave a wharf and head downstream with a tow by coming full ahead with hard over rudder, unless you want a thrill. The tow will swing rapidly until about 45 degrees across the river. Then the ratio between advance (the turning circle) and transfer (the distance required to turn) will become very large. The tow will drop sideways in the current much faster than it turns downstream and will become difficult to manage and to get headed straight downriver. This is particularly true if in the notch of a large empty ocean barge. In addition to the current, a strong wind blowing upriver will cause a problem in turning a light barge. An assisting tug may be required.

Departure for Sea with a Tow

"For sea" refers to a tow bound out of a harbor, bay, or river into the ocean whether on a coastwise, interisland, or long ocean voyage. Both the owners and the tug captain must take a number of complex considerations into account before departing on such a voyage. Many companies routinely send their tugs with what is called an "outside tow" up and down the coast or ocean and through other seas of the world.

CREW TRAINING FOR UNMANNED BARGES

Before leaving port with an unmanned barge that has no assigned personnel riding on the tug, responsible tug crew members should have complete familiarity with and understand how to use all features of the barge. The tug master, mate, and deck personnel should understand how and where the barge's bilges can be sounded and how its ventilation, anchoring, mooring, and ballast systems operate. All watertight closures should be made. The towing bridles, shackles, pendant, retrieval lines, emergency tow cable, float, and pickup line should be inspected. Tug engineers should be familiar with the care and operation of all barge machinery. Complete operation of the tug's towing winch should be understood by all who will operate it. The more each man knows, the better the teamwork.

CHARTS AND SPARE EQUIPMENT

Every tug bound for sea should have up-to-date charts for the entire area she will traverse or pass even though she may not enter these areas. Emergencies do arise. Having no chart is a poor excuse for a grounding. Up-to-date *Notices to Mariners* should be kept, read, and applied to the affected charts.

SURVEYS OF OCEAN AND COASTWISE TOWS

Any unit that a tug is to tow "outside," which includes the waters of lakes, bays, sounds, along the coast, or into one of the many oceans and seas, should be carefully surveyed before departure. This is a primary and outstanding duty of the tug master or his reliable designated representative. It is suggested very strongly that this person be one of the tug's officers. Quite often, an outide surveyor of a salvage association may have accepted the unit to be towed as seaworthy.

Such statements or certificates of seaworthiness do not preclude the tug master's responsibility if it is found that any casualty en route can be traced to some fault, defect, or condition that was overlooked before departure.

All conditions regarding bilges, drafts, listing, on-board pumps, navigation and other lights, their source of power and how they are operated, should be made known to the tug master and at least one member of the tug's crew. This should be someone who could board the vessel if the occasion arose.

BARGE RETRIEVING LINES

Any seagoing barge or other unit that is being towed by its own tow bridles should have at least one stout retrieving wire or line running from its towing end to a point near where the bridles are joined at either a shackle or a flounder plate.

Frequently and particularly in the Pacific Northwest and on transocean tows, two retrieving lines are rigged. The second one runs from the tow pendant back to the bow of the towed unit. These two retrieving lines should be led in such a way that they are separated and will not become entangled. Wire ropes that will stand the chafing from towing and that have the strength to retrieve the weight of chain bridles must be used.

All units built or rigged to be towed in outside waters and that have been accepted by a classification society will also have an emergency towing cable rigged for quick release. The end of this emergency towing cable into which the tug will eventually connect may have a heavy messenger coiled on the end of the unit and a trailing pickup line with a buoy on its outer end. There also have been several other experiments such as a radio-controlled projectile being fired toward the tug that is attached to this line.

If the unit is manned, the tug master or his representative must check all of these items with whoever is in charge on the tow. All inspections of tows and the findings should be entered into the tug's log with the date and time.

WEATHER CONSIDERATIONS

Assuming the tug is fully equipped with shackles, pendants, spare hawsers, fuel, engine stores, and galley provisions, the next consideration should be the weather.

Tugs are, or should be, built to stand rugged weather. Rugged weather at sea is to be expected. Winds and seas up to gale force can usually be handled through prudent seamanship in reducing speed, heaving to, if necessary, and the choice of more sheltered and favorable routes. The deciding criterion whether to go or not to go should be the weather expected at the point of departure from sheltered to open waters. This is the point where most tows suffer extremely high possibilities of a casualty.

Marine Weather Reports

It should be remembered that weather reports as broadcast are forecasts based on the best available data gathered over a relatively large coastal area. They do not take into consideration tidal current effect on seas, swells breaking on bars, or the fact that wind velocity may increase or double over open waters. A good formula to follow on the U.S. East and Gulf coasts where weather fronts often come from off the land is, if the forecast is a Small Craft Advisory for winds of 15 to 25 MPH, add these velocities. The result will frequently be winds of 35 to 40 MPH at sea.*

Wave Heights

Wind velocities as given at various offshore buoys and light towers should be compared with what the barometer is doing. The maximum intensity of a fast-moving storm may last for only four hours. Large, slow-moving fronts with strong winds may last for several days.

The tug master then must consider if his towed unit is well secured. Can he safely get out of the river or over the bar? Each harbor or exit to the sea has its own peculiarities. It may be smooth at certain times and treacherous at others. How much headway can be made? Will it be better to wait a few hours until the weather changes? How much wear and tear will his towing gear receive? If the unit is manned, how much weather can they stand? Can they communicate with him via VHF?

These questions are commonly discussed between tug masters as they approach the point of proceeding to sea. A consensus is usually reached whether to sail or to remain at anchor or moored until the weather improves. Many a would-be hero has departed or tried to depart when the weather was deteriorating, only to face with chagrin a parted hawser, a barge swept clean of gear, leaking pollutants, or running aground.

WHERE TO CONNECT UP TO THE TOW BOUND FOR SEA—CHOICES

There are a variety of choices as to where and whether the tug should make up to the tow alongside, or in the notch, or on the hawser. Much depends on how the unit to be towed is lying or moored. Here are a few examples:

If the unit is an empty or loaded barge and is at anchor, the tug may lie in the barge's notch or alongside and assist in easing ahead as the anchor is heaved in. Once the anchor is secured and the barge is nearly motionless and drifting, the tug should run ahead and either pass her bridles to bargemen or pick up the barge pendant if the barge has her own bridles.

If the unit is a dead ship, an assist tug or two may be required to hold the ship after the anchor is up and while the lead tug is connecting her hawser. If the anchorage area is restricted, connection of the hawser

---

*See Beaufort scale of wind velocity and sea description in appendix V.

should be done when tide, current, and wind will allow the towing vessel to safely turn and clear the tow into the channel. If this is not safely possible, an assist tug should be used.

CONNECTING TO THE DRILL RIG

If the unit to be towed is a drill rig about to be "floated out," much planning as to time, tide, and traffic must be done with local authorities and any relevant VTS. As more than one tug would be used, any suggestion as to which one should connect up first, and where, would be premature. Most rigs are towed from certain legs. The pendants that tugs connect to will be pulled up by a bight and secured to the rig's platform walkway. Tugs must back under this in such a way that the wind or current will not turn them crossways while they connect up.

After her hawser is connected, each tug must slack enough cable to clear room for another tug or tugs so that they may also back in and connect. To maintain position while waiting, single-screw tugs may idle ahead but not strong enough to place a strain on the hawser or move the rig. Twin-screw tugs may maneuver to walk sideways; Z-peller and tractor type tugs can most easily maintain immobility while holding a slight strain on the tow cable.

LEAVING PORT FOR SEA WITH VARIOUS UNITS

If the unit to be towed is a loaded notched petroleum barge moored on the face of a wharf, there should be no difficulty in getting in the notch and clearing the berth using a back springline. If the same barge is moored in a slip with her bow in, the tug can make up in the notch and back out slowly if all is clear. A securité call should be made advising traffic control and others of the movement. A careful estimate of the tidal current effects should be made. If there is the possibility of the current causing the barge to turn and strike the other side of the slip or another vessel, help from an assist tug should be requested.

Even though the touch may be slight, the concentrated weight of a loaded oil barge is sufficient to make a slight crack in a pier timber or piling. At today's prices, the repair or replacement of broken wharf timbers or pilings may exceed the tug owner's profit for the voyage about to be undertaken.

If the above type loaded barge is lying in a slip bow-out, the tug may either get in the notch or, if there is not room enough, make up alongside and push the barge out of the slip using similar precautions.

HAWSER LENGTH ON INLAND WATERS

In most U.S. East and Gulf coast ports, it is neither safe nor prudent to tow loaded barges on the hawser through a harbor's narrow and trafficked channels, the reason being lack of control. Loaded barges towed on the hawser in shallow or narrow channels tend to sheer.

For years, the length of hawsers between vessels on U.S. inland waters was limited to 450 feet (137 meters approximately) or "as much shorter as weather or sea will permit" (33 CFR 84) (84-10). While this regulation does not appear in the Navigation Rules (International, Inland, CG-169) of 23 August 1982, it has been implied that it has been absorbed within other of the above rules.* On the U.S. West Coast, in the Ports of San Francisco and Tributary Bays and in the many rivers, loaded barges are towed on the hawser. Frequently, a helper tug is made up alongside to assist in steering.

On the inland waters of the U.S. East and Gulf coasts, most loaded barges except trailer and container barges are towed alongside or by pushing to an area inside a river mouth, bay, or harbor entrance where there is sufficient room to safely get on the hawser.

Between the point of departure in a harbor and the point where it is necessary to get on the hawser, tugs should avail themselves of any counter or favorable current. Low-powered tugs should await slack water or a fair current. If a tow is being pushed and a sharp bend is encountered, it is wise to reduce speed so that control may be maintained until the tow is steady on a new course. This is particularly true in narrow channels. If it is necessary to engage a pilot, the tug master must reveal to him any and all mechanical, electronic, or other deficiencies the vessel may have.

When the tug master puts his hawser on a unit and heads toward the sea in a river, down a long, buoyed channel, or out of a bay, he should remember that he, not the barge, is in charge. Unless he is towing an unwieldy unit that requires special handling, assistance, and the use of the whole channel, his tug has no special rights or privileges over normal traffic. If, on the other hand, the tow should be difficult, he should be routed via any alternate channel after getting clearance from port authorities and should broadcast advice of his movements to warn other vessels. Occasionally, such movements are announced ahead of time in local *Notices to Mariners*.

AREAS COMMONLY USED TO GET ON THE HAWSER

While there is a great increase in the push towing of deep-notched barges, both coastwise and at sea, many other tows require the use of the tug's hawser. Some of these frequently used areas are: in Maine off Owls Head in Penobscot Bay, and between Spring Point and Portland Head in

---

*See Commander, USCG, Chief Marine Information and Rules Branch ruling, July 22, 1986, Chapter 13-7-a.

Portland Harbor; in New Hampshire between New Castle and Kittery Point at Portsmouth; in Massachusetts at President Roads in Boston Harbor, at Cape Cod Bay off the Cape Cod Canal fairway buoy, and at the other end of the canal in the anchorage above the Cleveland Ledge Light in Buzzards Bay. From Providence or Fall River in Narragansett Bay, tugs get on the hawser above Gould Island or off Castle Hill. Entering or leaving New York using the East River, barges are picked up or put on the hawser off Whitestone or Execution Rocks according to the weather. Leaving New York Harbor or entering below The Narrows, Craven Shoal lighted bell buoy off Hoffman Island is the area. If bound in or out of Raritan Bay, an area southeast of Light #18 is used.

In Delaware Bay outbound tows may get on the hawser in the anchorage off Artificial Island (Salem Nuclear Power Plant), or in good weather between Fourteen Foot Bank Light and Brandywine Shoal or below according to traffic. From Baltimore below Seven Foot Knoll Light is the first comfortable area; and inside Old Point Comfort or Thimble Shoal Light in Hampton Roads, tows frequently leave Yorktown with loaded oil barges on the hawser. Great care must be exercised coming out of the York River entrance channel as the crosscurrent tends to set the tow onto the buoys.

Leaving Morehead City, the Cape Fear River, Charleston, Savannah, and Jacksonville, tows not already on the hawser should try to get in this mode at nearly slack water just inside the jetties where the widest area of deep water is available. Tug masters who regularly run to these ports are able to go from the notch to the hawser or vice versa at almost any stage of the tide or current.

At Port Everglades, Florida, a tug sailing with an empty or partially discharged oil or cement barge frequently starts out of the slip made up head-and-tail, with the hawser already connected to the barge pendant and bridle. She is assisted by a local harbor tug. The port and the state require tows to take a local pilot even though he serves about 15 minutes or less on a nearly straight channel. Off the Port Everglades pierheads, the sea tug breaks away as the assisting tug pushes the barge's stern. The tug captain familiar with this port should give all orders to the assist tug as his experience tells him, with concurrence by the pilot. Otherwise, some confusion may result if the pilot gives orders to the assist tug without the knowledge of the lead tug's master.

At Miami, barges are normally taken away from the pier on the hawser, with a harbor tug assisting until out of the jetties.

Tows bound for sea out of Tampa frequently undock with the tug head-and-tail and break away in the inner harbor off Southport Terminal or below Black Point Channel. Some may remain in the pushing mode out into the Gulf or go on the hawser inside Egmont Key.

From St. Andrews Bay, Pensacola Bay, and Mobile Bay, there is plenty of room inside to connect or disconnect hawser. At Horn Island Pass, off Pascagoula, Mississippi, it is necessary to push out clear of the islands.

In the Mississippi River, loaded barges are not towed out of South Pass. In good weather, they are pushed and let go outside the sea buoy if the tug must get on the hawser. Loaded tows outbound in Southwest Pass change to the hawser below Burrwood, about 1 mile inside the jetties, if they cannot push out of the pass.

In Texas, loaded tows from Lake Charles get on the hawser just inside the Calcasieu River jetties, and from Port Arthur, Orange, and Beaumont, at the Sabine Anchorage several miles inside the Sabine jetty.

Bolivar Roads anchorage is the area used by tows in or outbound from Galveston and Houston. From Corpus Christi and Brownsville south, ports around the Mexican perimeter of the Gulf and Caribbean require departures from or off the dock with the tow on the hawser.

Most Central American and Caribbean harbors are small and have some swell rolling into them. Their areas are so limited that it is prudent to connect up with the tow at the wharf and maneuver carefully out toward the entrance. Tows leaving or entering at either end of the Panama Canal change modes in the anchorage at Cristóbal or Balboa.

There is little or no pushing of tows on the West Coast of the United States. Because of the depth of the harbors, and rough river bars, tows are rigged with extra-heavy chain bridles and surge pendants. They normally pass out to sea on the hawser.

San Diego, Wilmington, and San Pedro (Los Angeles) harbors open up to plenty of sea room for switching to or from the hawser. In San Francisco Bay, tows not already astern are put there off the east side of Angel Island or in the anchorage area below the bay bridge.

Exit with tows from the various harbors and rivers from Eureka, California, to Cape Flattery, Washington, is normally done on a good length of hawser, as the bars are treacherous much of the time. Tugs usually connect their tow cables before leaving the dock, mooring, or anchorage.

HAWSER TERRITORY

The ports and waterways of Puget Sound, the Strait of Georgia, the Inside Passage to southeast Alaska, and, of course, the Gulf of Alaska to the Aleutians en route to western and northern Alaska are considered hawser territory from dock to dock or anchorage to anchorage. This also applies to Canada's ports of Victoria, Vancouver, and Prince Rupert, British Columbia; also Saint John, New Brunswick, and Halifax, Nova Scotia. Also included are the St. Lawrence River below La Malbaie

(Murray Bay), Saint John's, and other Newfoundland ports, and the Northwest Territories including Hudson Bay.

CONNECTING THE HAWSER IN PORT

If connecting up to the tow in port is done at a berth and there is room for the tug to lie alongside, ahead of the unit's bow, the method is fairly simple. With the officer piloting, usually at the after controls, the tug is backed or moored close under the unit's towing end (normally the bow).

Power should, of course, be on the towing winch. A length of the tow cable should be laid out on deck ready to receive the pendants and bridles of the unit to be towed. The mate or deck crew on watch should have the tug's connecting shackle (50- to 75-ton) passed through the thimble or socket of the tug's tow cable. The nut or nuts and shackle pin should be close at hand in a safe place.

Depending on the type of bridles the unit has, the tug personnel will throw a heaving line to the men on the unit. If the barge has a set of heavy chain bridles and a surge pendant, the tug may have to back close enough so that a portion of the chain pendant can be reached and secured by a messenger. Careful heaving from the tug and slacking from the barge will pass the connecting link onto the tug's deck. In such rigging, spring lay or "swede" wire is used to slack the chains down from the barge.

If the pendant to the bridles is wire rope, a stout messenger of suitable length attached to the heaving line will be thrown to the barge. This will be attached by a shackle to the barge pendant so as to leave the pendant's thimble free and clear.

Many companies have their barges rigged so that there is a thimbled retrieving line that runs on the pendant. When the messenger is fast on the pendant, its other end is led to the tug's capstan and heaved aboard until it is conveniently close to the open shackle in the end of the tug's tow cable.

Getting the tow shackle in line with the pendant thimble is sometimes a struggle and requires the strength of more than one person, par-

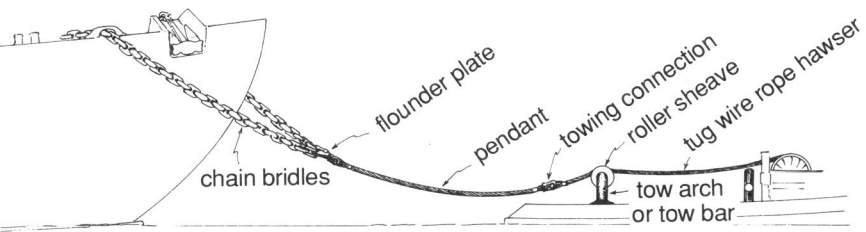

Connecting a sea hawser in port

ticularly in inserting the pin. The pin may require a tap with a maul to get it all the way through. Great care must be exercised so as not to damage the shackle pin threads. When the nut (or nuts) have been screwed on and tightened, a cotter key, pin, or pieces of heavy welding wire must be inserted through the hole on the end of the pin and bent over. Normally, nuts are given a last-minute tug with a wrench.

For reasons mainly theoretical, it has been found that on wire rope hawsers, it is best to insert the shackle pin from left to right so that the right-hand-turning nut will tend to tighten when dragged through the water. With the pin in the opposite direction, nuts have been found to be missing.

A slight strain may be taken on the tow wire when connected. The pendant, now having a little tension, will cause the messenger to slack. It can then be cast off and unshackled. When clear, the tug may start ahead slowly and pay out hawser.

These hawser towing connection methods are commonly used in port before leaving the wharf or pier with empty oil barges, car barges, and container and RO/RO barges, unless the slip, basin, or channel is restricted or well trafficked. Then, the choice would be to ask for a helper tug to steer the tow. These connections are also used in making up head-and-tail.

CONNECTING HAWSER WHEN HEAD-AND-TAIL

When this method is chosen, the tug is first secured on the outboard bow of the barge opposite to the way that the barge is heading. Thus, the tug's stern and the barge's head or bow are side by side, or head-and-tail. To best attain this position, a bight of the tug's bowline is passed around either a recessed cleat in the side of the barge or a cleat on its deck. The cleat must be located far enough ahead so that when the tug has connected her tow cable, she may move ahead and still use this as a bowline. In the latter position, the tow cable is heaved tight and becomes a combination tow- and stern line. The tow winch should hold this stern line by its brake only. This method has been used very successfully with empty oil barges of up to 250,000 barrels capacity (520 feet, or 158.5 meters).

GETTING AWAY FROM HEAD-AND-TAIL MAKEUP

After having sailed as one would with a barge on the hip (alongside), it is quite simple to get away from this head-and-tail makeup and to be on the hawser. A spot large enough to make this maneuver should have been chosen and the tug should approach at reduced speed, nearly dead slow. Make sure the tow is making very little way through the water. As in all maneuvers, "easy (usually) does it."

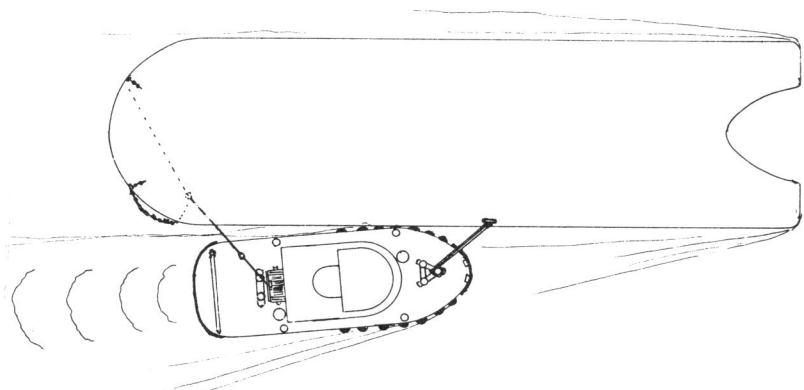

Tugs handling petroleum barges frequently make up to them in what is known as head-and-tail. Here the tug's towline acts as a stern line with the barge pendant. The offshore leg of the bridle would be under the turn of the bow. The tug's bowline has been hove tight. While the barge is being towed stern-first, her skegs aid greatly in steering and maintaining course. Proper lights must be shown as if stern were bow.

The reason for allowing the barge's way to be almost killed is so that it will not pass the tug as soon as the tug's bowline is released. Remember, an empty barge may draw only 4 or 5 feet, while the tug may draw 16 or 20 feet. As the ratio of resistance due to the barge's deadweight and windage vs. that of the tug is much in favor of the barge, the barge will tend to hold headway. All of this is stressed because suddenly putting the tug's throttle on stop from one-half or full ahead and letting go the barge's bowline would cause the barge to continue on its way and jerk the tug's tow cable tight. It might catch under the deck edge and temporarily place the tug in irons or a girding condition. Personnel could easily be injured.

By correctly running at dead slow speed for some minutes, always observing other traffic, and choosing a good area, a pilot can perform the maneuver very smoothly. A deckhand should be standing by forward to release the tug's bowline. (A tug that is on a regular run with the same barge usually has a soft poly line permanently secured to the barge. This is used as the tug's bowline in these maneuvers.)

With the tug almost stopped and in a position such that neither wind nor current will affect its turning, the tug's rudder is put hard over toward the barge and a slight touch ahead is given. As the barge swings, the tug is stopped and the bowline cast off. The tug's rudder is quickly shifted back to amidships or a few degrees away from the barge. As the bowline is cast off, another person aft slacks the brake of the tow winch, allowing the cable to run free as the tug moves ahead. Understanding, cooperation, and coordination between pilot and deck crew are imperative.

Everything must pay out smoothly on deck and the pilot must keep an eye on the barge's bow lest a corner of overhang swing in to strike the tug. As the tug angles away from the barge on a steady course, the pilot must watch the tow cable to make sure it does not run out too fast or get under the tug's counter.

Using the winch brake, a good deckhand will tenderly feed out cable. When the space between the tug's stern and the barge opens up, the tug may steady up on course. Speed should not be increased until the barge has fallen astern with the tug's tow cable leading directly aft, and both barge bridles have an even strain. Otherwise the barge will swing rapidly as one bridle leg suddenly tightens. The result will be a sudden jerk to the tug, rolling her enough to throw a person off balance and upset things in the galley.

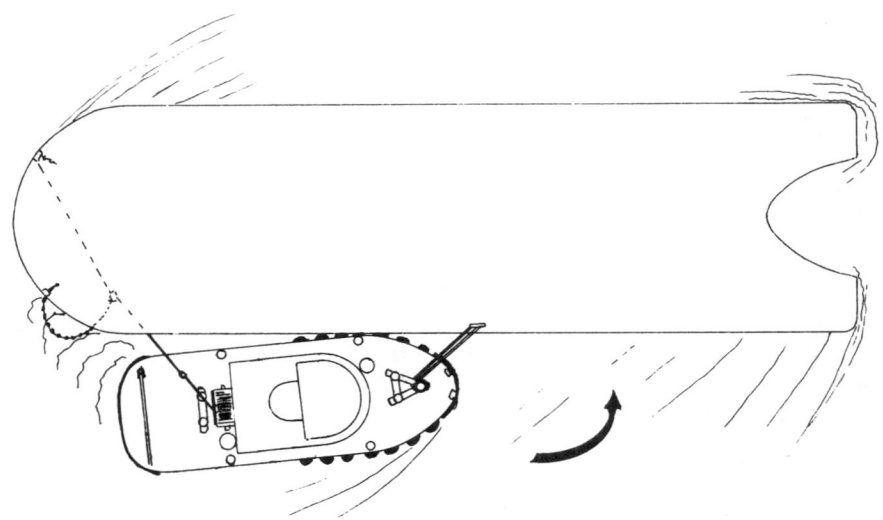

Head-and-tail swinging to let go

Opposite, above, the tug is still pushing gently against the swinging barge while the bowline is released and the tow cable slacked. *Below,* the swinging barge is on its own as the tug maneuvers slowly to keep the hawser directly over her stern, not allowing the barge to jerk and roll the tug.

UNDER WAY WITH THE TOW AND AT SEA 287

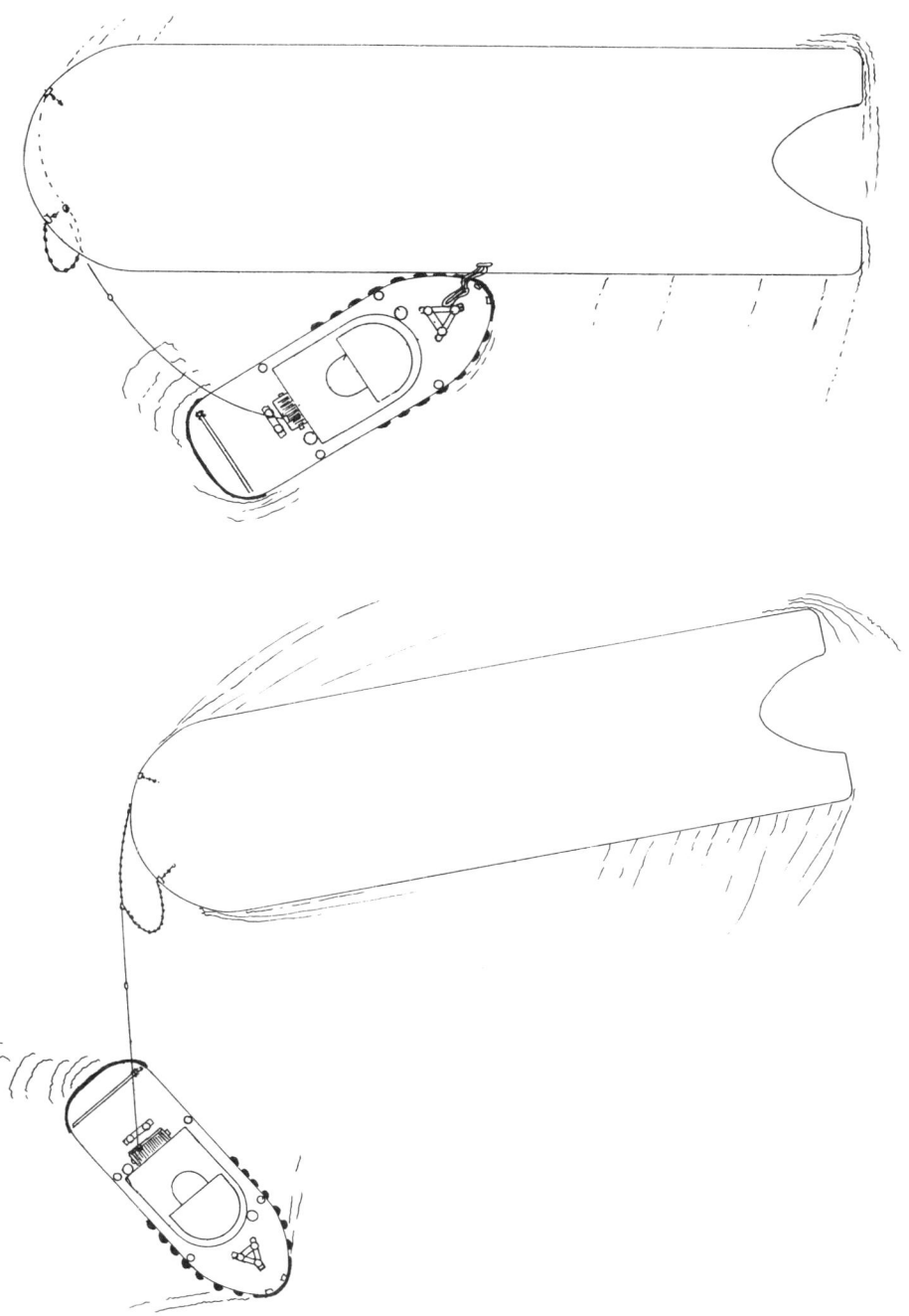

CONNECTING HAWSER AT TRANSITION POINTS

When tugs that are pushing or towing alongside arrive at the usual point for changing to the hawser, certain preparations must have been made. Any extra personnel required should have been called out. Needed tools and gear should be laid out. The tow winch and deck machinery should be energized and any protective covers to the after controls should be removed. If running at full power, the engine should be slowed to dead slow for about five minutes prior to arrival. When the barge has lost some headway, the propellers should be reversed, using such power as necessary to get the barge to come to a complete stop. Try and keep it on a steady heading. Some lines should be let go while backing. It is wise at this time to give a securité call so that VTS and other traffic that may be several miles away will know what you are doing.

When all lines and push cables are in, the approach to connect up will depend on how the barge is loaded and rigged. If the barge is loaded and manned, the tug usually proceeds slowly up the weather or windward side—slowly—because if the tug captain rushes and then has to back down heavily, he may end up with the tug out of shape in relation to the barge. The idea is to arrive near the barge's bow and stop the tug so that its stern quarter is about 20 feet off. From there a heaving line with a messenger is thrown to the bargemen. They will secure the messenger to the barge pendant and connection will be made as described earlier.

If the barge has no bridles, the tug must provide them. The tug should stop and back so its stern is close ahead of the barge in a position so she can drift downwind to pass up the bridle leg, which goes on the side she is closest to. Each bridle leg should have a 3- to 4-inch-circumference messenger spliced to it behind the eye.

It normally takes two bargemen to pull bridles aboard. The tug should lie still if possible, just maneuvering slightly to swing its stern to the opposite side to aid the passing of the second bridle. When the bridles are fast, the tug may then start ahead slowly and pay out hawser.

Opposite, above, a small manned barge. The tug *Mariner*'s crew has already passed the starboard bridle to the tankerman. He is presently walking the port bridle from the tug to the port chock and bitts. The tug will swing her stern in that direction to assist him and to be directly ahead of the tow. *Below,* starting ahead, the tug's stern is being swung to port for a minute so that her tow cable will jump into the towing arch roller. The roller is now on the port side with the cable resting against it. As soon as the cable is in the roller, the mate at the controls will swing the tug so that the roller is amidships and then an A.B. will slack out cable until directed to stop or until the mate tightens the winch brake by using the hand wheel to his right.

On connecting up to unmanned barges at these transition areas an entirely different approach is used. These barges, whether loaded or empty, are rigged with two pickup lines. On the flat side near the bow, about 30 feet back of the rake or sheer, a good line is fastened to a cleat and hung down if the barge is empty so that the tug crew can reach it and secure it. If the barge is loaded, this pickup line is left coiled and lying near the deck edge, where it can be pulled aboard the tug. It is picked up first and used as a tow spring so that the tug can work ahead or lay on it, allowing the stern to swing in and the second pickup line to be retrieved.

The second pickup line goes to the barge's shock line or tow pendant and bridles. It should be rigged so that just enough pendant or shock line can be heaved aboard the tug to connect. The barge pendant and/or shock line should be stopped off at various places on the barge so that it will not suddenly drop between the tug and the barge while connection is being made (see "Preparing and Rigging the Ocean Tow," chapter 9).

The shock line's pickup is trailing in the water. The crew of the *Sarah Hays* will grab it with a boathook. At the same time, another crewman will place the eye shown dangling from the barge over the tug's forward starboard side bitts. (Courtesy Watkins)

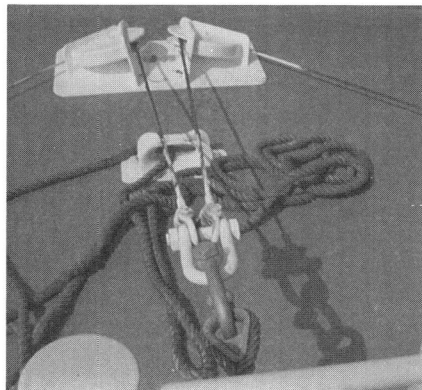 

The tug *Sarah Hays*'s push cables would have been slacked and disconnected at the galvanized shackle. The tug's hawser with the shackle still in it would be slacked and run over close to the starboard bulwark rail.

Here on the tug *Libby Black,* whose bow is higher, Mate Howard Newton is standing by the tug's holding line while the crew connects the shock line at the tug's stern. The shock line can be seen stopped off along the barge's edge.

 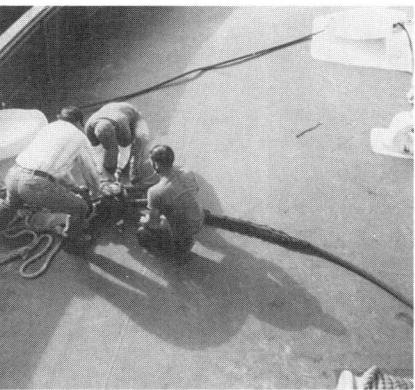

The crew pulls the thimble of the chock line aboard.

Connecting the hawser

Tug moves ahead slowly as the shock line breaks free. The pendant and bridles will also follow suit in a moment.

Once connected, stop the tug and cast off the forward pickup line. Then steam ahead slowly as the shock line and/or pendant break away and pull the bridle clear.

BARGE PICKUP LINE

A floating polyurethane or similar pickup line may be streamed from one corner of the barge's stern. This line of 8 inches in circumference and about 200 feet in length, with a small buoy on its end, is attached to the barge's insurance wire, which is lying stopped off along one edge of the barge's deck. This insurance wire runs to the bow and is connected to its own chain pendant. If the barge breaks adrift from the tug it may be retrieved by the tug picking up the buoyed end and taking it to the tug's capstan. When the thimbled end of the barge's insurance wire is heaved aboard, a new connection can be made allowing the tug to proceed carefully to sheltered waters or a port. The reason for that length of floating line is that a shorter one will become involved with the barge's wake and get sucked under the rake and entangled with the skeg. If the barge is not manned, this retrieval line can be streamed by the tug crew before they let go.

In actual practice, it has been reported that as the parting of towing hawsers is not a frequent occurrence, the 8-inch poly pickup line deteriorates rather quickly. Also, the messenger and float often become entangled with the barge skegs when towing has stopped. In some cases, the poly line on the barge stern had been stowed incorrectly and the line would not run out freely. It became fouled on a cleat, around a bitt or deck vent. Because of this, it was necessary for the tug to come close to the tow and grapple for the insurance wire or land a man to pass a line back to the tug (refer to "Retrieving and Reconnecting the Tow," in this chapter).

Having found that this trailing retrieval line is neither satisfactory nor a foolproof method in reconnecting to one of their own large seagoing barges, Crowley Maritime has devised a new approach. This is based on the time-honored Lyle gun method in which a projectile and small line are fired. The bazooka type launcher has a 3-foot screw-on barrel. It is secured on the barge's bow at an angle so that the projectile will traverse an arc of 300 feet. The tug must position itself stern-to and just ahead of the barge. The charge is fired electronically from the tug with a remote control garage door opener.

Attached to the projectile is a very small line about ¼ inch in diameter which has an exceptionally high tensile strength of 16 to 18 tons. This small, extremely strong line, when run out of its container, will be attached to a larger one of similar characteristics. When placed on a deck capstan or gypsy, these lines will have the strength to pull the barge's insurance wire to the tug for reconnection.

TAKING CHARGE OF THE BARGE

The tug captain should have confidence in his ability to take charge of his "friend astern" and know what to do to make it behave. Assuming the barge is following well, and has a favorable drag, the amount of hawser to put out and the power to apply depends on what type of transition area he is leaving. If leaving an open bay with a good exit channel, the hawser length should be limited to the amount required to keep the tow inside the channel and to successfully meet any incoming swells.

This is easy to regulate. At moderate or half speed, the bight of the hawser between the tug and towed unit should remain underwater. If on meeting swells it begins to break the surface, slow down, and pay out some more hawser, if possible. A good tugman can feel the strain of any surges on the hawser. When the tow is clear of the channel entrance or sea buoy, it may be slowed to pay out the balance of the hawser necessary for this portion of the voyage.

If the tug and tow are exiting a narrow river mouth where the tug must pull at an angle and where the length of the hawser must be limited to the shock line and/or pendant and a short length of hawser, the chafing gear should not be installed until the tow is outside and clear.

When to Rig Chafing Gear

This should be rigged as soon as the hawser length to be used is secured at the tow winch. The tug should be operating at a speed that will cause the tow wire to lead out over the center of the stern (roller) or to ride in the tow arch sheave.

The tug captain or pilot on watch should keep the tug from swinging and concentrate on the safety of his men working to install the chafing gear and any signals they may give.

KEEPING A GOOD LOOKOUT

During all of this transition from alongside or in the notch to the hawser, the captain or pilot must keep a good lookout for traffic. If he is working at the tug's after controls, it is a good practice to carry a hand-held VHF set and monitor the local bridge-to-bridge channel. Particularly at night, a pilot on an incoming or outbound ship may call to inquire as to what is going on and to request meeting or passing arrangements.

APPLICATION OF POWER

When the tug starts to pull ahead on the barge with a full hawser out, power should be minimal. The feel and the cavitation of the propeller(s) and the appearance of the hawser will indicate what speed to use. Don't start off pulling like a dog jerking on a leash. You will be wasting fuel and power. Increase speed slowly, keeping the hawser in the water. If you try to use full speed at once, the resistance of the nearly motionless barge to the yanking tug places undue strain on the tug's towing gear. When acceleration is properly done, the barge will quickly assume the same speed as the tug until a full throttle can be used and both are making the best way attainable from the tug's bollard pull. Getting a heavily loaded barge up to speed usually takes about 20 minutes.

BRINGING THE TOW OUT OF SWIFT RIVERS

Coming out of swiftly flowing rivers such as the Mississippi or over bars at jetties with a tow on the hawser can be dangerous. Tugs frequently use just enough hawser and power to keep control. By using the river's current and slight touches of power, the barge can be floated out. If too much power is used and the barge sheers, nothing is left to pour on to straighten it. The result can be catastrophic, with the barge trying to pass and gird the tug. Once outside in the clear, the length of sea hawser required can be paid out and the chafing gear installed.

WEST COAST SURGE CHAINS

On the West Coast of North America, most barges rigged for sea, whether coastwise to Alaska, or through Puget Sound and the Inside Passage, have heavy chain bridles and chain pendants. When outbound over bars such as ports and rivers of the northern California, Oregon, and Washington coasts, tugs will put out more hawser and pull harder than those

The thimbled jackstays restrict side travel of the tow hawser. If the rag moves, hawser is slipping. Brake must be tightened. The cable is held in the sliding sheave by a short chain with a screw pin shackle on each end. (Courtesy S. Lang)

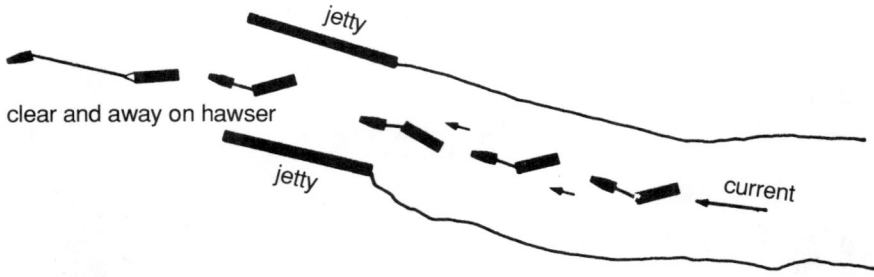

Floating a loaded barge out of a river on a short hawser

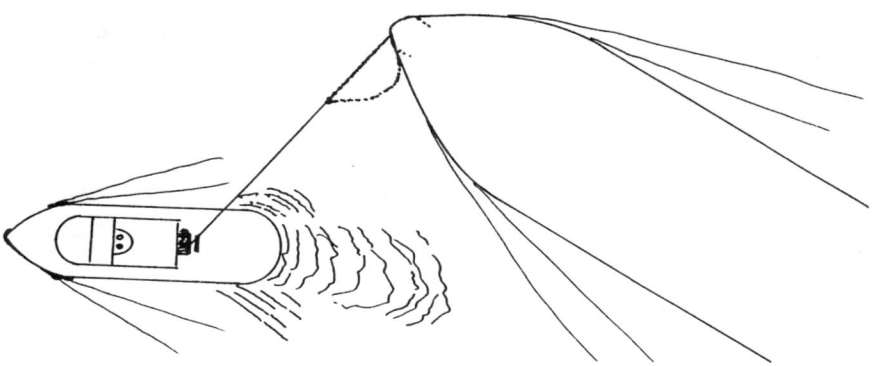

Heavy tow sheering

tugs exiting Gulf of Mexico ports, where nylon shock lines are used on barge pendants. The reason for these measures is the ever-present moderate-to-heavy Pacific Ocean swells, which break on the bars, rather than heavy swells; in the Gulf entrances it is strong tidal flows and crosscurrents which require shorter hawser lengths and easier pulling to maintain good control. Surge chains have also been adapted to some East Coast tows.

Surge chains are used on the huge trailer barges that exit the St. Johns River in Florida and work well on tows leaving other ports such as Charleston and Savannah. However, unless the tow can safely follow from its undocking to its exit to the ocean, the use of surge chains on any bridle may cause problems in narrow channels such as the Cape Fear River and other neighboring ports because good towing speed must be maintained or the surge chains will drag on the bottom.

Tug *Ellena Hicks* has cleared Mississippi River with the barge *Thelma Collins* and is under way across the Gulf.

Surge chains, as commonly used on the U.S. West Coast, are of various lengths. They are kept on the after deck of many tugs and are also referred to as the pigtail, when they connect to the tug's towing hawser. Due to their weight and the catenary they offer, their use at sea is fine. In restricted waters due to their length and weight it is difficult to heave them short. This leaves too much distance between the tug's stern and the towed unit. It is very dangerous without a good assist tug handling such a tow through narrow channels and drawbridges. When the tug's speed is reduced these heavy surge chains will hang down and drag on the bottom.

The Crowley Maritime super TMT RO/RO barges that are 730 feet LOA and carry 510 units have two separate single $3^{1}/_{8}$-inch surge chains. They run from two vertically separated points on the barge. One is connected to a heavy fitting in the barge bullnose. The other is from a heavy pad eye in the barge's stem about halfway between the loaded waterline and the deck.

These surge chains are connected to two independent tow wires running from the tug's double-drum towing machine. The purpose of using two individual tow hawsers tug-to-barge is safety. As Crowley's 9,600-

Trailer barge outbound in St. John River. Steering tug's pilothouse just visible over barge stern. Note single chain legs from barge bow, which run to individual drums on the towing tug. (Courtesy Crowley)

HP Caribbean tugs tow these barges at speeds of up to 12 knots, the possibility of a hawser parting exists. If that occurred, the tug would still have the barge under tow.

One of these hawsers is always kept with a little more tension than is placed on the other. Until recently, the actual load factor placed on this hawser with the greater tension was an unanswered question. To find such an answer, a load-cell was devised.* This load-cell is placed in the first outboard link from the barge after that link's stud has been removed and the link smoothed and faired. As any open link and some stud links will stretch or squeeze together slightly when under strain, this microcaliber load-cell will electronically register the chain's slight-

*Load-cells are the combined product of Coordinated Equipment Company of Wilmington, California, and Specialized Testing Services of North Hollywood, California.

Seaspan 251's bridles reveal a common arrangement used in British Columbia. Notice retrieving line running to bridle and its pigtail. (Courtesy Seaspan)

est distortion. This is transmitted to the tug and appears as a readout on her bridge. By studying the results, the tug captain can learn at what power he can get the greatest speed without overloading his towing chains. In this way, he can also find out the best ratios among power, speed, and fuel consumption.

Load-cells are not a permanent part of towing gear. Once the desired results are in, they are removed and made available for use elsewhere. Foss Maritime has found that a 140,000-bbl. barge loaded at 16,000 tons deadweight can be towed at 8 knots by a 9,000-HP tug. (No dimensions or draft of the barge were given).

TANDEM TOWING

From San Diego to Alaska, the tandem towing of medium- and large-sized barges is a common practice. Most of the tugs built for use in this area are like many in other parts of the world, equipped with a double-drum winch. They can tow more than one unit at a time as long as the total resistance does not exceed their power output.

Connecting up to such a tandem tow requires the assistance of a second tug. The first barge is brought out to or by the lead tug. It may be placed on a mooring buoy until the second unit is about to arrive, or the lead tug may hold it alongside. As the second unit is seen to be approaching, connection to the cable from one of the lead tug's tow winch drums is made and the tug very slowly moves ahead, paying out some of this cable. With the opposite side of her tow winch and stern being clear, the second unit is placed so that this cable may also be connected. This is done with either the tug's propeller(s) stopped or turning at medium revolutions.

The cable to the first barge should now be placed between a pair of hydraulic norman pins at the stern roller and slacked so that the barge it holds is three or four times its length away from the pulling tug. The second barge, now on its own cable, will fall in line between the first barge and the tug.

When the tug is under way, these tow cables are separated at the tug's stern by the norman pins and are gobbed or tied down by their own stoppers, a short line with a shackle that supposedly keeps the tow cable from jumping off the stern roller and from getting out from between the norman pins.

The tug master can adjust the length of hawser to each barge as weather and waterway conditions allow. Because of this individual hawser arrangement from the tug to each barge, the tandem tow can be safely taken out over some rather tricky river bars in Oregon. Large and normally smooth Elliott Bay off Seattle is the scene of frequent departures and arrivals of tandem tows. Occasionally, tugs in British Columbia will tandem tow three barges.

THE BIG TUG—PUSH VS. PULL

The relative merits of push towing have been known for some time. Since the mid-1950s notched barges of 500 feet or more in length have been designed. It was the intention that they be pushed by a conventional tug, designed especially to fit the notch of the particular barge. The notches were rather shallow at first and the tug's bow ran in until about even with the forward end of her deckhouse. It was felt that this method would allow the tug to push between many U.S. East Coast ports. It was soon proven that this was not to be. The tug's displacement was much less than that of the barge whether loaded or empty. The tug suffered. In rough weather her push cables and lines continually parted as she battered the barge notch. Owners were most unhappy as tug captains began going on the hawser before leaving transition areas.

Because pushing increases hydrodynamic efficiency, resulting in fuel savings with an average 25 percent increase in speed, tug designers

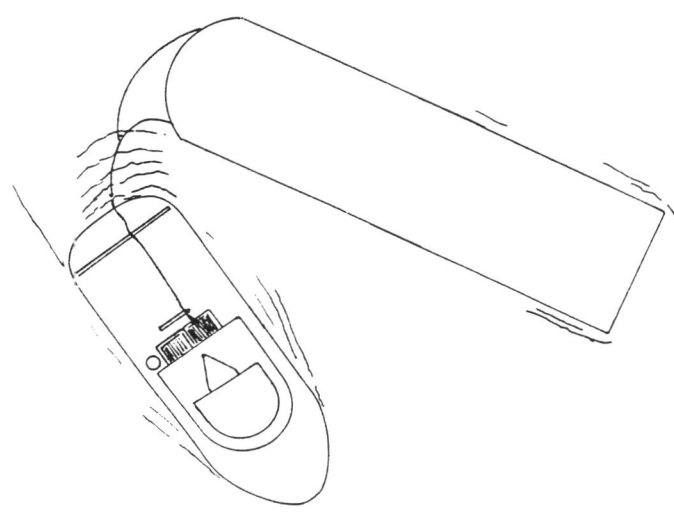

Streaming the first barge of a tandem tow

have come up with various methods to improve the notch-push towing system. Barge design has been improved. Tank tests have been done at various research institutes simulating random seas up to 25 feet and winds up to Beaufort Force 10.

From the shallow-notch and wire-rope system, deeper notches were introduced. Having the main push cable running to or from the tug was experimented with and the notch depth increased.

It was found that on mated tug-barge combinations, if the push cables were shackled into a pad eye on the barge skeg, better control resulted. This was especially noticeable in turning when the barge was empty. The next step was a much deeper notch, allowing the forward third of the tug to be admitted. More flexibility was tried in the connection system. Those now in use include the pinhole connection, where a sliding coupling between tug and barge, adjustable for any draft, allows the tug to roll or pitch. It is not a pleasant system to be aboard in quartering seas.

The pinhole system has been improved and notches in more recently built barges allow from half to three-quarters of the tug to enter. An entirely new variety of connection system using rubber side fenders allows the tug to squeeze tightly into the notch. One system requires the

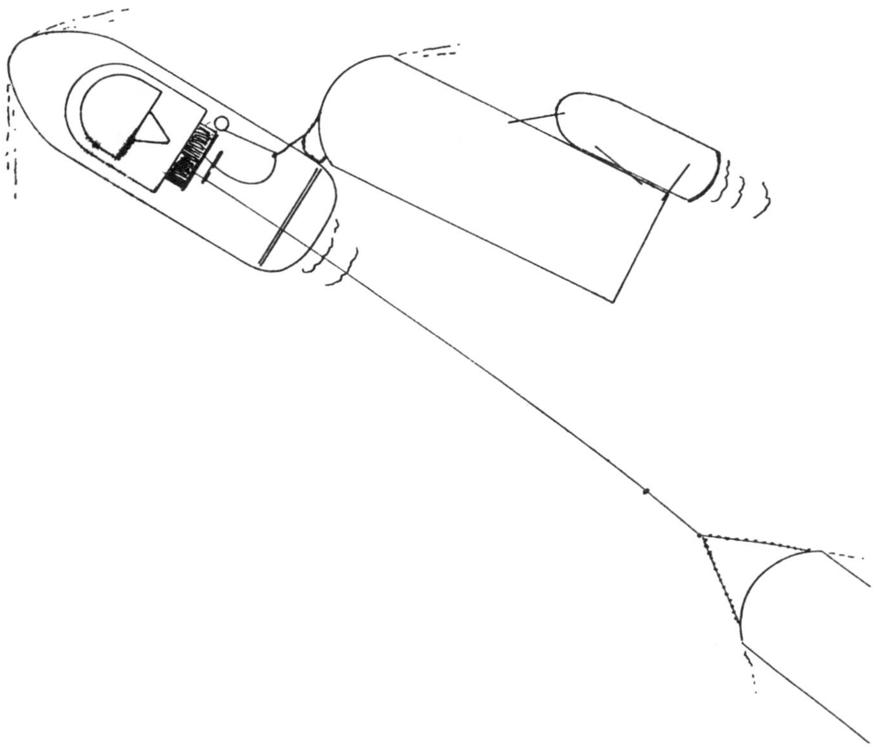

Connecting to the second barge using a tow cable from the alternate winch drum. An assist tug is made up to keep this barge clear as it drops aft.

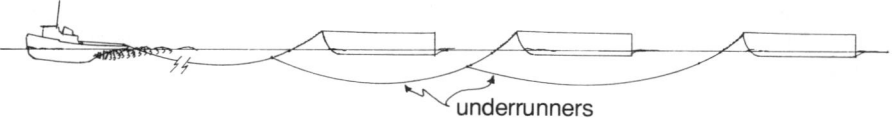

A triple tandem tow using underrunners. This method has been used for long ocean tows by tugs with a single-drum winch.

barge to have heavy raised bulwarks around the notch, and the tug has added padded blisters above the waterline on each side.

Other configurations include high vertical rubber-faced buffers built onto either the tug or the barge. One connection, Hydropad,* consists of hydraulically inflated pads. Other systems offered are: the Fletcher arrangement of rigid arms which hinge between tug and

*Hydropad, by ACB Marine, 44040 Nantes Cedex, France.

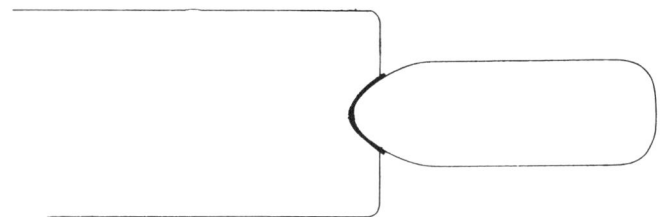

A shallow-notched barge

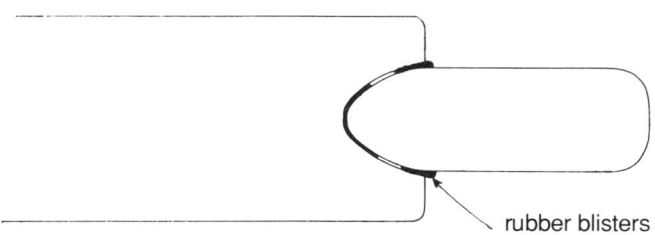

rubber blisters

The deeper-notched barge was introduced by IOT in the late 1960s for the tug *Clipper,* and others.

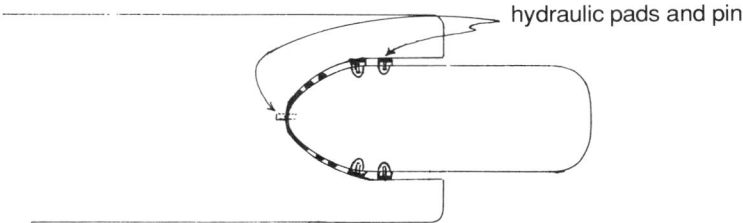

hydraulic pads and pin

The second generation IOT notch arrangement. In this so-called stinger method, a hydraulic pin and chain is connected to the barge from the tug, which was also given oversized fenders.

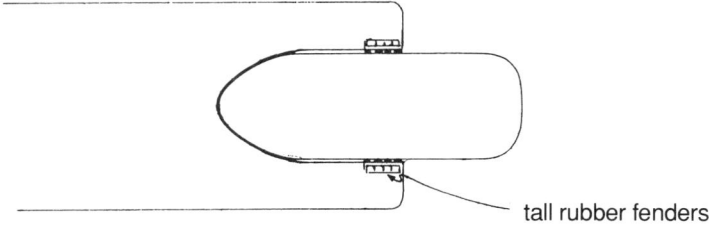

tall rubber fenders

The latest generation of deep-notch barge towing in which no push cables are necessary

barge, the Murvicker linking mechanism; and several Japanese winch-and-rope arrangements.

Barge units involved in any of these systems may or may not have movable skegs and/or a bow thruster operated from the tug.

PUSHING THE BIG BARGE—DOCK TO SEA

When pushing the loaded big barge from its berth toward the sea, or in the opposite direction, several precautions are necessary that differ from those of a similar movement with a smaller barge. The large barge of 500 to 600 feet LOA when loaded with about an 18-inch drag will draw between 33 and 35 feet, depending on the type of cargo. Immediately, it is obvious that this tow must stay within the deep water of the main ship channel. Realistically, the tug-barge notched together is a ship and must be handled as such with certain exceptions, the first being speed. Understandably, the 10- to 12-knot tug-barge cannot compare with a ship's minimum sea speed of 16 to 18 knots. Another exception is that some fast ships when meeting or passing will throw a large wake, which can part push cables. A lookout for these should be kept and a request for reduction in speed is common if necessary. Many state-regulated and commissioned U.S. ship pilots will call on VHF bridge-to-bridge regarding their ship's speed and wake prior to meeting or overtaking tows.

As the barge will displace from 20 to 25 times the tug's displacement this difference of draft and displacement slows down the results of rudder action: The water coming from under the barge contains air and is disturbed and in motion.

Settled down on a steady course in relatively wide channels or open bays, the tow will track well. On turns in a narrow buoyed channel, however, the water leaving the barge hull on the inside of the turn will follow the barge to replace the volume pushed away on the opposite side. The effect is, on any channel bend where the course change is 20 degrees or more, for the barge to run in that direction if too much rudder is given.* It may take hard-over rudder to start the turn, but the minute the barge's head starts to swing, rudder angle should be reduced to the minimum required. When the heading is within 10 to 15 degrees of the new course, opposite rudder may be required to check and finally steady up.

Particular care in lining up the barge must be used in poor visibility or at night. Try and find an object such as an aid to navigation or shore light ahead in relation to which the speed of your swing can be noted. Reliance on the gyrocompass and autopilot alone can be misleading when changing course. As the tug becomes a pivot point and is some 600

---

*Good examples are various cuts in Tampa Bay, Florida.

feet back from the barge's bow, the barge's bow may swing 10 degrees before a 1-degree course change will be noted on the tug's gyrocompass.

If the large barge being pushed is empty, the speed over the ground in still water will normally be within 1 knot of the tug's own speed when running free. Handling will be entirely different than when the barge is loaded. Hopefully, the barge will be ballasted with an 18- to a 36-inch drag.

On straightaways with a crosswind it may be necessary to crab sideways to make a course good. Plan to give channel buoys and markers good clearance. On turns and bends, the barge may slide sideways unless speed is reduced. Moderate to slow speed should be used in restricted waters or if approaching a situation where it may be necessary to back down.

DANGERS IN PUSHING AT SEA

Advocates of each of the various tug-barge systems proclaim that their system will work satisfactorily up to certain sea heights. Usually the direction of these seas is not mentioned. No suggestion is given as to what the tug master faces once the height of the sea goes beyond these limits. Obviously, the marketers of the systems have never tried to pick up a barge when seas have forced the tug to get out of the notch. At that point, both the tug and the barge are in extremis. The tug deck crew will be endangered in the attempt to recapture the barge on the hawser. The tug will be rolling and diving with her decks awash.

Notch pushing has its time and place. A careful tug master must calculate when that time is up. He must choose certain criteria that indicate it will soon be time to get out of the notch. He should not wait further. Heavy damage has been done to both tug and barge through indecisiveness and waiting too long. If it is the question of a too-rough night ahead, hawser connection should be made before sundown.

Never forget that Father Neptune and Mother Nature have never been lovers. Occasionally when they meet, a tempest results and there can be seas and winds in which no ship, let alone the tug-barge combination, even an ITB (integrated tug-barge), can hold together.

THE ITB

This is the so-called final solution for pushing at sea. It consists of a barge hull with a parabolic bow fitted with a thruster and a floating power unit solidly married into it. The term "tug-barge unit" is incorrect. There is no tug. IPB (integrated power-barge) would be more appropriate and realistic.

The power unit, which is detachable, has no other use except being joined to a similar barge with identical connection characteristics. This floating powerhouse cannot tow independently. It has no tow winch for pulling astern, only a pair of H bitts mainly for looks or for securing a mooring line. Structurally, it is not built for any alongside towing. Its configuration and marrying appendages, which protrude, make unfeasible any towing attempts other than when snugly secured to the barge for which it was built.

The Ingram ITB is mated by the power unit sliding on a plate extending down from the barge stern. The configuration at the forward end of the power unit consists of vertical wedge-shape extensions on each side. The unit is also held by a central series of hydraulically tightened cables and steel plates bolted between the main deck of the barge and the pushing unit.

When disconnected, most of the ITB power units have great difficulty maneuvering even in the calmest of weather. An exception is the Bloodworth-designed* *El Gaucho* which, when detached from its barge hull *Pampas,* does run to fuel docks or for provisions.

Catug power units with their catamaran hulls can detach and safely maneuver. *Seabulkmagnachem* seldom disconnects from its barge hull.

One problem with ITBs is the necessity to maintain a draft deep enough to keep the power unit's propellers(s) fully immersed. This requires constant ballasting of the barge unit.

All of these ITBs have successfully negotiated voyages between the Gulf of Mexico or Caribbean ports to the New York and New England areas, and from the Pacific Coast to Hawaii. One of Occidental Petroleum's ITBs did break up in the Atlantic when bound from the Gulf to the Black Sea. Bloodworth's *El Gaucho* ran successfully from Delaware to Brazil for some time.

These so-called integrated tug-barge units are actually ships, tankers, or bulk carriers. In the United States, they were built to evade certain crewing and other U.S. Coast Guard requirements. Now, automation and advances in electronics for navigation, communication, and cargo handling, which have led to a considerable crew reduction on a ship of equal or larger size with better speed, may soon make the ITB economically obsolete.

SKEGS

Most modern barges have skegs, heavy fixed fins extending at a slight outward angle from the barge's fore and aft centerline. Their purpose is to keep the barge from sheering or tracking back and forth when towed astern of the tug.

*Bloodworth of Pasadena, Texas.

Early on, studies were made to ascertain how much they retarded the tow's speed. One such study at the University of Michigan was eagerly accepted as it revealed the relationship between angle of skeg deflection and speed, and resultant operating costs. Because of the various findings regarding the deflection angle, owners immediately began incorporating removable skegs on new barges. Movement and adjustment were by crowbar, tackles, and, later, hydraulic rams. Some were even controlled by radio. At first, the results appeared economically favorable.

However, maintenance of movable skegs has proven to be costly. Heavy seas striking, or the grounding of, that portion of the barge has necessitated expensive repairs. Movable skegs have since become a luxury and, at times, a costly nuisance that most tug and barge owners have forgone.

One company, in an effort to increase towing speed and fuel economy, gave their tug master a set of guidelines in relation to the movable skegs on their 250,000-barrel barge. As the barge was unmanned, the skegs had to be set in port and remain in that position until arrival at the next port. It was the company's desire to use as little deflection as possible. The effort was concentrated on towing the barge when loaded at a mean draft of 32 feet. The tug was nearly 5,800 HP and her tow cable was connected to a 2½-inch wire pendant of 90 feet, which was shackled into a single 50-foot bridle leg of 2½-inch stud link chain.

This bridle leg ran from a bullnose chock in the center of the barge's bow. A single bridle leg from such a position gives a towed unit, particularly one with a ship-model bow, very little resistance against any hydrodynamic force that would cause it to sheer. When this force is not countered by some skeg resistance, the barge may fail to track directly astern of the tug.

Of course, if the barge had been rigged with two bridle legs secured one on each side of the barge's bow, it would be checked like a horse when its one rein (or bridle leg) tightens. This would occur even if the skegs were set at 0 degrees, which is amidships or fore and aft in line with the barge's centerline. With the barge in loaded condition, various angles were tested. At 0 degrees, the barge ran off to 45 degrees on the lee side of the tug. Any increase in expected towing speed was nullified when the tug's 2¼-inch wire hawser, which was vibrating heavily on the stern rail, not only listed the tug slightly but laid the barge so far off that it was like a paravane being towed by a minesweeper.

At the other angles of deflection, 20, 30 and 35 degrees, 20 degrees was acceptable with a partially loaded barge. The best control when loaded was obtained at 30 degrees. When empty, the skegs could be set at 0 affording a ½-knot increase in speed over that at a setting of 35 degrees.

Another skeg improvement, patented in Canada, is the Hydralift skeg, consisting of various vertical foil sections rigidly attached under the light draft waterline at a barge's stern. With one set of these foils on each side, the barge gains directional stability through the creation of a lift rather than a drag, thus it is an energy-saving innovation, as well.

Seatronics Technologies Limited of Vancouver, British Columbia, an independent firm set up by the British Columbia tug and barge giant Seaspan International Ltd., has developed and marketed the Hydralift skeg. Fifteen of these Hydralift skegs are in service all over the world on long-distance offshore routes. A reduction in fuel consumption of 20 to 30 percent has been claimed, with a 10 to 15 percent increase in towing speed.

Canada, particularly British Columbia, has shown alacrity in improving towing modes and conditions. British Columbian tugs are small, modern, and powerful, with a seaworthiness beyond the requirements of their classification.

Because of the great distance of Pacific Northwest towage of wood products, studies have been made to increase towing speed by reducing skeg resistance, which varies from 20 to 40 percent drag.

With Canadian Department of Energy backing, a North Vancouver group associated with Seaspan has developed a barge-steering system. It is known as the TNT* Active Flap Rudder and Barge Steering System. This rudder with a flap is mounted at the middle of the barge's stern. The flap is fitted on its trailing edge. In a test using the *Seaspan Pacer* with the flap rudder installed on the barge *Seaspan 821,* an improvement in the tracking of the barge gave an increase of 1¼ knots.

Operation of this system is accomplished in three steps. An angle sensor attached to the hawser will send a signal. When the tug's towing hawser runs at an angle from the barge, even though the barge's bow, the towline, and the tug may be in one straight line, the barge may be tracking at an angle. To rid the tug of this barge resistance, the sensor on the hawser will send a signal through a potentiometer, which relays it to a microprocessor. This stored information causes a small hydraulic pack on the rudder head to turn the flap. The barge then tracks directly behind the tug. Actuation of this system may be controlled either manually from the barge or through radio telemetry from the tug. This rudder flap system is particularly adaptable to the barge with a square and flat or raked stern, with or without skegs.

---

*Total Naval Technological Associates Inc., 4039 Violet St., North Vancouver, British Columbia V7G 1E3.

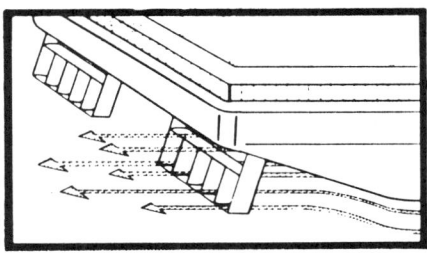

An artist's conception of the Hydralift skeg as installed on a square-stern, rake-ended barge. Its effect reduces drag of the barge under tow and increases towing speed. Hydralift II can be retrofitted on existing barges. (Courtesy Hydralift)

CHAFING GEAR—TYPES AND RIGGING

Material and methods used to protect tow cables over towing arches and at the stern taffrail vary greatly according to the tug's construction, the sea distance of the tow, and the area of operation (also see "At Sea with the Ocean Tow—Chafing Gear" in chapter 9).

For the short sea tow of two to five days, the simplest proven method, if the vessel is so provided, is the free-running sheave on the towing arch which holds the hawser clear of any obstruction between it and the tow winch, and itself and the stern rail.

Many large East Coast and Gulf of Mexico tugs use this method. There are still a few tugs that use synthetic hawsers and are protected at the stern with a hawser board or protective sleeve of rubber, etc.

What kind of chafing gear is best is a matter of experience and improvisation, resulting from the trial and error of seamen. Some of the more successful types are the split 6-inch steel pipe filled with a channeled wooden trough to fit the hawser, or a 3- or 4-foot piece of 6-inch aluminum I beam fitted with cable clamps.

Putting chafing gear on the tow cable is a dangerous operation and, if possible, should be done in relatively smooth waters with the tug holding a steady course. Whoever is on watch in the pilothouse should always be aware when men are working on the tug's stern with the hawser out. Particular attention should be paid when they are installing or working with hawser chafing gear. Men have been injured trying to remove a hawser chafing device that has turned or jammed over the stern rail. Wooden hawser boards are prone to tip due to the twisting action of ropes when they are slacked out. The split-pipe type chafing gear with tapered ends can be easily slacked over the rail or roller. It will roll as the tug pitches and rolls. It cannot catch on stern rollers. The alumi-

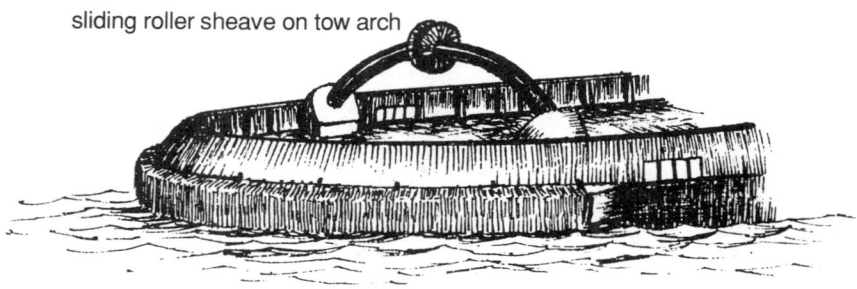

The single towing arch or "rainbow" is a popular method of avoiding use of hydraulic norman pins, stern rollers, or chafing gear on many U.S. East and Gulf Coast tugs.

When soft lines or nylon hawsers are used, it is necessary to place wood chafing boards where they pass over the towing arches. In this photo, the Dutch tug *Scheldt* is towing two barges from Rotterdam to Montevideo. (Courtesy P. Van der Hidde Collection)

num I beam type should have its outer edges rounded for easy passage over the stern rail or roller. None of these rigid antichafing devices should protrude more than 12 inches over the tug's stern, or the weight of the tow cable will cause them to bend.

In continuous rough weather, the point where the cable leaves this chafing gear and drops toward the sea is the weakest. As the tow wire is held flat and rigid in the chafing gear, this point is the center of all up-and-down bending. After continuous bending, heat and crystallization of

Chafing gear such as this is made from an extra-heavy steel pipe shaped and open to admit hawsers. Hawser is held by heavy wire rope clamps welded near each tapered end of the chafing gear. This gear can be placed and secured inboard and slacked to the stern roller.

The Irish are great seamen, as shown by the chafing gear on this tug's ship line both at the arch and in the tow hook.

inner strands and the core can lead to breakage of the entire cable; thus it should be changed at regular intervals.

Tugs on the U.S. West Coast have a different method of treating the hawser where it passes over the stern roller. They are fitted with hydraulically operated norman pins and hold-down bars on deck.

ROUTINE ON THE SHORT SEA TOW

When under way with the tow and passing out of a river or any area where water is liable to come on deck or where the tug is liable to roll heavily, all watertight doors to outside decks should be closed. Sudden and unexpected heavy rolls or near-girding conditions have caused several tugs to sink because a watertight door had been left open and unattended. During daylight hours and in good weather, watertight doors on the main deck may be left open.

Even though the duration of towing from sea buoy to sea buoy may only be two to five days and may have been accomplished repeatedly without any problems, this fact should not be the basis for any gradual relaxation of safe operational procedures or extension of the normal lifespan of equipment for just one more trip. Such pitfalls lead to casualties.

Prudent seamanship requires a consciousness of eventualities and a plan to overcome them as they challenge. This extends from the pilothouse to the deck, galley, and engine room.

The tug watch officer at sea is a busy man. Clearing the departure buoy or lighthouse he should note and log the time and course and check the gyrocompass heading vs. that of the magnetic compass. Generators have failed, usually at the wrong time. Overreliance on the gyrocompass has often resulted in neglecting to compare it with the magnetic compass.

The tow's position should be checked by radar, Loran, satellite, or visually, and placed on the chart at least every hour. Half-hour intervals are better. A good lookout should always be maintained both visually and by radar. In poor visibility and at night, good practice is to have the seaman on watch in the pilothouse as the lookout.

This watch-standing seaman should know or have been trained in how to steer the tug both with hand gear and by autopilot; what gauges to monitor if the vessel has an automated engine room; and how to stop the main engines in an emergency. He should also know the location of the general alarm and when to sound it. With this knowledge, he can safely relieve the watch officer for short periods, if necessary. The wheelhouse should never be left unattended when the tug is under way. When conditions are normal, the seaman on watch should be sent below at regular intervals to check things both inside the tug and out on deck. Particular attention should be given to watertight integrity, fire, and the towline and tow winch. If the tug has an unmanned and automated engine room, this area should also be checked regularly to see that conditions are exactly as shown on monitors.

On some tugs, it is customary to tie a rag or piece of rope on the tow cable midway between the tow winch and the stern where it can be seen from the pilothouse and any slackening of the cable can be noticed. Under no circumstances should deck personnel slack or adjust the tow cable without first notifying and getting permission from the person on watch in the pilothouse.

Navigation and Tow Lights

These should be checked hourly. If the barge lights seem to be weak or cannot be seen, it may be necessary to keep the tug's searchlight sweeping over the tow in order to warn other vessels, particularly small craft, that you are towing. Small pleasure craft and fishing vessels have been known to try to cross between tug and tow. Several sinkings with fatalities have resulted.

Radio Watch

In the pilothouse of the modern tug, VHF channel 16, the international calling frequency, as well as the area bridge-to-bridge channel, must be monitored. At sea, tows frequently identify themselves to other traffic and, if it is necessary to cross a traffic lane, other traffic should be kept advised until the tow has cleared.

The Log and Chart

Any positions obtained visually, by Loran, or by satellite should be marked on the chart. A good log should be maintained listing courses steered, weather, compass headings, barometer readings, condition of tow, length of hawser, and any other details of importance that will indicate the tug's protection of and service to the customer whose product is being towed. The same protection should be extended to the tug's owner and the watch officer when writing the log.

If the tow is manned and radio-equipped, a regular communication schedule between tug and tow should be held. Acknowledgment of this should be given under the "Remarks" section of the tug's log.

The log should be truthful, complete, and continuous; list personnel on each watch; and be initialed or signed by the writer. Attorneys are quick to note erasures, discrepancies, or lack of good logkeeping habits as an indication of poor seamanship and unseaworthiness of the tug. Many towing companies have guidelines for operations, etc., over certain towage routes. Nothing in their guidelines should suggest any restriction of log entries or priority of engine, fuel, or other data over that of normal overall recording. Some companies print their own "log sheets." These should be a supplement to the regular tug log, as they usually do not contain all of the material that would protect the owner, customer, or tug captain.

Maintenance at Sea

In good weather at sea, it is frequently possible to carry out a daily maintenance program. Hopefully, crew members will, in addition to keeping their own quarters and heads (toilets, washbasins, and showers) clean, extend this cleanliness policy throughout the vessel. When practical, watertight deck and navigation lights' glass lenses should be wiped clean of accumulated moisture. All grease fittings on fairleads, sheaves, rollers, tow winch, and watertight door hinges and dogs can be regularly filled. Pilothouse interior and windows should get daily cleaning. All loose gear should be kept lashed. Spare shackles should have their threads lubricated and spare towing hawsers should be protected from the weather. The noise associated with rust removal, if necessary, often creates a problem to those resting off watch and this procedure can only be accomplished for short periods when all crew members are up, or when laying over in port.

Disposal of Waste at Sea

Modern tugs are often fitted with trash compactors. Nonbiodegradable waste should be crushed in these and stored in plastic containers until arrival in port, where they may be placed in a disposal container. Biodegradables may be dumped at sea, away from shorelines, bays, or rivers. Body wastes are usually treated and pumped overboard at sea or into pump-out facilities in port in accordance with local regulations.

UNFORESEEN OCCURRENCES WITH THE TOW—
SHEERING—TOW ATTEMPTS TO PASS

Eternal vigilance should become the guiding principle to be instinctively observed by the tug's crew. This requires teamwork among the pilothouse, deck crew, and engine room personnel, particularly during maneuvers.

Some tows will follow very well behind the tug. Others will sheer or lay off to one side. A careful adjustment of speed and/or hawser length may be necessary to reduce this to a minimum.

Most troublesome are powerless light draft ships with a narrow knifelike stem under a flaring bow. In certain areas of towing, such as east- and northbound in the Straits of Florida where the full force of the Gulf Stream and a fresh southerly breeze often occur, a light draft ship or a large barge will tend to run to one side, usually to starboard. If no corrective action is taken by shortening the tow cable, the tow will sail up on that side trying to pass the tug. The tug's hawser will lead off on that side giving the tug a list in that direction. The situation will become dangerous due to the possibility of the tow cable jumping over and in front of the quarter bitts, cutting into the deckhouse, and finally girding the tug.

Quick action must be taken when a tow begins to behave in this manner. Traffic in the area must be warned that the tug *does* have a hawser running from it to the vessel well off to one side. An attempt to correct this situation must be made by first taking a relative bearing from the tug to the tow to see how far it advances forward of the tug's quarter before it stops its passing attempt. If it appears to continue past the tug's quarter, an attempt to speed up the tug may be futile. Instead, a slow turn away from the tow, keeping the tug's hawser as nearly directly over its stern as possible, may get the tow to slow and turn until both tug and tow are headed into the wind and current. Then, with the hawser shortened somewhat and towing speed reduced to steerageway, the tow will drift backward in the original direction until the elements allow it to again proceed on its desired course. All of this of course requires sea room to turn in.

GROUNDING AND THE REPORT OF MARINE ACCIDENT, INJURY, OR DEATH

There are several types of groundings that relate to tug and tows. Intentional grounding for a few moments due to local conditions has been an acceptable practice. These conditions would be when a waterway is restricted and leaves little turning room and/or there is a swift current in which the tug and tow must be turned. Also in certain areas it may be necessary for a tug with or without a tow to bounce off the bottom several times while successfully crossing a bar during a slight swell. These types of groundings are accepted as common practice in certain areas. The person in charge during one of these maneuvers should be aware that while some practices in life are common, they are not always legal. Any casualty resulting from an intentional grounding maneuver may bring the charge of negligence against the tug master.

Intentional grounding due to extreme tidal conditions has been necessary in many places. Tugs built specifically for work within such areas usually have heavy twin skegs that allow them to ground out nearly upright on an even keel. Otherwise the tug or vessel must have a flat bottom or be strongly supported against a solid wharf or other stable vessel. Mooring lines must be kept tight so that the vessel will not list away from the supportive side. Even if the vessel, particularly a tug, does not ground out at low tide but is subjected to rapidly changing tidal levels, a continuous watch must be kept to prevent the vessel from becoming hooked on an obstruction during her descent or ascent as the tide falls or rises.

Accidental groundings are all considered logable, reportable, and due to negligence. These include such occurrences as the temporary grounding outside a dredged channel, the striking of rocks, reefs, or any area of the sea bottom over which the vessel could not freely float.

Most tug captains respond to the problems of their own seafaring society when trouble arises. This is particularly true when a tug loses power, catches fire, or goes aground with or without power. Regardless of rivalry or competitive owners, tug A afloat will often help tug B which is aground. This is sort of an unwritten gentleman-seafarer's agreement throughout the towing industry. When carried out successfully there is rarely any remuneration. This in itself does not establish any precedent for not logging the incident. It also does not permit the master of tug B to avoid reporting the grounding to his owners and such other authorities as his vessel's nationality may require.

In the United States reporting is accomplished by filing a Department of Transportation, U.S. Coast Guard form CG-2692 as revised June 1982. Instructions accompany this form. CG-2692 coverage of accidents has been extended to include barges and mobile offshore drilling units (MODUs). A tug captain should report any damage sustained by unmanned barges under his control.

Logging and reporting of accidental groundings protects the tug owner if any previously undetected underwater damage is revealed on dry-docking and survey. It also identifies the person in charge of the tug when the damage occurred. Any injury or loss of life must be fully logged and carefully recorded both with the tug or vessel owners and authorities.

NARROW PASSAGES AND RAPIDS

It is very dangerous to tow light draft or empty oil, chip, or high-sided barges or ships through narrow passages or out of rivers when the current exceeds the tug's normal towing speed. In many of these places there are rapids with whirlpools and eddies. With a hawser tow, the towed unit may get into one of these and race ahead suddenly, swinging radically, jerking the tug and possibly breaking a shackle or parting the tow cable itself. If the tow is being pushed, it may get completely uncontrollable and strike rocks. (See chapter 10, "Notes on Towing in Specific Places.") Plan to transit such areas close to the time of the least or slack current. Transiting such passages with loaded barges should not be attempted on the hawser, except at the very last of a fair flood or an ebb head current that will change to slack water. In this way, the tow will clear before a heavy head current or strong fair current occurs. This timing with the current direction also applies to towing inbound or outbound over river bars.

NARROW DREDGED CHANNELS AND CANALS

In taking a tow into a narrow, dredged channel that has tidal mud flats, banks, or hidden ledges on either side, great care must be exercised. Inbound with a loaded barge the tug master can usually see over his tow. He can relate his progress to the channel markers, ranges, or buoys, and stay within the confines of the channel.

When operating a low-powered tug, or one that has a loaded barge on a hawser, the tug master may have problems if there is a sharp bend or an angular slip to enter. Barges either loaded or light tend to hold their headway for a long time. If they are loaded, they will not respond at once to the tug's change in direction. If the tug captain does not order his hawser slacked, the tug may be run over or girded. If using "gate hawsers," the slacking of one leg will free the tug to come around to head-and-tail. This may result in the tug and tow being out of shape. The empty barge may follow the tug if a little extra power is applied. But at this point, an increase in speed as one approaches a turn or mooring may not be desirable. A solution is to have a helper tug made up on the barge's quarter on the side toward which the barge is to be turned. When this helper tug backs and the lead tug keeps a gentle strain on the hawser, the turn will be negotiated safely.

If no helper tug is available to assist with an empty barge, the tug should make up alongside before starting in the narrow channel. If the tug's pilot cannot see over his tow, knowledgeable personnel with a walkie-talkie should be stationed on the head of the towed unit to advise him of aids to navigation. In avoiding them, care must be taken not to swing the tug out of the dredged channel at a point where she may go aground. If this does happen, the empty barge, which is heavy nevertheless, usually continues on, parting all the lines from the tug. If the tug was made up head-and-tail with her tow hawser used as the stern line, the barge in such a situation may cause the tow cable to cut into the tug's deckhouse as the barge starts to pass.

If pushing a loaded oil barge in the narrow land cuts of the Erie or Northern Canal of the New York State barge canal system, remember that at least half of the dredged cut belongs to the loaded tow when it is meeting another. The canal's bottom rises abruptly, often with a stony edge. Aids to navigation are only used to mark bends, points, and individual rocks or shoals. Otherwise, the rule is, as in most narrow channels, "keep in the middle." It may be necessary when meeting another tow to drift by using "dead slow" RPMs with only a foot between each vessel. Personnel should stand by with hand-held fenders.

OCCURRENCES AT SEA

Parting the Hawser

When at sea, perhaps the most exasperating, if not distressing, occurrence for the entire tug crew is the parting of the hawser. Immediately, the tug's speed should be reduced but not stopped. The tug should not be turned back toward the tow as the tow cable may slide to one side and be pulled into the tug's propeller. That portion of the towing hawser still remaining and running from the tug's stern must be heaved in to see where the break occurred and what remaining gear, if any, is still safe to use.

As the weather would undoubtedly be rough, the tug must be held head to the wind and sea at reduced speed or steerageway so that men on the afterdeck can safely work and clear away chafing and holding-down gear. These men should wear life jackets and/or survival suits if seas are liable to break on deck.

Taking time and care should be the keynote. The drifting barge cannot be picked up without proper preparations. If the cable breakage happened in poor visibility or at night, a warning and position report should be given to approaching traffic on the radio emergency channel. Tracking of the drifting barge by radar should commence at once. If necessary and possible, illumination of the barge should be maintained.

Unless someone has been injured, or the tow unit has been damaged or is leaking, or assistance is needed, it should not be necessary to advise anyone else if recovery of the tow seems imminent. This decision must be made at once by the tug master. It should not be forgone, however, if he has any doubts about reconnecting. His position and status should immediately be passed to the owners or to the nearest organization that can render assistance.

Retrieving and Reconnecting the Tow

As soon as it has been ascertained whether or not the tug's tow cable is usable, such articles as shackles, lines, etc., should be laid out for reconnection. If the tug's main wire tow cable is damaged or broken and cannot be shortened, clamped, and reused, the tug's spare hawser, usually a heavy nylon or other sythetic line of equal tensile strength to the main wire hawser should be roused out and made ready for running.

A plan should be formulated as to how the reconnection will be made. If the barge is empty and is fitted with an auxiliary tow wire with a buoyed floating pickup line trailing astern, this method would seem the safest. Empty barges will have some way on as they drift in the wind and roll in the swells. They are dangerous to get too close to, particularly at the bow in attempting to retrieve a dangling set of bridles or at the stern near the rake.

Lying off the barge's stern as it drifts, the tug master should spend some time noting its action and direction of movement. If the buoyed floating pickup line is available, the tug should be brought up parallel to the barge's lee side. In this way, as the tug makes her creeping approach, she will tend to be blown away from the buoyed line, and the backwash from the tug's hull will keep the line far enough off to be reached by a boat hook. An attempt to pick up the buoyed line from the windward side usually results in the tug running over it, or running onto the line where it will be so close to her side that hooking it is difficult.

Any of these maneuvers and those that follow will test the tug's skipper's coolness and ability in shiphandling. Once the pickup line is aboard, the tug should proceed ahead slowly at an angle to safely clear the barge's stern. If the lee side of the pickup line is opposite to the lee side of the barge, the end of the pickup line should be walked forward on the tug and passed around the stem before being taken aft to the capstan. In this way, the tug will not run over it in getting itself free so that it can pull the auxiliary barge cable aboard. It may be advisable to have a long, strong messenger available to secure to the pickup line, thus giving the tug more freedom to maneuver.

Once the pickup line and/or messenger are at the capstan and have sufficient turns to hold, the tug should be located in such a position that the thimbled end of the auxiliary tow cable can be heaved aboard and connected without the barge blowing down onto or off too far away from the tug. As soon as the thimble end of the barge cable is aboard, it should be shackled into the tug's hawser and the flotation line removed. Towing can then be resumed.

If the drifting barge is loaded and low in the water, the same method for retrieval and reconnection may be used. If it is not awash and rolling and seas are not breaking over its decks, the tug may be able to work up to it close enough to land several life-jacketed crew members so that new bridles and pendant may be passed over if necessary for reconnecting. This is a very slippery and dangerous task.

Barges rigged for pushing at sea frequently have a flotation line rigged from the regular bridle-pendant-shock line arrangement. This runs back to the notch with ample line to spare so that if the tug is forced to break away, this float line can be held on the tug as it backs out of the notch.

Helicopter Reconnection

If, for some reason, none of the above methods is practicable and it is necessary to get men aboard a barge that is adrift, but it is unsafe or impossible for the tug to land them, the services of a helicopter may be considered. This has been safely done on several barges that were adrift (afloat or aground) in water too shallow for the tug's draft.

The helicopter ordered must be large enough or have the lifting capacity to fly the bight or end of a suitable towline to the barge. It also must be able to transfer men back and forth. It must have fuel capacity to make the round trip from its base and remain on the scene for one or more hours.

Good communication between copter and tug should be arranged. The tug's deck should be cleared for the landing of a basket or cable. No one should touch items dropped from the copter until instructed by its pilot. No lines from the copter should be secured to the tug or barge or anything other than that which is free and about to be hoisted. If a line is to be run from the tug to the barge, it should be completely slack and free. Nothing should strain, tie down, or restrict the copter in its flight. With care, copters can accomplish amazing missions.

The Tug Stops—Lying on the Hawser

One of the most disheartening occurrences when towing at sea is either to have the engineer tell you that he must completely stop the propeller(s), or to have the engine(s) suddenly come to a grinding, banging, or silent stop.

Almost immediately, the pilot on watch should turn the rudder so as to use the tug's remaining headway to carry her away at right angles from the previous course, thus avoiding being run down by the oncoming tow. A heavily loaded barge will continue on its course at 5 or 6 knots for several minutes. If the barge is loaded, turn the tug away from the wind. If the barge is empty, turn the tug into the wind. Tugs catch more wind and drift more rapidly than loaded barges. The opposite is true of the empty barge.

The next consideration is the tug's hawser—keep it clear of the propeller(s) and slack it at once so that the barge will not pull the tug stern-first behind it. When the barge has finally stopped and the slackened tow cable begins to sag, consideration of the depth of water is the next item. Presumably, the tug and tow were proceeding at full speed just prior to engine failure. The tow cable would have been taut with a slight catenary. Now, its weight will slowly pull the vessels toward each other as the cable forms a huge U-shaped bight. If the depth of water is 50 fathoms (300 feet, or 91.44 meters) or less, the hawser will rest on the bottom. It may catch on something or it may drag.

If reports from the engine room indicate a long delay in, or impossibility of regaining propulsion, a call for assistance should go out at once. Preparations should be made for the many problems that may occur in the meantime. Among these are the cable catching around something and pulling the tug backward or fouling itself by crossing between the barge and the bottom. The tug and barge may wind up side by side or nearly so. Heavy fendering may be required. If such is the case, all sorts of seamanship can come into play and a constant lookout must be made for sudden changes in either vessel's position. When a relief tug arrives, instructions should be given to her master as to the course he should take to clear the crippled tug's cable from the bottom without fouling or kinking.

Anchoring the tug would, of course, have been the first feasible action when the tug was clear of the barge's course. But many tugs, and nearly all of those in the United States, do not have any readily accessible anchoring gear.

The Sinking Tow or Tug

One of the most horrible experiences a tug captain can face is the sinking of his tug or his tow. Never having personally experienced either, the author can only offer precautionary advice based on the close observance of others. Tugs sink mainly due to the entry of water, girding, nonwatertight integrity in heavy weather, collisions, or by being pulled under by a sinking tow.

Girding and watertight integrity have been covered elsewhere. Being pulled under by a sinking tow can happen at any time, particularly if the tow is unmanned. For this reason, tugmen from captain to cook, engineer, and deckhand should, when on watch, conscientiously keep a sharp lookout not only ahead but astern. At night, beware if the towed unit seems to list or cause the tug to slow down. It has either fouled something or has increased its draft. It soon may become critical.

By day, particularly in rough weather, the lead of the tow hawser should be frequently checked along with all of its appendages. If it is rough, the hold-down gear may be wearing or may break, allowing the hawser to jump off the stern rollers or the towing-bar sheave. If this happens and the wind is aft, the tow may start to pass the tug. The tow cable leading off at such an angle as to roll the tug's rails under can trip her before rudder and power adjustments can save her. In any of these situations the tug should be able to quickly free herself by letting the hawser run completely off the tow winch drum. If the water is deep, the hawser should be jettisoned, its bitter end flying overboard.

On many tugs, the inner end of the hawser is held by a large clamp with nuts which are rusted. They should be kept lubricated and handtight. In an emergency, such as a suddenly sinking tow, there is little time to get a wrench and slack the winch drum so that the nuts are accessible.

Because of this lack of quick hawser release (a flaw in winch design) and the fact that those in the pilothouse are unable to control or release the tension of the tow winch, several tug casualties have occurred.

The tug *Eagle* was lost when her hawser hold-down broke. The wire hawser jumped free of the stern norman pins. Her sole survivor of a crew of eight stated that "her two barge tow overran her on one side and rolled her over before her tow cable could be released." Her tow winch required two operations in order to free her cable. In another casualty, *Eileen C*, towing on a short hawser, was pulled under when her rudder jammed hard over and her barge passed in a head-and-tail direction. The tug's afterdeck was pulled underwater before anyone could get to the tow winch to release the cable.

As these tugs were U.S. flag, the National Transportation Safety Board has made recommendations to the Coast Guard and ABS "that standards be established for towing systems on ocean tugs including the means for releasing the brake of the towing winch remotely from the pilothouse or any other steering station."

The Orville Hook

On barges that are adrift and equipped with a surge chain bridle and pigtail, retrieval of the chain has always been most difficult. Now, a new system has been designed by a Sause Brothers Ocean Towing Company's tug captain. This is known as the "Orville hook." The Orville hook, the invention of Captain Orville "Bud" Fuller of Coos Bay, Oregon, is a proven retrieval system. It is fully patented and carefully guarded by the company; at present its use is restricted to their vessels.

Basically the Orville hook is a piece of 2-inch steel cut in the shape of a huge hook 2 inches wide with an 8-inch opening. A slot similar to a devil's claw is cut in the back to accept a chain link. The hook, which weighs between 200 and 400 pounds, is suspended by cable from a large fisherman type buoy. Its depth may be adjusted to whatever is needed to snag onto the barge's chain pendant. The hook has a pair of ears welded on each side. From these, lines lead back to the tug so that the hook may be turned or adjusted as it strikes the barge's chain. This entire assembly is floated down the side of the barge on about a 200-foot line. The tug is then maneuvered to drag the hook around and across the barge's bow. In actual tests, it has worked every time.

APPROACHING PORT

Most short sea tows proceed from sea buoy to sea buoy without suffering any mishaps. The tug captain's concern then is directed to the condition of the entrance channel: Can he proceed in? Is the bar rough? Must he

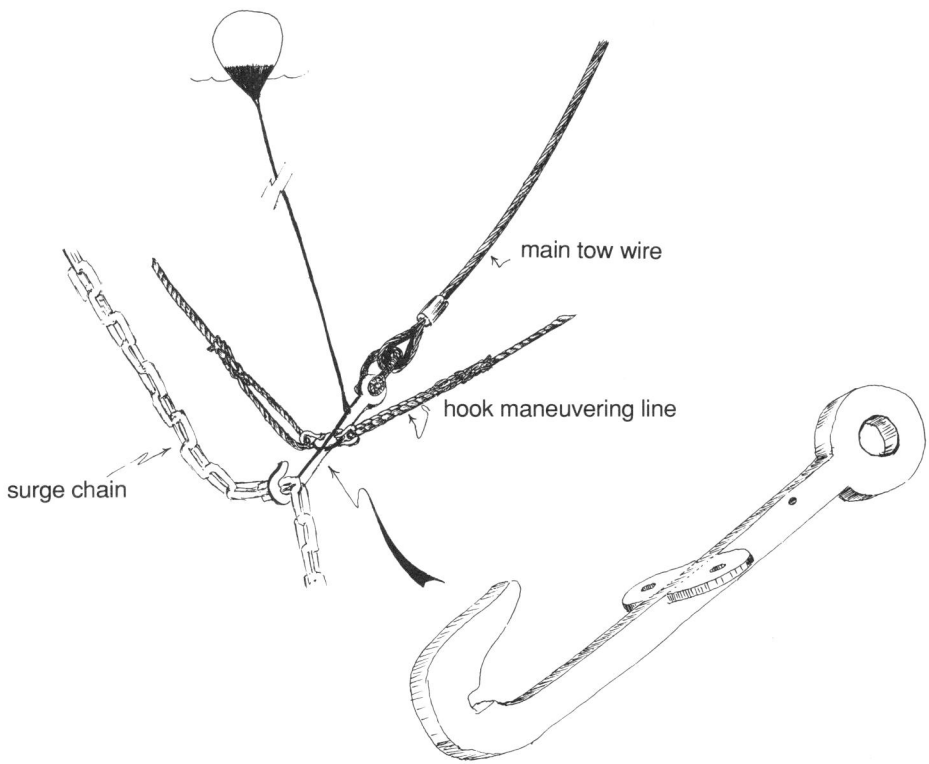

The patented Orville hook for retrieving chain bridles or pigtails

await outbound traffic? Must he enter on the hawser and shorten, or pick up his tow in a transition area such as Gravesend Bay, New York, the Solent off Portsmouth, England, or below the Bay Bridge or Angel Island in San Francisco Bay?

If the barge is empty, a shortening of hawser usually allows good control even in entering such openings as the South Pass of the Mississippi River, the Calcasieu River, and the small river mouths of Mexico.

On the northern California, Oregon, and Washington coasts below the Strait of Juan de Fuca, tows nearly always enter on the hawser provided the bar is not breaking. They are started in at the very end of outgoing current so that the swells on the bar are at their lowest and the current will begin to assist them as they slowly drag their burdens into safer waters.

In other areas, where it is both possible and desirable to get into the notch and push from the sea buoy, a long period of approach at reduced speed is required. It takes 20 minutes at dead slow, the lowest RPM of the tug's propeller(s), to get a 550-foot loaded barge drawing 30 feet or more to slow down enough so that her tow pendant can be easily disconnected and the tug relocated at the notch. If this is not possible, it may be a slow, hard pull, adjusting speed and hawser for sea and current before arrival in smooth water where disconnection can be made.

Letting Go

Letting go of the barge from the hawser is a delicate maneuver requiring training and communication between deck crew and the captain or pilot. Calmness and cooperation without any shouting of orders will bring success. Crew members on the stern must constantly watch the tow cable and avoid being struck by it. The slightest tap of a taut tow cable results in serious injury.

As the tow approaches the area chosen for disconnecting, it should be slowed as much as possible while heaving in on the tow hawser. Crew members assigned to assist in disconnecting should be standing by aft with such items as a maul or heavy machinist's hammer, cold chisel, pipe wrench, pinch bar, and a long rope stopper. Different connection methods require some, if not all, of the above, and the procedures are similar.

With the tow now shortened so that the shackle of the tug's tow cable is about to come in over the rail, the tug's centerline should be as nearly as possible in line with that of the barge. Loaded barges are slow to turn and, due to wash from the propeller(s), may lay off a little. If so, at this point the captain may order heaving of the cable to be stopped until he has brought the tug and loaded barge to a more compatible and steady position.

With an empty barge, the tow bridles and cable lead downward, and it is easier to get the tug and barge lined up. It is also easier to get fouled up if done with too much haste. Patience and calmness are required.

When the barge, loaded or light, is in line with the tug and not swinging, the tug's cable is heaved in until the tow shackle at the pendant is about halfway between the stern roller and the tow winch rollers. With the tug's engines still turning slowly, the rope stopper is secured for quick release, outboard of the pendant thimble. Cotter keys on the shackle nuts are cut or welding wires removed, and the shackle nuts are loosened and can be taken off.

The tug is stopped and the tow cable backed off until it falls slack on deck while the stopper holds the pendant. A tap on the shackle pin may be necessary to knock it free. Many tug crews have had a bail loop welded onto the rigid end of the tow shackle. This helps in quickly pulling the shackle pin free.

When the pin is free the end of the stopper is released as the tug starts ahead. The stopper should fly loose and the pendant drop over the side. The tug then proceeds swiftly into a position to tow either on the hip (alongside) or in the notch.

To Notch from the Hawser

It is not unusual for a barge released by her tug to begin swinging slightly. When the tug enters the barge notch, safety lines should be put out from the side toward which the barge must be headed when on its new course. A tug in a snug notch can work ahead at dead slow speed and still steer a loaded barge. Turning and advance will be slow. With all wires and lines fastened, the tug should be able to go from hauling hawser to full ahead in the notch in 30 to 40 minutes.

Picking Up the Light (Empty) Barge—Head-and-Tail

If the tug arrives from sea towing an empty barge and the tug master wants to pick it up alongside head-and-tail, consideration must be given to wind, current, and available room so that this tricky maneuver may be quickly and safely accomplished. If the captain has never done it before, he would be wise to choose the first smooth and open area. With practice and confidence in one's ability, this maneuver can be done smoothly in rather restricted areas.

To begin the maneuver to bring the barge to a head-and-tail position, the tug master should reduce speed until the barge has lost nearly all headway and the tug's hawser is slack and hanging down at the tug's stern. The tow should be heading in such a direction that the wind and current are ahead. If wind and current are opposite in direction, the captain should decide which will have the stronger effect, the wind on the barge or the current on the tug. The next choice is which side the tug must make up to. This will relate to which side of the barge will go next to the pier or to some other consideration regarding the passage to the landing.

If the barge is unmanned a heavy length of deck line would have been rigged on either side from one of the barge's near-amidships cleats. This line should be securely fast on the barge and hanging overboard above the water's edge. In that position, a crew member can hook it when the tug comes alongside.

If the tug is a twin-screw, the final maneuver is easier than with a single-screw. When ready, with the tug and barge nearly stopped and headed so that the barge has the wind slightly on the side opposite to

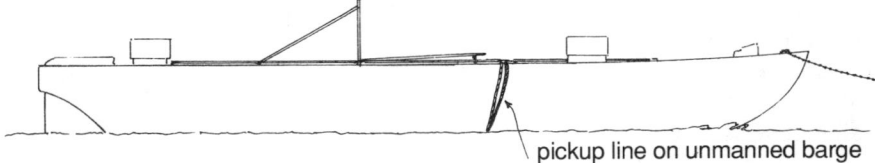
pickup line on unmanned barge

A good doubled soft line is often rigged and hung over the side of empty barges so that the tug may come around and make up head-and-tail.

which the tug will land, the tug is stopped for a moment. Enough hawser must be out to allow the tug to swing well clear and ahead of the barge, but not enough to allow the hawser to drag on the bottom. Practice will tell the correct length.

Two crew members must be standing by on deck, one at the tow winch and one with a boat hook on the bow. With the tug stopped, the rudder is put hard over in the direction to which the tug will go alongside the barge. Without taking a strain on the tow hawser, the tug should be quickly worked away from the barge's path and turned until it is parallel to the barge but heading in the opposite direction.

This is the tricky part of the maneuver. The tug must not be too far off or the pickup line from the barge will be missed. Being too close before the barge's bow or rake passes will cause damage to the tug's rails and deckhouse. If the hawser is not slack enough and takes a strain, the barge will swing away, and, like a dog chasing its tail, the tug will never catch the barge and a new attempt will have to be made.

With the tug quickly turned and in the right spot with some allowance for the barge to blow toward it slightly, the pickup line from the barge should be easily taken aboard and secured and the tug's hawser heaved tight. In this position, head-and-tail, the shackle of the barge pendant should be close to the tug's rail and the barge now alongside should handle very well. If it is dark, do not forget to change the barge running lights. Its stern is now the bow and should exhibit the correct sidelights.

TOW LIGHTS AND DAY SIGNALS

With the 1981-83 adoption of the International and the Unified Inland Rules, the navigation lights required on tugs and tows have changed considerably. Pilothouse watch standers and lookouts should be thoroughly familiar with each display as required for the length of the tug, the towline, the position of the tug, and the area of operation. When a different area is entered or a change in the towing mode occurs, these lights should be changed at once to the new display required. At daylight, if the weather is clear, they should be turned off, a habit that many tug masters have at times failed to carry out.

The diamond shape, Rule 24 (a) (v), required under International and Inland Rules when the length of the tow exceeds 200 meters (656.2 feet), need not be shown on inland waters but no ruling forbids ocean-going tugs from having such a diamond permanently displayed whether with or without a tow. See "U.S. Coast Guard Ruling Regarding Length of Towing Hawser" in chapter 13.

9. Special Types of Towing

LONG-DISTANCE TOWING—THE TUG

TUGS specially built and classed for long-distance ocean towing are larger than those multipurpose tugs regularly servicing routes to Alaska, Hawaii, and the West Indies. These large ocean tugs must be able to tow without replenishment for a minimum of 30 days. Some often are at sea for 90 or more days. They all have the features of a modern ship compressed into a smaller and heavier hull. They are maintained by some of the world's largest towing companies, as described in the first chapter.

These long-distance tugs may vary in size and horsepower from the multipurpose 120- to 145-foot, 3,500- to 9,000-HP U.S. offshore tugs, to the 165- to 300-foot, 11,000- to 26,000-HP big deep-sea tugs of Bugsier, Selco, Safmarine, Smit, or Wijsmuller.

Due to the size, construction, and layout of these mammoth tugs there is a vast difference in storage space and endurance over that of the multipurpose ocean tugs. This is quite true regarding those built under U.S. Coast Guard and American Bureau of Shipping requirements and is reflected in the rather strange interpretation of U.S. admeasurement statutes.

Because of rigid U.S. Coast Guard inspection requirements, all "power-driven," U.S. vessels* including tugs of over 300 gross tons are considered inspected vessels and come under full annual U.S. Coast Guard inspection, manning, and other regulations regarding licensed officers. Those of between 200 and 299 gross tons are termed uninspected, and the requirements are fewer. Tugs of under 200 gross tons have to meet the fewest requirements and are the admeasurement target of naval architects in the designing of most U.S. tugs. To attain this low tonnage in a large hull fit for ocean towing requires a great deal of jiggery-pokery within the hull structure. The result is several unused void spaces. Crew quarters are limited in size, and storage space is at a minimum, holding only enough spare equipment and stores for one replace-

---

*Power-driven is a U.S. Coast Guard analog for a steam or motor vessel.

Sea King is an example of the smallest type multipurpose tug built on the U.S. Gulf Coast for use in the North Sea. Presently, she has been altered for ship-assist work in the San Francisco Bay area. (Courtesy S. Lang)

Ajax, a 369-ton Swedish icebreaking tug, serves the Baltic (2,500 HP). (Courtesy Neptun Salvage Company Limited)

Abeille Languedoc's 23,000 HP is provided by four engines coupled to twin variable-pitch propellers. At 220 feet, this 1,600-ton ocean tug has 16-ton bollard pull and can run at 17 knots without tow. (Courtesy Les Abeilles International)

ment or for a set number of days at sea. If there is any question of the tug's voyage extending beyond her normal endurance period, extra equipment, fuel, and stores must either be stored on deck or on the unit to be towed.

This can-do type of ocean operation has worked even though it is a surveyor's nightmare. There are also those who have for certain considerations assumed that what the underwriters do not know will never hurt them. Success for such an operation is never assured. Unless the tower and the unit and equipment to be towed have a financial or ownership bond wherein delays for refueling, repairs, or replacements are acceptable, it is more satisfactory to contract towage with a large and reputable ocean towing company. These companies have huge tugs of from 500 to 1,500 gross tons, fully crewed for months of comfortable living and fully stored for worldwide towing from the Arctic to the Antarctic.

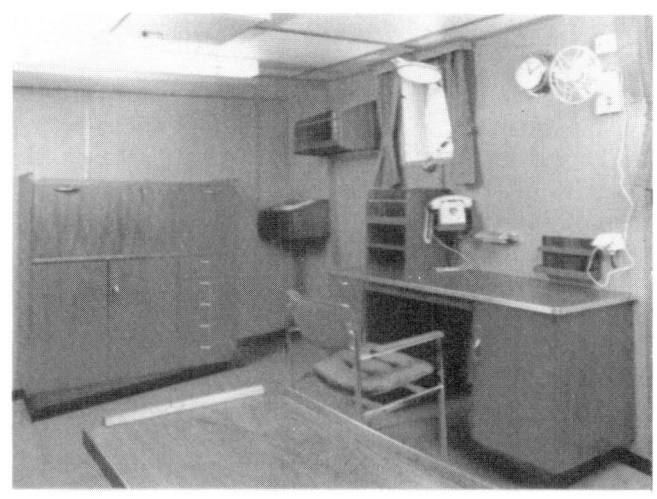

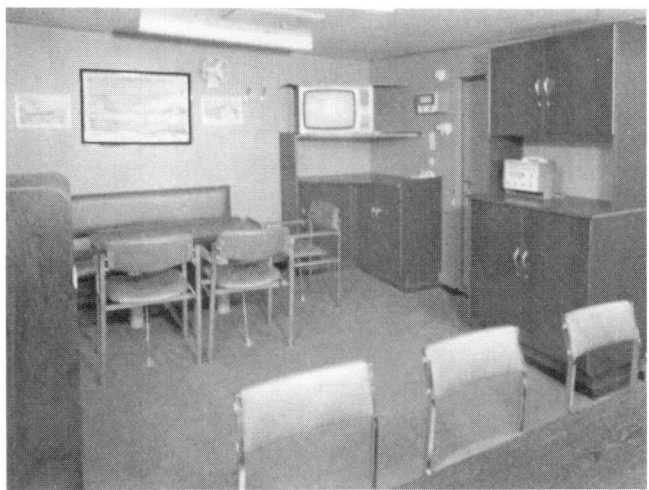

The quarters and mess rooms on deep-sea tugs are similar to those found on ocean ships. These are aboard *Abeille Provence* and accommodate a crew of from 12 to 15. (Courtesy Les Abeilles)

THE TOWAGE MARKET

As with other commodities, all towage—whether for continuous short sea hauls or one or more long sea tows—is placed either on the world towage market or through various marine brokerage firms. The brokers, in turn, endeavor to get their customers the best rate possible. Towage brokers are well informed as to the abilities of the worldwide towage companies, their equipment and record.

MARINE INSURANCE AND THE OCEAN TOW

On short sea tows, owners of barges or other units usually carry insurance coverage that allows them reasonable latitude to shop around for a competitive towage contract. Coverage for long-distance ocean towing differs somewhat. With the increasing value of such units as drill rigs, CIDs, and usable ship hulls, marine insurance coverage will be difficult if the market value of the tug or tugs to be offered is less than the value of the unit to be towed.

To protect marine insurance underwriters, brokers will usually stipulate the size and bollard pull required for certain tows and may not accept offers if the tug or its operator's towing record is questionable. For these reasons, the tug whose record is good but whose age has greatly reduced her market value may be able to tandem-tow several ships across the ocean for scrap. Yet the same tug would have difficulty in being accepted to tow a new drill rig on a similar voyage.

THE TOWAGE CONTRACT

This document specifically outlines the service offered by the towing company in every respect; it should be studied carefully and fully understood. It will be the basis of arguments in any litigation resulting from a possible failure of either party. To fulfill their acknowledged responsibilities anyone or any company unfamiliar with the parameters of responsibility and liability between tug and tow should scan *The Law of Tug and Tow*.* They should also go over it with the tug master who will be in charge of the tow.

PLANNING THE TOWAGE

In planning any long-distance tow, the tug master should be included as the major consultant. If he is to shoulder the responsibility of safely

---

*Published by The Continuing Legal Society of British Columbia, University of British Columbia, Vancouver. The standard work in the field is Parks, Alex L. *The Law of Tug, Tow, and Pilotage*. 2d ed. Centreville, Md: Cornell Maritime Press, 1982.

delivering a multi-million-dollar tow to its destination at a profit, his position and confidence must be included with that of top management.

To arrive at a profit, consideration of all eventualities must be reviewed. Alternatives such as storms, the possible breakage of tow gear, reduction in speed, time lost, and cost of assistance at fuel stops, if necessary, must all be considered.

If the tow is to commence as many do at an outport, one away from the tug's and company's base, a conference should be held between the tug captain and his operational persons covering all aspects of the towing contract. If the projected towage happens to be a repeat movement of the same or a similar unit and/or for a prior customer, the tug master with his local agent can often conclude all the necessary details.

The Route

First priority in planning long-distance towage is usually the choice of the route. Looking at profit, the company will select one. Looking toward safety, the underwriter's surveyor may choose another, and his choice may not agree with the route thought to be reasonable and safe by the tug master.

A compromise may be necessary. The tug master should not be intimidated. If he is a real and proven master, his reasoning should be accepted. He will know that the old adage that "a straight line is the shortest distance between two points" does not always hold true on the water.

So, in considering what route is to be taken, prevailing winds, currents, and weather patterns must be compared with those of an alternative route that may be more advantageous. If the unit to be delivered is bound for Arctic waters, the seasonal ice movement will be the criterion for study.

When requested or otherwise involved, Lloyd's Register has always issued a towage certificate to certify that the hull under tow is seaworthy. They also provide comprehensive towage certification covering towage arrangements and towing equipment. This certification is entitled "Fitness to be Towed."

Other classification societies may provide a similar service. Many individual heavy tows emanating from U.S. ports are certified by either the American Bureau of Shipping or the U.S. Salvage Society.

If the tow planned is to run into known ice areas, consideration of ice classification should be given. Finland was the first to give such classification for ice navigation. Lloyd's and A.B.S. also issue the same (see Ice Classification in appendix I).

BeauDril's class IV Arctic icebreaking tug *Terry Fox* was built by Burrard Yarrows Corp. in North Vancouver, British Columbia. This rugged craft is named for Terry Fox, the teenager who ran across Canada from coast to coast to raise money for cancer research. He was terminally ill at the time and ran with an artificial limb.

Speed and Estimated Towing Time

The type of tow, draft, and speed also will enter into the final decision as to whether it will be faster and safer to take a longer and more sheltered route than a direct transit. A fair target formula is to find out, if possible, at what speed the unit has been previously towed and by what tugs. Comparing their bollard pull or horsepower with that of the tug now proposed should produce an average speed that can be made good.

If the unit has never been towed, the designer's estimated towing speed may not be accurate. A 10 percent reduction is a safe application. Applying this speed estimate to the total towing distance will give the number of days required in good-weather towing. To this add 10 percent for bad weather plus the number of days estimated for slowing down to repair or refuel. This number times the tug's daily towage rate is the lowest figure that can be safely offered by a towing company. Before the offer is made, a further check with a marine weather specialist would be advisable in confirming the right choice of route.

PREPARING AND RIGGING THE OCEAN TOW

The ocean tug that has been accepted should, if possible, arrive at the site where her tow is moored in plenty of time to go over requirements with the surveyor, to take on stores and fuel, and to accomplish any customs or other port entry or clearances.

If the country from which the the tug is departing is not the same as that in which it is registered, a formal entry will be necessary. Crew lists, stores lists, deratization certificate, load line certificate, radio license, health and effective classification documents, and the tug's registry are among the many items the master must produce.

On well-run tugs, lists of these items along with all stores, spare parts, and equipment with their current status are kept updated. When computers are available, this task of a readily available inventory is simplified. Updated printouts of these items should flow between the tug and its operational office. An established shipping agency is usually appointed to handle all of this entry and clearance paperwork. If not, the tug master must do so.

While all of this is going on, the tug master or chief officer and the surveyor representing the owner of the tow or its insurance carrier will go over a conditional towage survey and its recommendations. If the tow is to be made up of one or more ships bound for a scrapyard, or a single ship or large hull without power, the surveyor's recommendations will include the interior blanking off of all sea valves and through-hull fittings, securing of the steering gear with the rudder in an amidships position, sealing of all watertight closures, securing the propeller and shaft, and tightening the stern gland. All voids and bilges will be dried and the interior of fuel, potable water, and other tanks for liquids will be checked for proper leveling.

Welders are usually in attendance and are involved in much of the above as well as in any stiffening of deck fittings on the bow where heavy chain bridles are to be secured. Shackle nuts and bridle connections may be tack-welded in addition to being fitted with cotter keys. When checking for wear, particular attention should be given to shackle pins and the flounder plate. The latter is usually checked by the ocean tug's chief mate.

Due to past breakages, the surveyor may require a certification of the chain bridle and shackles by a certification society. Where chain bridles pass out of chocks on the vessels to be towed, grease should be applied and the openings stuffed as tight as possible as a precaution. These wedges may not hold but will stop some wear for a while. Draft and Plimsoll marks should be repainted and a prominent short white stripe should be painted at the waterline on either side of the bow and near the stern so that the tug can see them.

All loose gear on the tow must be secured and the towing day signal rigidly installed. A set of long endurance navigation lights must be provided. If the tow is a loaded barge, the tug master and mate should make a very careful in-depth inspection of its vents, sounding pipes, watertight closures, and cargo lashings. Draft marks and load lines should be clean or freshly painted. These tug officers should act as independent surveyors and refuse to accept any conditions they feel are unsafe regardless of whether the surveyor(s) of one or more interested parties has/have indicated their approval of seaworthiness. Several cases have occurred where loaded barges and other tows have been issued a statement of seaworthiness only to have cargo lost at sea or an unclosed vent allow sea water to enter and sink the barge. Such occurrences are at once directed against the tug master's responsibilities.

CONNECTING TO THE OCEAN TOW

This is done in port at the pier that the tow may be lying at or in an anchorage. If the tow is of one or more ships or barges, they should be moored bow out, side by side so that the tug can back in to make the connection on her deck. If it is a drill rig, the tug may have to anchor and run her small boat with a messenger to pick up the rig's heavy chains from its legs.

Single barges are connected to the tug wherever they lie, either by directly backing up to the barge's towing end or by lying head-and-tail if the barge is not headed out.

DEPARTURE WITH THE OCEAN TOW

Leaving port with an ocean tow is not just a matter of letting go lines at any time. Planning and coordination are necessary, particularly if the harbor is small and/or there is considerable other traffic or if the tow is to be classed "as restricted in its ability to maneuver." The latter designation is usually applied to drill rigs, CIDs, and damaged or loaded ships.

A time of departure must be chosen that has favorable current and tidal conditions. It should also, if possible, be at or during a cycle of good weather. Winds should be no greater than force 5 or 6 and not expected to increase at least until the tow is well at sea.

Advice of the expected tow's departure should be given to and agreed upon by port and pilot authorities. Any special notices due to restriction of maneuverability or draft should be announced well in advance. Assistance from local tugs will be necessary in holding the tow against the wharf while lines are taken in and the tow clears its berth, in streaming tandem tows, and in helping to steer through congested harbor waters.

Prior to getting under way from the wharf, the lead tug should invite the captain of the helping tug(s) and the pilot to meet and confer on any plans necessary to get the tow away from the wharf, streamed, and headed out to sea. Assisting tugs usually remain with such tows until they are clear of jetties or buoyed exit channels.

AT SEA WITH THE OCEAN TOW—CHAFING GEAR

With the hawser slacked out to its sea length, ocean tugs, most of which have several tow arches, will protect their tow cables at these points by adding sleeves of neoprene or similar material that can be slid over the wire cable. Before the tow winch is secured it should be stopped in the pay-out mode so that the tow cable will lead from the center of the tow winch drum and directly through the tow winch vertical threading rollers.

At the stern, a chafing arrangement (see chapter 8) of split pipe, H beam, or other protective devices may be used. However, many ocean-going tugs use a combination of stern roller with hydraulically operated norman pins and a hold-down device. Some use a heavy shackle on a short heavy wire strap shackled into a pad on deck. All of this chafing gear should be watched hourly and attended daily by heaving in several feet or letting out several feet so as to change the friction and bending points.

TOWING WINCHES AT SEA

The winches should be set for the appropriate tension if tension devices are included. The fine U.S.-made Almon-Johnson winch has such a device. Others have springs that give and take as the added strain of swells are applied. Those with tension regulators usually have drums that unlock separately, similar to the wildcat of an anchor winch. When using such a winch at sea the drum should be set and held by the tension gauge, not by engaging a claw dog, because the drum moves back and forth slightly. The dog will hammer itself in and out of the securing teeth and eventually break off.

Maintaining a constant and proper tension on the tow cable is of utmost importance and should be thoroughly understood by the master and every watch stander both on deck and in the pilothouse. Actual readings in the pilothouse or on the bridge of seagoing tugs showing towline hawser tension have been provided by manufacturers for some time. The A. J. Hoyle* tow winch first introduced a mechanically activated dial mounted in the pilothouse, which registers the amount of

*A. J. Hoyle Manufacturing, Iron Mountain, Michigan.

Above, this example of a tow cable hold-down is on *Smit Singapore.* Tension is adjusted by a separate winch. Her hydraulic norman pins are recessed, and guards have been built over rollers and valves. (Courtesy Smit Tak) *Below,* a Safmarine ocean tug tows the disabled Danish containership *Adrian Maersk.* Notice that the tug's cable hold-down is a socketed wire held by a safety shackle. (Courtesy Safmarine)

towline out and the tension. The huge modern deep-sea tugs now have LED readouts of both.

Markey winches do not have springs and are provided with automatic or electric brakes. Other tow winches offer a hand-set tension-control indicator. The cable will pay out once the preset limit is past. The cable will reel in when this amount of tension is released. This is an automatic pay-out, heave-in feature. Another, the Hydraulik-Tencomp System used on some anchor-handling tug/supply vessels in deep-sea towing also pays out and heaves in the towline as the towing load increases and decreases. On tow winches that are fitted with automatic tension devices, the range of the tension is the greatest when only one layer of towline remains on the drum.

One of the problems with these systems is the requirement to keep power on the winch at all times when under way with the towline out. Another is the continuous wear on tow winch rollers, tow bars and arches, and the stern rail, all of which makes the use of chafing gear futile without its continuous adjustment. Some drastic results have occurred through the incorrect setting of tension devices. Tow cables have payed out to the bitter end without retrieving. Towed units also have been heaved up very close to the tug's stern.

Tugs with dual-drum tow winches or waterfall winches (where one drum is mounted over the other) may use a different size tow wire on each drum. Winch drums are usually manufactured to receive so many turns of a certain size tow wire. This may range from 1-inch diameter as used on a 1,000-HP tug to $3^1/_2$-inch diameter for a 22,000-HP tug.*

An incorrect-size tow cable should not be wound on a winch drum. The result will be a half turn that will not fit or fill it. Next, the threading device that guides the cable to form smooth, tight layers will, on rewinding, get out of step and out of line with the wire's turns, making frequent adjustments necessary. This difficulty is hard to correct if there is tension, as is usual when shortening tow cables. Other causes for misalignment can be paying out too fast and wear on the threaders' vertical rollers.

On some of the largest deep-sea tugs, the tow winch is located under deck and aft of the engine room. The cable enters a slot in the deckhouse and passes over a large sheave. Then it runs down at right angles to another set of in-line movable sheaves, one of which comes from a spring-loaded ram. This dampens the strain on the cable, which, due to the great weight of the tug descending on a swell, can build up quickly.

The Murdock tension damper is another installation offered, which, through sheaves and counterweights, relieves sudden strains on the haw-

---

*See "Tug Power versus Towing Wire Size" in appendix II.

These are synchronized towing winches on *Abeille Languedoc*. They are fitted with side-by-side tension drums and springs. The tow cable is wound on the drums on the lower deck. (Courtesy Les Abeilles)

ser. The Murdock tension damper can be installed on deck between the tow winch and the long stern of the tug/supply vessel or below deck if the tow winch is so located.

Care of the Tow Winch and Tow Wire

Maintenance of the tow winch and tow wire should receive top priority. Failure by either renders a tow useless. Properly cared for, tow winches will serve faithfully for years. Simplicity in operation, rugged construction, and the ability to stand hard use are challenges the many manufacturers of these devices must meet.

Among the most successful, the Markey tow winch* has proven very popular. It is hydraulically controlled from pumps in the engine room and operated through a valve on deck at the winch-operating station or by torque-converter transmission mounted on the winch's base. Three very positive actions are possible through sliding gears. When engaged, the winch can heave in or pay out under power. Disengaged, the cable can free-spool in paying out and may be held by a tightened brake band. The latest models offer power replacement for hand operation of the levers. There are many other fine tow winches provided by this and other firms, designed to serve whatever special purpose is required.

A program of lubrication and protection against deterioration of the tow winch must be followed. Drains must be provided to allow standing water to escape from around and under tow winch beds. This is the first collection point of rust and crud. All grease fittings should be kept full and pressed up if possible when heaving the tow wire in. This allows the lubricant to circulate through moving parts. The traveler and rollers of the threading device should be kept greased. Tug mates and engineers bear the responsibility of overseeing and carrying out tow winch maintenance.

Amount of Towline on the Winch Drum

Very few tow winches have an indicator that shows how much cable is out or remains on the winch drum after some has been slacked out. Lacking this, each tug master, mate, and winch operator should know how many complete layers of tow wire are on the winch drum. As an approximate example, a tow winch built for use with $2^{1}/_{4}$-inch diameter, 6×37 wire rope tow cable of 2,200 feet would have a 30-inch-diameter drum on which the cable is spooled. There would be about $10^{1}/_{2}$ layers of cable, which should always be tightly spooled, leaving no space between each turn.† Otherwise when a heavy strain is placed on the tow cable, it

---

*Markey Machinery Company, 79 S. Horton St., Seattle, Washington 98124.
†See tables in appendix II.

will cut down and wedge itself in the space between turns and will not free itself. The entire length of cable may have to be run out and tightly rewound to correct this.

As many captains regulate the length of tow cable they use in certain waters by the number of turns let out, it is important for the winch operator to watch the winch drum, counting each layer as it rotates. The wire should be braked in time to hold when the final turn is leading from the exact center of the drum. This will avoid any bend in the tow cable between the drum and threading device rollers. The tow cable should run straight from the center of the layer in use through the rollers and follow the tug's center line to its stern.

INSTALLATION AND CARE OF THE TOWING HAWSER

Care of the tug's main towing hawser is an absolute necessity. It is the most expensive item of tug equipment. If used frequently it must be regularly replaced. Eighteen months is a recognized limit of safe usage. On tugs where much of the daily or weekly towing is done on the hawser this time will be shortened if proper care is not exercised.

Three progressive steps are included in the normal care of a wire rope hawser. They are installation, usage, and preventive maintenance. Wire rope is shipped on heavy wooden reels. When a reel of 6×37 improved plow steel cable is delivered alongside the tug or to the owner's yard or warehouse, it should be handled carefully so that it is not dropped or the reel bent or damaged. Such damage will make winding

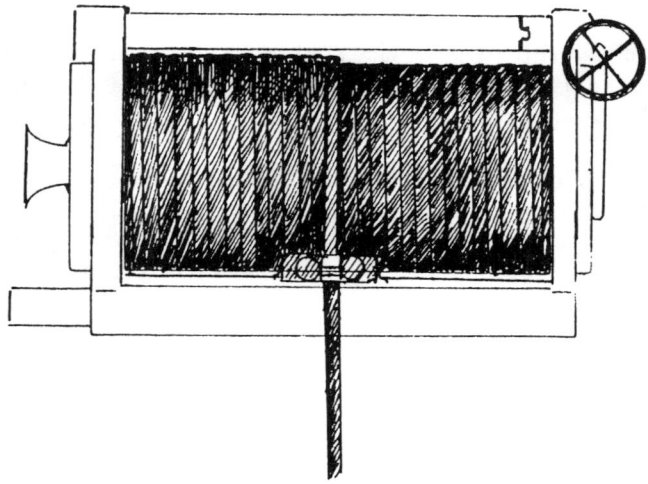

When towing from a single-drum winch, the cable should lead straight from the center through the threader.

the wire rope onto the tug winch most difficult. While being run to the tug, it should not be dragged over the ground or any surface that will cause removal of the wire's protective coating or allow it to pick up grit or harmful liquids.

If the cable is to be stored on the reel for future use it should be covered and if possible placed out of the weather. If the wire rope is to be installed immediately on the tug, good suppliers often provide riggers with a set of steel jacks and pipe so that the cable can be run from the wooden reel to the tug's winch drum without touching anything other than the tug's deck. New wire rope will have the manufacturer's soft, thick, protective coating, which will tend to flake off as the hawser is run aboard the first time.

Securing the bitter end onto the winch drum is usually done by passing it out through a hole in the side of the winch drum shell or through the eye of a wire rope clamp on the inside. The hole that some winch manufacturers provide and the wire rope clamp arrangement

Red Circle Towing's *Gayle B* has a single-drum hydraulic Markey TYS 36 tow winch. With all the turns filling the drum, the wire leads directly off its center. Notice the tow cable ends in a solid thimble fitted with a heavy-duty bushing.

with the nuts on the outside shell are usually on the right hand or port side of the winch drum when facing aft. The nuts on the clamps are kept greased so that if at any time it is necessary to slip the towline they may be quickly unscrewed.

The wire is reeled onto the winch by guiding it through a threading device so that each turn rests tightly against the previous one. The turns run first over the top from port to starboard on the first layer and reverse the direction on the second so that every odd layer goes from port to starboard and every even-numbered layer from starboard to port.

As a 2,200-foot coil of $2^{1}/_{4}$-inch 6×37 wire on a 30-inch drum will result in $10^{1}/_{2}$ layers, the last one will end in the middle after having run from port to starboard. Remembering these facts and applying them to other size cables, operators can apply a quick rule of thumb solution to determine how many layers are out.

Tow hawser usage requires a constant effort to keep the wire rope from being dragged on the bottom, jammed or pinched in any way, or kinked or worn at a particular spot through lack of chafing gear. In rewinding the hawser, attention must be given to see that the threading device directs the cable so that it will lie tightly against the previous turn and that the threading device does not start to return too soon in the opposite direction. On some tow winches this can be corrected by stopping and paying out a little to take the strain off the cable as it passes through the threader, which can then be reset by turning a hand-operated linkage.

After its first use IPS* wire rope cable will have exuded some of the oily fluid contained in its hemp core. This protective lubricant should be replenished from time to time—once a month if necessary, as wire rope will first deteriorate from the inside strands. Without advance planning, replacing the lubricant, while necessary, can be messy and difficult. Warm, sunny weather is the best time for this operation, although a cold, dry day of above-freezing temperature is acceptable if the lubricant to be used is heated. There are many such products offered by major oil companies and wire rope manufacturers. The consistency of the lubricant and the possibility of diluting it must be considered as application is accomplished by using either a long-handled turk's head brush or an overhead tank with a spray pipe or pressure sprayer. After first soaking the layer(s) still on the drum, each additional layer should receive a coating as the wire comes in. IWRC† hawsers should receive the same treatment.

The greatest wear on the towing hawser occurs within the outer third, which is connected to and nearest to the pendant and bridles of the towed unit. If the tug's normal assignments do not require the use of more

*IPS = Improved Plow Steel.
†IWRC = Internal Wire Rope Core.

than half to three-quarters of the length of her tow cable, turning it end for end after the first year of use may extend its life for another year.

Sockets and Towing Hawsers

Improved plow steel tow cables have always faced two enemies. These are deterioration of their inner strands and deterioration at the splice near the thimble on the towing end. To overcome these problems, some towing companies have changed to galvanized cables. They also have used sockets, which have always been considered a costly but superior method of treating this connecting point at the bitter end of the hawser.

Recently, new resins have made the cost of pouring sockets financially feasible. To pour a socket, the strands and wires are opened after the end has been passed through the socket. Then the resin, which has been heated and treated, is poured into the opening and runs among the strands and wires. When it cools or is set, a wedge-shaped mass of wire and resin is encased in the socket, and if the resin was properly poured, it cannot be pulled out.

CABLE PAY-OUT AND CATENARY

Paying out the tow cable should be done at an easy speed. Rope hawsers properly coiled on deck will run out smoothly. If too much speed is used, one turn of the hawser might grab several others in a half-hitch or knot—a knotty situation to have to stop, heave in, and untangle. With wire towing hawsers coming off the drum of a tow winch, the drum should be in free-spooling or pay-out mode with the brake open. Running an electric tow winch drum too fast while in pay-out can cause a reversal of polarity within the electric power motor.

When paying out tow cable, the tug captain or pilot must watch astern to make sure that his deck crew is clear. He must not make any radical course changes without warning his crew. If it is necessary to run the cable out in a hurry, the tug may have to back full for a moment to slow or stop the pay-out. Once the tow wire is out it should be held when the turn coming off the top of the drum is in the center and leads directly between the vertical rollers of the cable threader. When the cable is held, the tug should start ahead slowly, not at full speed. The tug master should work up gradually, not suddenly placing too much of a load on the hawser or other parts of the towing gear. The feel and sound of the cavitation will reveal to the tug master how the tow is responding to the tug's power. As the tow's hydrodynamic resistance reduces and the tow picks up speed, the tug's propeller cavitation and vibration will reduce until finally a smooth momentum with no cavitation is reached at full speed. The tug should then be pulling at her best.

COMMUNICATIONS AND THE OCEAN TOW

Whether the tug or her owners are in Argentina, Australia, the British Isles, or Canada, each owner will appreciate daily contact with his tug and its tow. For U.S. owners, if their tugs are within the country's continental limits, voice conversation through the use of pilothouse-installed voice high frequency radios (VHF) or single sideband (SSB) are commonplace. These frequencies have become very overloaded, however. To circumvent these systems, new ones have been introduced. Each bears its manufacturer's registered trade name. One, using the name Watercom, is a marine automatic direct dial system covering the Gulf of Mexico as well as some inland waterway areas such as Ohio and Illinois.

For the ocean tug, such a specialized system would have to be justified through constant usage. As the twenty-first century approaches, numerous electronic manufacturers continue to offer equipment that interfaces with satellite communication and navigation and weather reporting systems. Competition is great. Satcom is one such system available for deepwater towing vessels. It is a lightweight, compact satellite communication system offering telephone connection worldwide, telex, word processing, and facsimile services. It meets all International Maritime Satellite Organization (INMARSAT) standards. So swiftly are new developments appearing that some systems are already earmarked for obsolescence. Nearly all of these new communication systems are computer-linked with shipboard displays and onshore printouts. As most are tied in with other features, a look at the electronic navigational bridge of the modern seagoing tug is in order.

NAVIGATION AND THE ELECTRONIC BRIDGE OF SEAGOING TUGS

The seagoing tug is not restricted from accepting modern marine electronic technology. On many, the simple individual gyrocompass, autopilot, Loran, radio, and radar units have been retrofitted or replaced on the bridge by systems which either incorporate or are interfaced with one or all of these components, which themselves have been updated. Radar has never legally replaced the human eye as a lookout, but in supplementing it and greatly increasing the scope of vision, it has no equal. The basic radar set has been linked with the gyro and vastly improved for collision avoidance through computerized circuitry in order to meet the IMO regulations of Automatic Radar Plotting Aids (ARPA), a specification which takes effect in 1991.

In addition to ARPA, color resolution has been introduced, offering targets in different shades so that various objects can be identified. Furuno* is a popular color radar, showing targets in blue at various

---

*Furuno Electric Company Limited, Japan.

intervals running from 15 seconds to 6 minutes. Superimposed ranges, bearings, plotting, and pulse length appear at the four corners of the display. Another of these systems is Viewnav,* a highly sophisticated interactive computer system. It features three distinct displays. These are shown on a set of electronic charts for each harbor. They have been precisely digitized from a similar U.S. NOS chart. Landmasses and bridges appear as a light color. Bare shoals appear as light blue and shoal water as a darker blue. Navigational aids such as buoys and light beacons stand out in the natural color. Tugs and tows as well as other moving targets are revealed in true motion. An immediate readout of targets and one's vessel's course and speed are shown, and the time of next course change (now referred to as "way point") is given.

Naview has been found to be very bulky and expensive compared to other products being introduced. New electronic charts are becoming a valuable navigational tool at the average small vessel chart table or nav station. Navtex is another new and internationally adopted automated radio system. It is operated by the U.S. Coast Guard on the East and Gulf coasts. The onboard components are a smart receiver and printer. An operator may punch in and receive continuous reports on ice conditions, navigational warnings, weather, and local notices to mariners. By 1991 the International Maritime Organization (IMO) plans to have all merchant vessels over 200 gross tons carry a Navtex receiver. This will include many tugs of all the IMO member nations.

Weather Recorders

Tugs equipped with weather facsimile recorders are now common. The equipment is small and can be purchased for a limited area or for worldwide reception. Weather charts of every description are broadcast over 50 different transmitters covering 8 worldwide areas, including the navigational portions of both polar regions.

The data received include surface and frontal analysis, satellite weather pictures of cloud patterns and storm developments, surface wind data, and wave and ocean-current analysis with sea temperatures. For polar regions, ice location and its characteristics are given. Alden† and Furuno are early specialists in weather facsimile receivers.

Navigation and Piloting

This is a field in which big improvements have been made. No longer must the tugboat navigator wrap a leg around a stanchion on the boat deck as he tries to bring the sun, or a star the clouds are trying to hide, down to a dancing horizon, hurriedly read a stop-watch, and rush in to

---

*Viewnav, Navigational Sciences Incorporated, Bethesda, Maryland 20815.
†Alden Electronics, Westborough, Massachusetts.

work out the results on a small chart table where everything is sliding back and forth. The same results may be achieved much more easily today.

A sextant, a good timepiece, and *Bowditch* or other navigational tables are still great standbys. For many masters and mates who enjoy maintaining their proficiency in celestial navigation, it is nice to practice this form of old-fashioned position finding. It is interesting to note that navigation position finding by the celestial method remains the basic requirement by licensing agencies.

Celestial navigation in some areas was partially replaced by Loran or Decca and dead-reckoning sensors and radio navigational aids. Electronics and the microprocessor have opened up their application to new, rapid, position-finding systems. Omega was introduced in the U.S. through the Navy and Coast Guard. Most of these satellite systems have been linked with earth stations. To make position fixes more precise, Loran-C was introduced and installed on worldwide tug fleets.

More recently, the powerful integrated navigational computer appeared. Bowditch Navigation Systems* have introduced one that accepts input from all available navigational sources and aids. Satnav by Magnavox, Star Trac by RDI,† and Furuno are others. Several of these are designed to accept GPS (Global Positioning System) when it becomes available in the late 1980s.

Electronic Navigation for the Offshore Tug

GPS was introduced by the U.S. Air Force as the latest navigational tool for commercial use. It should surpass current navigation satellite systems just as they have surpassed the sextant. Also known as Navstar, it consists of 18 satellites in 6 orbits 20,200 kilometers high. GPS will continually provide latitude and longitude worldwide within approximately 262.4 square feet of the vessel's position.

In conjunction with the activating of GPS, the U.S. Coast Guard will phase out Loran-C by 1992, and the U.S. Navy will discontinue their Navigation Satellite System by 1994. Although the cost at breakthrough was tremendous, GPS has dropped to that of the other systems it has replaced and will continue to drop, as its circuitry is composed of the digital chips used in many other integrated circuits.

Other Electronic Devices for the Tug

With all of the communication, radar, and navigational systems linked to a central bridge computer, there is still room for others, if necessity requires. Here is one: Tow Watch by RDI (Mark X-T-W) is very useful in

*Bowditch Navigation Systems, Portsmouth, New Hampshire 03801.
†RDI, Radar Devices Inc., San Leandro, California 94577.

the fog or poor visibility. It allows the tug watch officer to set up a tow watch zone around his tow with a collision warning guard as well. As long as the tow remains in the preset watch zone, a digital readout of distance and bearing from the tug will appear. Through interface with the tug's radar, the watch zone will appear on the radar scope, and an automatically tracking and bearing marker will also track the tow. Should the tow drift outside the assigned zone, a lost target and audio-visual alarm is activated.

Other electronic equipment includes automatic speed transducers, which are improved nonfouling six-bladed propellers of alloyed plastic with Lexon housing to eliminate electrolysis problems. Autopilots, most of which interface with the Navigation Satellite System have been improved by Sperry's Adaptive Steering Modules, which reduce heading error and rudder angle. An ASM fits any Sperry Gyro Pilot.

Compasses have also changed. Master gyros are smaller, and one new automatic calibrated compass services Satnav and other satellite navigating systems. This is the Magnavox MX-35 Flux Gate Compass.

In addition to Tow Watch, there is the RDI Radar Watch, which provides a visual and audio alarm if a radar target should enter a preset zone.

Watch Standers

The watch stander must concentrate on whatever he is assigned to watch. The bridge watch officer will be responsible for continuously watching the tow astern, the way it rides, and if any unusual change in draft, list, or appearance occurs. He will also maintain a tight lookout by radar and visually and frequently check his position by whatever means available. All of the results of these observations should be entered into the log, stating wind, weather, and sea conditions, with courses and changes.

A good and truthful log shows psychological alertness and protects the watch stander, the tug master, the owners, and the customer. It is prudent practice for the tug master on long sea voyages to leave written instructions of some sort so that the other officers have a guide as to what they should do, if the weather deteriorates, or a change of course should be made, without having to first call the captain.

Conversely, the watch officer should always feel free to call the master without hesitation, for any situation he may feel is other than normal and which the captain should see or act upon. A conscientious tug master will welcome such responsible acts provided he is called *before,* not after, it is too late.

The same responsibility extends to deck and other tug personnel. They must be always alert for anything, however slight it may seem, which may escalate and lead to a major problem. Some of these are leaking fuel lines near a hot engine, overheated electrical circuits, an open door or port that allows spray to get on the main switchboard. Others might be large pieces of equipment getting adrift, bilges full of flammable liquid, or a broken pipeline. All of these situations have been experienced on large tugs and have led to fire and a sudden loss of propulsion. Keep in mind that unlike an auto, a tug cannot pull off to the side of the road and stop to change a tire or await a mechanic or the fire department. The tug crew, all hands, make up the repair team, the fire brigade, the protectors of their vessel, tow, and destiny.

Beyond daily cleaning and sanitary care, some scheduled maintenance must be carried out, weather permitting. The hawser and chafing gear should be adjusted each day by either paying out several feet or heaving in a slight amount in order to change the wear point. Continual bending will create heat as in bending a nail. Inner strands will break first and the hawser will be weakened and may part at some future time if it does not do so on the voyage itself. Just before the noon watch change is a good time to make this required adjustment of the hawser wearing point as both deck crew members are usually available.

Engine personnel will keep a close watch on the level of fuel oil tanks, lube, and potable water. Some trim adjustments of the tug may be necessary as the voyage progresses in order to get the most efficient bollard pull for the tow as the tug's draft lessens due to the weight lost.

DRILLS AT SEA

Among the first instructions usually given by the chief mate to crew members is the location of individual life jackets and survival suits. It is also within the mate's duties to draw up a station bill showing the location of fire-fighting equipment, lifeboat and equipment, life rafts, EPIRB, and any other emergency gear. This bill will show where each crew member is to report in the event of fire or an abandon ship order. If such a bill has already been drawn up, it may require updating as equipment is changed or added.

Weekly drills should be held. These are required on certain inspected vessels and are a recognized safety procedure for all ships regardless of size. Fire drills usually are announced by the sounding of alarm bells and appropriate whistle signals and/or over a public address system. Such drills should not be a repetitious mustering at a fire station and the stretching out of a fire hose with the ultimate production of water pressure. After familiarization with this function has been accomplished, coping with special types of fires in various areas should be practiced. Fire drill and abandon ship drill should be held frequently at night, preferably near a change of watch.

Consideration should be given to holding a drill on how to approach one's tow and fight fire if one should occur on the tow. This might require going close to or alongside with the hawser still connected.

The tug's station bill also should specify who brings what to the life raft or boat, and how it should be launched. When opened, inflatable life rafts will sail away from a tug like a balloon. Thought should be given to how to quickly attach a drogue or sea anchor. Other weekly drills can cover how to throw lighted ring buoys, how to turn the tow, and how to retrieve someone who has fallen overboard.

Onboard engine personnel must transfer fuel so as to always keep a steady flow to the engine(s) or its day tank. Cooling, seawater, and lube oil temperatures must be checked. Seawater temperature should be passed to the bridge whenever a noticeable change occurs. Filters and strainers should be changed on schedule and a continuous flow of electrical power maintained by alternating generator sets. The latter changeover should never allow the current to suddenly drop and then surge as the alternative unit takes over.

All of the incidences described have occurred on seagoing tugs. To experience any of these situations is unnerving; however, the best safety device to handle an emergency on a tug is a well-trained crew.

INVENTORY AND THE COMPUTER

The chief mate and chief engineer should maintain a running list of all equipment and spare parts. If the vessel has a computer, such a record can be updated very simply as items are used and replaced. Many satellite communication systems will allow voice or telex messages between the tug and its office ashore. Several companies offer computer ship management; however, this may be too expensive for tug application. Besides, as successful towing is really a reflection of the tug master's competence, individuality, and touch, it would be difficult to foresee his acceptance and application of someone else's program.

EATING AND EXERCISE

One of the most thankless jobs in the world is that of the tugboat cook. Up at 0430, and working through until 1430; then up again at 1630 until 1900, he must satisfy the tastes of from 8 to 14 men, often having different ethnic eating habits. Frequently, their tastes do not follow the dictates of good health. To assuage their appetites, he may unwittingly offer daily snacks that load his crew with unneeded triglycerides, fat, and cholesterol, all enemies of the human heart and body.

Although tugs on long sea voyages seldom offer areas long enough to jog on, part of a crew member's personal routine should include 15 to 20 minutes of good exercise. A spot on deck can usually be found where membership in the big-belly club can be rescinded (see "Health and Exercise," chapter 11).

RECREATION AND ENTERTAINMENT AT SEA

For years one of the most popular, yet poorest, forms of entertainment on tugs has been sitting in the galley, smoking, drinking coffee, and eating sweets while watching television. Crew members cannot be blamed for creating this situation. Neither tug designers nor tug owners offered any alternative area for those off watch.

On the U.S.-built tug this lack of recreational area was due to the desire to keep the vessel under 200 gross tons. Some ocean tugs of recent vintage have introduced a recreation room with a library, comfortable lounge chairs, and television. The latest addition is the onboard entertainment system such as SES,* which provides videotapes to over 1,000 subscriber vessels including tugs. Exchange facilities for this system have been set up in 150 ports worldwide.

ROUTES, CURRENT, AND WEATHER

With a good set of ocean current charts and area current tables, backed up by facsimile weather charts or reports from other sources, the ocean tug master should be able to take advantage of whatever nature provides in his favor. Sometimes, this may appear as nothing, when for days speed must be reduced to several knots. A search should always be made for an alternative route or course change in order to avoid the obvious bad weather. In some areas such as the Baltic, Mediterranean, Red Sea, Persian Gulf, Gulf of Alaska, and similar bodies of water, a diversion to avoid bad weather is not possible. Sometimes, shelter in a lee is good seamanship and the only recourse. On large ocean voyages, early detection can lead to safer, smoother, and shorter towage.

There are certain spots worldwide that should be avoided if conditions are not right. Care must be taken in the entry or passage through them must be made at the correct time. The English Channel, Straits of Dover, and the North Sea can offer the tug master nightmares if he arrives or is caught under way in bad weather. There is no solution in handling the tow other than using every lee possible and slowing progress while awaiting the best possible break in the weather and tidal current when preparing to enter port. Assistance by other tugs may be necessary.

The inland route along the west coast of Scotland and through the Minches should coincide with nearly slack current and low wind conditions, as there are many whirlpools and vicious eddies in which a tug or

---

*SES Ship Entertainment Systems, Inc., Thousand Oaks, California.

tow will outrace each other. Pentland Firth at the north of Scotland is another stretch of exciting water with a terrific tide rip and sea if the current is running east and meeting a strong easterly wind. It is best to pull against the last of a head tide so as to arrive off Stroma Island at nearly slack water.

Towing across the North Atlantic has proven successful by tugs that leave the North American coastline between Cape Hatteras and the Delaware Capes on a course north of Bermuda and 50 miles north of the Azores before hauling toward the European shore or Gibraltar. If westbound, this course or one a little farther south in the northeast trade winds is favorable if the tug bound for the Gulf of Mexico or the Panama Canal via Mona Passage.

The northeast trade winds mix with the southeast trade winds off the Leeward and Windward islands. Tows bound from West Africa or South America to the Panama Canal, the Gulf of Mexico, or the East Coast of the United States can find favorable current directly to the Windward Islands and across the Caribbean north of Aruba. Here, they may divert to Colón or proceed directly toward the center of Yucatán Channel. They should stand off Cabo San Antonio at the tip of Cuba, as inbound traffic may be using a countercurrent that runs in a southwesterly direction close off Cuba's west coast.

From the middle of the Yucatán Channel, the current begins to form the vortex of the Gulf Stream, with a branch running into the Gulf of Mexico toward the Delta of the Mississippi River. The vortex itself swings northeast toward the Straits of Florida and, keeping in their center, curves easterly northeast and north to follow the U.S. coastline from 5 to 50 miles offshore until passing close to Diamond Shoals. It then bends northeasterly to pass off Nantucket Shoals and Lightship before crossing the Atlantic to diffuse into three branches: one running north of the British Isles, Faroe, and Norway; another taking off toward the Bay of Biscay and the English Channel; and the third curving off Spain toward Africa to again join the Northeast Trades.

The use of the Gulf Stream by tugboats is often a disputable subject. As a good chef will point out, the secret of good sauces is the delicate blending of various ingredients, properly mixed at the correct time and temperature. This analogy can be applied when considering the use of the Gulf Stream when towing north off the eastern U.S. seaboard.

Unfortunately, some tug owners and/or their managing personnel have the idea that the Gulf Stream is always nature's bonus to any good navigator who can find its axis. Based upon actual experience, rather than theory, quite a few tug captains will strongly disagree. In good weather, the Gulf Stream's favorable current gives a bonus in speed and fuel consumption. If strong northeasterly winds are predicted, however, it is best to pull out of the Gulf Stream, as a terrific sea can build up quickly.

If towing in the opposite direction from the upper East Coast to southern Florida or to Gulf of Mexico ports, the route most frequently used from Diamond Light Tower south is inshore of the Gulf Stream passing up to 10 miles off Lookout and Savannah Light Towers. This course curves southward off the Georgia and north Florida coasts until passing Cape Canaveral. From there until Jupiter Light the course finds deep water and a countercurrent 2 to 5 miles off, while southward from Jupiter to Miami, it is safe to navigate 2 miles offshore.

From Miami to American Shoal Light, and Sand Key Light off Key West, stay from 1 to 2 miles off each of the various lighthouses, paying particular attention to the tide, which sets onto the coral ridges they mark.

In this area, ships will overtake tows on either side. If it is winter and a norther is predicted and the tow is bound for Tampa or ports in Alabama, Mississippi, or New Orleans, it may be desirable to take the passage between Rebecca Shoal and Dry Tortugas, staying on and following the 20-fathom (36.6-meter) curve off the Florida coast until 29°25′ North. From that point, a direct course to the Mississippi or intermediate ports can then be made and the heavy seas that exist in the 150-mile-wide path of Desoto Canyon with depths averaging 1,800 fathoms (3,292 meters) can be avoided. This canyon runs from the coast of Cuba to a point approximately 80 nautical miles east of South Pass.

Tows bound for Lake Charles, Sabine-Beaumont, or Galveston-Houston should head for the shipping safety fairway lane, which runs southwest from Southwest Pass. Navigators should be wary of errant ships or other vessels which for reasons unknown occasionally stray from the confines of these safety fairway lanes. Due to the ever-changing locations of oil-exploration and drilling equipment, tug masters should make more frequent checks of their position than normally done when in open waters. This lane runs through the maze of oil well platforms and drilling rigs. It has a branch leading off toward Corpus Christi and Brownsville. Tows bound for the Mexican ports of Tampico, Veracruz, or Coatzacoalcos cross the Gulf well north of Campeche Banks and must be alert for anchored shrimpers and drilling platforms.

Off Cabo de São Roque in Brazil, there is a north-flowing equatorial current. It has been found to run up to 6 knots as far north as Trinidad and the Windward Islands.

On the Pacific Coast south of Panama, the current inshore on the Colombian coast sets northerly. Forty miles off it flows south southwest and turns westerly off Ecuador.

Northbound from Balboa, the currents are mainly northerly but the strongest winds are also from that direction. The Great Circle Route from Balboa to the Hawaiian Islands is favored by both fair current and the northeast trade winds, which continue through the Pacific Islands to the Philippines.

In the Gulf of Alaska, ocean currents flow counterclockwise from the Queen Charlotte Islands around toward Yakutat, then over to the Kenai Peninsula before turning southeasterly off Kodiak Island to complete their circular movement back toward the Queen Charlottes.

In the Bering Sea, some currents run northerly, while on the balance of the Pacific Rim and the Orient, currents run in various directions in various latitudes, mainly southerly, southeasterly, or southwesterly. Throughout the world, these ocean currents, which may be used to help a tow, are subjected to climatic changes and the sudden seasonal gales, hurricanes, and typhoons, which should be avoided.

GOING BACK TO LOOK AT THE TOW

On long sea tows, occasions have arisen where the tow appeared to be listing, taking on water, or having its cargo adrift. The tow bridles or their connecting shackles may look as if a nut has come off. From the big deep-sea tug, these problems are usually given a closer inspection by launching a heavy, inflatable, powered workboat when weather permits. In this way, it may be possible for men to board a unit, transfer a small pump, start a barge ballast pump, or tighten lashings. It is better to leave the shackle that has lost its nut alone until smooth sea conditions will allow it to be heaved aboard for replacement.

When weather permits, ocean tugs that do not have a heavily protected work or rescue boat may circle back close to or actually go alongside a towed unit, if necessary. In doing this, it is best to reduce the speed of the tow until it is almost drifting. With the towed unit nearly stopped, it must be decided which way it will drift or be blown if it is not headed into the wind. The tug captain must then choose a side from which he feels he can safely depart after his inspection without damaging his tug. In deep water of 185 feet or more, it should not be necessary to take in or even shorten the tow hawser. By turning the tug slowly, the cable will assume a dangling loop dropping off the tug's stern.

After inspection of the towed unit, the tug can ease ahead and away from it and continue to tow once the cable is again taut. If it is necessary to repeat this operation several times, the tow should be turned so that the opposite side can be used for landing. Otherwise, a twist is eventually introduced into the tow cable.

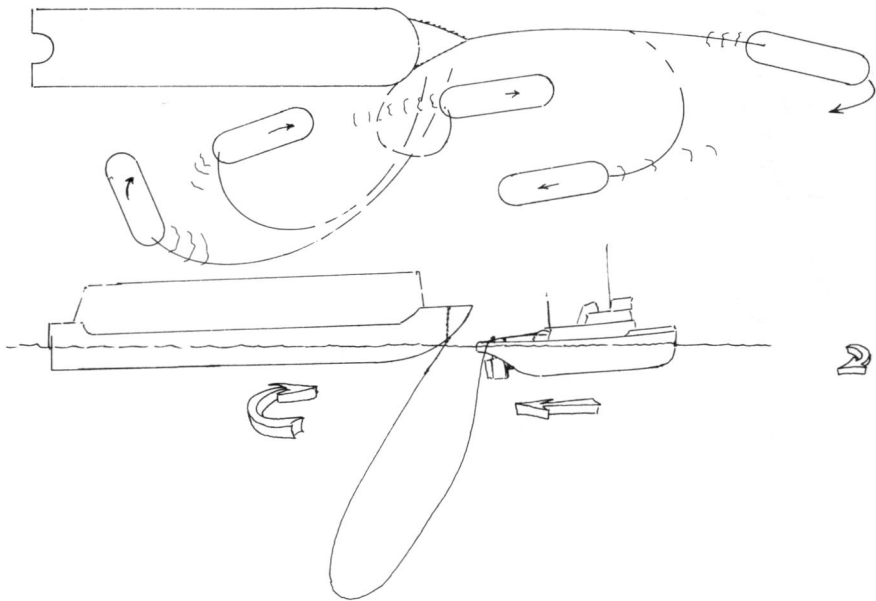

This is a method by which a tug may circle back to inspect her tow without disconnecting the hawser or heaving short, if the water depth allows. The hawser will hang down in a loop and should be held by norman pins so that it will be clear of the propeller.

SALVAGE AND RESCUE

For years the word *salvage* has been misinterpreted as the right to take possessive ownership of a vessel or floating property found adrift, aground, in dire distress, or abandoned. This is not true. Rendering assistance in such cases is perfectly legal, but any remuneration must come from a salvage claim against the property owner filed in the appropriate court. A favorable salvage award is usually based on the condition and value of the salvor's equipment, the time spent, the ultimate success or failure, and the value of whatever was salved.

Based on these facts, and both international and local laws, some companies have built, maintained, and stationed huge seagoing salvage tugs at strategic places throughout the world. These tugs carry extra portable pumps, diving equipment, fire-fighting apparatus and other material to sustain themselves for months. The names of Bugsier, Neptune, Safmarine, Selco, Smit, Switzer, United Towing, and Wijsmuller

The USS *Safeguard* is one of the Naval Sea System Command's new type of rescue and salvage vessels. They are 255 feet with a beam of 51 feet. Their draft is 16.7 feet and they are equipped with bow thrusters and controllable-pitch propellers, giving a speed of 14.5 knots with nothing in tow. (Courtesy Peterson Builders, Inc.)

are as commonly associated with salvage, as they are with deep-sea towing. The USSR also maintains a new fleet of 14 ice-strengthened salvage tugs.

Salvage is a tough game. It is an art in which the initiative and intuition of man is challenged by forces of nature trying to hold, take away, or destroy that which is in jeopardy. Salvage is conducted by a special breed of men not quite as perplexing as described by the chief mate of one of Europe's well-known salvage tugs:

> Ship salvage is a science of vague assumptions based on debatable figures from inconclusive instruments, performed with equipment of problematical accuracy by persons of doubtful reliability and of questionable mentality. Captain of a salvage vessel is said to be a man who knows a great deal about very little and he goes along knowing more and more about less and less until finally he knows practically everything about nothing. The chief engineer on the other hand is a man who knows very little about a great deal and keeps on knowing less and less about more and more until he finally knows practically nothing about everything. The salvage master

Smit's famous (older) *Zwarte Zee* stands by to run a line to the tanker *Jaguar*, which has broken in two in mid-Atlantic. (Courtesy Van der Hidde)

starts out knowing practically everything about everything and ends up knowing nothing about anything due to his association with captains and engineers.

Ship salvage of ships aground, sinking, damaged, or on fire has been well covered by several fine books;* therefore, it will not be dealt with here. That part of salvage involved with rescuing ships at sea will be touched upon.

Should an independent tug not specifically assigned for salvage respond to a distress call at sea, or be sent to assist, or come across a ship in distress, some form of agreement should be made either between the tug master and the person in charge of the distressed vessel or their offices ashore. This is usually the revised Lloyd's Open Form of Salvage Agreement (LOF 1980). Acceptance with the date and time, including the name and identity of the accepting parties, should be entered in the tug's log.

Marine Salvage Operations, by Edward M. Brady, published by Cornell Maritime Press, Centreville, Maryland; *Commercial Salvage Practice,* by Captain David Hancox, published by Thomas Reed, West Sunderland, England.

Bugsier's ocean salvage tug *Titan* stands by to run a line and connect her hawser to the fire- and explosion-damaged tanker *Puerto Rican*. The ship has been abandoned by her crew. Salvage crewmen from *Titan* are seen on the *Puerto Rican*'s forecastle head. (Courtesy Alex Rynecki, Inc.)

The vessel to be rescued should be contacted as soon as possible so that its general condition and ability to haul the tug's hawser aboard and to steer are known. Arriving on scene, the tug should make a slow approach, circling the casualty to observe its condition. It is good practice to stop and drift ahead or astern of the casualty to note how it is drifting in relation to the tug. Using this information, the tug master can choose the best possible spot for connecting up. Often, this will be in the lee of a bow near the break of the forecastle head unless the ship has broken in two.

In such a case, the stern section should be considered for towage if it is afloat. Many stern sections of ships have been safely towed to port. If the ship has been in a collision, or has been damaged forward in any way, such as by a fire or explosion, it may be necessary to commence towing stern-first. If the vessel and her cargo are of extremely high value and pumping or fire fighting is necessary, one or more able tugs should be called to assist. The qualified on-scene tug should take charge of all rescue and salvage operations until relieved by higher authority.

If the vessel is manned, a shoulder line throwing gun can be used to fire a projectile with a small line across the casualty's foredeck. If the ship has power, messenger lines of increasing size can be secured to this until a messenger heavy enough to support the heaving of the tug's hawser and pendant on board is used.

If the vessel has no power, it may be necessary to send over a snatch block first with the bight of a small line. The tug crew will then heave all subsequent lines through this snatch block until the hawser arrives on the casualty.

If the casualty is unmanned, the tug may be able to place some crew members on board. To safely retrieve them may be difficult. Large tugs will use their workboats, if necessary.

When the hawser pendant is rigged to be passed to the casualty, the messenger should be secured beyond the eye splice and then seized to the eye so that it can be heaved through a chock on the casualty without jamming.

If the vessel is on fire, control of the fire will be the first priority. In using the tug's water or foam monitors, it will be necessary to hold the tug in position by thrusters or propeller alignment if the tug has Z- or rudderpropellers. If the tug has neither, it may be necessary to play water from a monitor or hose nozzle opposite to the one being used to extinguish the fire on the casualty. Otherwise, the force of the fire-fighting monitor's water will push the tug away from the casualty. Once connected and ready to tow, movement should start at the slowest speed in order to observe how the tow will handle.

Two tugs, one U.S. and one Canadian, tow the stern section of a U.S. T-2 tanker into Block Island Sound. (Courtesy WorldWide Photographs)

SPECIAL TYPES OF TOWING 361

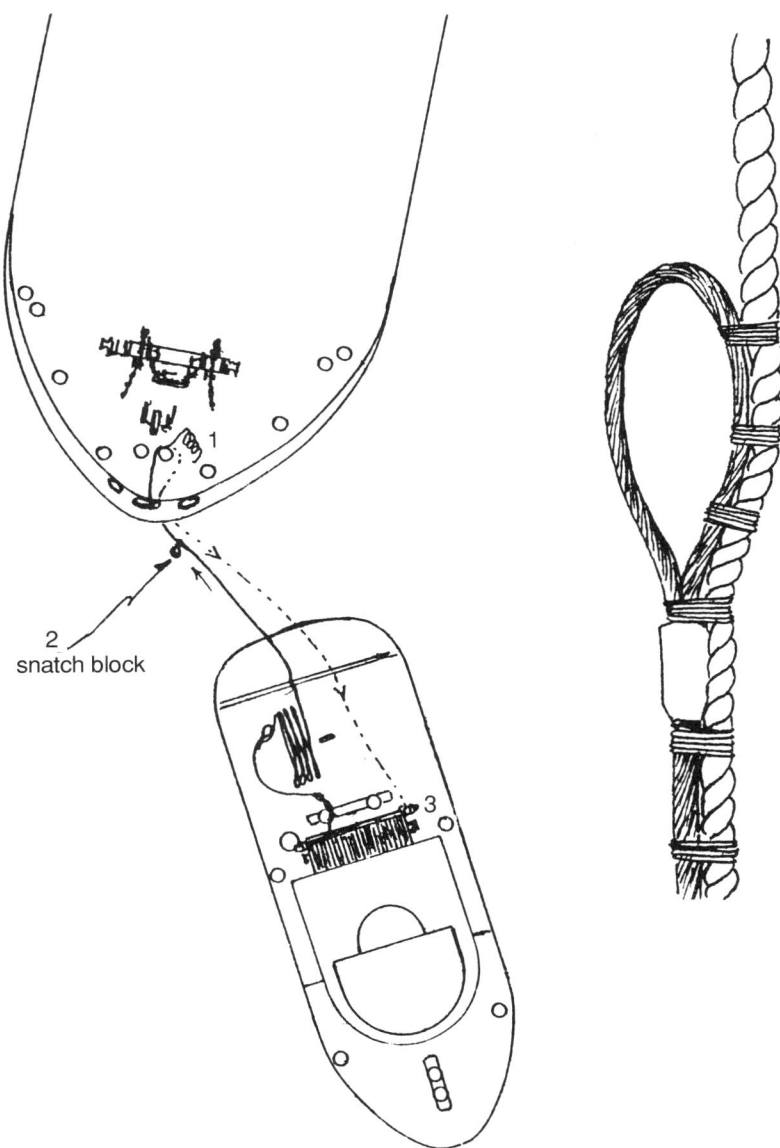

Left, steps in passing a cable to a powerless or unmanned ship: (1) tug fires line and sends bight of messenger; (2) tug heaves snatch block to ship; (3) with snatch block secure on ship, tug heaves hawser to ship, and ship's crew or crew from tug secures cable to bitts. *Right,* when passing messenger to a casualty at sea, seize the eye of the messenger to the tow cable; after the eye clears the chock on the casualty, the crew of the ship may cut the messenger loose after the cable is secured.

Smit-Lloyd anchor-handling tugs fight fire on Norwegian tanker *Thorshavet* in the Persian Gulf. (Courtesy Smit Lloyd—Smit International)

When a vessel is stranded or aground, the use of tugs begins when the vessel is thought to be ready to float. This may be from either an increase in the tide or a decrease in tons grounded. Tugs pull on vessels that are stranded in sand or muddy bottoms. They may free such vessels by positioning themselves so as to wash material by using their propellers. This will frequently clear out some of the bottom on which the ship's hull is grounded.

ANCHOR HANDLING

Anchor handling for the offshore petroleum industry requires large anchors and anchor-handling vessels. To cope with this, towing companies such as Smit Tak International have built special anchor-handling tugs with dimensions running from 129 to 208 feet and having horsepower from 3,000 to 10,000. Tugs with horsepower of 8,000 and upward can handle anchors used in up to 2,000 feet of water. Their stern taffrail is of concave construction, which allows an anchor and its chain to run freely over the stern roller.

The anchor and a little chain are landed on deck by the rig's boom and secured by a quick-release stopper. The tug then runs out slowly on the bearing given by the rig drill pusher or superintendent. Most anchors are accompanied by a large marking buoy attached by chain. These are slacked overboard prior to dropping the anchor. To pick up the anchor, the buoy and its chain are retrieved first. The anchor is held over the stern or on deck until the tug moves to a new location or it returns to the rig with the anchor as the chain is recovered.

Many anchor-handling tugs have a gate in their stern that allows anchor and chain free access to run out when being dropped. For towing use, this gate is closed.

RIG TOWING AT SEA

Between the 1960s and mid-1980s, there has been a great deal of drill rig towing. Voyages of over 12,000 nautical miles have not been uncommon. These voyages have been successfully accomplished by tugs of from 10,000 to 26,000 HP. Routine at sea aboard these tugs when towing a drill rig is much the same as described earlier. The speed, however, is less than that when towing a barge.

In port, to connect up, the tow cable of the tug is run to the rig's tow bridle, which normally is fast on its leg or floating base. To do this, the tug often uses its workboat or a smaller tug. Getting out of port with a drill rig requires the assistance of about four or five other tugs to keep the tow straight as it creeps through inland waterways. On arrival at a work area, the rig is usually turned over to smaller tugs while it is being spotted on location or anchored.

TANDEM TUGS TOWING AHEAD

Frequently, the situation arises where it is necessary to place more than one tug on hawsers ahead of a unit to be towed. Each tug will pull separately on her own hawser. This is common in moving drill rigs either in floating out to location or moving to a new one or bringing them in. There are no limits regarding the size of the tugs to be used except that they should be nearly equal in construction and length if at any point they are to pull while moored side by side. The very heavy deep drilling rigs used off Norway and at other sites are often moved by two or three of the world's largest deep-sea tugs pulling in tandem while maintaining position of about 300 yards from each other.

As each of these tugs will be towing from a separate point on the drill rig, steering and keeping from working sideways is not too difficult. Tugs involved in such tandem tows will each put out the same amount of towline. When course changes are necessary, the tug nearest to the new heading may have to reduce speed slightly until arriving on the new

course. When she has steadied on this new course, the other tugs may make their turns. When all tugs are turned on the new heading, full speed may be resumed.

If the rig to be used is one built to drill in shallower water and is to be moved by multipurpose tugs, up to four may be employed. Steering should be continually checked. Autopilots have the embarrassing habit of unexpectedly getting out of order or suddenly causing the rudder to go hard over.

When two tugs are towing ahead of a dead ship, the connecting hawsers are often run out of the ship's bow bullnose chocks. This will cause the two tugs to slowly come together unless a slight amount of opposite rudder is held. If they are to tow side by side with a line out, their tow bitts or tow hooks and propellers should, if possible, be even. This affords easy course changes and reduces cavitation.

MUD TOWING

This is the movement of self-dumping scows from alongside a dredge or loading wharf to a spoil area and return. Where this type of towing still exists, the disposal area may be in sheltered waters or one outside at sea. The so-called dumping grounds, as marked on charts, are deep holes well offshore and require certain towing and handling precautions for

Smit New York and *Smit Houston* pull from a common point on the barge astern. (Courtesy Smit Tak)

Opposite, Oceanic leaving Matsiu's Tano Shipyard in Japan with the world's largest semisubmersible rig, *Henry Goodrich*. Attending tugs assist the tow toward the sea. The departure date was August 5, 1985. Arrival date in Stavanger, Norway, was September 15, 1985. Towed distance via Cape of Good Hope, was 15,500 miles, which illustrates the endurance of the huge deep-sea tug. (Courtesy Bugsier Towing & Salvage)

the dump scow. The use of some of these dumping grounds has already been canceled. Those remaining in service require a permit for each load of material deposited. In the United States, permits are issued by the U.S. Army Corps of Engineers.

Most dump scows when loaded with 2,000 cubic yards of material have about a 1- to 3-foot freeboard. They are not designed for use in excessively rough weather where seas would break over their coaming and wash out material from the scow's open pockets. All of the discarded material is supposed to reach and be deposited only in the allotted dumping ground. For this reason, tug captains assigned to this type of towing are very careful to look at the weather before proceeding to sea.

The loaded scow is usually picked up at a mooring buoy where it has been placed by a smaller shifting tug. If the loaded scow is at the dredge, it may be moored alongside, or hanging on a line near the stern. The sea tug may be returning with an empty scow, which if it has an exposed shaft mechanism should be landed with the shaft away from the dredge. This maneuver will avoid the shaft being struck by the dredge's bucket each time it swings over the scow. Remember "shaft-side away from the dredge—shaft-side to the tug."

Landing a light scow alongside a dredge is not too difficult, depending on where the dredge is lying. If it is dredging in a tidal river or canal,

Canada Dredge & Dock's dredge *Monarch* is digging in the St. Lawrence Seaway at Lake St. Francis. Notice that the mud scow's shaft is away from the dredge and the two balls on the outriggers indicate to shipping traffic that this is the passing side. (Courtesy Canada Dredge & Dock Company)

it may be necessary to pass the dredge and turn the tug so that while the shaft is still outboard, the tug and tow will be heading into the current.

If the loaded scow has to be passed, it should be done at a very slow speed with the light scow parallel and about 3 feet off. When three-quarters of the length of the light scow has passed the loaded one, a very short burst of half speed with the rudder hard over usually swings the light scow's bow toward the dredge. As the scow's stern passes the loaded one, the tug backs, and the light scow will drop in alongside so that the dredgemen can place their mooring lines on it. The dredge operator will often drop his bucket in one pocket to aid in holding and placing the scow in position. Then loading commences as the sea tug makes up alongside the loaded one.

If the disposal area is inside in smooth waters, the tug may make up alongside or push. Arriving at the site, the tug should position the scow so that the wind will be blowing across the scow and away from the tug. In this way, any splashes or spouts of material will not land or blow on the tug or dumping personnel.

Old wooden scows, if any still survive, have a shaft running on one side of each pocket. More modern scows have a different mechanism. Those with doors at the bottom of pockets have chains that run up vertically to a sheave and through horizontally to the shaft on which they are wound tightly. The doors are held closed by a hinged dog, inserted into a tooth of a small heavy wheel, which is part of the shaft. To release an individual pocket, the dog is struck by a maul, the chain unwinds, and the mud, clay, earth, or rock starts disappearing. Whether a spout of water and mud shoots into the air is governed by the type of material that is in the pocket. Soft watery mud goes fast and innocently, while heavy clay often sticks before rolling down the sloping side of the pocket, creating heavy upheaving splashes that can rise as high as a tug's pilothouse. Gravel, cellar dirt, and rock usually run out quickly. For this reason the tug captain, if alongside or pushing ahead, will often direct the scowman as to the order of release of the pockets to avoid being covered with mud.

Modern steel scows have similar arrangements, but the pockets are controlled by hydraulic rams, which release or tighten the bottom doors. Another type of dump scow is split lengthwise and is hinged, opening in the middle. If any of these scows are to be towed to a disposal area at sea on the hawser, the weather in the dumping area must be such that a man can get out on deck to operate the dumping mechanism. The tug captain must be absolutely sure he is in the assigned disposal area and so indicate this in the tug's log with bearings and how they were obtained. The tug should be slowed and turned so that the wind will blow away from the shaft side and the area where the scowman must work.

The small tug *Frank B. White* returns with her empty scow from the inland dumping ground off Alcatraz Island in San Francisco Bay. (Courtesy S. Lang)

At the dumping ground, tugs usually blow a whistle signal notifying the scowman it is time to dump. They will then circle slowly, keeping their hawser well off the bottom. The scowman will signal the tug by hand or flashlight when all pockets are empty. At night, searchlight flashes from the tug are used to signal the scowman, and a flashlight acknowledgement from the scow has been used when no VHF radio or hand-held sets are available.

Dumping areas off some ports have been moved farther out. There has been a suggestion that the Port of New York's be extended to 100 miles southeast of Ambrose Tower for certain materials.

Speed in returning with an empty dump scow will not be greatly improved unless the open doors hanging down from each pocket are closed. Water entering each pocket retards progress as if the tug was pulling an open egg crate. If power is available, pockets may be wound up as the tow returns from sea. Scowmen should not be working on deck if the scow is rolling. If the doors of the scow are still opened when inside in shallow water, care must be taken so that they do not strike bottom as they may become damaged or have their hinges twisted or their chains broken.

A Great Lakes Dredge & Dock Company tug crosses in front of a Staten Island ferry as she returns from the ocean dumping ground off Ambrose Light Tower. This type of disposal at sea has been curbed.

TOWING TOXIC-WASTE BARGES

For years, specially constructed medium-size barges for the carriage of toxic waste have towed from some U.S. East Coast ports to special disposal areas at sea. The handling of these barges is done as routinely as it would be with any other oil type barge of its size. There is one exception: These barges are never manned. Therefore, special arrangements must be made for getting on the hawser after clearing the loading berth, and for arriving at a transition area. This may be accomplished by running the barges' wire pendants to the tow bridles down the edge of the deck toward the tug's bow and fitting it with a strong pickup line, which can be hooked and hauled aboard the tug. The tug will be provided with a portable radio transmitting panel that, when plugged into the tug's power circuit, will energize operations on the toxic-waste barge. The tug will be given an operating manual and special dumping instructions covering the exact area, its limits, and possibly, the course, turns, and time to take in disposal of the material. These should be observed meticulously, as air or radar surveillance by plane or patrol craft frequently occurs in order to spot check on tugs and barges proceeding to and in toxic-waste disposal areas.

These special waste barges are fitted with a Christmas tree yardarm mast with lights of different colors, which when lighted indicate to the tug that the operation requested is being carried out. On one such barge an orange light appears when the barge generator has been started. Another colored light will show that the air compressor or hydraulic pump has started. These will be followed by lights indicating when each compartment valve is opened, and others showing their closing, the washing out, and final system shut-down.

If any position fails it is possible to swing back and go alongside to put a man aboard to open a valve by hand. All captains, mates, and engineers on a tug assigned to tow such a remote-control barge should be instructed in the use of the system and should be able to troubleshoot simple operational failures.

In October of 1986, the 99th U.S. Congress renewed the Clean Water Act, which includes specific provisions to close New York Bight, an area extending approximately inside a line from Jones Beach, Long Island, to about Asbury Park, New Jersey, prohibiting all dumping of waste material after December 1987. A new dumping site for certain materials will be established 106 miles off the New Jersey coast. Boston also will curtail use of its dumping ground.

10. Handling Tows in Special Situations

PARTIALLY LOADED BARGES

FROM time to time, the tug captain may be confronted with moving or turning a unit that is not seaworthy. Both caution and some estimate of how the unit will respond are necessary. These movements are usually made from one berth to another, or turning at a wharf or for short distances in a narrow or crowded waterway.

Partially loaded barges that have one end much deeper than the other are hard to handle. If the tug makes up alongside, they are difficult to maneuver. If the high end, with less draft, is ahead, getting away from the pier will present a problem, as the pivot point is near the low end, which will resist sideways movement. The high end will swing easily. In such a condition it is better to place the tug behind the low end and push.

A barge half full of liquid cargo with a slight drag may follow well on a short hawser, but it will not swing quickly on bends in the channel. When it does swing, it may overrun the course change by sheering heavily due to the liquid cargo free surface effect. To counter this, an assist tug may be required.

Ocean barges rigged with a single-chain towing pendant from a center bullnose chock will not follow at sea as well as those rigged with bridles connected to an additional pigtail and pendant. Ocean barges with a single pendant should not be towed behind a tug in this manner in inland waters without an assist tug made up alongside or astern where it can aid in the steering.

Ocean barges partially or fully loaded will stay with a moderate current such as the Gulf Stream, if the tug makes a 30- or 40-degree course change for any reason. Such a course change, if the tug is about to enter port, should be made gradually over a period of up to one-half hour so that the barge will not be heading at nearly 40 to 50 degrees from the tug as it commences its new course.

LASH AND SEABEE BARGES

These barges are also very awkward to handle, particularly when a standard harbor tug is taking them to or from the mother ship's stern cranes. A pusher-type tug is better suited for this maneuver. As the deck fittings on these barges are small, special towlines are necessary. When they are to be made up as a tow, the loaded barges should be placed in one tier and the empty or lighter barges in another tier or behind the loaded ones.

PASSING THROUGH DRAWBRIDGES

These obstructions to navigation were never placed for the convenient passage of vessels, only land craft. They should be approached cautiously with some trepidation and knowledge of the way the current flows and sets on their abutments and fenders. It is also wise to remember that bridge fenders are not built for tows to rest against or slide through on. As these fenders are not always kept in good condition, the tug or tow that rubs against a bridge fender may be the unlucky one charged by the bridge tender with damages done by some previous traffic. The condition of drawbridge fenders and responsibility for their upkeep have always been a bone of contention between bridge tenders and tug masters.

Drawbridge operation is another. The standard U.S. whistle signal to request an opening is one long blast followed by one short. Many drawbridges where openings are frequent have 24-hour operation and monitor a bridge-to-bridge radio watch. The draw tender is required to respond either by repeating the whistle signal or by VHF, if he is prepared to open at once. If he cannot open, a signal of four blasts will be given.

In the interim between requesting an opening and seeing some action, tug masters should approach at a speed that will allow the tow to be stopped, if necessary. Drawbridge machinery has been known to stick or blow a fuse. Personalities also influence their immediate response. In some cities, bridge operation and maintenance comes under a highway department whose officials can be pressured by irate citizenry, who have often been delayed during rush hours while the bridge is open to allow a slow-moving tow or a tug-assisted ship to pass.

Railroads seldom altered the course of their tracks in order to cross a waterway at right angles to the channel. As a result, many railroad drawbridges are at an oblique angle to navigation channels. Passage through railroad drawbridges requires extra care. In the approach, the tug master should not allow himself to get in extremis with the tow before the bridge tender acknowledges that the bridge will open at once. Openings on some railroad bridges are regulated by an automatic signal system. Once the train's engine has passed the last bridge safety signal light, the bridge may not be opened. Some railroads have extended this safety bridge lock system to include the time it takes the train to reach the bridge as if it was already open and about to close. If a freight train is seen approaching or crossing a drawbridge, the tug captain should prepare for a long wait and choose a spot where there is room enough to drift.

Alignment in passing through drawbridges is sometimes difficult. If the bridge has a center island or abutment with a passage on either side, the right-hand passage should be used if the channel depth, the direction of the bridge swing, and the fendering condition so indicates. If the bridge draw span swings toward the tug, it is a good indication that the draw tender desires the tug to proceed through the other side; or that traffic is close to or entering from the opposite direction. By shielding oncoming traffic, drawbridges that are on or near bends in channels often present some surprises. A securité call on bridge-to-bridge VHF should be made to advise other traffic of your presence and intention.

Lining up the tow to pass through a narrow drawbridge is easy if the tug is in the notch. With the tow alongside, it is best to favor and steer for the portion of the bridge that houses the draw span machinery, such as the center island if the tug is made up on that side. If the tug will pass on the short-fendered side, the bow of the tow should be held close to it, allowing it to fall off so that the tug will clear without touching. The outboard corner at the stern of a unit when towed alongside usually comes closest to touching the bridge's fender work.

Towing through drawbridges on the hawser requires short individual towlines if possible, as response is better than with bridles. The tow can be so close to the tug's stern that there is just enough clearance for the tug to swing. By looking aft, the tug skipper can bring the barge through nicely. If towing a ship without power, on a short line, care must be exercised constantly to check any attempt by the ship to sheer. Presumably, another tug will be alongside or astern to assist in steering. If towing a ship with power stern-first through a drawbridge, the use of the ship's propeller and rudder at reasonable revolutions can help straighten her without endangering the tug.

If assisting a ship with power through a bridge, tugs may be placed alongside if there is clearance. Otherwise, the tugs may line the ship up and let go, running ahead or holding back, if the tug is on the stern. Tractor tugs, with their ability to apply power in a 360-degree direction without changing position, are excellent in assisting ships through drawbridges.

At night, drawbridges in U.S. waters are marked by privately maintained fixed red or green lights.* All bridge piers and abutments adjacent to the navigation channel are marked on all channel sides by red lights.

Code of Federal Regulations—Title 33, Parts 114–118.

Five Crowley-Red Stack tugs of their Puget Sound Tug & Barge Company assist the 770-foot tanker *Baltimore Trader* through Blair Waterway Bridge, Tacoma, Washington. (Courtesy Crowley)

On all types of drawbridges, one or more red lights are shown from the draw span and are higher than the pier lights when the span is closed. When the span is opened, the higher red lights become obscured and one or two green lights will appear, also higher than the pier lights.

To find out the clearance required at any particular stage of the tide, clearance gauges as regulated by the U.S. Coast Guard are provided on some bridges. Vertical numerical scales reading from top to bottom show the actual vertical clearance between the existing water level and the lowest part of the bridge over the channel.

In areas with a large daily tidal range, many tugs have hinged masts that allow them to pass under drawbridge spans when they are closed. Captains should be aware of and make note of whatever mark or reading will safely allow their tug to pass under a bridge without requesting an opening.

Never start to enter a single-span drawbridge until it is fully opened. There may be some vessel approaching on the other side that is not aware of your presence. If the bridge swings, it might stop before its roadway structure is completely clear of the channel. Such a target can tear a hole in a steel barge or tug deckhouse. On single- or double-leaf bascule bridges, the tug master should request more clearance if necessary by quickly getting the draw tender's attention on the VHF or sounding the tug's whistle if he feels that the leaf should go up farther.

Avoid letting the tug or tow rub on bridge fender work. These structures are not always in good condition. Occasionally, old breaks or cracks exist and the slightest pressure may cause timbers to fall or stick out to get hooked on the tow and be torn off. Protruding bolts have been known to scratch the side of a steel barge. The resulting sparks caused a disastrous fire. In another incident, a barge was being towed through a bridge on which the creosote fenders were hot and dry. The friction from the steel hull of the barge immediately started a fire. Fortunately at that time the barge was pumping ballast and the discharge water put the fire out. Beyond the possibility of disastrous results are the heavy claims that can be lodged against the tug's master and her owner.

SHIFTING PERSONNEL FROM BOAT TO BOAT

Tug captains, mates, and crew members who have been assigned to one tug for a long time become so accustomed to its response to their particular operating touch that when they are transferred for any reason to another tug, they are often momentarily ill at ease. The pilothouse of the new assignment will seem strange. It is like going into someone's home: things are arranged differently; engine controls given a different response; on deck, tow winch and other equipment may be entirely foreign.

No two vessels respond exactly the same. There will be changes in the location and types of equipment. Unless the whole crew from the other tug goes with an officer it may take a few days for an understanding to develop between him and the deck crew. It will take several days before the relieving captain feels at home. If it is only a temporary change, those men or officers relieving should try to adhere to the well-established work pattern in force on the tug. As tug men are innovative individuals, respect for and understanding of someone else's methods may lead to different or improved ways of doing certain things.

NOTES ON TOWING IN SPECIFIC PLACES

Every waterway should be looked upon as a challenge to the tugman's towing talent. There are a few places throughout the world that must receive more than a passing glance. If they are not approached correctly, they can turn and pounce like a tiger. Here are a few of these well-known areas through which tows can and do pass safely if the area is approached correctly:

One of the first of significance is the BAY OF THE SEINE and its bar above Honfleur. The tidal range may run to 30 feet above mean low water. This causes a very strong tidal current. Sand that has come down the River Seine has built up a bar with many fingers, which shifts daily. The incoming tide meeting the river's outflow causes a 6- to 8-foot wall of water or tidal bore to run up the river to Caudebec. This penomenon is locally referred to as "le mascaret."

Pilotage for the River Seine and bar is compulsory. Tows should pass over the bar at whatever time the Le Havre-Rouen pilots advise. This is usually for the period after the flood has run for an hour or more. Consult the latest pertinent pilot guides for communication and boarding instructions.

Outbound from the Seine with a tow requires careful timing. The tow should not be too far downriver when the bore begins. The tow will have a fair current and must meet the bore head-on. The place to do so should be far enough upriver from the bar so that the tow can exit the river after the bore has passed yet before the depth on the bar has dropped to a point where the possibility of grounding exists.

Conditions on the treacherous Seine River bar have improved recently. A dredged and well-marked channel is constantly maintained through a self-flushing system. Not far from the edge of the channel, however, the rest of the bay bares itself for several hours before and after low tide. The same type of condition exists in the Bristol Channel of Wales.

The BAY OF FUNDY, particularly near its head, is another area with a notable tidal bore. Any tugs with tows must come into such ports as Digby, Windsor, and Parrsboro, Nova Scotia, at the top of the tide and make fast to a stout wharf. The tug should be listed slightly toward the dock or object she is lying next to, if the tide will leave her grounded out. Some wharves have a timber grid for vessels to lie on.

Truro and Moncton have a large tidal bore that occurs daily and should not be considered for entry by any vessel without local knowledge.

At SAINT JOHN, NEW BRUNSWICK, the river waterfront and piers have great depth and the current is swift. Local pilots and assist tugs are necessary to land tows due to the various eddies. When the current is slack at the Reversing Falls, it is possible for tows to pass through.

PETIT PASSAGE, NOVA SCOTIA, between the Bay of Fundy and St. Mary's Bay, has very swift currents through its narrow pass. Tows must be prepared for a very strong tide rip when the current runs out into Fundy. There is a ferry crossing in the middle of this passage. In bad weather, an alternative route would be outside via Lurcher Shoal Buoy.

LUBEC NARROWS between Lubec, Maine, and Campobello Island, New Brunswick, is narrow, shallow, and swift and should only be considered as a high-water, clear, smooth-weather passage.

Because of the swift current in the Piscataqua River at PORTSMOUTH, NEW HAMPSHIRE, tows are assisted to the various wharves by local pilots and tugs who will advise by VHF radio as to what time they want to start up with the tow from Newcastle Point.

CAPE COD CANAL movements are governed by the Cape Cod Canal dispatcher (U.S. Army Corps of Engineers), who monitors channels 16, 12, and 14 from his office—call letters WUA-21* This office is located at Buzzards Bay on the west bank between Buzzards Bay Railroad Bridge and the Massachusetts Maritime Academy's pier. Canal traffic is also regulated or advised by traffic lights, which are displayed on poles at Sandwich breakwater and at Wings Neck in Buzzards Bay.

Tugs with tows are usually admitted for transit with a favorable current commencing about half an hour before the current turns fair at whichever end of the canal they are approaching. Ship traffic usually gets priority, and the canal dispatcher will judge whether he will allow other traffic in the canal at the same time. Large tows may require tug assistance. Tugs are available from New Bedford. If the canal is fog-bound, tows may be anchored off Sandwich breakwater in Cape Cod Bay. If there is an easterly gale and heavy seas or breakers occur at this end of the canal, tows can find shelter and good anchorage in the outer bight of Provincetown Harbor near Wood End Lighthouse.

At the Buzzards Bay entrance to the Cape Cod Canal, tows may be anchored outside Cleveland Ledge Lighthouse or in the small anchorage next to the buoys just past the lighthouse. Shelter may also be granted upon request of the canal dispatcher for tows to moor at the dolphins just east of Hog Island and on the north side of the canal opposite Sandwich.

The passages between BUZZARDS BAY and VINEYARD SOUND, although tricky, are used by tugs. Woods Hole passage should not be attempted except at slack water. Robinsons and Quicks holes may be used, but the current and available depth must be carefully considered.

*See "Cape Cod Canal Regulations" in appendix VI.

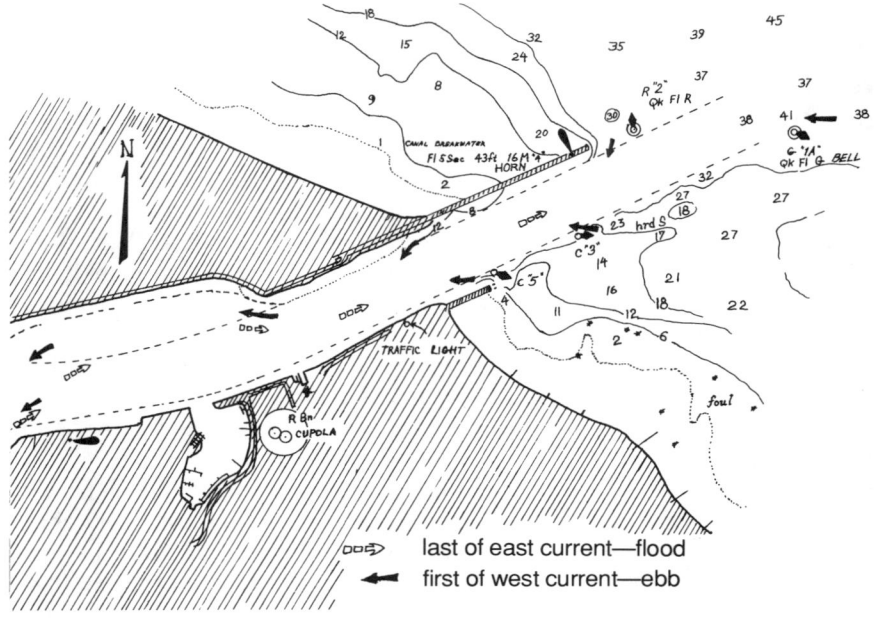

East Entrance, Cape Cod Canal

At a speed of about 9 to 10 knots, it is possible to carry a fair current from the east entrance of Cape Cod Canal to The Battery in New York Harbor. To do so, the tow must enter Sandwich breakwater between the time the current begins flowing west and up to 1½ hours after it has been running westerly. This westerly current will be carried as an ebb out through Buzzards Bay, where it will meet the beginning of a flood also running west in Block Island Sound off Point Judith. This fair current can be carried as a flood in Long Island Sound to Execution Rocks Lighthouse. Between Stepping Stones Light and College Point, it will still run westerly but change in name from a flood of Long Island Sound to an East River ebb. This will continue to run fair through Hell Gate to The Battery and The Narrows. Occasionally, a tow will be fortunate enough to run this entire distance and then pick up the beginning of the flood off St. George, Staten Island, and carry it into Arthur Kill.

Current charts for Buzzards Bay, Nantucket and Vineyard Sounds, Long Island Sound, and New York Harbor are very helpful in navigating these waters.

Running THE RACE between Fishers Island and Little Gull Island is frequently done incorrectly by strangers. This is especially dangerous if it is done with a tow. In the middle of The Race is Valiant Rock, a hard

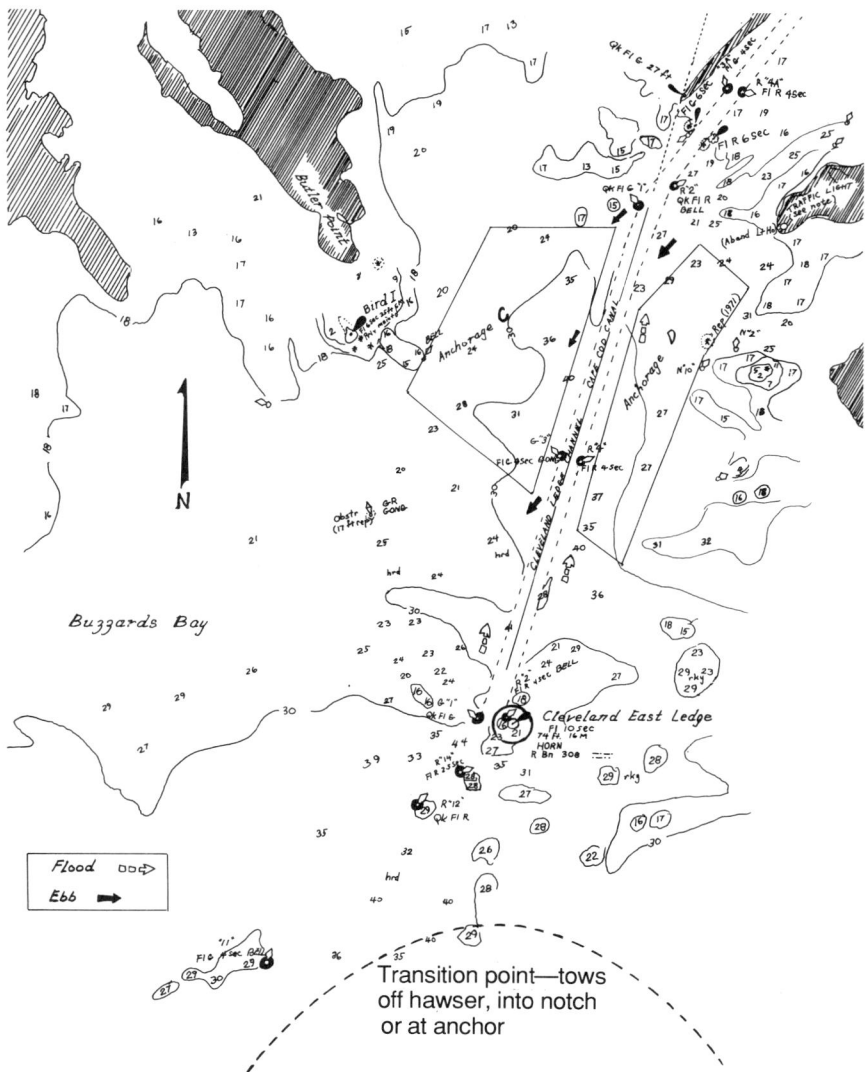

Buzzards Bay Entrance

pinnacle with 18 feet or less over it, depending on the stage of the tide. One or more barges and other vessels have had their bottoms torn on it. The tidal current runs between 4 and 6 knots over it as it fills or empties this portion of Long Island Sound. Approaching The Race from Long Island Sound, tows pass close to Cornfield and Plum Island Fairway

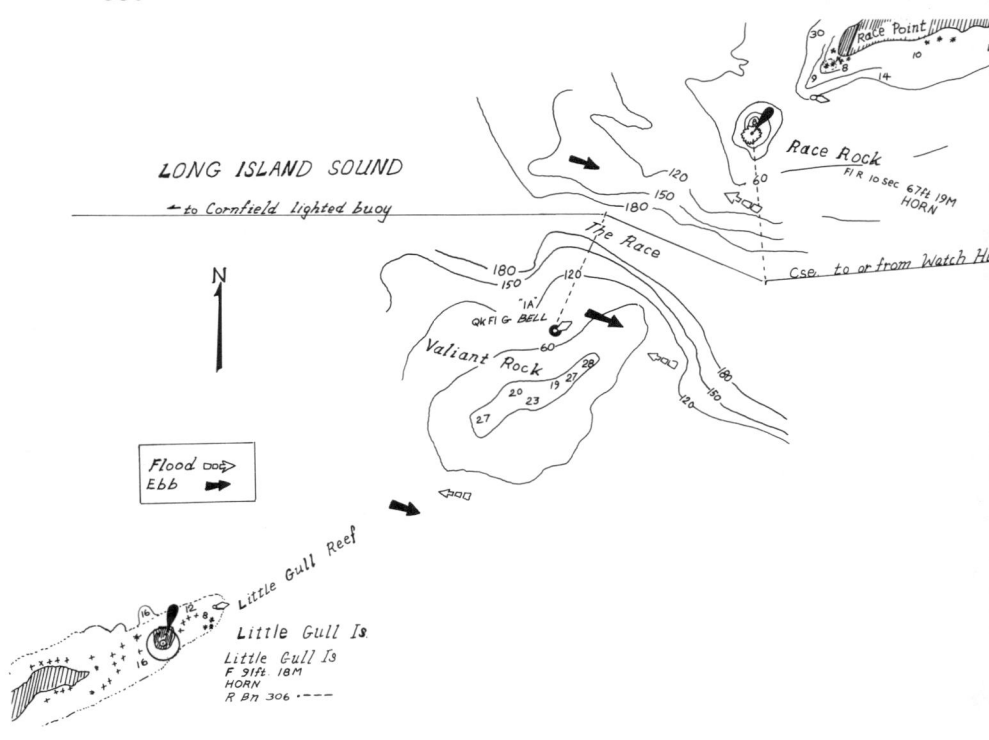

The Race

Buoys and then head on Race Rock Lighthouse until abeam of Valiant Rock Lighted Buoy which should be to starboard at no less than half a mile. Then a course of southeast will take the tow clear until a new course may be taken for Watch Hill Fairway Buoy and Point Judith Lighted Buoy. When approaching The Race from Block Island Sound, tows should head on a northwest course, which will take them midway between Race Rock Light and Valiant Rock Lighted Buoy. They should then resume their normal westerly Long Island course after Valiant Rock Lighted Buoy is abeam, unless they are bound for New London or the Connecticut River.

Entering the CONNECTICUT RIVER from the west with a strong ebb tide flowing easterly, care must be taken to start the 90-degree turn early enough so as not to be set onto the shoal water to the east of the entrance breakwater. The current coming out of the river, if in a freshet stage, and the ebb current of Long Island Sound meet here off the breakwater and act as a powerful jet stream on the port side of a loaded barge that is being pushed or towed alongside as this turn to port is being made.

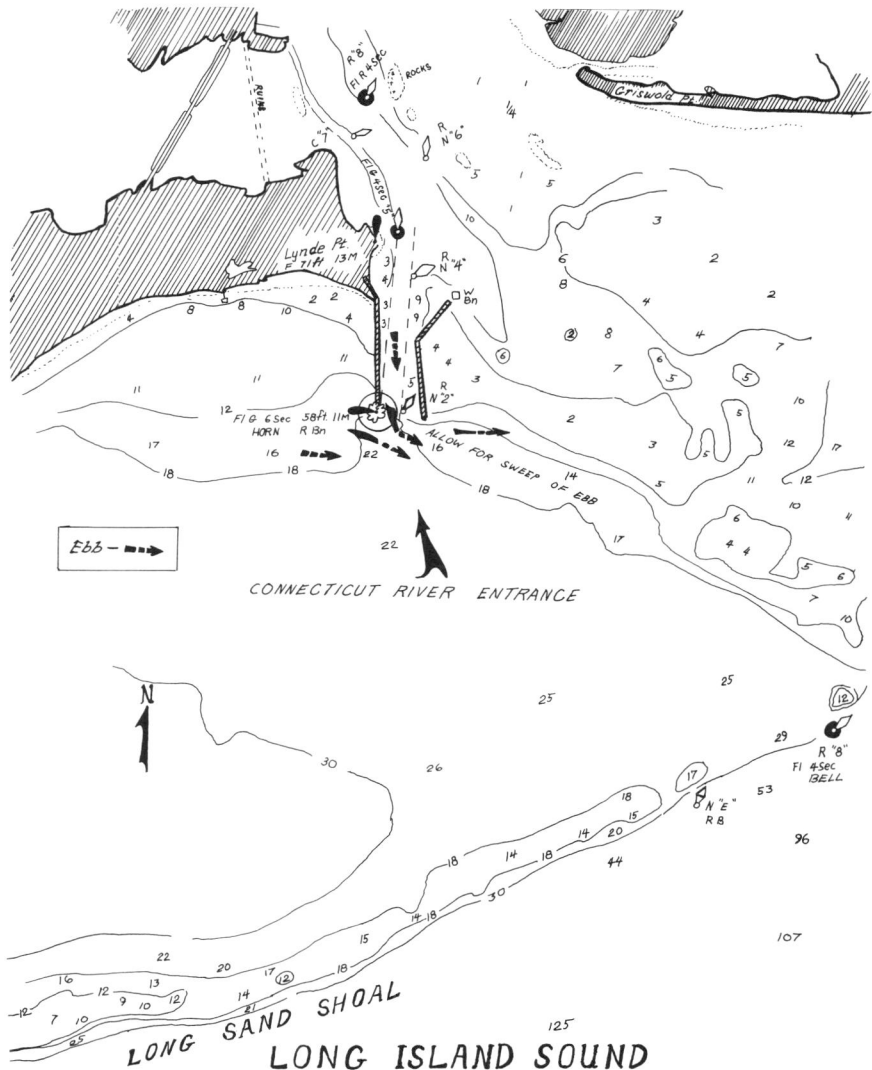

Connecticut River entrance

The current in HELL GATE in New York's East River, known locally as "the Gate," has always been an exciting experience for tugboat skippers. For years, the so-called sound steamers ran the Gate at any stage of tidal flow. Not so with tugs and tows. A normal passage through Hell Gate has always been made with a fair current. The time for safe passage with a tow ends about 10 minutes after the current turns in the opposite, or ebb direction, if north- or eastbound on the flood, or 15 minutes before it turns flood if south- or westbound.

This holds true today for several reasons. While it may be safe to push a light or loaded barge, or pull an empty one on the hawser, up the narrow reach from Man O'War Rock toward Hell Gate at from one-half to three-quarters of an hour before the current turns fair, such a tow in bucking the still-swift ebb is difficult to steer and is an undesirable target for any descending traffic. No tugs with tows should attempt an earlier passage.

The time span most experienced tug captains use, if eastbound up the East River to Hell Gate, is departure from The Battery or Buttermilk Cannel with a tow when the first of the flood has just started on the Brooklyn shore at Red Hook. In any event, it should be no later than the time it will take to traverse the same passage through Hell Gate as far as North Brother Island before meeting the ebb or head current.

In running the East River and Hell Gate, most tug captains announce their location frequently, using VHF bridge-to-bridge channel 13 as a security call. In this way, meeting in Hell Gate with either another tow or ship can be planned for or prevented as the size of vessel and prudence dictate. Common practice has been for tows of several barges or sand or stone scows held on gate hawsers to proceed through either very early or very late on the tide. This will avoid the extreme flow in the almost figure S bend under the Triborough Bridge and the Hell Gate railroad bridge.

Eastbound tows proceed upriver from off the United Nations Building using the center of the channel between Manhattan and Roosevelt Island (formerly Welfare or Blackwell's Island) heading for Hog Back Light until off Gracie Mansion.* They never go beyond here at full speed with a fair current. Tows are slowed so that the correct use of the current will aid in staying in the middle. Off Gracie Mansion at the mouth of the Harlem River, tows swing slowly to starboard. This swing must be checked before heading on the center of the Triborough Bridge. Then a turn to port must be started until the tow is headed for the middle of Hell Gate Bridge. If no traffic is headed in the opposite direction, the tow should be held to mid-channel by staying slightly to the left. The flood current coming around Hallets Point will push the tow to port, but as the tow starts under the Triborough Bridge, a strong current running along Wards Island shoreline will hit it. This current is running east from the Hog Back. If the tow is too far to the starboard hand, it may get set too close to the ragged rocks on the Long Island side.

---

*Gracie Mansion, the single white mansion behind a high wall on the Manhattan shore; the New York mayor's residence.

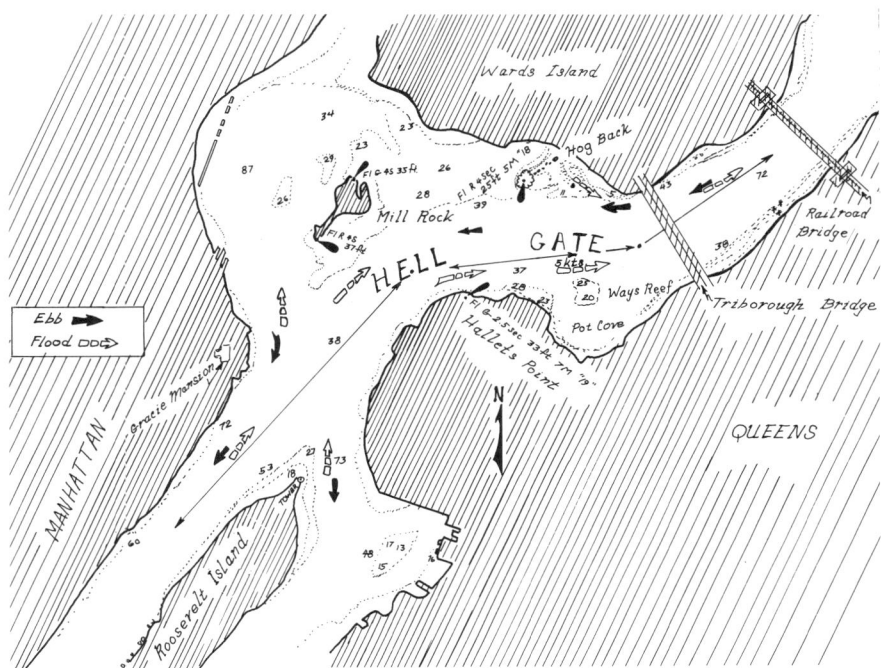

Hell Gate

This same steering and sailing procedure applies if pushing or towing alongside. When bucking the flood it has been common practice for westbound, downbound tugs pushing or pulling an empty barge, to slow and hold back on the Bronx side of the area between Sunken Meadow Light and Hell Gate Bridge if another tow or ship is eastbound and will be met before passing Hallets Point. Westbound tows with a fair current should slow down enough so that the tow will not sashay as the turn is made under the Triborough Bridge and again below Mill Rock when the Harlem River ebb flow strikes it.

There are two things to remember when approaching Hell Gate. First, while it may not seem practical due to outside noise, it is still a legal requirement under the Inland Navigation Rules, revised in 1982, as it was in the old Inland Rules of the Road, for vessels or tows to sound the bend signal as stated in Rule 34(e). The other is that ships do pass through Hell Gate either way. If they are bound for a pier in the Bronx, or Steinway terminals, they usually have tugs accompanying them. If so, the docking pilot may be on the air using the tug company's VHF frequency. Momentarily, he might be off the bridge-to-bridge frequency.

Whether the tow is east- or westbound in the upper East River, care should be exercised in the choice of whether or not to pass between the Brother islands. This channel has good depth in the center. When eastbound there are dangerous rocks and a crosscurrent above the powerhouse at Steinway. Westbound, the current at times sets onto the rocks off North Brother Light. Passage through North Brother Channel should not be attempted if other vessels will be met. Also, a careful lookout for cross-traffic coming around the main channel on the west side of North Brother Island or bound in or out of Greater New York Terminal should be made.

All of these details may seem to be too explicit, but there have been a great many collisions, groundings, hawser partings, and sinkings of tug and barges in the Hell Gate area of the East River. The only other areas in New York Harbor where similar care is always necessary is rounding The Battery, in Buttermilk Channel, and the Kills. The latter covers one of the nation's busiest shipping lanes commencing at Constable Hook. Here at the entrance to Arthur Kill between Robbins Reef Light and St. George, Staten Island, is a very dangerous spot for tugs. Deep-draft ships come in on the range lights while tows enter and exit from the New Jersey bulkhead channel or head for the main channel either bound up the bay or across the ferry lane for Brooklyn and out of the narrows. Shipping has been aided by the Coast Guard in the recent addition of a single visual range marking the center of Constable Hook channel. This is a single station lighted screen,* which when viewed from a vessel will indicate which way the vessel must change its course to be on the centerline of the channel. It is used mainly by ship's pilots and their riding pilots who maintain control assisted by tugs.

The waters between the Exxon Constable Hook piers and the buoy on the turn are frequently congested, as this is one of the narrowest parts of Arthur Kill. Great care must be used by tows in keeping well to the right without striking this buoy. The next critical point is after passing under the Bayonne Bridge at Port Richmond. A very short way to the west is Bergen Point, whose rocky outcropping is marked by a light. Here, the channel divides, with one leg going north into Newark Bay and Port Elizabeth's great container terminals. The other, equally well used, channel runs westerly as Kill Van Kull. This is also heavily trafficked by deep-draft ships and tows. It is in this area where the tidal current changes. Flood tide coming around the back side of Staten Island through Kill Van Kull meets that coming in from Arthur Kill. Together, they flood into Newark Bay.

*Inogon Leading Mark System, New York.

There are very good tidal current charts in booklet form covering all of New York Harbor and tributary waters.

Using SANDY HOOK CHANNEL when entering or leaving New York Lower Bay on the hawser can be interesting. If inbound with a very nearly calm sea condition the surprise may come in the form of some very heavy swells as the tow progresses up to the turn to port toward Sandy Hook itself. Slow speed may be required with a fair amount of hawser, otherwise the tug may run and lift the tow cable out of the water. When off the tip of Sandy Hook itself, the tow may want to set to the west and run over the red buoys.

When outbound with an ebb tide the set will be in the same northwesterly direction for a short period. After passing the tip of Sandy Hook, the current will set easterly toward the green buoys that mark the top of the False Hook Channel Bar.

Inland and Intracoastal Routes

Tugs running with or without a tow and drawing 6 to 8 feet can proceed up or down the United States East and Gulf coasts in relatively sheltered waters. The tow's draft, dimensions, and vertical clearance required under certain fixed bridges, and the crossings of high-tension wires are criteria for passage. Small tugs running independently will find many safe inlets and shelters if running along the coast of New England, New York, and New Jersey and as far south as the Delaware Capes. All of these HARBORS OF POSSIBLE REFUGE should be studied for possible use if this is to be the skipper's first trip through these waters. The latest up-to-date small craft or harbor charts should be carried. They should be updated through the local *Notice to Mariners*. If the small tug or light-draft tow is caught out in bad weather in unfamiliar local waters off an inlet, her skipper should try to contact a local inbound fishing vessel or the Coast Guard for advice.

From the DELAWARE CAPES to CAPE HENRY, VIRGINIA the small tug must watch for good weather and estimate her running time. This is of the utmost importance. There are few places to enter, so it is foolish to take a chance in questionable weather.

It is possible and at times profitable for even large tugs with heavy tows to take the shipping route up Delaware Bay and through the Chesapeake and Delaware Canal to the upper Chespeake Bay if bound for Baltimore, Washington, Yorktown, or Hampton Roads.

On the outside, the small tug should forget the route beyond Cape Henry to Diamond Shoals and around Cape Hatteras. Several large tugs have been lost and many hawsers have been parted off Diamond Shoals.

Towing Routes for Small Tugs—
Norfolk, Virginia, to Brownsville, Texas

Those tugs drawing 8 feet or less can take their tows into Norfolk Harbor, Virginia, and proceed via the South Branch of the Elizabeth River, which is the upper end of the Intracoastal Waterway (ICW). This ICW leads its winding way through sounds, across swift-running rivers, close to all the major ports and inlets as far south as Miami. This route is heavily trafficked by pleasure craft southbound in the fall and northbound in the spring. There are long navigation cuts through the Carolinas. As they often join tidal rivers, it is not uncommon to find the currents changing directions as they are crossed. The tidal range increases considerably in Georgia. As a result, there is a sort of jet stream effect, which hits tows as they cross some inlets at right angles. In parts of this wide and uninhabited marshland, there are many sloughs. At night, pilots must take great care not to make the mistake of heading up the wrong passage. The day markers and lighted beacons can be seen from afar. What may appear as the next lighted beacon may actually be across a grassy islet and around a bend in the channel. The main channel of the route is fairly well maintained as far south as the ST. JOHN RIVER below Jacksonville, Florida.

As there is much less commercial traffic on this southern or Florida section of the ICW, tows may find some difficulty in the shoaling near various inlets as well as hard-to-see day markers and light beacons on various turns.

At STUART, FLORIDA, a tow with a 6-foot draft may take the St. Lucie Canal and a dredged route around Lake Okeechobee to Moore Haven and the Caloosahatchee River. Two locks bring traffic to Fort Myers and the Gulf of Mexico, where a choice may be made at Sanibel Island. If the weather is fine a trip along the coast is possible. The alternate route is via the ICW, passing Boca Grande, Venice, and Sarasota and across Tampa Bay to Clearwater where the ICW is interrupted at Anclote Keys.

It is about 140 nautical miles across the head of the Gulf of Mexico from Anclote Keys to Apalachicola Bay, where the Gulf Intracoastal Waterway resumes again. This is a very heavily trafficked towing route, which extends around the north perimeter of the Gulf and down the Texas coast to Brownsville at the Mexican border. It offers the normal pitfalls of some difficult bends, open bays, and bayous. There are crossings of swift rivers and busy marine traffic at New Orleans and the Atchafalaya River at Morgan City.

The MISSISSIPPI—GULF OUTLET is a well-marked dredged channel. It is frequently used by tows as it leads from the Industrial

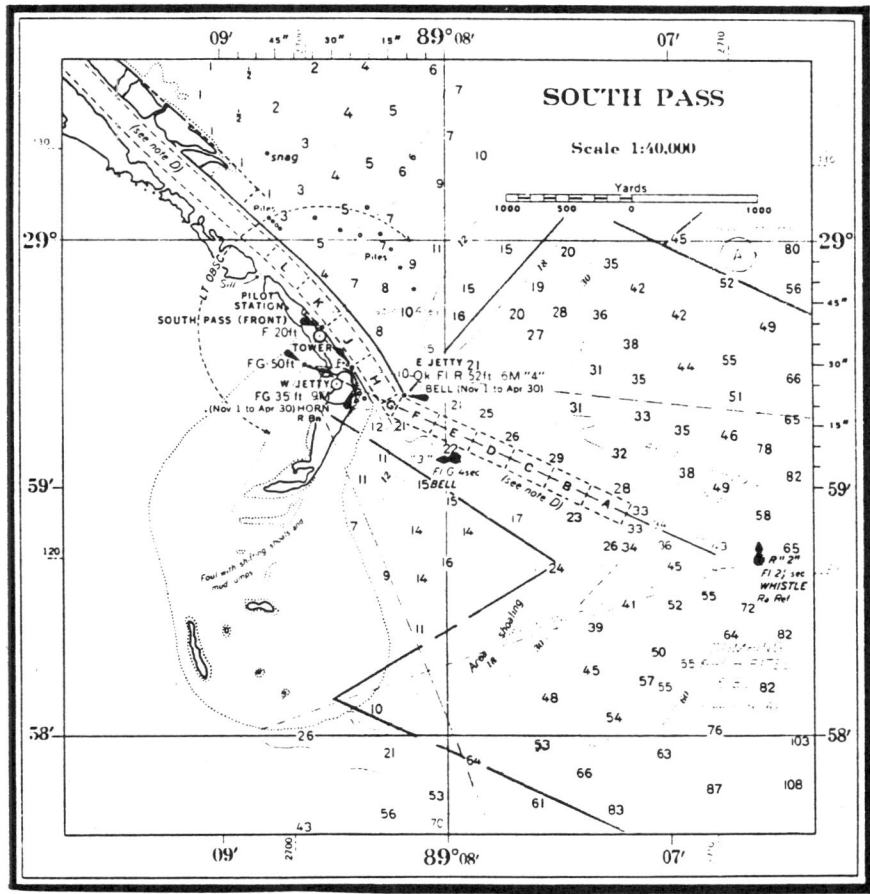

South Pass (Courtesy National Oceanic and Atmospheric Administration)

Canal and Gulf ICW near New Orleans and offers relief from bucking the main river current when inbound.

The Mississippi River Passes

The two main entrances to the MISSISSIPPI RIVER are South Pass and Southwest Pass. Navigating them with a tow, for the first time at least, is in itself an adventure. Both passes are well marked and have lighted ranges covering their approach channels. Their entrances, however, and the method of using them are quite different.

SOUTH PASS should not be considered if it is necessary to enter on the hawser with a loaded barge or any unit drawing over 20 feet. There are depths averaging 25 to 30 feet in some places, but the river builds up mud balls or sludgy water near the bottom. This makes steering and control of a tow difficult if its draft is too close to the maximum allowed.

Entering South Pass with an empty barge on a short hawser, the tug should keep to the right of the center quarter of the range, that is, with the range opened a little to the right. As the tip of the breakwater is abeam, begin a slow turn to starboard. If pushing, the turn should be started at the same point, in order to allow for the set of the river current, which does not follow around this turn at the end of the breakwater. Instead, it runs diagonally across the channel in a southeasterly direction, toward the shallow bank the river has built up. Outbound tugs and tows do not use South Pass unless they are pushing a loaded barge or pulling an empty one.

It should be remembered that the current normally flows out of both passes except during late summer. Occasionally, when the river's stage is low a flood current has been felt upriver as far as Pilottown. There is, however, a rising and falling of tide each day, on the bar, and up the river according to the lunar cycle.

Coming out of South Pass with a light barge on the hawser, the tug should start the turn when inside the breakwater so that it will arrive on the course to the entrance buoy and be as nearly as possible on the center line of the range. If pushing a loaded barge, the tug should slow down and start this turn heading as if the barge's port bow is going to just clear the end of the breakwater. Then, as the current sets the tow to the right and onto the range, speed should be increased to full. A few moments later, the excitement ends as the swift choppy brown water of the river turns to the smooth green of the Gulf and the tow is clear.

SOUTHWEST PASS is the main entrance used by ships bound into the Mississippi. Because of the difficulties in navigating the short entrance channel and staying on its ranges, pilots and tug captains adhere to the local custom of avoiding a meeting between the sea buoy and a point about $1/2$ mile inside Southwest Pass Lighthouse.

It has been agreed that no ship or tow shall attempt to enter Southwest Pass if another is outbound and below Burrwood. Conversely, no traffic is to pass Burrwood if a ship or tow is inbound from the sea buoy off Southwest Pass.

Entering Southwest Pass with a tow is not easy. A check should be made with the Captain of the Port, New Orleans, Louisiana, regarding any requirement for an assist tug when entering Southwest Pass with a tow. If the river stage is high, there will be approximately a 6-knot ebb or outflow passing Southwest Pass Light and running directly onto the

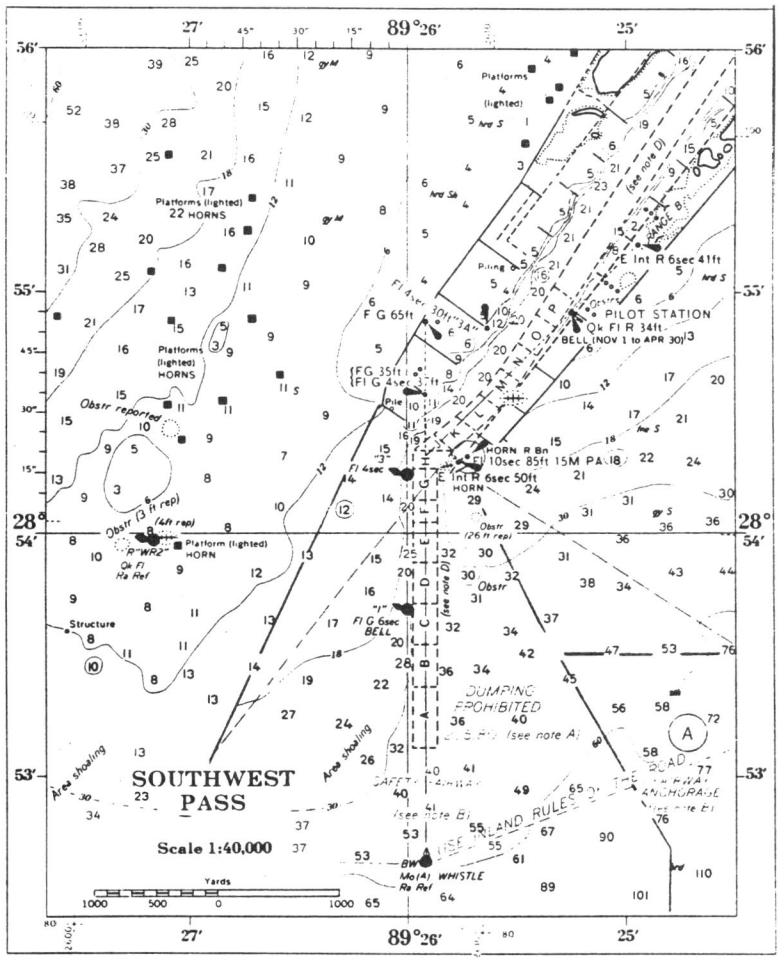

Southwest Pass (Courtesy NOAA)

mud bank on the west side of the entrance range. It should be noted that the U.S. Coast Pilot #5 is a bit misleading in stating that the current runs out of Southwest Pass "straight between the jetties." It does not state that there is a sharp turn from southwest to south onto the entrance range. Straight ahead leads onto a treacherous mud bank which has been a sort of barge graveyard.

Tugs entering with a loaded barge on a short scope of tow cable should be headed onto the set of ranges marking the east limit of the entrance channel. They should maintain about half to three-quarters of their pulling power while on this range. It may take a short burst of full speed to start the barge to break around Southwest Pass Light and turn into the river current. The tug should be kept as far to the right as possible and then pour on the power once the barge is straight astern.

If the tow seems to have stopped, keep pulling. Do not slack hawser or zigzag. If no headway is noted, try to work the tug close to the lighthouse and get the barge astern so that the entire tow may drop back with the current until clear, and an additional helper tug arrives. Inbound with a light barge on the hawser the same procedure is followed, but the danger of the tow grounding is much less.

Outbound in Southwest Pass with a loaded barge is even more of a thrill. If the tug is able to stay in the notch, her speed should be reduced to a point where control can still be maintained. As the lighthouse is approached, left rudder should be applied so that the barge appears to be diagonally across the channel. It will be still going on course, but sideways. As soon as the first green buoy is sighted clear of Southwest Pass Light, speed should be increased and the tow steadied up to slide on the ranges. After passing clear of the light, it may be necessary to head more southeasterly to clear this buoy and the mud bank it marks. When it is necessary to go out Southwest Pass on the hawser, many tugs go through the transition from pushing to astern pulling at a spot about 1 mile inside Southwest Pass Light. At this point, the barge should be on a short hawser. This would be with the barge's wire or chain pendant or shock line almost reaching the tug's stern rail. The connecting shackle would be just outboard and no norman pins would be up. The barge should be making very little headway. Just enough power should be applied by the tug so as to keep a slight strain on the short hawser and shock line if one is used. The tug and barge are actually floating out of the pass. With the tow headed down the last river course above Southwest Pass Light, it should be kept a little to the left on the channel's center line.

As the tow approaches the lighthouse, the tug should start turning left as soon as lighted buoy #3 appears clear of the lighthouse. The tug should then pass about 100 feet off Southwest Pass Light and begin to add a little power to start the tow swinging. As the tow crosses onto the range the tug should start pulling full ahead. If the tug heads southeasterly out of the channel the barge will turn and have the current on its quarter. The barge will resist but eventually follow and settle astern in a few minutes. Then the length of hawser required at sea may be slacked out.

The passes of the Mississippi are not the only places where a continuous outflow or tidal ebb runs between jetties at from 2 to 6 knots over a narrow shallow channel. When outbound in such an area with a loaded barge on the hawser, certain things should be kept in mind. First, to maintain control use as short a hawser as possible. Second, use as little power as possible—just enough to keep a slight strain on the towline. Third, let the current do the work, floating the barge out. Do not be misled if the barge seems to swing sideways and heads for the bank. It usually is still following the tug. To prove this, take a quick line of sight bearing by looking astern and noting the angle at which the towline crosses the tug's taffrail and runs to the barge's bow. Note how far the angle increases before it stops. It may come back to amidships slowly and go to the opposite side. All the while the barge's heading may give the illusion that it is wildly running back and forth across the channel. Unless the barge continues to run well off to one side, it would be a great mistake to increase the tug's speed. A short burst of power might be necessary to correct this swing. By unnecessarily increasing speed, the barge actually will start to sheer. This is a very dangerous situation.

Due to these possibilities it is also good practice to have all the watertight doors on the main deck closed. Men working on deck at such times should wear life jackets. Someone should be stationed at the towing winch ready to slack cable as necessary.

Lightering in the Gulf of Mexico

The lightering by barge of deep-draft oil tankers has been going on at anchorages in New York Harbor and Delaware Bay for some time. This has now been extended to the open waters of the Gulf of Mexico. Most of this lightering takes place at the LOOP,* where ULCC and VLCC moor to pump their cargo ashore. While such ships will act as breakwaters and lie almost motionless during discharge, an empty 600-foot oil barge will rise and fall with any existing swells. Landing such a barge on the hawser without the use of a second tug to hold back is very risky.

A deep-notched barge mated with its own tug also will have difficulty. The tug's push cables are frequently parted. To improve upon and overcome this problem, Sonat Marine, whose expert tug and barge fleet do much of the East and Gulf Coast lightering, have changed their push cable arrangements for this Gulf of Mexico lightering. They have had heavy doubler plates installed port and starboard underneath the rake of the barge. Into these doublers, heavy pad eyes have been welded. The push cable arrangement is connected at these points. They consist of a short $2^1/_2$-inch wire pendant connected to a 12-inch circumference nylon

---

*LOOP = Louisiana Offshore Oil Port.

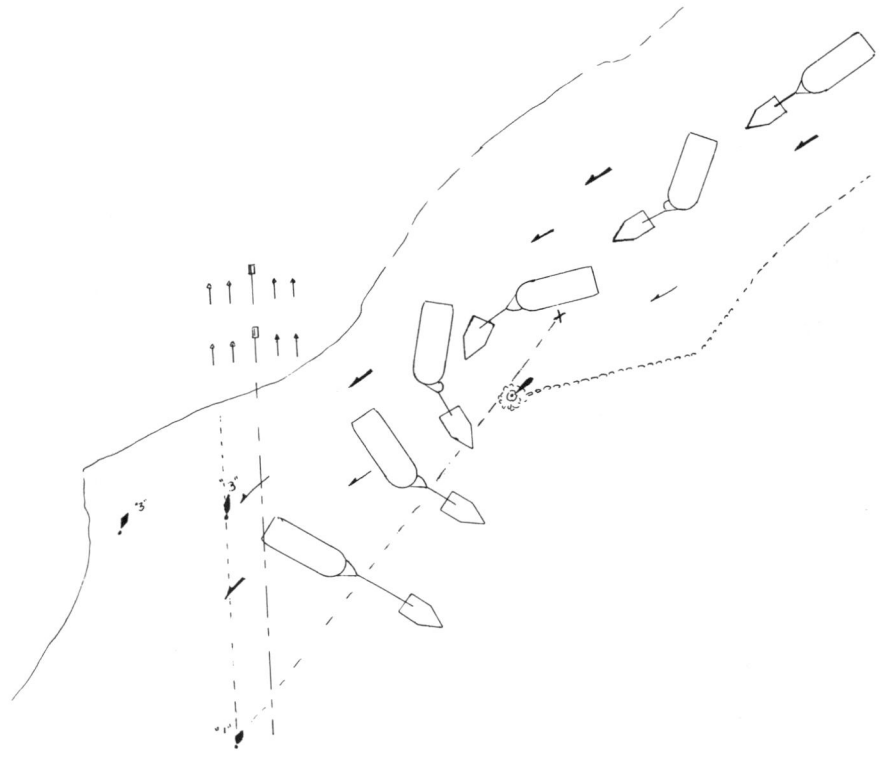

Recommended course for towing a loaded barge on the hawser out of Southwest Pass.

strap that runs partway to the tug. This in turn is connected to another wire push cable, which runs through a sheave in the tug's bulwark, where it is connected to one from the opposite side. Both are joined to the tug's main tow hawser, which when heaved tightly will hold her in the notch and still allow some elasticity.

It was found in using this 12-inch nylon strap, which is inserted into the normal push cable arrangement, that each strap must be compatible. They must both be cut from the same coil in order to have the same amount of stretch and recovery. This is one of the peculiar aspects of nylon. If the straps are not cut from the same coil, one side will stretch more than the other, causing the tug to be off balance in the notch. One push cable would be slightly slack. To avoid this, both nylon straps must be replaced when one wears or requires renewal. To ensure that they are even in length the cutting and splicing is done at the cordage factory.

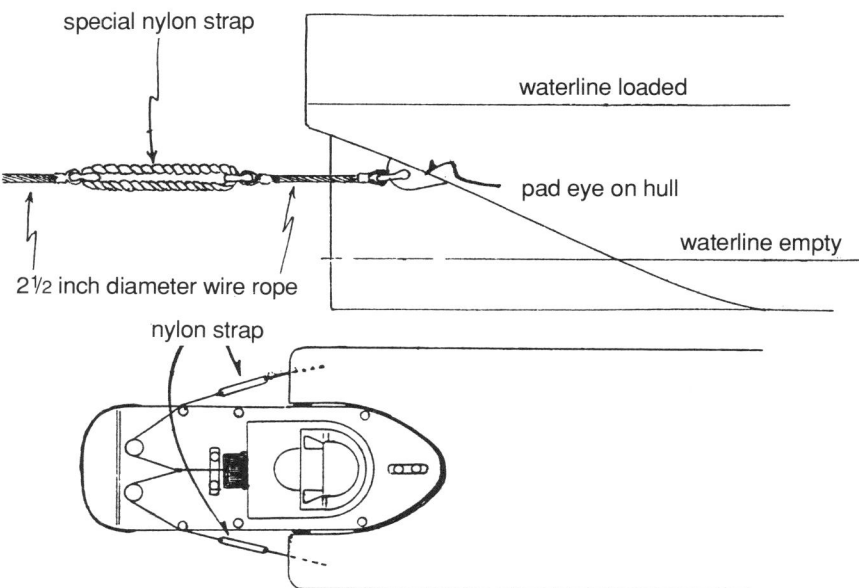

Lightering from large tankers in the Gulf at Louisiana Offshore Oil Port (LOOP) is often done in rough weather. In order for the tug to remain in the barge notch, special nylon straps have been introduced and connected between the push cable from the barge and that from the tug.

The East Coast of South America
This area has three rivers frequented by tugs. The Orinoco of Venezuela has been improved by constant dredging but should be entered with the assistance of a local pilot. This also applies to the Amazon and Tocantins above Belem in the State of Pará, Brazil. The Rio de la Plata is shallow and buoyed as it runs between Argentina and Uruguay past Buenos Aires to Paraguay to Asunción.

A strong equatorial current with velocities up to 6 knots runs off the coast of northern Brazil, extending at times to the Windward Islands.

The Pacific Coast
Several troublesome spots for tows exist on the Pacific Coast. The STRAITS OF MAGELLAN are a wild and windswept series of deep fiords and swift currents. Williwaws of 70 MPH will suddenly descend. This passage is only used by tows too large to transit the Panama Canal. The engagement of a Chilean pilot will be a great aid.

The area of the Gulf of Farallon off Golden Gate Bridge is hazardous at all times except when the waters are smooth. There is little out of the ordinary north of this for tugs until arriving at the COLUMBIA RIVER. As its gorge is approached above Corbett wind becomes a problem. At times, it will reach 60 to 70 MPH, making a tow very difficult to handle. Above Bonneville Lock, winds are still strong but erratic and may blow hard from either east or west, making approaches to the lock or its mooring dolphins very difficult.

British Columbian Waters

Probably among the world's busiest towing areas, these waters also have some of the most treacherous rapids, which, like the Reversing Falls at

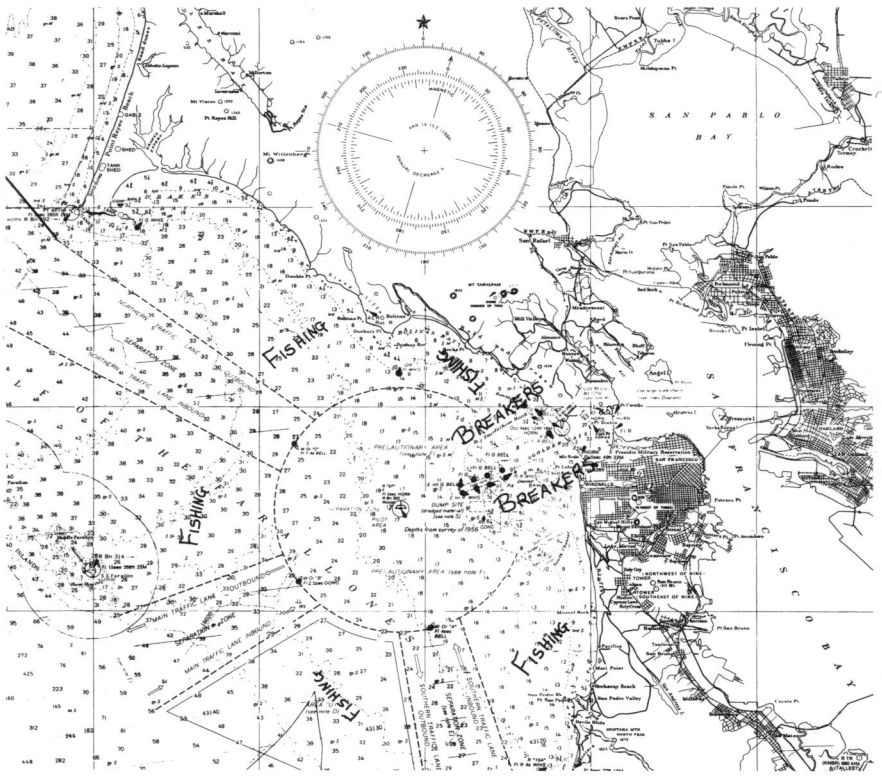

The entrance to and exit from San Francisco via the separation lanes is very dangerous. Tows from sea should time their ETA at the San Francisco pilot station area at slack to flood and stay within the eastbound 3,500-yard traffic lane. It is 11 nautical miles to Golden Gate Bridge. When outbound, follow the western traffic separation lane. To deviate from these lanes could spell disaster. (Courtesy NOAA)

HANDLING TOWS IN SPECIAL SITUATIONS

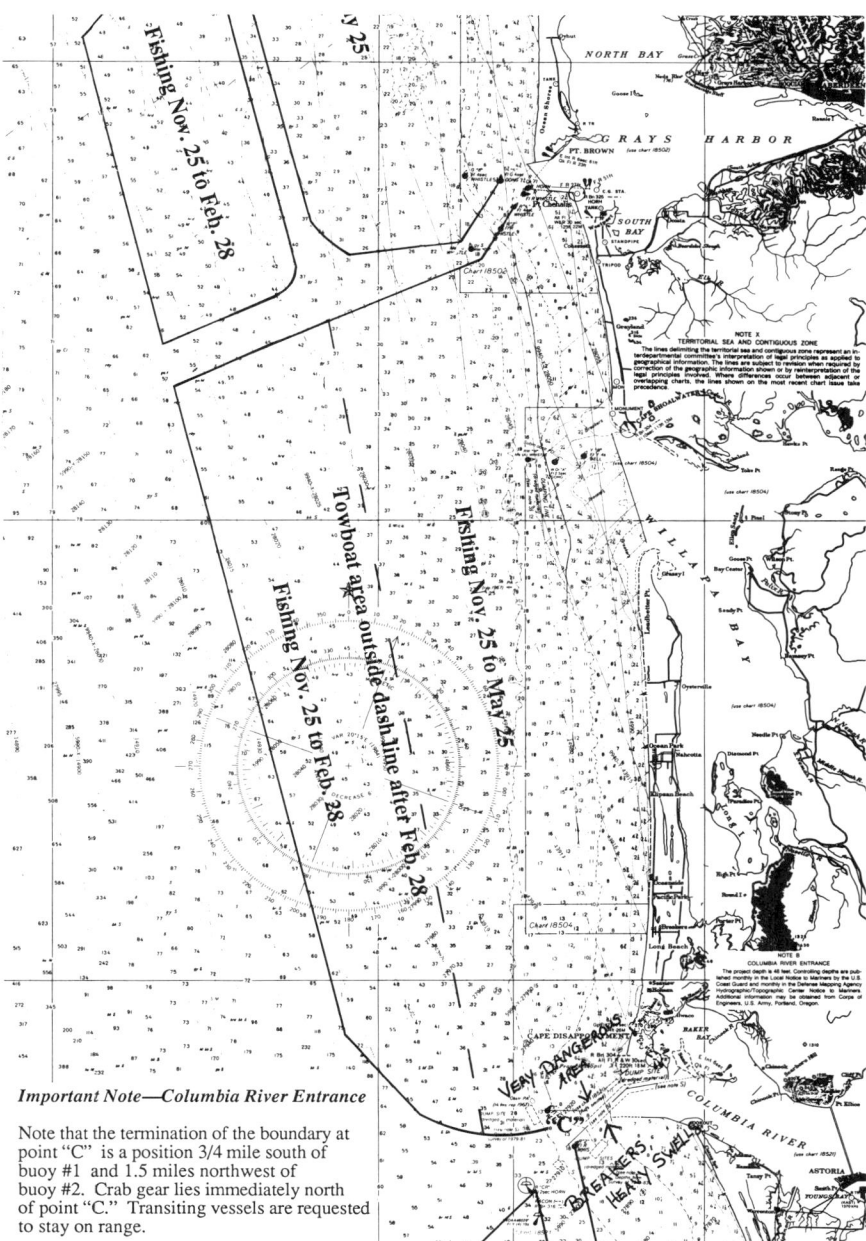

Important Note—Columbia River Entrance

Note that the termination of the boundary at point "C" is a position 3/4 mile south of buoy #1 and 1.5 miles northwest of buoy #2. Crab gear lies immediately north of point "C." Transiting vessels are requested to stay on range.

Grays Harbor and Columbia River entrances. Tows should approach the bars cautiously. Slack to flood is the best time to cross bars inbound; slack to ebb, outbound.

St. John, New Brunswick, can be traversed during the short period of stand when the current is slack just before and during high water. The major and most frequently used rapids are Seymour in Discovery Passage. It is at this point that the tidal currents sweeping around the north of Vancouver Island meet those coming from the south via the straits of Juan de Fuca, Haro, Rosario, and Georgia.

All traffic movements through these areas are within Vancouver Traffic Zone's Vessel Traffic Movement, which has regular calling in points. VHF Channel 11 is used for this purpose.

Due to the difference in distance of this area from the Pacific Ocean, it will be high tide in Discovery Passage a few miles north of Campbell River when the tide is already ebbing at Cape Mudge. This draws water from Discovery Passage in the north toward the Strait of Georgia through Seymour Narrows at up to 15 knots. When the conditions are opposite, the direction changes. These changes occur approximately every six hours.

Tows that are northbound hold back below Cape Mudge so as to pass Campbell River and enter Seymour Narrows no sooner than ten minutes before slack water.

Southbound tows wait off Plumper Bay. One or two tug captains have made the mistake of going through Seymour Narrows with a fair current.

When either the flood or ebb are running at strength, there are great eddies, swirls, and upwellings, which occur about every five seconds. They increase in size and will suddenly shove a tug or barge rapidly ahead and then release it, as they turn into whirlpools. The reaction on the tug and tow is hectic. The tow will sail off to one side and race, almost passing the tug, whose hawser may be slack. Then, the tug will suddenly pick up speed and snap the hawser tightly. Don't even think about trying to pass through Seymour Narrows when the current is at strength.

There are many more rapids in the side channels of these British Columbian waters. They are used at slack tide by tugs towing log rafts. In some of these rapids, which are very narrow, whirlpools will form with such a swirling vortex that they have sucked fishing vessels of up to 65 feet under the surface.

TOWING LOG RAFTS

This is one of the slowest types of towing. Although its volume has greatly diminished, it is still carried out on the upper Columbia, Willamette, and other rivers of Oregon; some areas of Puget Sound, and throughout British Columbia. On lakes in northern Ontario and other parts of Canada, small, specially built tugs push rafts of logs and herd them into storage ponds at pulp mills. To form a raft, frames of flat, square sections of single logs are held together by boom sticks. Boom sticks form the ends, sides, and crosspieces between each section. Without logs inside them, they might be compared to a huge, short ladder floating on the water. When each space between its rungs is filled, the raft will hold from 25 to 30 logs across. There are usually two such groups butted fore and aft so that when the log raft is towed, the ends of the logs are facing the tug and offer the least resistance.

These rafts are originally made up in a remote marshaling area where the end of a road or chute runs down to an inlet, cove, or river. Heavy wires hold the boom sticks together and a wire from each outboard corner is provided for the tug to hook onto. Some rafts are towed by a single wire from the center of the raft.

Because of the size of these cables and the resistance of the raft, tugs of low horsepower are used. Log rafts will not stand any jerking by the tug, or any seas or chop over 3 feet. In a slight swell, the raft astern of the tug resembles a rippling carpet. The slow progress of 4 to 6 knots and precautions regarding weather make timing for towing a critical matter.

Towing is so slow and the areas from which it is done are so remote that replenishing stores for the tug and crew changes are often accomplished while the tow is under way. A seaplane, the auto of northwest

Catherine Foss with a small log tow. (Courtesy Foss)

British Columbia, will land alongside the tug, which slowly holds her raft in tow. Log raft tow movements are planned to use the most favorable tidal current and to pass through rapids at slack water. It is not difficult for the tug to pull at a very reduced speed in order to hold the raft almost stationary above or below rapids while awaiting slack water, which will, after passage, change to a fair current in the tow's favor.

When tugs depart with a raft in a wide inlet, progress will seem almost nil. The tug must take a very slight strain and slowly work up to what the tug captain feels is a safe towing speed. In daylight, captains and pilots of vessels that throw a heavy wake will normally slow down. If it appears to the tug captain that this is not going to happen, he should call the oncoming vessel, requesting a reduction in speed.

As there is frequent ship and pleasure boat traffic through British Columbian waters, log rafts should be well lighted* and swept by searchlight† if other craft are approaching. Pleasure craft have been known to mistake the significance of tug and log raft lights. They have run into rafts and have even jumped over boom sticks to land in the raft itself.

The tug master of a log raft tow normally knows the route very well and must be aware of various points and turns where he can safely use eddies or steer clear of them. He will know other points where the current will set him into shoal water and the maximum wind and current through which he can safely manage the raft. In towing log rafts from rivers such as the Fraser or narrower waterways such as the LaConner Canal in the state of Washington, the lead tug will have a small tender tug on the tail end. This helper tug will straighten the raft, aid in making narrow bends, and ensure safe passages through drawbridges or into locks. This tender will also help in the landing at the log storage area or booming ground.

Deckhands who regularly work at log raft towing must be agile and wear proper shoes as they are required to scramble over the logs to handle mooring or towing wires.

In some areas of Puget Sound where the distance is great between log assembly and delivery to a mill, sudden changes in weather have caught tugs in exposed areas where seas have caused logs to jump out of rafts. Until they ground on the beach, these unmarked floating missiles are very dangerous. They are nearly perfect excuses for a damage claim if struck by an unsuspecting vessel.

Tug masters would be prudent to choose the most sheltered route with a safe holding area, which may be used if the weather turns bad. If,

---

*International and Inland Navigation Rule 24 (g)(i-ii-iii-iv).
†Rule 24—Inland (g)(v).

HANDLING TOWS IN SPECIAL SITUATIONS 399

at any time, he feels that his tow is "restricted in ability to maneuver," he should display the lights and/or day signals of International and Inland Navigation Rule 27.

ICE CONDITIONS

Ice conditions exist on three levels: the fall freeze, the winter deep freeze and thaws, and the spring breakup. For the tug and tow, ice conditions in the fall and spring can be the most dangerous. This is due first to the unexpected appearance of ice and secondly to its great movement as a body as it breaks up.

Ice and Towing

In the Arctic, a ring of solid or broken fast ice continually exists. It runs from Novaya Zemlya Islands, the Kara Sea, Laptev Sea, eastward along the Siberian coast across Alaska and Canada's North Slope through the Beaufort Sea and the channels and islands above 70 degrees north. This is icebreaker territory. Open water is seldom seen until arriving in Baf-

Arctic might. The 260-foot, 14,600-HP Canadian-built *Miscaroo* is a specially built icebreaking tug. Here she is stopped to have her picture taken. (Courtesy BeauDril Limited)

Ikaluk, Miscaroo's Japanese-built sister, is towing the conical drilling unit (CDU) off Deadhorse, Alaska, bound for the Beaufort Sea. These specially constructed Arctic tug/supply vessels have significantly extended the exploratory system. They are equipped with four diesel engines geared to twin controllable-pitch propellers. They are designed for continuous movement through 4 feet of solid ice at 3 knots. In addition to towing, they set anchors and supply material from Tuktoyahtuk, and winterize at Herschel Basin in Thetis Bay, Northwest Territories, Canada. (Courtesy Gulf-Canada Resources, Inc.)

fin Bay or Bering Strait. Any Arctic towing except around Alaska's North Slope is done by specially built icebreaking tugs and supply vessels.

With increasing activity in the North American Arctic, special icebreaking tug/supply and anchor-handling vessels have been built in Canada. Their work is mainly tending Concrete Island Drilling Systems (CIDS) and other rigs in the Beaufort Sea. For perhaps 30 days a year, they may work in open water. Because of this, their normal operation is considered ice navigating. To carry out their work, they are fitted with engines of up to 15,000 HP developed through four main power units driving twin controllable-pitch propellers. These propellers are in nozzles and, like the rudder, are protected from ice damage by large ring

Miscaroc's wedge-shaped forefoot shoves the ice away leaving her wake as clear open water. (Courtesy Gulf-Canada)

bossings. To add to their maneuverability, thrusters are installed fore and aft with fast ballast adjustments that allow the tug/supply vessel to operate in temperatures of minus 50 degrees Celsius while breaking solid level ice up to about 4 feet (1.2 meters) thick at a steady speed of 3 knots. The design in their hull form aids in this; a wedge-shaped forebody deflects ice from propellers and rudder. Most deck equipment is under cover.

As all icebreaking tug/supply vessels as well as icebreakers themselves are rigged for towing, some comments on their methods in ice-canopied waters must be considered. Most of these icebreaking craft have powerful towing machines with pulling and holding power of up to nearly 400 tons.

In St. John, Newfoundland, many large oil rig tugs and tenders connected with Canada's offshore drilling are based. One of their additional procedures is constantly watching iceberg movements. When icebergs move in any direction that threatens collision with a drill rig, one or more of these huge ocean tugs will lasso the berg and pull for hours. Movement is hardly perceptible, as the course is changed away from any drilling structure. To date no rig has been touched by an iceberg.

Finland has led the world in icebreaker building and design. It has provided the world market with Antarctic, Arctic, and Baltic icebreakers of all types including those for use in lakes, rivers, and harbors. The propulsion power of these Finnish icebreakers ranges from 3,400 to 36,000 HP. Nearly all are designed, laid down, and powered in and by Wärtsilä at their Helsinki building yard. Finland was the first country to draw up specifications for different levels of "ice class"-(ification).*

The prime objective of these icebreakers is to keep an open track in the ice so that shipping can pass. As aids to navigation are impractical in waters above the Arctic Circle and other ice-canopied areas, the ice-broken track, unless the whole ice area moves, may be considered a safe path to navigate. Tugs with tows will stay in these tracks. If their speed is such that they can barely move, the icebreaker captain may offer, or local regulations may require, that he assist the tow.

Most tugs that regularly operate in these icebound waters are capable of handling their assignments without assistance. Tugs that are occasional visitors are the ones that may require help. Those tugs with tows assigned to the annual replenishment of North Slope oil production, such as Crowley's Sea Lift,† employ an icebreaking and salvage unit backed up by a helicopter. They wait for an open track around Pt. Barrow before proceeding toward Prudhoe Bay.

Farther south are the seasonally iced-over areas of rivers and harbors. Some are in western Alaska, eastern Siberia, Canada's large northern lakes, the Swedish and Finnish inlets from the Gulf of Bothnia, the Gulf of Finland, and their ports. Towing and ship assistance in these areas is carried out by local ice-strengthened tugs.

Finally, there are the open waters of the northeastern United States and Canada, outside the Gulf of St. Lawrence; the upper Illinois, Mississippi, and Ohio rivers; and portions of the Great Lakes, all of which succumb to some ice covering each year.‡

It is in these waters that the greatest care must be exercised. All tugs are not ice strengthened or equipped, nor are they powerful enough for ice navigation if suddenly caught in a seasonal freeze-up. In many cases, it is not economically feasible for ordinary towing to attempt to continue in some of these waterways that freeze for a part of every winter. Some are closed to navigation.

*For ABS Ice Class, see appendix I.
†The annual Crowley Maritime Convoy of up to 16 tugs with tows plus their salvage and rescue vessel, *Sea Salvor.*
‡See appendix I.

HANDLING TOWS IN SPECIAL SITUATIONS 403

Above, Bugsier salvage tug *Titan* breaking ice ahead of Bugsier salvage tug *Simpson* with mobile Arctic caisson *Molikpaq* (not shown) in tow, Beaufort Sea, August 1984. (Courtesy Bugsier) *Below,* Helsinki harbor tugs tow the new *Royal Princess* through the ice from the launchway to the outfitting dock. Ice like this is common during the winter in most Baltic harbors. (Courtesy Wärtsilä)

Icing the Tug

Like icing on a cake, the icing of a tug with or without a tow can be beautiful. It can also be disastrous. As the earth's orbit causes the sun's annual trek southward, the ice-free navigation season swiftly closes in the Arctic and sub-Arctic areas. Thereafter, it is only safe for specifically constructed icebreaking and ice-repellent vessels, including tugs and tug/supply vessels, to operate in these now-frozen areas. Ordinary tugs should stay out. Winter can commence after early September in some latitudes. It can be hazardous for an unprotected tug, with or without a tow, to be caught in one of the many open-water areas late in the fall when a northerly driven sea spray in temperatures of 24 degrees Fahrenheit (minus 5 Celsius). The tug's hull above the waterline, rails, and superstructure will quickly build up with ice. Watertight doors, wheelhouse windows, and navigation lights all become encased with ice. The tug's stability is immediately threatened. Crew members may not be able to get out on deck. Not only are all fittings and lifesaving apparatus solidly frozen in place but the tow winch and deck machinery may be so covered with ice that they are temporarily disabled.

Crowley-Red Stack tug *Crusader* is already iced up leaving Alaska with tandem tow. Ice is heavier on the starboard side of the tug. (Courtesy U.S. Department of Commerce, Pacific Marine Environmental Division)

Tugs and tows should not be dispatched on long pulls across such waters as Hudson Bay, Davis Strait, the Strait of Denmark, or the Bering Sea in the late fall unless the tug is large and classed for icebreaking. When under way in areas between 44 degrees and 60 degrees North, including the Gulf of St. Lawrence, the U.S. Great Lakes, Canada's freshwater lakes, and the Gulf of Alaska, a watchful eye should be kept on satellite weather patterns. Quickly dropping or rising barometric pressures can give an early warning of sudden changes that will produce the cold air blasts, or "Siberian express," so often experienced from the Gulf of Alaska through mid-America to the Gulf of Mexico.

Eventually, such a cold front will break into a northeasterly track on the U.S. East Coast. There, such fronts usually occur after a protracted period of slowly dropping barometric pressure and overcast humid weather with light southerly winds. As the barometric pressure stops dropping and suddenly rises, its speed will warn of the wind's velocity. Land observations will give adequate warning through Canadian or U.S. weather broadcasts of the approach of "cold Canadian air."

This occurrence on the U.S. East Coast any time after early November presages the possible icing of tugs at sea anywhere north of the Virginia Capes. Icing is unusual before the first of December and may not occur at all during some winters; however, it is better to be prepared than to be caught unaware.

During some winters, tugs in New York's Lower Bay and New York Harbor have been coated with ice and required a trip into a safe slip for relief and removal.

In many ports it is necessary during the ice season to clear a slip in order for the tug to berth her tow. To do so, the tug may have to let go and break the ice and then wash it away with her propellers. If the pier has a solid face the tug may have to put a line out and work her wheels. Often it is necessary to have an assist tug if a ship is to be docked, as ice tends to float right back once the clearance stops. Trying to push a ship broadside into a berth with heavy broken ice between it and the quay is most often not successful.

The same possibilities extend to Long Island Sound waters. It is amazing how much and how quickly a tug with a tow can ice up between Cape Cod Canal and Boston Light or Cape Ann. For a tug to proceed toward Portsmouth or Portland while ice is still forming on the hull is a difficult decision for the master. To continue across the Gulf of Maine might place the tug in jeopardy. In all of these waters, the freezing spray of a northwest gale will make ice on the tug's port side, resulting in an increasing and dangerous list. All of the tug's courses when northbound from Cape Cod Canal would be northerly or northeasterly. This would

Piers and tugs in New York Harbor may have changed since this picture, but slips with winter ice have not. (Courtesy Turecamo)

make it difficult to ice up evenly on both port and starboard sides. The icing on one side of the tow can also be dangerous.

Tugs operating on the upper Baltic, Caspian, Sea of Okhotsk, Barents, and Beaufort seas and their tributaries are built to withstand freezing sprays. Many have heated decks and handrails and are structurally designed for work in both open or frozen waters.

It is in the Gulf of Alaska, Aleutians, and Bristol Bay, where the standard ocean tug is used year-round, that icing has become a major winter problem. Since 1980, studies have been made jointly by the University of Alaska's Arctic Environmental Information and the Pacific Marine Environmental Laboratory of NOAA. These studies have been taken from 85 documented icing cases included in 200 reports from oil rigs and other vessels in the area. Many of the tabulated results were submitted by the crews of Foss Launch and Towing Company's tugs, *Justine Foss* and *Sandra Foss*. One of the findings was that seawater temperature alone was not the major determinant in vessel icing. Another result of this study is an easy-to-use table* that can predict the rate of ice buildup based on wind speed and sea and air temperature. As an example, a surface wind of 35 knots with air temperature of 14 degrees and sea temperature of 34 degrees will cause icing of from $1/2$ to $3/4$ of an inch every three hours.

*See table in appendix V.

Breaking ice in Stockholm Skagard is a normal winter requirement. Some tugs are now fitted with a bubbling system around the hull, which aids in icebreaking. (Courtesy Neptun)

Towing in Ice

Handling a tow in ice can be a difficult, noisy, and frustrating experience. Much depends on the type of ice you are trying to crunch through, its formation, and the waterway and area in which it is encountered. For the unsuspecting or inexperienced skipper who is meeting ice for the first time with a tug or tow, lesson number one is to slow down before entering it. Look for a crack, open water, or a broken track that appears to be in the channel. As many aids to navigation are removed in ice areas, great care in piloting is necessary. Towing through ice requires many different approaches by the tug and tow according to the type of ice. Freshwater ice, which is found in Canada's northern lakes, is so hard, thick, and fast that towing ceases in the late fall. Marine equipment is hauled out on marine railways until ice breakup in the spring.

Farther south, on the upper Mississippi, tug-barge traffic attempts, not always successfully, to exit to below the Chain of Rocks Locks before freeze up. In the Illinois and Chicago Drainage Canal, however, winter traffic buffets its way through, if possible. Lockages are a problem. Too much ice in a lock will block the lock gates and prevent closure or clog valves and obstruct pumping facilities. Pusher tugs with a single tier load ahead can break through ice reasonably well if a track has been kept broken. In cold weather, below 20 degrees Celsius, a broken ice track will refreeze in 20 minutes. If stopped, tows will freeze in quickly.

To resume headway, the tug's propellers may have to open up some water astern to allow the towboat to back and then thrust ahead to gain headway. It may be necessary to repeat this maneuver several times before steady progress is attained.

Meeting or passing other tows is a risk that must be carefully assessed. If the track is not broken wide enough, both tows may have to stop and break a turnout on their side of the channel.

Unbroken virgin freshwater ice is as hard as glass. When first broken, it is just as sharp. Tugs, barges, or any other craft with a wooden hull making an original track through it will be shredded at the waterline. In one hour's steady progress through such sharp freshwater ice in the Dismal Swamp Canal, a vessel's 2-inch planking was cut through.

Once a channel is broken through this type of ice, its cutting effect is greatly diminished. If a channel is cut open by an icebreaker, its cutting effect is reduced even if it refreezes at night. If it is snowed upon, ice loses some of its brittleness. All of this must be kept in mind by the tugman as he approaches a field of ice. The fresh water that makes ice as described is usually found in rivers above the tidal or saline boundary. Here, ice remains fast to the shore and out to the broken track or channel. If the weather is mild, the ice in the broken track may move with the tidal current. If it is cold, the broken pieces freeze almost as soon as the tow passes by or stops. It is a common practice in some areas, particularly the upper Hudson River above Esopus Island, for tows in this situation to stop in the track at sundown. They lie in the broken ice, which quickly freezes and will try to start ahead at daylight if possible or wait for a U.S. Coast Guard icebreaker to open up the pass. These conditions and actions also occur on upper Chesapeake Bay. The route through the Chesapeake and Delaware Canal has at times been solidly frozen and closed for several days. Here the ice, of course, is mainly from a mixing of brackish waters from the Delaware River via the C & D Canal and the salt water of Chesapeake Bay, topped off with the fresh water from the Susquehanna River.

Ice outside the broken track will remain relatively stationary in the cold weather. But when the thaws of late winter start, the whole field of ice lets go from shore and can give the navigator and tugman great problems. Ice that has built a thin sheet extending out into bays or unusually wide areas in rivers during cold, calm weather, may be broken up by wind and sea and driven into a hard pack in the bottleneck of a bend in a river or the head of a bay. This jumbled pack is called "windrowed ice," and may be very much thicker than the ice behind it. This condition is the usual occurrence around the breakwaters of Buffalo Harbor, New York, the Welland Canal (which of course, is closed), Toronto, Ontario, and Montreal Harbor. Not too many vessels and few

The towboat *George T. Horton* slowly pushes her tow through broken ice on the Illinois River. (Courtesy U.S. Coast Guard)

tugs challenge these points, but in the Hudson River between Storm King Mountain and Constitution Island, tugs frequently become fast until broken out. In upper Chesapeake Bay off Tolchester Beach is another spot where windrowed ice may be encountered.

In the Baltic Sea, the harbors of Stockholm and those entered via the Gulf of Bothnia or the Gulf of Finland all have their ice barriers built by nature. In addition to hard winter ice there are several other types that tugs and tows might meet or be forced to work in. These types are frequently confronted in the late winter and early spring.

When towing through broken Arctic field ice, tugs with tows look for leads of open water and proceed slowly, stopping if necessary to avoid striking heavy compressed flows. There are places in Hudson Bay, Hudson Strait, Baffin Bay, and Greenland where tugs occasionally take tows. Icebergs and their smaller pieces, which have calved off Greenland's centuries-old glaciers, must be strictly avoided. A large portion of any glacial ice may run out as a shelf under the surface. When struck even by a glancing blow, glacial ice is hard enough to put a hole in a vessel.

Starting from the East Coast of the North Atlantic, there is a great field of broken ice containing large icebergs, growlers or smaller bergs, and bergy bits. This mass covers much of Baffin Bay, and the Labrador coast. It runs through the Strait of Belle Isle to the Gulf of St. Lawrence and around Newfoundland. Unless on a priority tow or engaged in ice management—the breaking up and diverting of ice flows threatening drill rigs—the tugs and tug/supply vessels or anchor handlers should keep out unless they are built to Ice Classification 4 or better.

Any tow proceeding necessarily through this area and meeting these ice conditions when coming from open sea will need icebreaker assistance. The dangers are great. Even the smallest bergy bit is hard as granite. To strike one with tug or tow can be damaging. To become fast in an ice flow subjects the tow to the whim of the weather. Gales in the Gulf of St. Lawrence have carried ice flows and vessels in them into shoal waters, over rocks, crushed them against ledges or driven them ashore.

Winter navigation on the Great Lakes is often hampered by extremely heavy ice. Studies are ongoing concerning the length of the navigation season and feasibility of extending it for the eleventh month. This would entail some ice navigation. One Great Lakes company has refitted its tug *John M. Selvick* with air bubbles, which greatly aid in

---

Opposite, above, Sisu is towing a ship out of windrowed ice in the Gulf of Bothnia. *Sisu* is 340 feet × 79 feet × 27 feet, 7,900 tons. Her diesel electric plant of 22,000 HP is connected to two propellers forward and two aft. Her speed through open water is 18 knots. (Courtesy Wärtsilä) *Below,* Canadian icebreaking *Atomic* crashes through 12 feet of hard freshwater ice in the Detroit River off Amherstburg, Ontario. (Courtesy McQueen Towing Limited)

HANDLING TOWS IN SPECIAL SITUATIONS

breaking the track around her. The U.S. Coast Guard has several new icebreaking tugs stationed in the Great Lakes area.

Farther south, moving salt water or sea ice can be encountered in bays and rivers. As in towing through any ice where the hawser is used, care must be taken so that the tug does not suddenly hit a cake or pan so heavily that it will have its progress retarded. When this happens, the tow wire becomes slack or lies on the ice. The tow astern continues to come and may run into and damage the tug. It also may just slow down and stop, while the tug frees itself. This sudden movement by the tug will snap the hawser dangerously.

If it is necessary to tow through large sheets of moving ice in a bay or river estuary with a barge or other unit astern on the hawser, proceed through open water at slow speed unless the tow is still in ice astern. When the tug's stem comes against the next pack, increase speed slowly until the hawser comes taut. Then, pull to break through, and slow down again when the tug comes out into the clear if another solid patch is ahead.

These conditions have been experienced on Buzzards Bay, Massachusetts; Long Island Sound; Haverstraw Bay; and Tappan Zee on the lower Hudson River. During extreme winters, Delaware River, Delaware Bay, and Chesapeake Bay may also be ice covered.

Alongside towing becomes difficult as pieces of ice build up between the tug's bow and the unit towed. If the unit's head end is square, it will also pile up enough ice to eventually bring the tow to a stop. Landing at piers, wharves, and in slips requires extra caution. Forcing a loaded barge into an ice-clogged berth may cause damage to wooden pilings. If pushing or pulling, the best approach to the berth is from nearly parallel and as close off as possible. In some cases, it may be necessary to clear the ice by dropping the tow and running the tug back and forth in front of the berth and then to tie up ahead of the berth and work the tug's propeller. In this way, ice will be pushed away from and not into the wharf piling.

Other ice no-nos include: Do not ram a loaded barge into the ice. The weight of the barge and cargo vs. the hardness of the ice will surely cause indents in bow plating and may result in hull cracks and leaks. Continuous pushing through heavy, thick ice can render a washboard effect on a barge's sides. The fact that the barge was able to deliver her cargo must be weighed against the eventual repair costs a survey may require. The winter's profits may be wiped out!

Tugs working in harbors where ice fills slips, or in breaking out other tows, have to be careful when coming alongside or passing anything. Ice will be dirty, and hard pieces may not be easily recognized from among the other jumble. The tug with some way on may be thrown into the wharf or ship by one of these unnoticed heavy cakes. Just such an occurrence has caused considerable damage.

Any tugs that have been iced up, particularly ship-assist tugs with fenders, should be handled gently when coming alongside anything. Soft cushioning rubber or other type of fenders will be rock-hard. Rubber-tire fenders may have a solid filling of ice. Fender chains and lashings snap easily when frozen. Bow fenders are no longer protective bumpers. If the tug has a rope bow pudding that has been saturated and frozen, her approach head-on to a ship or other object should be at a drift until contact is made.

All sorts of crazy things happen when towing in the ice. Shackle nuts have been loosened and come off after much dragging over the surface of ice cakes. Shackles have also become embedded into large pans of ice. This has slowed the tug as the ice was temporarily added to the tow. Wire hawsers become shiny and bright as they are dragged over and under the ice.

If the tug must let go her tow to break around it or to aid another vessel or tow, a warning should be given to the tug crew. Ice will throw a tug into a short, jerky movement that can injure crew members. If it is necessary to back into ice, the rudder should be placed amidships. Backing into solid ice when there is any angle on the rudder may cause the rudderpost to be bent. On a wooden tug, this spells disaster, as the rudder may be broken off or become unhinged from its gudgeon or heel pin. On a hand-steering gear tug, the rudder should be placed amidships when backing into ice. The steering wheel should be lashed or placed in beckets. A free-moving rudder, quadrant, and wheel rope can suddenly cause a backlash if the tug backs into heavy ice. Many a sore hand, arm, or fingers have resulted from a steering wheel spinning unexpectedly.

TUGS EQUIPPED FOR FIRE FIGHTING

Whether it occurs on a ship or at a waterfront property, fire is one of the most dangerous situations that can confront the tugman. There is always the potential for an explosion. The nature of the product fueling the flames is not always known. Yet many tug crews have bravely fought fires from their tug, nearly always successfully.

Many tugs are rigged for fire fighting. Some are general purpose harbor tugs fitted with a fire gun or monitor on top of the pilothouse. Water pressure is furnished by an auxiliary engine and fire pump.

These tugs often supplement local fire-fighting forces, and in some areas are under contract to provide this service. Nearly all tugs from Valdez, Alaska, to the head of the Persian Gulf that are assigned to handle mooring lines and assist ships at oil terminals are also fire

Dorc III and *Dorc IV* are a pair of Dutch-built ship-assist tugs assigned to oil terminals. Their fire-fighting capabilities are being displayed. (Courtesy Damen)

fighters. The American Bureau of Shipping issues a Fire Fighting Class rating for tugs.

Oil rigs have always been a fire hazard. For their protection, some tug/supply vessels have been provided with fire-fighting equipment and can spray seawater, light water (a fine powder that smothers flames), and foam.

In the United States, the tugs of the Sonat fleet, which tow oil barges of up to 260,000 bbl. capacity, have been fitted with a fire-fighting agent called Fire Boss. This upper-deck-mounted unit consists of a cylinder full of powder (light water) and a container of nitrogen. The flick of a lever dumps the nitrogen, and immediately the cylinders pressurize. A long hose on a reel reaches up to 150 feet from the unit and can be used when alongside another vessel. The results are immediate and fantastic. A blazing oil fire on a barge can be swept and put out in a few minutes unless the entire vessel is burning; then added water is required.

Oil tankers are most susceptible to fires. Explosions from cargo handling, tank cleaning, collisions, and bombings in the Persian Gulf have been the main causes. Often the engine room damage or heat is so intense that ships' crews are unable to fight the fire. Salvage companies have fitted many of their units for extensive fire fighting. While the

HANDLING TOWS IN SPECIAL SITUATIONS 415

Kodiak II, a Titan-class, Halter-built tug/supply vessel tests her fire monitors. (Courtesy Halter)

regular ship-assist tugs may hold or completely extinguish the small engine room fire, or single tank fire, it takes tons of water powerfully arched and directed to cool and snuff out a hotly burning oil tanker.

Salvage and anchor-handling supply vessels can throw between 4,000 and 20,000 gallons of water a minute. Smit Tak International has fitted a group of these anchor-handling tugs with a sprinkler system around the entire upper superstructure. This provides a shroud of water around the entire ship when it is approaching a burning vessel.

The huge salvage tugs of Bugsier, Neptune, Smit, and others are prepared for heavy-duty fire fighting and carry every conceivable type of necessary backup equipment. Smit offers transportable fire pump sets, which they will fly anywhere within 24 hours of a request.

If a tug is engaged for fire fighting, the operation becomes one of salvage rather than tow. The burning vessel may be moved if possible, in order to get it away from a pier or other vessels and to head it in any direction that will retard the flames. Meanwhile, the company's fire-fighting team is flown out from the home base with portable pumps, added foam, and inert gas generators.

These marine fire fighters are specially trained. The on-scene firefighting tug and its own trained crew will have approached the burning craft from the windward side away from the flames. They will attempt to cool the casualty's hull with water and at the same time work toward the flaming areas. Burning oil that has escaped over the side must be contained by a foam blanket.

Tankers and other vessels have been known to burn for days. In the United States, a serious shipboard fire occurring when a vessel is moored to a pier often becomes more of a confrontation among officialdom than a confrontation with the conflagration! Pier owners, port authorities, environmental commissions, and the U.S. Coast Guard will all disagree as to whether the ship should be moved, and if so, how, when, and to where.

The harbor or ocean and coastwise tug master who is called upon to fight fire or who volunteers to do so, must decide quickly whether to fight the fire or tow the burning unit clear of wharf or other vessels. Presumably, the best approach is to be prepared for fire fighting with the fire monitor and/or fire hoses on deck manned and pressurized. Most oil tankers have a wire towing pendant hanging from both bow and stern. This may be used if the ship must make a rapid getaway caused either by a fire on the pier or one on the vessel that threatens surrounding shore installations.

Oil companies and port authorities that have tugs under their control will have a fire-fighting and evacuation program in force. Private tugs that are called or volunteer to assist should ascertain who is assuming charge and quickly join that person's communication system. In the United States and Canada, the countries' Coast Guard may have an on-scene fire-fighting vessel and team, which may be in charge.

Tugs approaching a burning vessel should do so cautiously. If the fire is fresh, an explosion may occur. Fragments of steel and burning oil may fly and strike the tug or injure personnel. It is best to approach a portion of the vessel that is not burning where it is still possible to play water on the fire area. A tank vessel that is hot from burning for several hours may have a boil-over. This occurs when oil in the tanks adjacent to the fire is heated to a boiling point. In expanding, it may back up out of vents and on-deck openings and run off overboard to spread a sheet of flame around the ship's side. A tug threatened by this situation must blanket the water around her with foam and back away quickly. Water should be played on the tug's superstructure to prevent paint and other material from scorching or igniting. It will be necessary to use the tug's engines to maintain position at a close distance as the force of fire nozzle pressure when the water hits any object will push the tug away. Cycloidal or rudderpropeller tugs have a great advantage in holding their position when fighting fire.

If a tug master is asked to tow a burning ship away from her berth and the dangling wire pendant is to be used, another short wire pendant from the tug should be shackled into it and then connected to the tug's hawser or regular ship line. This is a precaution against the use of a synthetic line close to the ship, as the heat might cause it to melt and part. It must also be assumed that the ship's mooring lines have been cut or hurriedly let go and that someone in authority will advise the tug when she may pull the ship out. In the interim, while awaiting such advice, the tug once connected should have her engines working ahead dead slow with just enough strain on her line to keep herself in position.

The tug master who does pull a ship away should realize that he and his owners may be liable for any damage caused by such a movement. He should log all times and details as well as hours of service and what materials were used to aid the stricken vessel.

The recently built U.S. Navy Rescue Vessels of the A.R.S.-50 Class are teamed and equipped for fire fighting. The U.S. Navy and other organizations maintain fire-fighting schools in which merchant marine officers and crews, including those of the towing industry, are trained under actual shipboard fire conditions. Some towing companies and many maritime schools send students through this rather arduous and intimidating course. It is great for instilling know-how and confidence when a fire is confronted.

TURNING AN EMPTY OIL BARGE IN THE WIND

Among the remaining items of handling tows in special circumstances is turning an empty oil barge or docking it in the wind. Picking up an empty barge under normal conditions has been covered in chapters 7 and 8. Attempting to get empty barges under way and heading on the desired course in a moderate to fresh breeze is sometimes difficult. When the wind is 25 knots or above, either an assist tug (if available) should be used or an area large enough must be chosen to safely accomplish this maneuver. The latter is often the case in some sheltered bays or coves where the barge must be placed alongside a ship at anchor or when proceeding into a slip or wharf.

With the barge's way nearly stopped, the tug must let go her hawser quickly. Her master or pilot should know whether it is more expeditious to back down or turn to go alongside and get a line out on the barge. Under such conditions, the barge quickly turns or blows sideways downwind. The tug should chase from the windward side and, like a cowboy, lasso the fleeing steer with a line anywhere.

However accomplished, a good line on a barge cleat should be used and the barge's headway and drifting stopped. If the tug is getting in the notch, it may be necessary to work slow ahead to hold her position until all lines are out and tightly secured. If the tug is to make up alongside, the original line can be shifted, and after a second one has been added the tug can work her way to the correct spot and side to be used.

Because of the wind and the probable necessity of maneuvering using the tug's full power, all lines should be doubled and be without slack. When all lines are fast, the change from the tug and barge drifting to leeward to moving ahead on the required course nearly always reveals that the inclination to come full ahead with hard-over rudder will not work. This is assuming that the new course will require turning into or across the wind's direction. What happens is that the pivot point for turning is now close to the tug's propeller(s). As there is little headway at first, the response is that the standard tug has become a tractor tug. The barge seems to swing rapidly. Actually the tug is working around the barge in a transfer action with little advance.

As headway is established, the swing will diminish and may reach a point where, like a sailing vessel, the barge will not come any closer into the wind. The best way to get out of this dilemma is to slow and alter the tug's rudder in the opposite direction, which would be to leeward. Then, go full astern on the tug. The barge will swing rapidly in the opposite direction and may come around to the desired heading on the first

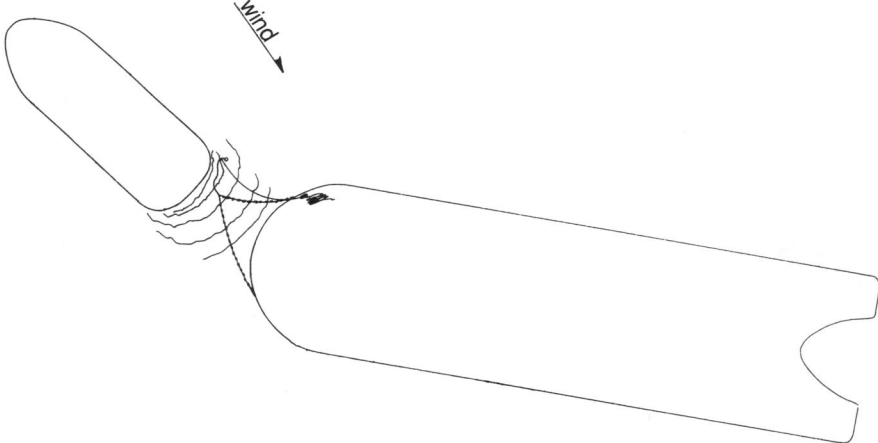

To pick up a light barge in the wind, the tug must head the tow nearly into the wind.

Opposite, above, if the barge is to go alongside a ship for lightering, it must be picked up so it will be landed bow-first. *Below,* once the tug is made up, it may be necessary to go full astern until the barge swings across the wind.

HANDLING TOWS IN SPECIAL SITUATIONS 419

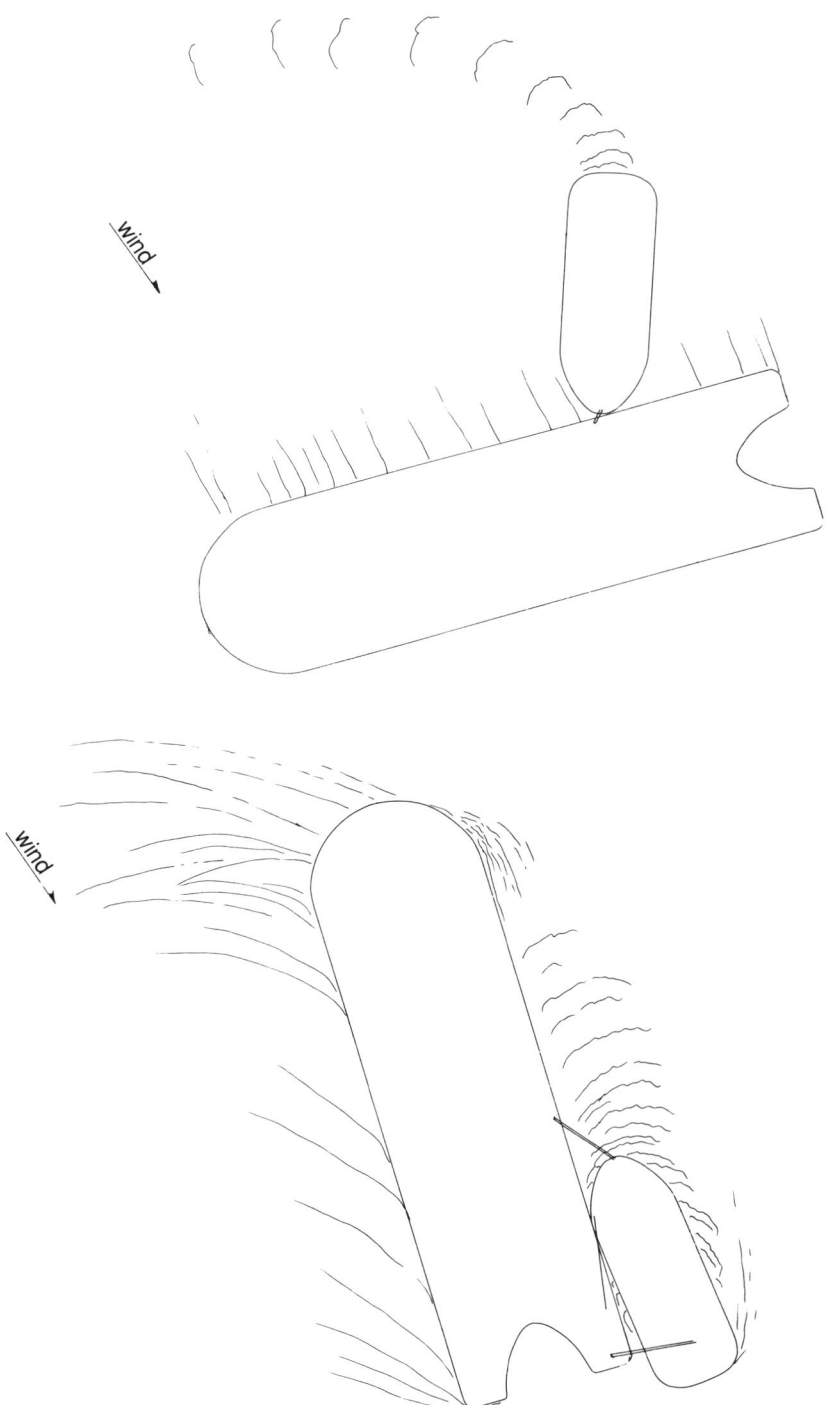

attempt. As it does, the tug's engines should be placed at full ahead and appropriate rudder used in time to catch the barge before it swings too far. If not successful at first, the tug should make a second try.

With the barge now under way, docking will be the next problem. If the wind is blowing on a wharf that is solid with a flat, long surface, a twin-screwed tug that can walk sideways may be able to approach and land the forward end lightly. Then she would be able to hold off the stern so that it comes in gently. This maneuver takes patience, coolness, and experience and is not recommended. Using an assist tug or waiting until the wind dies down would be the more prudent approach.

If it is absolutely necessary to get the barge docked immediately, and none of the above considerations are possible, one remains. If the barge has good ground tackle and there are no known obstructions on the bottom, such as cables, gas lines, or wrecks, the barge's anchor can be dropped, using just enough scope to allow it to have some resistance. This retarding force at the forward end of the barge will allow it to be worked in bow-first at an angle to the wharf. When lines are out, the stern can be held off by the tug and allowed to fall in gently by use of the rudder and occasional kicks of the prop.

If the wind is blowing off the wharf, the tug should head the barge directly into the wind. The forward end of the tow should approach the mooring bollards it will use when it is in place. If the wind is coming off at an angle that will blow the stern or entire barge off the berth, an assist tug may be required. If the berth consists of a short mid-section with a catwalk to pile clusters at each end, the windward cluster is usually the best end to try to get a line on or around. If the barge is much higher than the cluster pile, the bight of a mooring line might come off when under a strain. In such windy weather it will be necessary for someone on the wharf to make this line fast to a proper bollard, hook, cleat, or other mooring fitting. Then the barge can be worked in alongside.

It may be necessary and advantageous to have the tug made up so as to be between the barge and the berth as the landing is being made, particularly if the terminal or wharf has cluster pilings or caissons at each end. In this manner, the tug captain can observe the inside bow of his tow and keep it clear until enough of the barge is past the corner. Then, by backing, the bow will swing in gently and land on the corner.

If the tug is in the notch, it will not be possible for the tug captain to see any of this. A good deck person or mate will have to guide him in. If the barge is small and manned and does not have a high overhanging rake, it may be held close to the tug on two stern lines. As the berth is approached, the tug's stern can be backed against the barge to stop it. Then when the line on the side closest to the wharf is slackened, the bow of the barge will swing in close so a line can be passed ashore.

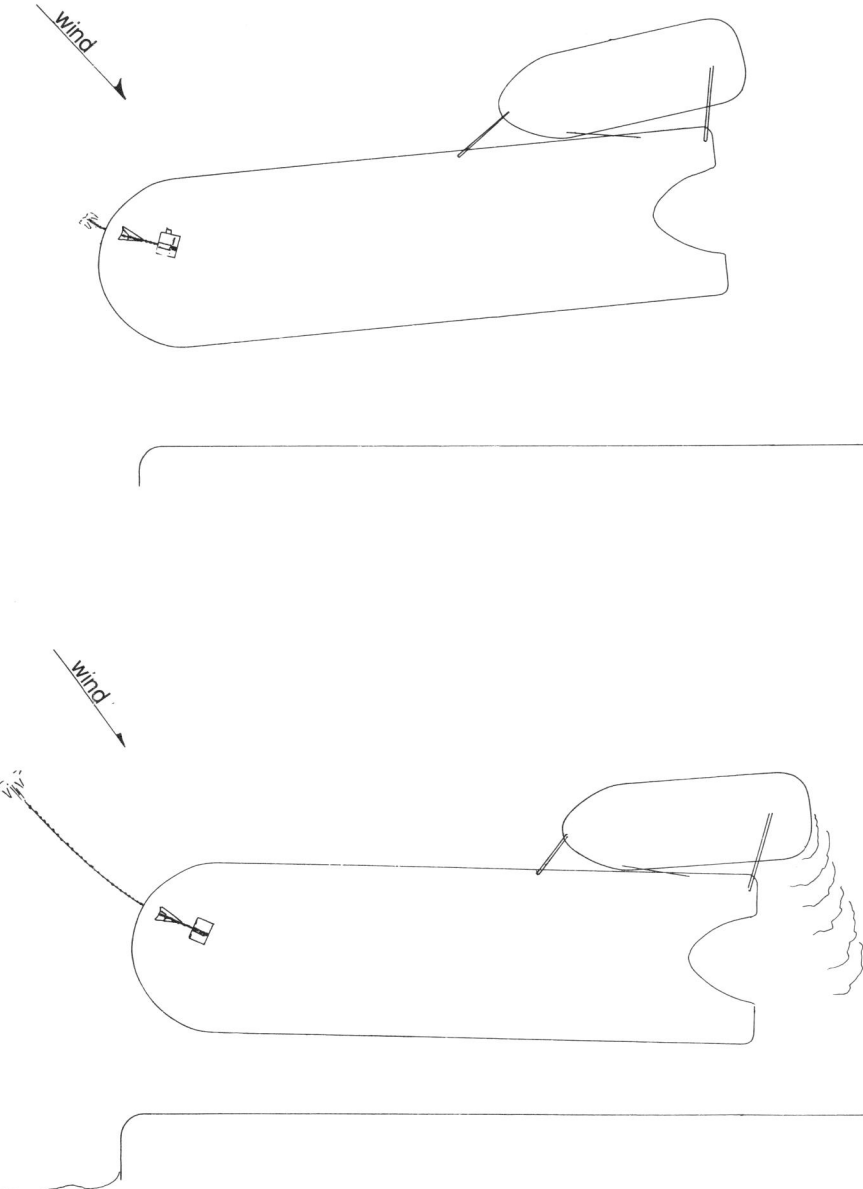

Above, dropping the barge anchor to aid in docking. *Below,* slacking the anchor to gently drop the bow as the tug holds the stern off and parallel to the wharf

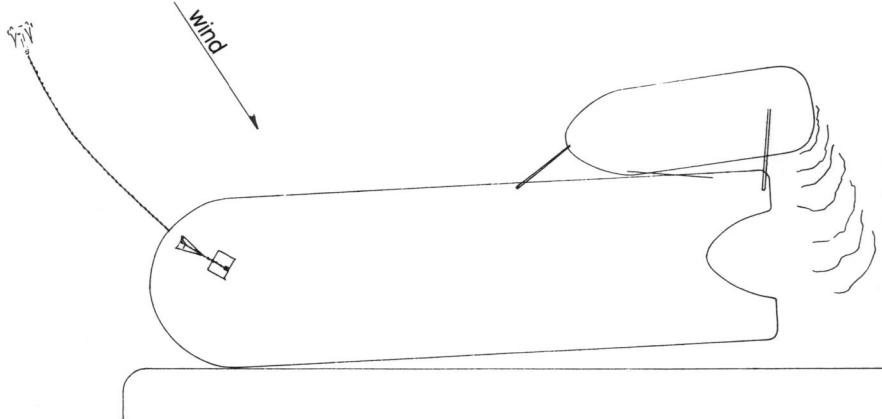

Coming alongside, the anchor chain should be held so that the barge's bow does not strike, and the tug should keep the stern off slightly until the barge lands.

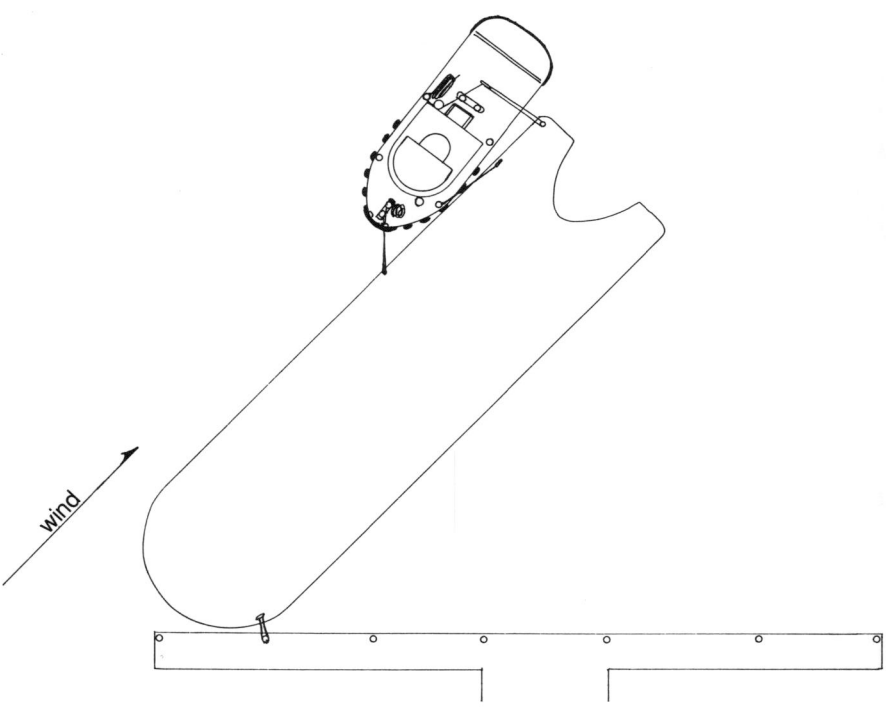

Docking a light barge with the wind off the pier

HANDLING TOWS IN SPECIAL SITUATIONS   423

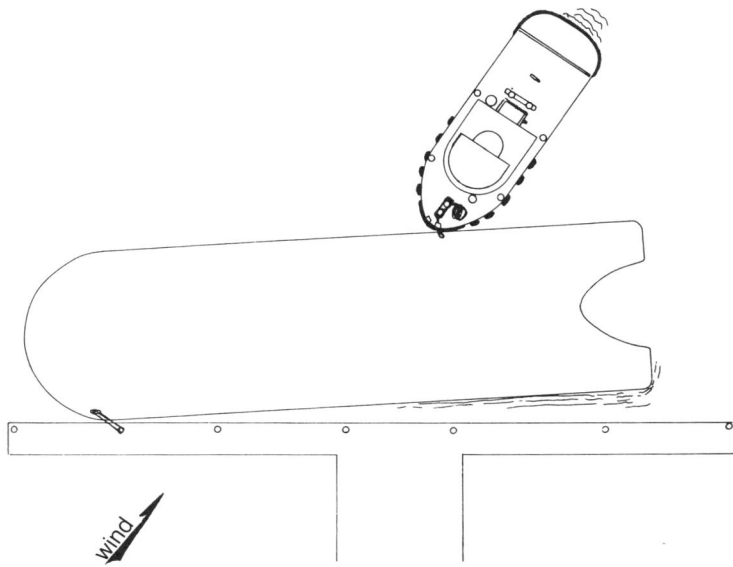

After getting the barge's bowline out, the tug holds the barge from striking the pier as it comes alongside.

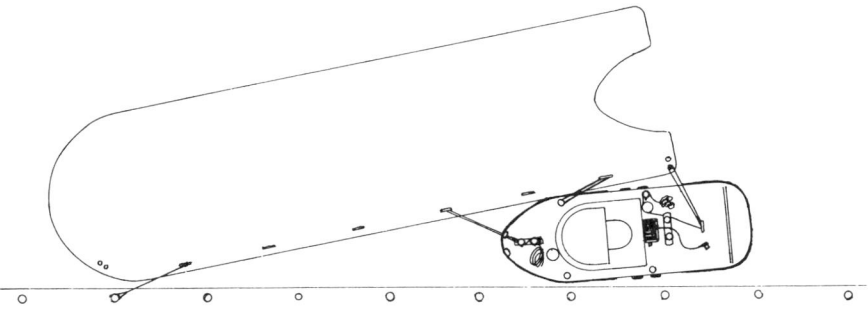

Docking a barge with the tug between it and the pier

USE OF THE ANCHOR WHEN DOCKING ON THE HAWSER

There are times when conditions require landing a unit by using its anchor. As mentioned earlier, docking a barge in strong winds is one occasion. Another more frequent use has been with barges that have a model or ship type bow, particularly if they are fitted with a hawsepipe and good anchor windlass.

The barge must have an experienced person standing by. The tug must have the barge on a short towline. There must be room for the tug to maneuver in, ahead of the berth where the barge is to be landed. The waters of the approach must be free of cables, pipelines, or other obstructions. The bottom must have sufficient depth at any state of the tide so that the barge, loaded or empty, can pass over her own anchor without striking it. The bottom must be sand, clay, or mud with some good holding qualities.

The maneuver usually commences when the barge is nearly stopped several lengths from the berth. On signal from the tug, the anchor is dropped and about one shot of chain released. This should give the anchor and chain a good lead under the barge and act as a brake holding the barge so that the tug must pull from slow to half-speed for the barge to move. Proof that this combination is adjusted correctly is for the tug to stop for a moment. The barge should start to drift backwards.

Once the anchor is down and the chain has a strain, the tug can work the barge up to its berth. By swinging the tug's stern in or out as necessary, complete control can be accomplished in docking. The tug should be handled from an after-control station. Once several bowlines are run from the barge to hold her in place, the tug should be let go and go back to push the barge into position and hold it until the barge is fast. Then the barge anchor should be heaved in. This same type of operation may be done with the tug in the notch. Good communications would be required between personnel on the bow of the barge and the tug's pilot. This type of operation has been successfully used in landing large empty phosphate barges in Florida ports and on huge cement and oil barges in Texas and other ports.

Any barge with the anchor held on a slide that allows it to run over the bow would be difficult and possibly dangerous to dock using this method. The wire anchor cable runs through a fairlead, which must be closed so that the wire cannot jump out and cut the deck edge or strike personnel.

ANCHORING THE TUG WITH A TOW ASTERN

Those tugs that have a standard set of ground tackle (anchor, chain running in a hawsepipe, and a windlass with a wildcat), can, and sometimes do, anchor, holding their tow close astern. The tug's anchor must be heavy enough to hold the tow if the tidal current or wind increases. The tow must be headed into the wind and current and stopped. This is not a common occurrence in the United States as few tugs, except those of the Navy and Coast Guard, are fitted with anchor gear. Elsewhere, for whatever the reason or necessity, the anchoring of a tow by a tug should be considered a temporary condition.

A good anchor watch must be kept and the tug's engines ready for immediate use. A change in weather or tidal flow may cause the anchor to drag or the tow to approach and strike the tug. This has happened. A large tug towing a ship became weather-bound and anchored in Delaware Bay with a dead ship astern. The tug's draft was greater than that of the ship. When the tide turned, the ship's stern swung to the north and the tug's stern to the south. The ship blew past the tug crossing her bow and the tug's hawser was wrapped around her wheelhouse, up one side and down the other. Another tug had to pull the ship back around to free the cable. In another accident a tug anchored in The Solent off the Isle of Wight in England was struck when the sharp-hulled unit she held on 200 feet of hawser astern rammed her amidships. This happened as the tide turned. The tug was swinging crossways when the unit on the hawser was pulled by its weight and the fair current caused it to strike a damaging blow amidships.

ANCHORING THE BARGE WITH THE TUG WHEN DELAYED EN ROUTE

This is a common practice when waiting for a storm to pass, or for fog to lift in a canal or waterway that authorities have closed. It is a simple matter of selecting a spot in an anchorage where the tow will clear other vessels and be out of the channel. Most anchorage areas are marked on charts. In many areas, it is necessary to get permission and/or notify authorities of this intention.

If the tow is on the hawser it should be headed into the wind and current and stopped. The tug may let go first if the tow bridles are hers. If they belong to the barge they must be cleared so that they will not foul the cable or anchor when it is being dropped and being heaved back aboard. This is particularly true if the tug is made up to the barge head-and-tail and is using her main hawser connected to barge bridles as her stern line. It is assumed that the anchor gear to be used is at the barge's bow.

In such a makeup, the tug, on arrival at the spot where the anchor is to be dropped, must back the tow in such a way that the anchor cable will not run under the tug or become fouled in the propeller or rudder. This is very difficult if the wind and current are different or opposing directions.

If the tug is in the barge's notch, anchoring is simple. After the barge is anchored, the tug, which was made up head-and-tail, may have to be moved. The yawing of the two vessels may cause the tug to swing over the barge's anchor cable or the tug may wish to go on the hawser when getting the barge under way again. In being released from an anchored barge when lying head-and-tail, a stern line must be put out from the tug to the barge and tightened before the tug's towing hawser is slacked and disconnected.

GETTING AN ANCHORED BARGE UNDER WAY

When an anchored barge has a hawsepipe and chain, there is little problem for the tug in assisting the heaving of the anchor. It is a matter of a slight touch ahead to relieve the strain on the chain and then following such orders and directions as given by the person on the barge's bow. The tug can do this from the notch or if made up from alongside on the hip. It has also been done with the tug head-and-tail.

Barges whose anchors when stowed lie in a slide on deck will give a bit more trouble when heaving up. Their anchor cable of wire runs off through an open chock. It will often jump out as it is being heaved in with a heavy strain on it. The person directing the tug captain must understand this. The anchor cable must be kept relatively slack and leading as nearly straight ahead as possible. Directing the tug to do this will be by hand signals or VHF radio. At times, the tug may have to back or use rudder angle to keep the barge cable in this desired direction.

When the anchor is at short stay, that is, still on the bottom and the cable is straight up and down, some captains may wish to get on the hawser. While this may be advantageous, it is also a delicate choice. The tug must put her bridles on the barge and start pulling gently so that the balance of the barge's anchor cable and anchor will come up and clear—one advantage being that an anchor full of mud can be washed off. If the tug has her own bridles, the tug must wait until the anchor is cleared and stowed before connecting. Then, she should proceed as described in chapter 8.

APPROACHING PORT IN THE FOG

With radar and position-finding equipment, a tug and tow should not have a problem in entering port in the fog. It is when the fog lifts that danger appears. Anchored ships that have been delayed start to heave anchor and proceed in. If the harbor entrance is a dredged channel between two breakwaters, as many in the Gulf of Mexico are, it is not a prudent place for the tug with her tow to be when the fog lifts. If it is necessary for the tug to have any amount of hawser out and she must use more than one sector of a channel range to keep the tow straight, the tug captain should wait until he can proceed with a reasonably clear and safe channel. In many places, vessel traffic control does not exist. Where it does, the outside approach may be beyond its limits. By contacting pilots on other vessels through bridge-to-bridge VHF, a safe entry arrangement can be made on most occasions.

PLACEMENT OF SHIP'S TOWLINES

When a tug takes a line from a ship's bow or stern, care should be taken in securing it. A tug that will work at the ship's stern using the ship's line may drop the eye over a bow bitt on the tug, but it would be much safer to take enough aboard to wrap it around leaving an end free, which can be slackened.

The safest method to use in accepting the eye of a ship's towline is to have a good piece of synthetic line to pass through it and secure to the tug's bitts. This line should have a good eye splice on one end that will fit over the cavel of a bitt. The other end should have a plain whipping. This piece of line should be long enough so that its end can be passed through the eye of a ship's line and back through the tug's bow bitts, repeated several times. Then, if the ship suddenly uses her engines and the propeller wash hits the tug and lists it radically, the ship's line can be let go before the tug heels far enough to girt and capsize. The record shows that there have been a few "telegraph happy" shiphandlers who have unwittingly girted tugs. Unfortunately, these tugs could not release themselves because the eye of the ship's line was over the tug's bitts. This holds particularly true when a line is placed on a tug's H bitts aft, and has been the cause of more than one capsizing, sinking, and loss of life. NEVER PUT AN EYE OF A SHIP'S LINE OR WIRE OVER A TUG'S AFTER H OR TOW BITTS. An eye may be placed in a quick-release tow hook.

TOWING "LIVE" SHIPS

Frequently ships that have power have to be started away from their berth by tugs. In such a situation, and in harbors and narrow waterways, it is very dangerous for the single tug to tow from ahead of the ship if: (a) the ship is loaded, as it will sheer if the channel depth is close to the ship's draft; (b) if the ship is light and there is a fresh breeze, as it will hang off to one side and try to sail; (c) the loaded ship will be affected by bank suction and will run across the channel; (d) the ship that uses her engines and steering gear will take control away from the tug.

Before starting such a movement, the tug captain and the ship's captain or pilot must agree on the best method. If only one tug is to be used, it should be made up near the break of the foc's'le head on the offshore side. Lines leading in opposite directions should be sent up from the tug so that it can either push at an angle or back as necessary. With the tug secured in this mode the ship can use her engines and rudder until clear. If a second tug is used, it should make up alongside on the ship's quarter. Then, the forward tug can either tow on the hawser or remain made up at the bow. If the forward tug tows from the ship's bow, care must be taken to keep clear of a bulbous bow, if there is one. This lead tug's propeller or "wheel water" will not affect the following of the ship if the tug is 150 to 200 feet ahead of her. If the ship must pass through a drawbridge, a shorter length of towline should be used. This is so that complete control can be maintained in counteracting suction from nearby seawalls, wharfs, piers, or other moored vessels. A ship, like a barge, can be pulled up to a wharf using her anchor as a restraining brake. There are also occasions where ships must be towed out of tight areas stern-first. Sometimes, a bow anchor is dropped to use as a drag. In all of these movements, a ship is free to judiciously use its engine and rudder without endangering a tug.

TOWING BARGES IN ROUGH WEATHER

One of the most powerful shocks in the world is that of the solid sea breaking aboard a vessel. Ocean liners have had their forward deckhouse shoved in from such a blow. Oil and other barges lower in the water take a terrific beating if pulled at any speed other than dead slow when in heavy seas. Pipelines, hoses, deckhouses, and on-deck stores get bent or washed overboard. Tug captains and watch standers must watch their tows. The tug may be riding well, while the tow may be burying itself in solid water. This may appear from the tug to be just heavy spray.

If a tow is caught in a storm at sea, speed and course should be adjusted to allow the tug to heave to. The towline should be regulated so that the tug and her tow rise together on each sea and drop together in the trough. This eases the strain on the towline and the pounding on the barge. In some areas, if the sea is not too great, and the tow is light of draft, or can stand to be turned in a reverse course, it may be prudent to run before the bad weather for awhile until it improves. Empty oil barges should not be pulled at full speed when seas are over 5 or 6 feet. The continuous pounding of their bow in rough weather causes indentations between frames and the loosening of welds on internal ribs. The monetary value of the time saved by the quick trip will not cover the cost of shipyard repairs at the next survey.

PUTTING A BARGE ON A MOORING BUOY

To put a barge on a mooring buoy is not quite as easy as landing one at a dock. Two things are required to start the operation: a long boat hook and a person who understands what response he will get from the directions he gives to the tug's pilot.

There are mooring buoys used by barges in such U.S. harbors as Boston, Narragansett Bay, New York, and Seattle. Some are privately

HANDLING TOWS IN SPECIAL SITUATIONS 429

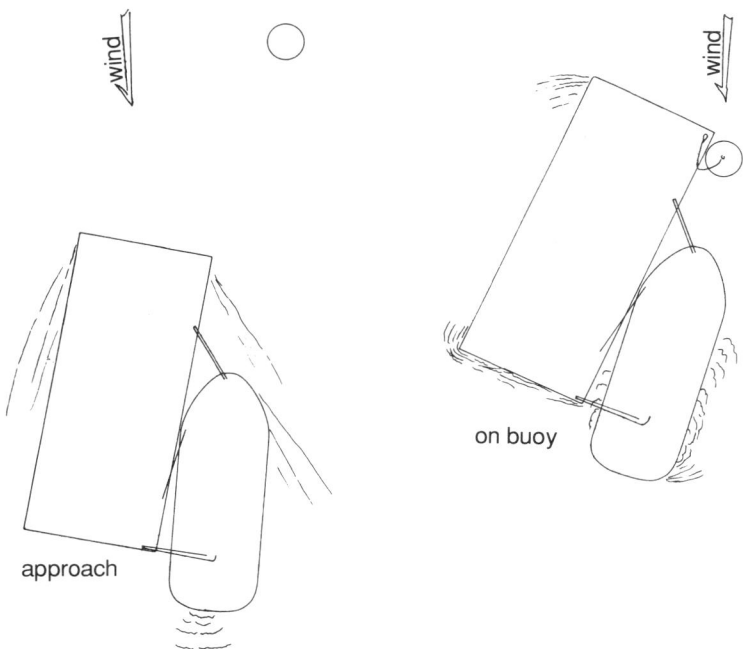

Tug approaching mooring buoy with barge alongside, and getting line on buoy.

owned.* Hanging a barge on the latter, except in an emergency, may result in an order to leave. Some of these buoys are round steel rubber-fendered cylinders or long wooden floating timbers with a steel loop or bale on each end. On the wooden-timbered type, there is usually a mooring wire with a small watch buoy floating nearby. This is attached by a rope messenger to the actual mooring wire, which would be hanging down. Steel mooring buoys may be fitted with rings or hooks to receive vessels' mooring lines. To make connection on either type of moorings, the approach is most important. The tug master must decide which of the elements, wind or current, will affect his tow. His tug should be alongside or in the notch. A loaded barge low in the water is easy to moor. The tug is positioned so that the bow is heading into the current and very slowly creeps up, keeping the mooring buoy dead ahead so that it will just clear the side of the barge. With the tug stopped and the barge's bow floating past the buoy, which should be in full view of the tug master, the two persons on the bow should hook the watch buoy and pull up the mooring wire.

*In New York Harbor near the Statue of Liberty and on either side of the Bronx Whitestone Bridge.

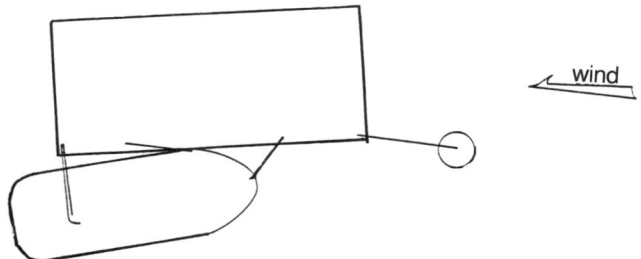

Barge lying on mooring buoy

If all is clear, the eye of the mooring wire can be dropped on a corner bitt of the barge. If the buoy is of the steel cylindrical type and has huge hooks, dropping the eye of a barge line over one of them and paying it out when the tug backs will serve as a mooring. If the barge has rings, mooring will be more difficult. The eye of a line along with some slack must be thrown onto the buoy. It must land near enough to a ring so that the ring can be flipped over on top of the eye. Then, reaching through the ring with a boat hook, the line must be pulled back to the barge. This will give a two-part mooring. It takes two persons to perform this feat. For safety, a third person is necessary to talk to or to signal the tug while the other two struggle with the line.

In approaching a mooring with a light or empty barge, the same procedures should be followed. Trying to pick up the mooring from the center of the bow is usually futile. The barge nearly always floats over the mooring buoy as the pickup is being attempted. If the tug backs, contact with the buoy is lost when the barge swings. A pilot of a tug in the notch must rely on his personnel on the barge for directions. A tug made up head-and-tail can carefully back up to the mooring so that the buoy is touching her stern. The end of a barge line can be thrown to her deck and passed onto or through a cylindrical mooring buoy fitting. This stern approach is not recommended with moorings that have their own wires and pendants hanging in the water.

If the tug must leave the barge on the mooring, her captain or mate should inspect the mooring. If it is to be left for any length of time, the lights of an anchored vessel should be shown on the moored barge as required in local navigation rules. In U.S. waters, this is Inland Rule 30.

HURRICANE, TYPHOON, AND TORNADO READINESS

There are no specific rules to follow if one of these weather conditions appears to be approaching a tow at sea. Common seamanship indicates that early action should be taken to direct the tow's course away from the storm's eye and dangerous semicircle. As modern worldwide weather forecasts give early warnings, there should be sufficient time to plan. That is, of course, unless the tow is unlucky enough to be in the area where the storm is originating.

The first consideration is to head toward a safe port, if one can be reached before the storm catches the tow at sea. In some areas, it may be safer at sea if there is enough time and room to avoid all but the fringes of the storm. On a tug with a tow that is about to be caught in a storm at sea—even if it appears that the storm's worst fury will miss it— preparations for the worst should be made. Any spare lines or gear that might be lifted by seas should be double-lashed. Lines swept overboard have tangled in tug's propellers. Loose lockers, gas bottles, and drums of liquid have rammed deckhouses, broken off fuel vents, and stove in watertight doors. All of this has caused disasters to tugs in a sort of domino effect. The propeller became jammed or an unexpected opening admitted seawater. Power was lost. Steering was lost. The tug became a helpless ship buffeted until it capsized, filled, and sank. Lives were lost.

Early action can prevent much of this from happening. The tug's position, condition, and action planned should be passed to other ships in the vicinity and to shore-based search-and-rescue facilities as well as to the tug's operational base. On board, the primary concerns are watertight integrity and a continuous flow of electrical power and fuel to the engine(s).

A check on the towing hawser to see if its securing nuts or clamps, which under extreme conditions might have to be released, are free and not frozen with rust. If the storm begins to affect the tug, off-watch crew members should be alerted. Added watch standers and lookouts may be required. As a safety measure, without causing alarm, the crew should be outfitted with survivors' suits or life jackets. The EPIRB should be checked to see that it will float freely. A plan should be made as to how a self-inflating life raft can be held to the tug's hull long enough for the crew to reach it and then release it. Courses on gyro and magnetic compasses should be compared in case power is lost. Any existing hand or auxiliary steering that might be required should be checked. Keeping the tug afloat, hove to with the tow intact, is the target. Beyond good seamanship, the rest is in the hands of fate. Hopefully, all of this will end as a good exercise in preventive seamanship and self-preservation.

When a tug with or without a tow is caught in port by one of these cyclonic storms, a good solid and sheltered mooring should be chosen. It is difficult to judge what the maximum wind direction will be. If this can be predicted beforehand, the tug's bow, which will offer the least resistance, should favor that direction. Every good line aboard should be run out to whatever the vessel is moored to. If the vessel is alongside something else that is afloat, the lines of that craft must be doubled and in good condition to hold the added weight. In rivers, water from torrential rains will raise the water to flood stage. In this case, a wharf with a very high caisson or piling should be chosen. Floats should be avoided, as they will rise and float in over dry land. After the storm subsides, any tug or vessel may be 100 feet or more up on the shore.

As on the tugs at sea in this weather, the radar and all radio antennas will be damaged or torn off. This occurs when the wind exceeds 70 MPH. A course of action should be planned to cover the possibility of the tug being blown away from her mooring. The adjacent waterway should be visually surveyed and a magnetic course to be used to steer up and down its middle should be plotted on the chart. Points where the tug should reverse course should be chosen. An estimate should be made of the normal time required at one-quarter to one-third speed to go from one end to the other. Under hurricane conditions, one-third of this time should be used in running before the wind and twice the time when running against it. If this cruising becomes necessary there will usually be torrential rain and flying spray with zero visibility. Fog signals will not be audible, only the wind's roar. Aids to navigation may have moved. Lighted aids such as range lights or shore beacons may be blown away. Other vessels and objects may be adrift and collisions are possible. A good lookout of several people should assist the captain or pilot who must watch and steer by compass. Seamanship and luck can bring a tug through such an ordeal unscathed. Then, there will be real work to do.

VESSEL TRAFFIC SERVICES

In order to help maintain a safe and steady flow of traffic in some estuaries, rivers, and harbors of the world, a system to advise vessels of the presence of other vessels and vice versa has been installed by various authorities. This is known as a vessel traffic service, or VTS. Through this system, some of the waterways are scanned by radar, while others depend on the voluntary information announcing their position, course, speed, destination, and type of vessel. This includes all vessels including tugs and tows.

In some ports, tugs running light, "looseheaded" (without a tow), do not have to report to "Traffic" (VTS). Many do out of courtesy. Vessel traffic control in the United States does not give orders. There is, however, traffic control by the U.S. Army Corps of Engineers over both the Cape Cod Canal and the Chesapeake and Delaware Canal (see chapter 10).

All other U.S. vessel traffic services are operated by the U.S. Coast Guard. Tranport-Canada controls all of the Canadian VTS, which are used in the following ports and waterways: the St. Lawrence River from the Gulf to Montreal; Sydney Harbor; Canso; Halifax; St. John, New Brunswick; Victoria, British Columbia; Vancouver, British Columbia; and the straits and ports of British Columbia's Inland Passage.

In the United States, VTS exists to a limited degree in the Port of New York. Full compliance is required on the Mississippi River to Baton Rouge, Louisiana*; on Galveston Bay and in Houston Ship Channel; in the approaches to the Golden Gate and San Francisco Bay as far as Sacramento and Stockton; and upon Puget Sound. A complete description of VTS will be found in the applicable area's *Coast Pilot*. The VHF channels presently used by VTS appear in appendix IV.

These vessel traffic services should not be confused with other traffic controls established by authorities that govern movement into locks or canals throughout the world, or the various traffic separation schemes.

TRAFFIC SEPARATION SCHEMES

Traffic separation schemes are an apparent outgrowth of a custom on the U.S. and Canadian Great Lakes. The maritime forefathers who were members in the International Ship Masters Association, together with the Lake Carriers Association, agreed to establish separate eastbound-westbound or upbound-downbound steamer courses throughout the Great Lakes system. They still exist.

Presently, this type of traffic separation exists at sea in the international waters approaching ports or through channels or along coastlines where there is heavy marine traffic (see International Rule 10). As tugs with tows operate in these areas, they are very much involved.

Here are some of the places where these traffic separation schemes are located: the English Channel and off some North Sea ports; the Strait of Belle Isle; Cabot Strait, and the Gulf of St. Lawrence; from Nantucket Light Vessel to Ambrose Light Tower; approaches to Boston from Nantucket Shoals and Cape Ann; approaches to Portland, Maine; the southerly approach to Ambrose Light Tower from sea; the southerly approach to Ambrose Light Tower from off Barnegat, New Jersey; Five Fathom Bank Buoy to Delaware Bay Pilot Station; and an adjacent one from the southeast. From Cape Henry Pilot Station, one separation zone runs northeast and another southeast.

The next traffic separation scheme is a single wide fairway through the Gulf of Mexico's oil patches and rigs. It runs southwesterly from Southwest Pass and then around in a west and northwesterly direction with branches running off toward Corpus Christi, Galveston, Sabine, and the Calcasieu River.

---

*Navigation regulations for control of vessels during high-river stages at New Orleans are given in appendix IX.

A vessel traffic separation scheme consists of two parallel shipping lanes laid out in deep water with a separation zone between them. Like a street with a median strip, traffic flows to the right. The median or separation zone may not be entered or used except as noted in International Rule 10 (e)(i)(ii). These traffic separation schemes are from 1 to 3 miles wide with lanes of from $1/4$ to 1 mile in width. The separation zone between them is usually two-thirds of the entire width. Each scheme is well marked on navigation charts.

On the U.S. West Coast, traffic separation schemes cover the approaches to Wilmington-Long Beach, California; the adjacent Santa Barbara Channel; and approaches to San Francisco; the Straits of Juan de Fuca and Georgia; and Puget Sound. Tugs with tows may use courses that are inshore of these traffic lanes, but must announce their intention and position when crossing and do so as described in International Rule 10(c).

FISHING VESSELS AND AREAS TUGS SHOULD AVOID

There are several types of fishing methods and areas from which tugs and tows must keep clear. Off the East Coast, particularly in Chesapeake Bay and from New Jersey and Long Island Sound northward, lobster and crab pot buoys should be avoided. This is sometimes difficult as some fishermen plant trawls of buoys across the navigational channel.

Fishing vessels anywhere that are dragging nets or engaged in any underway fishing should be allowed to pass clear with plenty of room. Menhaden vessels often appear in the summer and use large seine nets, which circle their catch. They need and should be shown the courtesy of plenty of room.

South of Cape Hatteras, the shrimp boats must be avoided, although they can and will turn away from crossing the path of a tow. Many shrimp boats are found around Dry Tortugas, Florida Bay, and off the Delta of the Mississippi.

On the U.S. West Coast since 1984, there have been annual towboat/fishermen's meetings under an Oregon State University Sea Grant Program, the purpose being to establish towboat lanes that will clear fishing areas north of the Columbia River and from there to San Francisco. In 1983, the captain of the Port of Seattle issued an order regarding restricted fishing areas and the gill net fishery with detailed instructions for tugs including whistle signals to gain passage through nets (see appendix VIII). Through this sea grant program free chartlets have been issued showing these lanes. They are frequently updated, as fishing areas change and are very valuable in assisting the navigation of tows in order to avoid fishing areas. There are seven charts in all, which extend from Point Sur, California, to the Strait of Juan de Fuca.*

Within the waters of Puget Sound and British Columbia, there are large seiners and gill netters, fishing seasonally. Tug pilots must watch out for seine net floats and gill net corks. Gill nets are usually set with the tide and run from the bow of small aluminum "bow-pickers" and from the stern of older wooden fishing vessels. All can easily be identified by the large net reel they carry. When they are working with their net out, they hold one end and an orange buoy by day or a "jacklight" at night is used to mark the other end.

When the salmon are running these gill net vessels will be found in packs with their nets floating across rivers and channels. In the Puget Sound area and the Fraser River entrance channels it may be necessary to slow and almost stop to find an opening between nets. The method of navigating through gill nets appears in appendix VIII.

*Fishing lane chartlets may be obtained from The Marine Extension Agent, Oregon State University, Corvallis, Oregon 97331.

PART III
Towing as a Business

11. The Shore Establishment

TOWING COMPANIES

MANY a towboat captain, in a local bar, has claimed that he is the best in the business. However, he may disagree with the argument that all towing companies are not solely managed by the male species. He is wrong! For years, the Bridgeport Towing Line of Connecticut was solely and ably managed by Mrs. Ruth MacDonald. The still-renowned Providence Steamboat Company tugs had the leadership of the late Katherine "Kate" Sutton. Her regal, Queen Mary-like appearance belied her ability to yell orders to tug captains on their vessels from the open door of the towing line's wharfside office. Presently, Mrs. Lucille Johnstone manages RivTow Straits Limited, one of British Columbia's largest towing concerns. There are probably others.

Today, numerous large towing companies are a diversified part of energy, construction, salvage, or other conglomerates. A few remain in operation under family ownership. Many of these are run by second and third generations of the original owner's family. One or two towing operations in the world are subsidized branches, or an integral part of, a government's transportation authority. There are also a great many single and proud tug owner-operators.

OPERATING COSTS

No towing line will publicly reveal its operating costs. Each is bound by certain basic costs such as amortization, hull insurance, fuel, crew wages, maintenance and repairs, stores (including food), shore establishment, marketing, and advertising. Using these necessary ingredients, the company comptroller must arrive at a reasonable cost ratio for each item in order to establish a daily towing rate in what is a very highly competitive market. To keep within this range makes any sudden increase in any category the cause for a complete budgetary revision. One of the most argumentative changes is crew wages. These are affected by cost-of-living increases but cannot be expanded to take a greater percentage of the company's gross income without adjusting other costs downward or increasing towage rates.

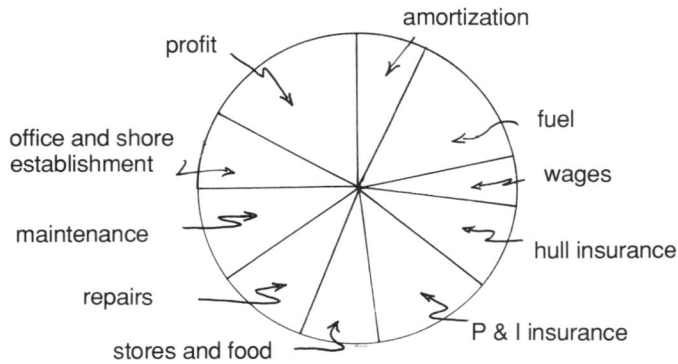

"The Wheel Of Fortune." A rough example of a single tug's operating costs. If any cost item increases, another must be decreased. When all increase so as to squeeze out profit, the tug should be laid up or disposed of. If the company's whole tug fleet fails to show a profit, filing for reorganization or bankruptcy soon follows. Many large towing companies have faced this prospect during doldrums in world shipping and economy.

MANAGEMENT

The large towing company, which may be a subsidiary of another corporate structure, will be headed by a chief operating officer. This executive will answer directly to the corporation, of which he may be a director. He may also be the chairman of the board of directors of the towing line if one exists. Vertically, the management team responsibility passes down to a president, vice-president, and the various necessary departments, each with its own specialist. The management team has two reasons for its existence. The first is to see that the tugs—their product—are properly manned and maintained. The second is to sell their product as towage at an attractive rate to their prospective customers.

PROTECTION AND INDEMNITY COVERAGE

In the United States, due to the unusually large awards made by the courts in personal injury cases involving mariners, marine protection and indemnity insurance coverage has become a problem. Coverage has been refused to tug operators by some insurers, while others have raised the deductible for a single case to $100,000. This is partly due to the difficulties insurers have in estimating the potential size of awards. Some smaller speculative insurers have been attracted to the tug market; however, any large loss may wipe them out without the ability to pay the entire award of the court.

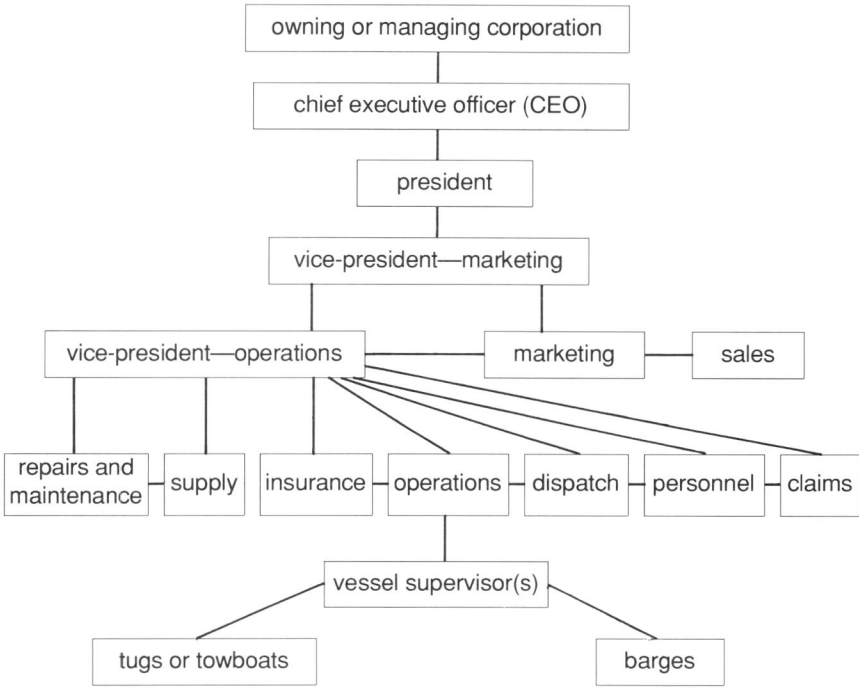

Corporate structure of a large tug and barge operator

Some protection and indemnity clubs in the British Isles and Europe have considered entering into this towage market, as their other types of coverage are so widespread as to support any risks. To regain adequate protection, towing companies in the United States must enhance their safety record if they wish to achieve any reasonable coverage.

RELATIONSHIPS

A good rapport should exist among all departments. An equally fine relationship must exist between those who work on the tugs and those who work ashore in supply, maintenance, dispatching, or in the administrative office. Recognition and appreciation must flow both ways. Those working ashore should realize that the tug and its crew members earn the monies that keep the company financially afloat. It is the shore establishment's job to serve the tugs. Tug crews should not forget, however, that the company provides as tools the tug and the contracts that enable them to work. Teamwork and respect between those afloat and those ashore make for a successful and profitable towing company.

COMPUTER MANAGEMENT

Several marine computer management companies* have introduced software packages to meet maritime small vessel needs. Towing companies are gradually introducing these helpful systems into various departments. Such microcomputer information services help reduce operating costs. By replacing many manual reporting and recording operations both afloat and ashore, a continuous flow of information is at hand about the status of any tug in the fleet and its needs, and the tug master or engineer can be relieved of time-consuming paperwork. Most of these microcomputer software packages are for IBM PC or compatible hardware. They cover vessel administration, offering voyage data and crew particulars (who is on board and their personnel records). Other support packages cover accounting, spare parts, inventory control, chart and publication corrections and inventory, voyage plan, preventive maintenance, consumable stores and shoreside spares, tug and personnel rotation schedules, and fuel consumption. Variable information within system programs can be updated.

In the United States, the Federal Maritime Administration (Marad) has sponsored a program for improving productivity in the towing industry through computer technology. Several simple programs for small vessel and tug operators have been developed by Phillips Cartner & Company together with the Great Lakes Towing Company of Cleveland, Ohio. One program, the Small Vessel Cost/Pricing System, breaks down costs into fixed and variable elements. A very sophisticated system has been aimed at inland towboat operations. For harbor fleets, a Harbor Tug-Port Performance System using computerized records from vessel operating data and crew productivity reports offers the tug operator an opportunity to revise costs.

OPERATIONS, DISPATCHING, AND VESSEL READINESS

The operations department is the overseer of all that affects the tug. It originates company policies and vessel procedures; schedules annual inspections, classification surveys, dry-docking, and repairs; and presents to top management any plans for tug replacement, lay-up, or new buildings.

Whether a towing company uses a partial or complete microcomputer system, its operations department, including tug dispatchers, should have an updated status sheet for each tug. This should show the tug's work assignment, fuel on board, and days endurance remaining. It

---

*For example: Marine Computer Management Inc., Antioch, California 94509; Transtema Kockumation, Malmo, Sweden; Marine Management Systems, Inc., Stamford, Connecticut 06902; Veson Inc., New York, New York 10006.

should also include a running inventory of the tug's towing gear and spare bridles, hawsers, shackles, and other operating necessities. It is the responsibility of the tug captain and mate to advise operations of any changes and replenishment of fuel and stores as they come aboard. Shore operations will be greatly enhanced if their personnel has served afloat on board the company's tugs. Prior to appointment to a shore berth, these employees should be shifted around to the various tugs and the types of assignments the tugs accomplish throughout the fleet.

TUG DISPATCHING

This is a separate yet integral part of the towing company's operations department. The number of personnel involved depends on the size of the fleet and its dispersal. If the owner operates from a base in only one port, his dispatching section may have three to six dispatchers, one of whom, because of greater experience, will be the chief dispatcher.

In major ports, such as Rotterdam, Antwerp, New York, and Philadelphia, where there is a mixture of ship and barge movements at all hours, the tug dispatchers work in shifts. Rush hours for dispatchers often occur when shore establishments open for the day and just before they close in the evening. When ten or twelve operating tugs have been already been assigned or programmed for their next job, any last-minute changes put a dispatcher to the test.

A tug dispatcher should be a diplomat and have a good phone and radio voice and diction. The basic art of harbor tug dispatching involves a thorough knowledge of every aspect of the port's waterways, berths, depths, channels, bridge clearances, and hours of operations and closures. The dispatcher should also be aware of the state of the tide and current as well as the maneuverability and suitability of a particular tug for a particular job. In an emergency, dispatchers should be able to change tug assignments quickly when the unexpected occurs and to coordinate short-term emergencies, which can be accomplished without inordinate downtime for the tug.

Tug dispatchers' relationships with tug captains and mates should be respectful, if not cordial. Dispatchers should not demand or threaten without good cause. Tug captains should not refuse assignments without adequate reasons. Dispatchers should also be wary of ulterior motives when analyzing the requests of some captains. There have been cases where dispatchers within certain towing lines have been unfamiliar with the depths alongside wharves or of their condition and have sent the wrong tug to do the job. In some cases, the tug captain should have known better but he may have been new and relied on the dispatcher. Tug dispatchers need to have a library or microcomputer system of data before them.

In assigning tugs for outside towing for which no particular vessel was contracted, the dispatcher must choose one that will normally accomplish the voyage in a reasonable speed and that is compatible with the unit it is about to tow. Compatibility requirements are adequate horsepower; the correct towing gear, including spares; and sufficient fuel, lube, potable water, deck, and galley stores.

There should not be a scheduled crew change within the time frame of the voyage if it is to be longer than 24 hours. If it appears that a crew change would occur in a port with limited transportation facilities (few daily flights or a distant airport that necessitates long taxi rides), a different tug, which does not require a crew change, should be assigned. Transportation costs and delays eat into profits. The crew change phenomenon, unless it results in a smooth transition of personnel, often creates ill feeling among tug crews, dispatchers, and vessel personnel supervisors. In the end, it is costly.

VESSEL SUPERVISORS

These personnel, who include port captains and port engineers, should assist in interpreting for vessel crews any of the company policies and operational procedures about which there is a question. They should also see that each vessel is provided with all instruction manuals; the latest navigational charts; tide, current, and navigational tables; light lists; notices to mariners; navigation rules and regulations; and such other data and materials as the tug master deems necessary in the operation of his assigned tug.

In many towing companies the operations department draws up operating procedures and policies for approval by top management and implementation by the tug's personnel.

COMPANY SEMINARS

In one U.S. company, annual seminars have been held alternately between management and captains and mates or the management and engineers. In these seminars, members of top management were introduced, and they circulated on a one-to-one level. Policy questions received clear-cut answers. The entire company policy was reviewed, and open discussions between operating management and tug personnel resulted in a positive approach to better understanding. When this was followed by a convivial refreshment hour and a pleasant meal, a feeling of teamwork resulted.

SUPERVISORS AFLOAT

An innovation in U.S. towing circles has been the sudden classification of tug captains, mates, and engineers as supervisors. This is a no-change situation. For years, there have been outspoken captains and engineers who have earnestly told owners how things were and how they must be done. These requests or demands have been quibbled at but were usually granted. The originators have been grudgingly respected. This demonstration of authority based on experience truly makes that person a supervisor. After all, if he is to be entrusted with millions of dollars' worth of vessel, machinery, or towed units, he should be considered a supervisor and treated as part of the management team.

The towing lines designation of supervisor to various categories of marine labor should be serious. Knowledge and treatment afforded them should be consistent with similar categories ashore. The name and classification of "supervisor" should not be used specifically to limit a person's relationship with a labor organization (see chapter 12, "Personnel and Management").

PROFIT SHARING

Another proposal recently introduced in the towing industry is that of profit sharing. While this is not an entirely new maritime concept—fishing vessels have for years operated under profit sharing—it is new to the towing industry. Under a profit-sharing program, tug crews become more flexible in their responsibilities. The tug captains become very familiar with engine room operation, fuel consumption, and stores inventory. Engineers have repair capabilities beyond main engine maintenance. Other crew members accept galley, engine, and deck responsibilities as needed daily. Together, these people sit down and figure out a budget for the tug. They manage their own ship's accounts. They order food and stores and decide how much will be spent for maintenance. They designate the priority of repairs and choose the facility to be used. All phases of the tug's costs are listed and the effort to stay within the budget will supposedly produce a higher profit for sharing with all hands.

SALES AND MARKETING

This department is one where personal contacts plus the towing company's reputation can be blended. To what extent, from a telephone offer to a luncheon or social gathering, a marketing approach should go depends on the scope of a possible towing contract. Personnel selected for marketing should be of the extrovert type with university-level intellect. Some business administrative ability and some maritime experience afloat and ashore should be included. A top-level merchant marine officer or maritime school graduate makes a good candidate.

In practice, one of the pitfalls of marketing has been the overestimating of the speed and regularity of tows. Some towing companies have entered towing contracts that were based on the tug's top performance, which both weather and conditions beyond their control rendered impossible.

The top management personnel of some customers is often unrelenting in the demand for ironclad fulfillment of a contract's specified speed per voyage. A "we-do-make-but-can-average" is a safe clause that should be included in a towing contract.

MAINTENANCE AND REPAIRS

In the late 1980s the average annual maintenance cost of a 4,000-HP tug was about $100,000. Of this amount $13,000 went for engine overhaul. Some towing companies have their own M & R facilities. These range from a shed on a wharf with a small machine shop and welding facility to a mini-shipyard with a separate shore gang equipped for electronic and all other types of repairs and replacements. Other towing companies may have only one or two shore personnel for immediate minor repairs. These companies may farm out all major repairs. In so doing, personnel should not always let price overshadow the quality of work and the yard's efficiency. If a tug is to be sent into a shipyard for a biannual overhaul, repair requests made by tug personnel should be approved by the vessel's supervisor or operations authority. If the cost estimate or yard quotation exceeds the vessel's budget, a review or rearrangement of priority of items is indicated. The originators of the repair request should be included in working up this revision. Tugs should not be sent to a shipyard without advice that the repair list that they submitted to the office has been greatly altered. Such a situation can be very embarrassing to the tug captain and shipyard superintendent if they have already concurred.

Once the tug is in the shipyard, the chief engineer and captain, if present, and the vessel's supervisor should establish smooth relations with the yard superintendent in charge. The expectation of what is to be accomplished should be reviewed and a thorough understanding should be agreed upon. A continuous casual survey of work as it progresses eliminates later claims of poor or unsatisfactory workmanship.

The temptation by tug personnel to get a workman to do a little personal job or project should be avoided, if it is not already on the repair list. Due to the rugged nature of shipyard walkways, with overhead cranes, air hoses, and welding wires underfoot, liability insurance may cause the yard to refuse to allow tug crews to live aboard while the tug is on dry dock or in the yard for repairs. Hard hats may be required in any open working area. Certain other requirements will be passed to the tug captain or person in charge for signature and dissemination. It should be understood that once a vessel enters a dry dock or marine railway and grounds on the keel blocks, the yard must accept full responsibility for safety, security, and conduct.

SURVEYS

If any surveys are to be accomplished while the tug is in the yard, the chief engineer, captain, or vessel supervisor should be present as an observer. Any deficiencies on the tug should have already been known by the tug personnel and supervisor. These deficiencies should either be in the process of improvement or covered as a work order for completion before the tug leaves the shipyard. Any recommendations by the surveyor must be taken as requirements, which if left unattended, might at a later date render the vessel to be judged unseaworthy.

Surveyors of classification societies will check and require that all aspects of the vessel's hull and machinery be in good order. They will not overstep these requirements and get into other areas in which a captain or chief engineer may desire improvements unless a blatant deficiency is apparent. Certain areas of safety, towing, and fire-fighting equipment do not fall within classification societies' realm of control. They do come within national regulations.

For years, it has been a custom of some towing companies in some ports to offer or give favors to surveyors, the goal being that the surveyor will overlook certain items that would normally require attention. Whether the form of gratuity is a fine meal and a few drinks before or after a visit to the vessel or an envelope with currency surreptitiously placed where the receiver will pick it up, this custom is a disservice to the organization the surveyor represents, the towing company, the tug crew, and the industry.

TOWAGE RATES

In many parts of the world, towage rates are based on the bollard pull of the tug to be hired. In others, horsepower is the criteria. Towing rates are based on many different types of towing; they vary from port to port and company to company. A general harbor towing rate may be offered by the hour, the day, or the job. Rates for ship-assist work are often quoted with a base scale covering ships of certain tonnage. Additional charges will increase as tonnage does. Included will be certain variables such as shifting berths, the rate for a second tug, and premium time for movements before or after certain hours. There are also holiday rates, which in the United States cover from 9 to 13 such holidays.

A towage price for a certain tow may be bid through a towage broker for single movements. In large, long-term towage, rates are often based on the product to be moved, such as barrels of liquid cargo in a barge or the number of containers on deck. In this type of towage bidding, a difference of a fraction of a cent per barrel can separate the winning towing line from the loser. Rates for ocean towing of oil rigs are extremely expensive. In quoting to the petroleum industry and other potential customers, some towing companies use the tug's horsepower or bollard pull as their price criterion. The rates for transocean towing of dead ships to scrap yards are made on a daily basis. Some flat towage rates are quoted for single movements.

SUPPLY

Large towing companies often have their own supply department: a warehouse or storeroom stocked with items that are frequently used throughout the fleet. These supply operations are normally at the company's tie-up and repair facilities. Every type of item, from hand soap to a new wire hawser for the deck and from rags to injectors for the engine room, are either stored or quickly available. Requisitions to draw supplies for a tug are usually approved by a vessel supervisor before they are issued. Control of each vessel's stores is one of the tug master's responsibilities. An onboard inventory should be kept by one of the mates or engineers. A computerized system is of particular help both for storeroom accountability and for other tug activities.

Grub

Supplying food has been a point of contention on some tugs. Food is ordered by the crew or the cook with the sometimes casual, sometimes serious, review of the tug captain. It is paid for by the towing company and furnished in some cases by so-called good grub suppliers of questionable quality. This may not be true in Canada and some other countries, but it has been a problem in many U.S. ports. The method of payment is often the evil genie in the tug's grub supply system. It would behoove the tug master to rectify this or any other unsatisfactory food situation unless he is involved personally in the grocery transaction. The latter has occurred. Captains have been advanced sums of money for the provisioning of the tug but have not used all of the funds provided for this purpose. In one U.S. area, where tug crews rotate, the "grub check" was mailed to the cook's home. By union contract, the amount of money was figured as so much per man per day. Sometimes, however, a great difference was apparent in the amount and quality of food between tugs in the same company with the same number of personnel in their crew.

Several major problems were inherent in such a system. The amount of the grub check issued to cover a 15-day period was large. Since immediate cashing by banks was prevented by clearance delays, tug cooks turned to bars where they were known or to the grocer himself in order to get cash. Most of these grocers are located in waterfront neighborhoods and depend on the tugs for their main business. These grocers could not compete with supermarket chains in either prices or brand names. The tugboat galley suffered, particularly if the grub was ordered by phone and picked up later. Unless the cook went to the supplier himself, some tugs received the oldest milk and bread, and the stalest baked goods. Supposedly fresh produce did not exist.

These grub suppliers should not be confused with ship chandlers, many of whom are scrupulously honest. Unfortunately, only a few are set up to service tugs. Provisions and galley supplies for offshore ocean tugs are generally provided from well-known chandlers whose reputation and integrity have been established.

Until recently, however, the provision situation for tugs in the New York area was often a rip-off. Tugs suddenly assigned to an out-of-town tow would need to top off their food stores. As the tugboat would already be engaged in some towing assignment, she could not be delayed. The cook could neither get ashore nor tie up the vessel's VHF with a long grub order. So the tug dispatcher would call the grocer and state that the tug needed three or four days' grub, leaving the type and amount of items up to the supplier. As the supplier knew the number of men on board and the daily allowance for each, he came up with whatever he chose to use the daily amount of money. When the tug made a quick stop to pick up her grub, it often appeared that a selection of much that could not be sold elsewhere had been wished on the tug's cook and crew. There is a lesson to be learned from these evils of the past—most of which have been eliminated. Now, many towing companies make arrangements with supermarkets to allow the tug cooks to make their own selections. A target of $9.00 per man per day is used by one large East Coast tug operator. Others have no specific allowance and depend upon reasonable purchasing. No grocery money passes through the captain or cook; the grocery supplier bills the company directly.

FEEDING AND HEALTH

It is sad to sit at a tugboat or towboat mess table and watch shipmates slowly begin to dig their own graves with their teeth. Yet that is what many have done, mainly by overeating and indulging in the wrong type of food. The nostalgia of a tug is often linked to the smell of fresh coffee, sizzling bacon, or some other delectable concoction. Writers never mention the ripples of fat that hang over some belt buckles or the sudden departure of shipmates with angina or heart attacks.

For years, the eating habits of tug crews were never challenged. Meal planning was left to the cook. To maintain peace between the galley and the crew, certain ethnic tastes—sometimes more than one—had to be satisfied. Tug owners wishing to mollify their crews because of the long hours and restrictive conditions of a tug allowed a great variety of junk food to be available between meals. Habitually, coffee was brought at frequent intervals to the watch standers. Often, it was old, too strong, and very harmful to the recipient. Those crew members who could pass through the galley would get their own coffee with a cookie, doughnut, or piece of bacon left over from breakfast. None of this was healthful.

A U.S. tug's normal breakfast menu offered cereal, bacon or ham, and as many eggs as one could eat, washed down with lots of coffee. Lunch started with soup, a roast with potatoes and canned or frozen veggies, plus dessert. The evening meal was much the same. Nearly all male tug cooks participate in this daily ritual. Some are lackadaisical food preparers whose creativity emanates from the use of a can opener. While they read the directions on labels of prepared and packaged food, they fail to read the cholesterol or fat content on the same labels. With a sprinkle of monosodium glutamate, they turn out a presentable and often tasty meal. Many crew members are satisfied with this. Health-conscious ones would not be. If the cook is of the chef variety, he will be conscious of healthful, nutritious diets and will go out of his way to purchase and serve a balanced and tasty menu. Unfortunately, not all tug men have been health-conscious. A day that starts with a cup of coffee and some inhaled tobacco may seem delightful on the outside. If that person is under 40 years of age, it may seem foolish to consider otherwise. If one's body could speak, it might say, "Hey, fella, what are you doing?" Like unpaid interest accruing on a long-term loan, the clouds of self-destruction are slowly gathering within the lungs and bloodstream.

How do we know what is good for our bodies? There is a measurement system. Like a tug's main engine, our body only requires so much daily fuel. This fuel is caloric intake measured as calories. As in a tug, the fuel for the human body must be of good quality. It should be very low in cholesterol, salt, saturated fat, and sugar (see appendix XI).

Our daily energy action is our propulsion system. It is measured as mets, the metabolic cost of this energy usage. When too much fuel passes through the injectors of a diesel engine, it smokes, becomes sluggish, and slows down. If this oversupply is not cut off, an explosion in the base or fire may follow. In the human body, this is angina, a stroke, or heart attack. Afloat and away from home, these can be fatal.

A few years ago, the *Seafarers Log*'s "Final Departure" column reported the deaths of many whose ages were between 40 and 60. The cause, of course, is heart attack. Why? Even though some of the deceased sailed deep water, the answer is evident. It was overeating and the wrong food. Recently, as a result of national attention in the U.S. toward nutrition, training programs have been given for cooks and stewards.

One of note is that of the Harry Lundeberg School of Seamanship at Piney Point, Maryland. In 1979, the District 2-MEBA introduced a physical fitness program based on both nutrition and exercise. Within the U.S. towing industry, a few companies have held seminars for cooks on good nutrition.

It will take understanding and cooperation between a tug cook and his crew to create a healthy nutritional relationship. On many tugs where cooks are still in vogue, they stand one continuous watch from 0430 to 1930, broken by a short afternoon nap. These tedious hours are often aggravated by off-watch members who leave unwashed cups and dishes or use the galley as a smoking room hangout. A cook or person assigned to cook is supposed to remain composed and eager to satisfy the discriminating taste buds of eight or nine crew members whose mothers may have come from eight or nine different ethnic backgrounds and cooked accordingly.

Basically, the person assigned to cook should know that the human body's worst food enemy is fat. Saturated fat is the most dangerous. Fat in food comes from animal and vegetable fats. Animal fat is found in fresh beef, lamb, pork, ham, processed bacon, bologna, sausage, hot dogs, and other products in which organ meats are used. Vegetable fat comes in dairy products, packaged cakes, pies, and cookie preparations. With this in mind, cooks should become contents label readers. They should watch for the amount of saturated fat content. They should recognize the use of liquid vegetable oils such as corn, cotton- and sesame seed, safflower, and soybean over that of solid or hydrogenated cooking oils. All of these liquid vegetable oils are better for use in cooking. Frying should be kept to a minimum. Eggs, organ meats, and some seafood are high in cholesterol (see "Words Vital to Your Health" in appendix XI). Not only is personal health of great interest to each crew member, it is a burden on the employer when it breaks down. Lost time, sick pay, and replacements are added operating costs.

HEALTH AND EXERCISE

In addition to good eating habits and a healthy menu, tug crews are faced with the challenge of how and where to exercise. There is a definite relationship between the daily calories one accepts and the mets one expends. Roughly, the ratio is : 1 calorie = .8 mets. Average daily calories for weight control are as follows: for a male 22 years of age—3,000; at 45—2,700; and at 65—2,400. For females the figures are the following: at 22— 2,100; at 45—2,000; and at 65—1,800.* For every 3,500 calories you get and do not use, you gain about one pound in weight. As an ex-

*U.S.D.A. Pocket Guide #364.

ample, 2,800 calories are about the value of the tug meals described above. For maintaining good health without weight increase 2,240 mets must be expended. To solve this obvious discrepancy, a reduction in the size of helpings at each meal will help. Low-calorie protein foods must be introduced and supplemented by exercise. This is the quickest way to burn up calories and expend mets.

How does one exercise on a tug? Find an area where you can lie down, stand, and swing your arms. Open deck areas are preferable. There are two kinds of exercise. Isotonic exercise involves the movement of arms and legs; walking and running are examples. This type is very good. Isometric exercise involves contractions of the muscles. Weightlifting is an example. This type of exercise does not help the heart and may add stress.

Before any exercise commences, a warm-up period of stretching should be taken. A good exercise program results in physical fitness. Research* has shown that a physically fit person is able to withstand fatigue longer, is better equipped to tolerate physical stress, and has a stronger, more efficient heart. There is usually better mental alertness and absence of nervous tension in a physically fit person.

Any tug owner interested in maintaining an alert, physically fit crew should encourage exercise by seeing that space is available and, if possible, providing equipment such as a small rowing machine, weight pulls, and other exercise apparatus. This would be especially appropriate on tugs on long runs and overseas voyages, where crews do not get ashore frequently.

To safely participate in an exercise program, a crew member should be checked by a doctor to see if there is any possible reason why he may not exercise. The extent of exercise can be measured by the results desired. From one-half hour to one hour daily is excellent. Twenty minutes, three times a week, is minimum.

A target heart rate checked by the pulse should be worked out, using 220 minus one's age as the maximum heart rate that can safely be attained. Eighty percent of this is the target heart rate to shoot for and maintain for 20 minutes. Example: 80 percent for a 25-year-old is $220 - 25 = 195 \times .80 = 153$ beats per minute. To check this during exercise, stop a moment at the peak. Take your pulse for 10 seconds either at the neck carotid or wrist pulse. Multiply the number of beats by 6; example: 18 beats = 108. Try to work up to your target heart rate when exercising. On a tug, skipping rope and running in place are among the many ways an ingenious crew member can exercise.

*Research: Nathan Pritikin, *The Pritikin Permanent Weight-Loss Manual* (New York: Bantam, 1981) and Royal Canadian Air Force, *Exercise Plans for Physical Fitness* (New York: Pocket Books, N.D.).

SAFETY

This is a broad and vital subject in the United States. In October 1980, Public Law #96-380, "Towing Advisory Committee," became effective. This law authorized the appointment of a 16-member committee whose members were to be from the barge and towing industry, mineral and oil supply industry, port terminals and labor unions. Together with the U.S. Coast Guard this committee is to review safety, navigational, and other issues relating to the towing industry as a whole. This law is partially the result of a 1978 Coast Guard "Safety Analysis of Towing Vessel Casualties." At that time, there were just over 6,950 documented U.S. towing vessels. In number of vessels, the U.S. Inland Rivers led with 35 percent, followed by the Atlantic Coast and Gulf Coast with 24 percent each. The number of tugs on the Pacific Coast and Great Lakes was much less. The total number of casualties reported to the Coast Guard at that time was over 1,600. Capsizing and flooding led the casualty list, followed by fire, collision, and drowning. Extending these causes and figures worldwide has resulted in the renewal of safety-consciousness.

In the United States individual towing companies have undertaken thorough personnel safety programs. Companies specializing in safety now issue newsletters.* In the United States, there is the National Safety Council, which includes The Maritime Towing Advisory Board. In Canada, there is a counterpart for the towing industry. Through these groups, safety has become an ongoing program aboard tugs. Still, there are casualties. In the 1980s, tugs have sunk and lives have been lost off both U.S. coasts. The apparent cause is the entry of water. So watertight integrity becomes a prime foe. Tug crews must be watertight-conscious. They must be made aware of doors, vents, and ports that are open, and be instructed on when to secure them.

Fire is another enemy high on the list. It should be common practice on any tug when under way to have a watch stander go through the entire vessel at least once every hour. He should watch for a possible fire or entrance of water. Fires can be caused by trash or oily rags in lockers or the failure of a thrown cigarette to go overboard. Another cause has been the spilling of overheated fat on a hot galley range when the tug rolls. More often minor fires occur on tugs due to overloaded circuits or motors. A potential danger in engine rooms is the leakage of fuel oil under pressure striking hot exhaust lines.

All of these cause vessel casualties. As a result of a strenuous long-term safety program, one large U.S. towing and petroleum barge operator has at last been able to report a noncasualty year. This does not hold

---

*Lifeline, c/o Ships Operational Safety, Inc. Port Washington Harbor, New York 11050.

true in many other segments of the industry. Aware of an attitude by some vessel owners to shortcut safety procedures, the U.S. Department of Transportation initiated as a trial the Marine Safety Reporting Program (MOPS).* Through it, potentially dangerous conditions on board and near-misses of vessel accidents can be anonymously reported. These reports are strictly confidential. The information received will not be used to charge owners with violations. Instead, the conditions found will be disseminated as a safety precaution to be taken throughout the towing industry and maritime unions.

In furthering safety consciousness covering watch standing officers, several large companies (Crowley and Maritrans), have sent their tug masters and mates to one week of collision avoidance training and bridge protocol at the Kings Point Computer Assisted Operations Research Facility.

Operational safety has been specifically noted in various sections of preceding chapters. Personal safety should also be reviewed in certain items. On board the tug, the slogan "Safety is the name of the game" should be stenciled in each crew member's mind. It should be evident in each act he or she performs. Crew members should be ever-cognizant of preventive personal safety in performing their duties. Here are a few reminders about the most common injuries to tug personnel, their cause and prevention:

> Slipping on deck caused by water or oil or a nonskidproof surface.
> Shoes with crepe or slippery soles and failure to hold onto a handrail.
> Carrying heavy objects when it is rough.
> Decks should have nonskid surfaces in work areas.
> The rule "one hand for yourself and one for the ship" should be observed.
> When holding the end of a line or wire coming off a capstan or gypsy, do not grasp it too close to the drum—stand back. If the line suddenly surges it may pull back and jam your hand.
> If rigging tow cables, wear heavy protective shoes. Many a toe has been smashed by a dropped shackle or heavy towing gear.
> Hook watertight doors, if open.
> Don't stand with or put your fingers on door jambs.
> Watch your shins when climbing ladders.
> Don't go on top of the pilothouse or aloft without notifying people on watch or if the radar antenna is turning. If no one is in the pilothouse, leave a note on the radar set.

*MOPS—Transportation Systems Center, Kendall Square, Cambridge, Massachusetts 02147.

Remember that the stern is the most dangerous work area. When connecting or disconnecting a tow hawser, play "heads up." If you don't, it can be "heads off." This horrible fatality has occurred several times. Wire hawsers are guillotines awaiting unwary suspects. Bights of line on deck are bear traps if stepped into as their coils are running out. To be jammed against a bulwark or deckhouse by a line or wire is an awful sensation to remember—if you live!

Engineers should not start the main engine without notifying those on deck or in the pilothouse. If no one is present, he must check to see that all main and remote controls outside of the engine room are off. If the propeller is to be tested, mooring lines must be checked and doubled if necessary and shore fittings examined to make sure they will hold against the tug's pull. Neither engineers nor any other personnel should enter any tank for cleaning or inspection without a rescue breathing apparatus and having someone else standing by. Men have gone into tanks without proper protection and collapsed. A shipmate noting this then unwittingly also entered and he, too, collapsed. It finally took a third or fourth party with a protective mask to recover those who had succumbed.

When moored alongside any other object, a safe ladder, gangway, or method of stepping over tug rails should be provided.

All these possibilities should be studied by the tug master. Drills and a review of preventive safety measures should be a weekly function. With overall responsibility for all that occurs on his vessel, the tug master must become a safety expert. After all, safety is the name of the game.

12. Personnel and Management—Labor Relationships

THE TUG MASTER OR CAPTAIN

As the title denotes, persons appointed to this position must accept all and full responsibility for what transpires on and with the vessel now under their command. The tug master should be chosen with great care. If the tug operates in a harbor or area where each movement is within its confines and is ordered by radio dispatch, the requirements for tug captain may be slightly less than for those who will serve on ocean or coastwise tugs.

There is much to consider. However denoted, the tug's skipper, captain, or master is not a demigod. Those who imagine they are often fail to impress anyone. The tug captain whose handling of his authority is rarely noticeable will reap rewards from his crew. Each company may regard the definition of tug captain differently. Some wish their captains to be robotlike puppets responding only to their directional pulls. They will pass down embarrassing nonoperational orders to be carried out without question. Other towing companies make no effort to maintain liaison with their captains. Communication stops at the port captain's level.

The tug captain whose license, experience, or choice limit him to rivers, canals, or the Great Lakes is no less important than one who can operate offshore. Each captain should be an expert in his field of operation. This necessary experience and knowledge should have been molded by some years of successful service on deck and later as mate.

A variety of people become tug captains. Some are born boatmen; others struggle toward excellence in boat and tow handling. The natural tugboat handler may be lean on education but inherently wise in seamanship and prudent action. The college- or university-educated tug master may possess equal talents or may strive unsuccessfully to perfect them. Usually, personnel with no long-term deckhand experience, although working under a good and experienced captain, tend to make decisions and tug movements based on preconceived notions that are not always prudent. This type of tug captain may possess all necessary skills except the ability to use them other than verbally. He will frequently fall into the most difficult and awkward positions with his tug or tow and somehow miraculously escape with little or no damage to either.

In advancing a person to the berth of tug master, the tug owner or his personnel representative has a responsibility to carefully analyze the candidate. Promotion should not be automatic. If the proposed new captain has served with the company for any length of time, his damage record, smoothness of operation, attitude, and ability to work with and lead his crew should be well known if he was a mate. What is not known is how he will respond to the new and overall responsibility he must accept. Many people such as tug mates work well under someone else. Once in charge, the person may find that the stress or worry becomes too much. Some persons make better mates than masters and will refuse to advance. Their choice should be respected.

If a berth of captain or relief captain is to be filled by a newcomer who has served in a similar capacity for another company, a careful study of the applicant's past is in order. Is he moving up because the equipment, pay scale, or working conditions are better or is he moving away from a questionable record, possibly tainted by alcoholism or other undesirable traits? Often, personnel department interviewers become impressed with appearance, the type of license, and the claims of the applicant. They should beware of those who boast about how good they are. These applicants usually prove to be show-offs and do the most damage. Those who claim that they have never had an accident have had little experience as a captain. It should be remembered that a merchant marine officer's license does not mean that its holder is a good tugboat man. He might not be able to correctly back a tug out of a slip. Some men have been hired as captains because they held an unlimited master's license. A few of them proved to be poor boat handlers.

Without becoming a psychiatrist, a person in charge of promoting or selecting tug captains should watch for certain telltale signs. Nervousness, insecurity, and lack of confidence are not the mark of a good tug or towboat captain. There are, unfortunately, a few captains who yearly fight the battle of docking, picking up a barge from the hawser to alongside, or passing through certain waterways. To offset these fears, and the resulting delays or minor damages, they will attempt bravado in other directions, such as never slowing down at sea when it is rough or checking weather reports before departing on a voyage.

In general, the performance of a tug captain can often be compared to that of an athlete. On some days he is at his peak and on others he may be sloppy. On this note, first-time captains should begin by riding for several weeks with a senior captain. When the senior captain is confident of the first-time captain's ability to assume command responsibilities, a probationary agreement between the new appointee and the tug owner should be prepared. Such a test period is not only fair to the owner but also to the prospective captain. He may find that he does not wish to be captain and would rather remain as a mate. In this way, he can revert to that position regardless of whether it is by his choice or his inability to satisfy his employer.

Before applying for or accepting a promotion to tug captain, the applicant should be aware of company policy regarding that position. Do they just give lip service to the title, and run the tug through orders from the office? Does the port captain accept suggestions? Does he back up captains or at least listen to complaints? Most companies have a booklet of policies and procedures that they expect to be carried out. This should be read and considered before making the step to captain.

Upon acquiring a good captain, a good towing company should treat him with such confidence and respect that he will be able to make decisions without first questioning the office. On coastwise and deep-sea tugs, this ability to make and act upon decisions is the difference between being just a captain and being a tug master. Such decisions coming from an able and trusted tug master are a great relief to tug operations departments. Operations departments cannot, of course, give tug captains at sea with a tow any direct orders. To do so would jeopardize the limit of liability of the towing company and the tug in the event of a casualty.

Each tug captain should be briefed on pertinent parts of the towing contract in which he is participating. The value of heavy tows or single units pushed by towboats or pulled by tugs often runs into millions of dollars. In this type of towing the tug master should be a part of the management team. If tug captains are to be classed as supervisors, they should be included in all management matters.

The designation of a tug master as a supervisor has been a long time in coming. It should be used honestly and meaningfully. It should not be a method used to curtail or cancel the tug captain's prior membership or accrued benefits in any labor organization. This should hold true whether or not the company had a contract with the employee's labor organization when he was acting as or promoted to captain. If this is not acceptable to a towing company, a compromise or crossover plan acceptable to all parties should be worked out. This is a very necessary solution in countries or areas where the tug employee has been a union member since entry into the industry. He has probably been given upgrading training by his labor group through the various levels, including licensing as an officer. He cannot be expected to turn his back on his sponsor—the union—when he becomes a captain. The union in turn should not penalize him for accepting a higher level of responsibility. Instead, the labor organization should train men for eventual management. In time, through such a linkage of labor-trained, industry-experienced tug master/supervisors, a much better labor management relationship will result. The salary and benefits of the tug master/supervisor should be commensurate with that of equivalent shore personnel. No deduction for room and board afloat should be made.

The tug captain should be a serious management representative afloat. He should possess the ability to lead and, through mutual respect, hold together his crew. Together, the captain and crew should make their tug an outstanding example of their owner's service.

Alcohol use by crew members is one of the most difficult nonoperating problems facing the tug captain. A close second is the harder-to-detect use of drugs. The use of either on vessels is forbidden by laws in the U.S. and many other countries. Using alcohol as personal fuel is an insidious problem. Chapters could be written on alcohol abuse and its dire consequences. Without the company's knowledge, captains have brought alcohol on board, shared it with crew members (or vice versa) until a nearly wholly imbibing crew ran an often tipsy tug. Until the one inevitable collision, grounding, or injury occurred, some operations departments overlooked these problems. Not so with labor organizations. They have tried to weed out the abusers and offer them help and cure. The Seafarers International Union (SIU) has a special facility for abusers at their Harry Lundeberg School of Seamanship at Piney Point, Maryland.

The captain who does not deal with the problem of alcohol or drug abuse by a crew member is foolhardy. He is not helping the addict. He may himself become an "accessory after the fact," if an injury or accident should be linked to the use of alcohol or drugs on the vessel under his command. He should meet the afflicted as a friend with a no-nonsense warning. If no change occurs, he should see that the offender is removed from the vessel. It is a hard choice, but it is a favor—and sometimes a lifesaver. The rest is up to the person's desire to help himself.

If the tug has a relief captain who is listed on the vessel's documents as alternate master, he must also bear the same responsibility as a regular master. If their duties are similar, the pay scale should also be the same, unless the senior captain is paid by the towing company as a supervisor.

As the seat and target of all responsibilities on his vessel, the captain's daily duties will vary according to the type of work assigned to his tug. Logbooks and record keeping are an anathema to most tug captains. Towing companies desire explicit records of towage movements, times involved, and crew payroll. This is basic to all tugs. Sometimes, that part of the tug log dealing with weather, tides, and other incidents may be kept casually. This is a mistake. A tug captain should protect himself, his owners, and their customers by keeping an accurate daily log (see chapter 7, "The Tug Log").

Tugs involved solely in mud towing, ship assistance, or marine contractor's movements may only record the time involved on paying assignments. However, if an accident or injury occurs, notes should be entered on all other items such as time, visibility, weather, tide, and precautions taken. All of these entries will be asked for in any hearing. Write logs defensively.

Coastwise tugs will keep much more extensive logbooks, and large ocean and salvage tugs will include navigational positions. If any of these matters are temporarily recorded on note pads, only the corrected information should be entered in the rough log. Tug captains often assign to mates the final smooth recording of the rough log. All entries in logs should be initialed by the watch stander. No erasures should be made. If a correction is necessary, errors should be struck out and initialed by the corrector.

Tug captains should always be aware of the general condition of their tug. A thorough checking of the interior from forepeak to rudderpost should be done by them once a week. Through their mates, a Station Bill and Fire and Abandon Ship Drill and Safety Program must be instituted. If the tug is involved in towage between different countries, the tug master must familiarize himself with the number of each document that each country requires. Such documents include crew lists, stores lists, bonded stores, declarations of crews' purchases, fuel, and water. The tug's draft, current certificate of inspection, load line and classification certificates, health and deratization certificate, radio license, waste disposal, and evidence of environmental pollution coverage should be at hand for port authorities. A supply of these necessary forms should be kept on board. In Canada and the United States, they may be obtained through the customs authorities or a shipping brokerage company.

RELIEF CAPTAINS AND CREWS

Tugs that work 24 hours daily have relief or second crews. The way they rotate varies in different work areas. In some, where the tug operates within a certain distance from its home port, they alternate every week. In other areas, it may be two weeks to a month on and a month off. On large deep-sea salvage tugs, crews are relieved on a schedule ranging from 6 to 18 months. The entire new or relief crew is flown to the tug's location abroad. Whatever the time span, the duties of the relief captain and crew are the same as those of the regular first crew.

Relief captains and crews may use slightly different methods and equipment arrangements than the other crew when making up to a tow. Before being relieved, unless this is done while under way, conditions on board should be left as they were when the relief crew took over. The work rotation of relief captains also varies. Normally, they work with the same complete crew. Some companies had arrangements where the relief of the captain was also done by rotating him with two other deck officers. In such an arrangement, the regular master worked for a period with the chief mate. At the end of this time, the chief mate relieved him as captain and another mate reported for duty. This second mate works half of his time with the relief captain and the other half with the regular captain. In this arrangement, the rest of the crew rotates as a single unit.

MATES

Persons assigned to the position of mate have arrived at the top rung of the ladder leading to a berth as captain. To get this far, they should hold the appropriate license and any necessary endorsement. In addition, they should possess a background of several years of tug- or other boat-handling experience. They should have other prior training including seamanship. A trial period of hands-on training on one of a company's tugs has often weeded out those who would not make good mates and future captains.

A perfect candidate for the position of mate on a tug is a young seafarer who has entered the industry as a cadet or ordinary seaman. By the time of consideration, he or she should possess papers as an able seaman or seawoman. This person should be enterprising, industrious, and enthralled with towboating. He or she should also have an enthusiastic desire to become a good leader. Boat handling is the one remaining qualification to be proven for service as a mate on harbor or nearby coastal service tugs. Some persons are born boat handlers. Others, unfortunately, don't have this talent.

Those who pass this hurdle and advance to mate for the first time will, if they possess any of the qualifications above, be eager to learn quickly everything about the tug and its work assignments. This can best be accomplished by starting a small notebook—a practice that is not limited to wheelhouse personnel on tugs. Apprentice, as well as many experienced bar pilots, carry and refer to their own little self-compiled encyclopedias of local knowledge. For the young, or new to the tug, mate, this type of notebook will be of great help. In it may be stored all of the tug's characteristics; any special methods used to make up to certain units; list of current documents carried; sketches of frequently used entrances, wharves, piers, or docks. The locations of piers and berths by number or name are also of great help. The various VHF channels of other tug companies and pilots will be frequently referred to. These are but a few of the possibilities in compiling a personal book of knowledge for use when on watch.

In addition to watch standing, the general upkeep of the deck department, its towing, fire, and rescue equipment, are among the mate's duties. The writing of the smooth log and the requisitioning of deck stores may also be wished upon him. He may be asked to prepare the station bill for Fire and Abandon Ship Drill. He should maintain a computerized or hand-recorded running inventory of all deck equipment, stores, or spare parts. Unless directed otherwise, the mate should be on deck leading or assisting in the connecting or disconnecting of the tow hawser or in docking and undocking of barges and other units. In the latter operation, he should assist the captain who is piloting, using either VHF or hand signals. In some cases, he may be asked to relieve the captain on watch in the wheelhouse while the captain goes out onto the unit to direct the docking or undocking.

SEAMEN (HE OR SHE)

Tug company personnel interviewers should use nondiscriminatory standards in hiring or receiving from a labor organization a person who wishes to serve on deck. Both the personnel department and the tug captain must realize that seafaring applicants with prior experience on deepwater ships or in the military, will require time to adjust to a new work ethic.

The first thing these seafarers should learn upon reporting to a tug is a consciousness of what is going on as it affects everyone aboard. The first consciousness should be of sounds, the nerve system of the vessel. They are something everyone is conscious of. Everything that moves makes a sound. Learning to differentiate among the sounds is the important skill. There are no bells or gongs ringing or telegraphs jangling between the bridge and engine room, or the hushed sounds of the steam tug, to aid the deckhand in handling lines.

Much of today's tug machinery and electronics run unattended. Any change in their sounds may denote trouble if they do not have an alarm system. When the tug is moored with the main engine shut down, the exhaust of a generator is usually the main noisemaker. If, in changing from one generator to another, the current fluctuates, this may set off the gyro or other alarms. New crew members must learn to recognize and know whom to inform when one of these sound changes occurs. When the tug is under way and maneuvering to pick up a tow, assist a ship, or perform any other movement, the tug deckhand must recognize engine sounds or the feelings that accompany them. There is a different sound and feeling for each change of speed and direction from ahead to astern. It is through the recognition of these changes that a person working on deck knows what the tug is doing. It will alert that person as to which line to adjust if there is slack; how to pay out the hawser or hold the tow winch brake; and most important, how to act when working on the stern.

Once consciousness of the meaning of each of the tug's noises has been understood, the deck person should be inquisitive as to the cause. If on watch when the tug is under way without a tow and the tug suddenly slows down or reverses, he should look out on deck to see the reason. If the tug has a tow, he *must* do so. A towline may have parted or the hawser may require attention.

While adjusting to these noises, new deck personnel should familiarize themselves with every part of the vessel and patrol it hourly when on watch. If, however, the visibility is poor or it is dark, the deckhand on watch will be posted as a lookout. The location will be assigned by the wheelhouse watch stander. At other times, the deck personnel should learn or be shown by an able seaman or mate all of the gear and equipment and how and when it is used. Deckhands should also be given wheelhouse instructions in steering, both by autopilot and by hand and should know how to slow or disengage the engine(s) in case of an emergency. If the tug is automated and there is no designated engineer on watch, the deck watch should visually check the engine compartment for any unnatural noise, smell, or fire. Such trouble indicators do not always register on wheelhouse monitors until the damage has occurred. An experienced or able seaman will have known about all of these items and incidents and how to react to them.

The otherwise normal duties of deck personnel are maintenance of the vessel under the mate's direction or program. This should include the care and stowage of lines and other deck equipment. There should be a place for everything and everything should be kept in its place. When handling barges, unmooring, or taking in any deck lines, deck personnel should coil the lines, making them ready for use and placing them out of the weather. Lines should be covered to protect them from spray, rain, or snow. Frozen lines are weakened and dangerous. Synthetic lines that are constantly exposed to the sun will be damaged. Lines covered without airing in humid climates will become mildewed.

A deckhand should be proficient in splicing, making heaving lines, and using stoppers. On the large deep-sea tug, a rigger's vise is often provided so that wire rope can be spliced.

Deckhands on harbor tugs that do not have a mate may be called on to steer and handle the tug under the captain's direction. This is excellent training on any tug and is the first step up the ladder. Other than regular watch-standing duties, an earnest deck person should take an interest in how the tug maneuvers and should be aware of the tidal current. This is particularly true when called upon to go aboard a scow or barge to assist in landing or letting go or in directing the wheelhouse during anchoring or approaching a mooring buoy or wharf. Good hand signals should be made from a position readily visible from the wheelhouse.

Although it may not be pleasant, vessel's sanitation is a deck duty. To help, all crew members should clean up after using bath and toilet facilities. Tugs have washers and dryers; the users should take pride in cleaning them after use. Spilled soap powder, loose paper tissue, and anti–static-cling papers can find their way into engine air intakes. All of these seemingly minor items have had to be addressed on various tugs.

One of the most desirable attributes of a deckhand is the ability to "ring" a bitt or cleat on another vessel or shore facility as the tug comes alongside to moor or make up. There is an art to this act of throwing a tug deck line so that the eye will fall over the intended bitt or cleat. The opening of the eye of the line must be from 3 to 4 feet in length. It must be held in the hand normally used for throwing, with enough additional turns of line to reach the object to be ringed. It often takes the novice quite a while to accomplish success. Knowing the necessary amount of line and arc to use versus the distance of throw only comes with actual practice. There is no better way to do this than by using the tug's bitts, particularly at the stern.

The other side of the coin of line handling is even more difficult for some. This is the flipping off of the eye of the line that is already fast on a bitt or cleat. This is necessary when releasing a tug or barge. When attempting this act, an inexperienced deckhand usually does not allow enough slack in the line. To release a line from a cleat, the first move is to shake the eye of the line open so that its position leaves the splice facing toward the thrower. If the seaman is higher than the cleat the splice end may be open and off one of the horns of the cleat. Otherwise by giving sufficient slack and then throwing overhand as in pitching, the slack will roll and flip off the eye. Of course, when the vertical distance to bitts and cleats is high, a heaving line or another person will be required to secure and release the line.

Ringing a round bitt head is done in the same manner as hitting a cleat. Releasing the line's eye from a round bitt may require an underhand flip. Tug bitts are a great place for practicing this trick. Tug sailors who have recently served on deepwater ships or in military deck ratings were rarely called upon to handle lines in this manner. To become proficient in handling tug lines may require time and patience.

All deckhands have to be adept in the art of lassoing dock bollards. This sailor is putting his all into it. Knowing that he was being photographed, he did the obvious, he missed.

ENGINEERS

With the advancement of marine engine technology, there has been a great reduction in engine room personnel. The tug oiler followed the fireman down the gangway. On some tugs, the assistant engineer was dropped in favor of automation. However, two engineers are still common on some 24-hour general towing tugs and towboats. There are also tugs upon which engine operation is covered by a person qualified to serve both on deck and in the engine room. In certain areas, harbor tugs (see "Manning" later in this chapter) have this arrangement, whereby the deck personnel start the engine. Thereafter, all monitoring of the engine is covered on a console in the wheelhouse. On ocean and salvage tugs a chief and several assistant engineers are carried. In addition to maintaining their own engine rooms, they may be called upon to maintain compressors for divers, and the tug's deck and other auxiliary machinery. These engineers are usually licensed by their country and have specialized training and machinist background.

In the United States, most of the large towing companies require engineers to be licensed. Their license should cover the horsepower and tonnage of the vessel. This guarantees a sound technical knowledge based on prior service. For engineers, credible service is sometimes difficult to acquire. Most tugs do not carry engine room personnel below the level of engineer. Service in an engineering rating in the U.S. Navy or Coast Guard has been credited toward engineer licensing. Men with such experience have proven to be good engineers. Many others have obtained engine room experience on large fishing vessels or at maritime schools.

Applicants for engineer should be screened as conscientiously as any other personnel. On U.S. tugs and towboats, many engineers are members of the Marine Engineers Beneficial Association (MEBA). Their duties on the tug are well defined. In addition to the main engine, generators, and auxiliaries, including the steering gear, they will be responsible for the tug's electrical, plumbing, sanitation, and refrigeration systems. At times, they may be called upon to start or make minor repairs to a barge pump or anchor engine.

On tugs that do not carry a qualified or licensed engineer, minor repairs and maintenance are done through a port engineer or manufacturer's representative. Many towing companies maintain a shore gang to do major repairs. In other cases, such as on towboats on the Columbia River, the entire crew turns to for engine room maintenance at the end of each round trip. The method of engine care differs greatly from company to company and port to port.

COOKS

Cooks or persons designated to cook should be hired either through a reputable agency or labor organization. Cleanliness, affability, and cooking ability are the most important attributes to look for. Past records and verbal references should be reviewed. Galley personnel should not smoke while they are cooking or near food. In addition to producing three healthy meals each day (see chapter 11 and appendix XI), the galley, all its equipment, refrigerators, and food storage areas should be kept immaculate.

RADIO OPERATORS

Large deep-sea and nearly all salvage tugs carry a radio operator. Under the equivalency rules of some nations, tugs equipped with Satcom are required to carry radio operators. On salvage tugs and some others, the radio operator stands a regular monitoring watch when the tug is in port if they are away from their home base. When the operator is not on duty an automatic alarm signal will alert him that a call exists for the vessel. Tugs assigned to salvage-and-rescue duties also have an IMO-approved monitor set for international distress signals.

MANNING

The required number and designation of tug personnel depends on several things: the country of origin or registration of the vessel, its tonnage, its service classification, and its type and area of operation.

On deep-sea and salvage tugs, a crew of twelve may be expected. This includes a captain, two mates, a radio operator, a chief engineer, an assistant engineer, three seamen, a diver, a utility man, and a cook. Most large salvage companies will fly additional salvage crews and divers to the scene once their salvage tug has become involved in saving a casualty. The regular crew of these tugs usually work for about six months without relief.

Ocean tugs tending drill rigs may have crews of five to nine persons who work up to three months without relief. The personnel assigned to U.S. flag tugs that are bound for foreign ports or are stationed in non-U.S. ports are usually signed on through shipping articles which stipulate the duration of the voyage or period of employment as well as point and method of discharge. These articles may require service of from 6 to 18 months. In an emergency the individual may be relieved by mutual consent.

In the United States there exists, it seems, a plethora of statutes covering manning, licensing, and documenting of tug and towboat personnel. These statutes also cover tonnage, waters of operation, distance of voyages, and hours of work or watches required. Much of this information was once available to the seafarer through the freely acquired U.S. Coast Guard publications. These publications were actual extracts of pertinent parts of the statutes. When the supply was exhausted, they were not reprinted. Instead, it is now necessary to peruse various portions of Part 46 of the *Code of Federal Regulations* (46-CFR), which was recodified in the late 1980s. In October 1987, the U.S. Coast Guard published interim final regulations in the *Federal Register*, which became effective on December 1, 1987.

Parts 1-40 of 46-CFR, which deal with towing related statutes, may be purchased as one volume from the Superintendent of Documents, U.S. Government Printing Office, Washington, DC 20402, or any U.S. Government Book Store (10/1/86-$13.00). They may be viewed also at various U.S. Government Depository Libraries, many of which are public or college/university libraries.

To clarify and reconstruct the application of the regulations for the towing industry, Captain Kirk Greiner, President of Maritime & Environmental Consultants, who served in the Coast Guard headquarters as the Executive Secretary of the Marine Safety Council, prepared a paper for his associates and clients. It appeared in his spring 1988 MEC newsletter. The following is a condensation of the facts.

For background purposes see 46-CFR Parts 7; 10.464; 12.02-7; 15.301; 15.705; 15.910; 15.840; 15.850; USC 8104(d); USC 8104(g); USC 8104(h) which excerpts those portions that are pertinent. They will not be quoted in toto.

Seaman's Document Requirements

In general seamen are considered to be all crew members who are serving in a nonlicensed capacity. Crew members who serve on towing vessels (tugs and towboats) operating exclusively on *rivers* are not required to possess a merchant marine document known as a Z card (46-12.02-7). Operating exclusively on rivers does not include venturing into what the U.S. Coast Guard regards as lake, bay, or sound waters. Crew members on towing vessels of *less* than 100 gross tons are not required to possess a Z card, unless serving on a towing vessel that maintains a two- or three-watch system (see "Able Seamen or Seapersons, Watches, and Lookouts").

Crew members on all towing vessels of *over* 100 gross tons *are required to possess a Z card*. In the U.S. tug and towboat industry it is often difficult to obtain a position afloat at the entry level. The basic requirement in most cases is the possession of a valid Z card issued by the U.S. Coast Guard. To apply for a Z card a person must have either acceptable letters of prior experience afloat or a "letter of committal." A letter of committal is not always easy to obtain. It is a promise of employment by a shipping or towing company. As the towing company may not know the applicant or need him it usually takes a personal connection to get that valued letter. When there is a shortage of seapersons, some maritime unions such as the Seafarers International have offered entry training programs. The SIU's Harry Lundeberg School of Seamanship at Piney Point, Maryland, has graduated many who have gone on to become able seamen and licensed tug officers. The United States Merchant Marine Academy at Kings Point, Long Island, New York and the various state universities' maritime colleges annually graduate cadets who turn to the entry level of towboating. As with all entry-level personnel they should pursue their career as ordinary and then able seapersons in order to obtain the vital experience they need to go on as a mate and eventually a tug master. The same is available for, and applies to, those who choose engineering.

The number of crew members on coastwise and harbor-short sea tugs varies by country. In relation to manning requirements in the United States, there are four regulatory measurements that apply to commercial towing vessels. These are the following: (1) uninspected towing vessels of up to 26 feet in length, (2) uninspected towing vessels over 26 feet and under 200 gross tons, (3) uninspected towing vessels between 200 and 300 gross tons, (4) towing vessels over 300 gross tons. These vessels are inspected, and all U.S. Coast Guard regulations and applicable CFR statutes for vessels of this size and type of operation are enforced.

Licensing

Currently personnel on towing vessels of up to 26 feet not engaged in commercial or assistance towing are not required to have any document other than a Z card. Tugs of up to 26 feet have been manned by a single person, but usually have a person in charge as captain and a deck person. The Coast Guard issues two different groups of licenses applicable to towing vessels. One group is the master/mate licenses; the other is the operator/second-class operator licenses issued under 46 CFR 10.464.

Effective September 15, 1988, all mariners engaging in assistance towing services will be required to hold either a license as operator of uninspected towing vessels; a license as master/mate of vessels greater than 200 gross tons; or a license as master, mate, or operator restricted to vessels of 200 gross tons or less that is endorsed as "authorized to engage in commercial assistance towing." The new regulations apply regardless of the size of the vessel providing assistance.

To understand licensing for uninspected towing vessels one must refer to the Officer Competency Convention of 1936. Under this convention documented vessels 200 gross tons and over operating seaward of the boundary lines must be manned by personnel holding licenses authorizing them to sail in that capacity. An officer designated as mate must hold a chief mate's license. A second mate must hold no less than a second mate's license, and a master must be the holder of the required master's license. No person in a lesser position may cover the master by holding such a license (see 46 CFR 69 for definition of gross tonnage). Boundary lines are considered to be those waters seaward of the high water shoreline. They should not be confused with the demarcation line between Inland and International Navigation Rules. Under the Officer Competency Act the U.S. Coast Guard's license categories for master/mate/engineer fulfill the requirements. The operator's license does not. Thus as an operator's license for uninspected has no tonnage restrictions, it is suitable for use on towing vessels up to 300 gross tons *on inland waters*. It is also valid on towing vessels under 200 gross tons not engaged in a foreign voyage.

A person 21 years of age or more holding a license as master or mate on vessels *over* 200 gross tons may serve as operator of uninspected towing vessels within any restrictions on the individual's license. A mate's license, however, is only good to cover as an operator on domestic routes (46 CFR 15.910). Article 2 of the Officer Competency Act defines "Chief Engineer" as "any person responsible for the mechanical propulsion of the vessel" and "engineer of the watch" as "any person who is for the time being actually in charge of running the vessel's engines." On tugs over 200 gross tons which proceed out to sea outside the boundary line, a person performing the duties defined above must hold an appropriate license and not merely a merchant mariner's document.

Able Seamen or Seapersons, Watches, and Lookouts

The U.S. Coast Guard interprets "watches" to be the direct routine of so many hours regularly stood "on and off." There is no statutory requirement for an unlicensed deck force on uninspected towing vessels. However, if unlicensed personnel are carried in the deck force in the capacity of ordinary or able seapersons, *and* if they stand watches, there must be the required number to meet the watch requirements when the vessel is at sea (operated beyond the boundary line). When a vessel is on a prolonged voyage and required to carry out a three-watch system so that crew members do not work over eight hours, the same vessel may on a short voyage of less than eight hours need to carry only one watch.

Even though there is no specific requirement for a watch of unlicensed deck personnel; there are requirements in Rule 5 of both Inland and International Navigation Rules for the maintenance of a lookout. The general rule is that a proper lookout may have no other duties that would interfere with his responsibility as lookout, and thus the lookout cannot be the person in charge of navigation of the vessel.

Unlicensed deck personnel who stand a watch must be divided into three watches if the towing vessel is over 100 gross tons and operates outside the boundary line. The exception is that even if the voyage goes outside the boundary line at any point, but is under 600 miles and not to a foreign port, a two-watch system may be used. Thus a foreign voyage of less than 600 miles requiring a three-watch system could be between a U.S. port and Caribbean or Central America ports. Hawaii, Puerto Rico, and certain Pacific Island Trust Territories are considered to be United States. Canada and Mexico, while treated as nearby foreign, are regarded as foreign voyages under the act by both their countries and U.S. customs and immigration authorities.

Length of Voyage

The courts have ruled that the length of a voyage shall be measured from the port of departure to the last port of call. Where a vessel calls at a port or ports before her final port, the length of the voyage for manning purposes is not shortened. Additionally, if a portion of the voyage is outside the boundary line, the manning for the entire voyage must meet the standard for the seaward portion of the voyage. Thus a vessel that travels over a portion of its route outside the boundary line (such voyage being over 600 miles in its entirety but which may have been less than 600 miles at sea) and which has a deck watch must have a three-watch system. Sixty-five percent of this vessel's deck force must be able seamen. This interpretation of the law is based on the parenthetical exception to the three-watch requirement found in the first line of 46 USC 8104(d) which excepts from the three-watch system those vessels that operate only on "inland" waters. It should be noted that on the Great Lakes and certain tributaries, unlicensed and licensed personnel shall not be permitted to work more than eight hours a day. Vessels authorized to have a two-watch system may have only 50 percent able seapersons (46 CFR 15.840).

The provisions of 46 USC 8104(h) permit licensed personnel on towing vessels of at least 26 feet in length to work not more than 12 hours in any 24-hour period. As there is no reference to the waters on which this applies, it is taken to apply to any waters including the high seas. Subject to exceptions, 46 USC 8104(h) requires licensed individuals to be placed on three watches when at sea on vessels of greater than 100 gross tons unless on a voyage of less than 600 miles—46 USC 8104(g).

A final and perhaps confusing situation is caused by the apparent conflict between sections 8104(g) and (h) of the U.S. Code. The Coast Guard has interpreted this to permit licensed individuals (holders of master/mate licenses or operator's licenses) who serve as operators of uninspected towing vessels which *are not subject to* the Officer Competency Act to be divided into a two-watch system regardless of the length of the voyage. These would be towing vessels under 200 gross tons which proceed beyond the boundary line as well as all inland towing vessels regardless of tonnage. Most recently built U.S. tugs have been scrupulously designed by naval architects so that they come under that unrealistic admeasurement resulting in 199.99 or fewer gross tons regardless of the displacement or draft of the tug when equipped and fully loaded with fuel, lube, and fresh water. Thus with the under-200-ton factor present some owners have played with manning in a manner that may be legal but may not always be prudent for safe navigation. Some owners feel that the monetary saving from having fewer crew members than are required by law will be greater than the cost of a violation penalty, if they are caught. Thus, they sail their tugs short-handed. What they fail to consider is the cost if someone is injured or killed or if the vessel is involved in a collision. A finding of unseaworthiness due to inadequate manning may result. In lieu of this and in an attempt to remain within the law, many towing companies have adopted their own manning arrangement. A sampling of those in force includes a system used with a tug-notched barge operating in a steady trade. The two tankermen of the unmanned barge were granted a liberal allowance of sea time for riding on the tug. Through this, they obtained a Coast Guard endorsement as able seaman and as second-class operator of uninspected towing vessels.

When this tug/barge unit is sailing between ports, one of these men serves as third seaman when the distance is over 600 miles. The other man serves as second mate. This fills out the tug's roster of required watch standers. The tug, while legally measured at 199 gross tons, actually displaces nearly 500 tons. The loaded barge she pushes at a 35-foot draft has close to a 25,000-ton displacement.

An ocean-towing operator on the upper U.S. West Coast covers the law on long voyages by staffing his tugs with a captain, two mates, and three others, who are classed as engineer/deckhands. Cooking is done by one of them.

The average harbor tug (day boat) in San Francisco has a crew of three, while the 24-hour operating pusher tugs on the bay and adjacent rivers have only two crew members. Some of them are licensed captains and mates as well as serving on deck. Not all of these tug crew members are male. The same is true in many other ports.

On San Francisco tugs of 2,000 HP that work 24 hours daily, crews number five. They break down as two in the wheelhouse and three on deck, one of whom is engineer and another is the cook. On the harbor's 4,000-HP tugs, six persons compose the crew as follows: a master, a mate, an engineer, and three deckhands, one of whom acts as the cook.

On the East Coast north of the Virginia Capes, four or five persons usually make up the crew of harbor tugs. From Wilmington, North Carolina, to Tampa and around the Gulf of Mexico, three is the number: a captain, a deckhand, and a deckhand/engineer. The same crew reductions apply in Europe. In Germany both seagoing and harbor tugs are crewed by three persons while working within the boundaries of their harbor. In Portland, Oregon, only two persons, a captain and a deckhand, assist ships from tractor tugs.

Exceptions to these manning numbers are the crews required by the U.S. Navy, the U.S. Army, and the U.S. Coast Guard on their tugs. Navy harbor tugs draw manpower from a shore pool and will put from 9 to 12 men on a tug depending on the size of the ship they are to assist. The U.S. Navy seagoing or fleet tugs are considered fleet auxiliaries. As they venture into most combat areas, their manning scale cannot be compared with that of any commercial towing vessel. Their navigation bridge watch officers are backed up by signalmen, radarmen, quartermasters, and lookouts. All officers and crew members on these Navy fleet tugs have multiple duties in damage control, fire fighting, antisubmarine warfare, and regular battle stations.

U.S. Army tugs are noncombatant craft and are usually part of a harbor craft company. Their manning follows the pattern of commercial harbor tugs. U.S. Coast Guard tugs usually carry about double the crew of a commercial tug of equal size. This difference in operation and thinking often becomes evident at those times when a commercial tug has been involved in a casualty and its master is confronted by a Coast Guard investigation officer.

TUG LABOR RELATIONS AND THE UNIONS

Many crew members of tugs throughout the western world are associated with some sort of maritime labor guild or organization. In some countries this has existed for years. In others, tugs are part of a government or port authority. Examples include Dublin, Ireland, where the ship docking tugs are under the Dublin Port and Docks Board. Tugs on England's Manchester Ship Canal, like those at Panama Canal, are operated and manned through the canal's authority.

One of the first questions young persons desiring to enter towing will ask is, "How do I get on a tug?" It is not easy. In the United States until the 1930s, the answer was in whom you knew and where you were. In small East Coast ports, it was practically impossible to get a job on a neat and steadily employed harbor tug. At that time, even though tugs were busy and plentiful, one had to be a relative or friend of a captain or engineer to gain a berth.

In some areas tug men were a clan by themselves, linked through religious or fraternal orders. In Cleveland, Ohio, it was said that membership in St. Malachy's Parish and hanging around a certain tugman/boxer's gym were the necessary formulas for consideration when an occasional opening occurred on one of the "Big G" harbor tugs. On the Great Lakes, the tugboat season ended in December and opened in April. In the winter, these tugmen worked in various unionized trades ashore. With the advantages of trade union employment in their minds, they became organized into two tug unions, one for licensed pilots,* as

*Licensed Tugmen's and Pilot's Protective Association.

the tugs were steam-powered, and one for unlicensed.* It is said that they stood shoulder to shoulder if any trouble arose.

Early on, no East Coast tugs were organized other than some in New York Harbor. As there were plenty of merchant ships and tankers prior to and through World War II, maritime labor organizations did not venture into the towing industry. Even the Masters, Mates, & Pilots (MM&P) locals were not interested in obtaining licensed steam tug deck officers as members. In southern states, tug owners treated their tug crews as a family fiefdom. In Norfolk, Virginia, the late Captain Joe Wood revealed his manning secret. He divided his tug fleet up, crewing one half with North Carolinians and the other with men from Virginia's famous seafarers of Mathews County. Both groups were very clannish and jealous of one another. "They each try and do better than the other," Captain Wood claimed with a chuckle. In Jacksonville, Florida, the famous "Doc" Coppedge kept his boys on low wages but never let them down if they needed a loan. With many crew members said to be indebted to him, few could afford to quit.

Even in New York Harbor, which later became a bulwark of maritime unionism amid its waterfront corruption, there were tugboat clans. You need not have asked Captain Paddy McGuirl, owner of Shamrock Towing, for a job on one of his totally green tugs unless you were Irish. From the East River, most of Red Star's captains came from College Point or Green Point and were also of Irish ancestry.

On the New Jersey side of New York Harbor's North River, the Lackawanna Railroad tugs were referred to as the Austrian Navy. In Brooklyn, the crews of both Olson and Dauntless Towing Lines had pronounced Scandinavian accents as typified by one captain, who gave his tug's radio call sign of KJVP as Kay Yah We Pee.

By 1930, nearly all of these fine tugmen were unionized, but not through organized labor's "strength in unity." Local maritime unions did not have any reciprocal agreements. They made no attempt to unify the waterfront industry afloat. This prevented the seafarers on tugs of one union from crossing over to another unionized employer. As an example: If you were a steady deckhand on a railroad tug which went in to lay up, you might be bumped back to part-time work. Unfortunately, you could not work on any of the other regular harbor tugs, the reason being that railroad tug crews belonged to a railroad marine branch of the MM&P. At that time, the MM&P was affiliated with the Brotherhood of Railway Trainmen. The rest of New York Harbor's tugs were contracted with the United Marine Division, Local 333. This local was a branch of the International Longshoremen's Association (ILA).

*Tug Firemen, Linesmen, and Watchmen's Benevolent Association.

On coastwise tugs, engineers also belonged to this union. Many of the tugs were steam, and the firemen were usually Spanish and came from the Catherine Street area of Lower Manhattan. Local 333 manned all these New York Harbor tugs vertically from captain to cook.

During World War II, as tugmen were drafted or volunteered for service in the armed forces, unions lowered their entry barriers. After the war, there was a flurry of U.S. ocean and coastwise towing. Many ships were moved to various lay-up fleets or sold for scrap. A whole new generation of men manned these tugs. They came from everywhere: Some were fresh out of military service; some were from the huge laid-up merchant fleet; many of these men were already members of a labor organization. Very quickly, the various maritime unions began to revamp their territory. Attempts were made to organize any new towing branch of an old towing company. For the next four decades the tugmen and tug owners faced a hodgepodge of noninterlocking maritime unions. These unions were often hostile to one another as they viciously guarded what they considered was their right to represent workers in certain areas. Some of these unions had no prior connection with the maritime labor industry.

The AFL-CIO, the recognized authority and charterer of many of these then-member-hungry groups, did not take action to direct all maritime workers into one or two unified maritime unions. The results left many groups adrift and ready to accept the best offer. Union members had little to say in such decisions. The ILA temporarily lost its charter with the AFL-CIO. The disenfranchised Local 333 eventually linked with the National Maritime Union. On the Great Lakes, the two tug unions that had been associated with the ILA were also affected. They joined with the new International Brotherhood of Longshoremen (IBL), which temporarily replaced the ILA in the AFL-CIO. They are now a part of the SIU.

In other ports the MM&P suddenly became tug-conscious. Not to be outdone, John L. Lewis's United Mineworkers initiated a catchall union for maritime workers. This, called Local 25, organized the harbor tugs in Philadelphia and other ports. In some places, the Teamsters Union, the MEBA, and Operating Engineers Union all claimed bargaining rights for marine and tug workers.

In the 1950s, the Seafarers International Union, an outgrowth of the Seamen's Union of the Pacific (SUP) and what remained of the earlier International Seaman's Union, opened up a subsidiary for harbor, coastwise, and inland tug and tugboat workers. Known as the Inland Boatmen's Union (IBU), the union negotiated contracts with major towing companies on both coasts, covering both unlicensed and pilothouse tugmen. This coverage was extended later throughout the Mississippi River system. Today, the SIU, SUP, and IBU are the key labor organizations representing unlicensed and many licensed tug and towboat men of the United States. Although in early 1986 the president of the SIU proposed a merger of all maritime unions, a confusing number still hold contracts with towing companies.

All of the tug unions belong to the Maritime Trades Department of the AFL-CIO. The Masters, Mates, & Pilots man the wheelhouses of tugs and include the bar pilots on the U.S. West Coast. Now a division of the rechartered ILA, they have an inland division covering some harbor tugs in Florida and on San Francisco Bay.

The Seafarers International Union, with a total membership estimated at 80,000, is affiliated with 18 unions, 3 of which are U.S. West Coast marine unions and 2 of which are in Canada. Outside of New York Harbor, they have contracts with all the major East Coast and Gulf Coast towing companies, as well as for tugs in the harbors of Los Angeles, Wilmington, and San Pedro.

The Seamen's Union of the Pacific, granddaddy of all deepwater seamen's unions, has been affiliated with the SIU since 1938. It furnishes unlicensed personnel on many West Coast tugs.

In 1979, members of the SIU branch, the Inland Boatmen's Union, voted for and were accepted into the regular SIU. The SUP also has an affiliated IBU (Pacific), which has contracts for unlicensed personnel on West Coast inland tugs and barges. IBU (Alaska) was a small affiliate of the SIU until June of 1985, when it became independent. Its members work on tugs and barges in Alaska and Hawaii.

ORGANIZED TUG AREAS

Not all the tugs in the various harbors of North America operate under a union contract. In some areas there are various combinations of all unions on the same tug. SIU contracts for tugs normally cover the pilothouse and deck force, including cooks. In most cases, the engineers are furnished by the MEBA. On the U.S. Great Lakes, tugs are mainly covered by contracts with the SIU. In Canada, the SIUNA has contracted towing and dredging companies. On the East Coast, tugs from Maine to New York and up the Hudson are manned vertically from pilothouse to engine room by New York Harbor's Local 333. Tugs in Philadelphia, Baltimore, Norfolk and Jacksonville are in the SIU.

As Florida is a "right to work" state, some tug operators have fought to prevent workers from revealing their desire for membership in any labor organization. Tugs in Miami and some in Tampa are unorganized. The tugs in the rest of the Gulf ports and towboats on the Mississippi River are mainly covered by SIU and MEBA contracts. Tug pilots working on Panama Canal ship-assist and dredging tugs have for years been MM&P members.

On the West Coast, the harbor tugs of Los Angeles are divided among several unions. Some owners are contracted by the SIU and MEBA. Others are under MM&P, SUP, or IBU and MEBA contracts. At San Francisco, the AMO branch of MEBA mans the pilothouse of one company while the Inland Division of the MM&P covers others. Deck and other personnel are furnished by the IBU. In Portland and on the Columbia River system, the IBU furnishes most tug crews. The pilothouse personnel of Puget Sound tugs and those in the various harbors are covered by an MM&P contract. The unlicensed personnel come from the IBU.

As many tug owners on the U.S. West Coast, and a few on the East Coast, have tugs stationed in various ports, they may have from three to six different labor contracts. These signed agreement cover wages, working conditions, and other work arrangements including benefits. They are usually for a three-year period.

WORK ARRANGEMENTS AND WAGES

Working arrangements differ by coast and port. On the U.S. East Coast, harbor tug wages are often hourly with a guaranteed minimum for each day. On tug fleets in ports with a lot of ship-assist work, tug hours are staggered. Some boats will come out early and tie up early. Others come out later and tie up later. Any work done before or after certain hours is classed as overtime. All kinds of overtime pay scales have been applied to Sundays and holidays. On 24-hour tugs this has frequently led to the senior crews arranging their work schedules so that they could be on during all premium pay weekends. To alleviate this complaint by the opposite crew, some owners have worked out an average total monthly wage for each crew member's position. By dividing this result, a daily average has become the new daily pay rate. Considering that there are 52 weekends, and in some states 13 holidays are recognized, this change has been a huge bookkeeping relief. Under such a method, crew members have not lost anything. Captains, mates, and accounting departments ashore now have simpler time- and record keeping.

Harbor tugs on the West Coast follow much the same work pattern as those in East Coast ports. The big difference from East Coast operations occurs on the 24-hour ocean and coastwise tugs. As voyages between coastwise ports of Hawaii or Alaska are long, a credit for one-half of a day's pay for each day worked is accrued.

During two decades of towing prosperity, tug wages and other benefits have escalated to keep up with those common to industry ashore. Here is a sampling of wages paid by some of the larger towing companies:

Year	Basis	Captain	Mate	Chief Engineer	Deck-hand	Cook	Union
1965	**	$900.00	$700.00	$850.00	$400.00	$300.00	Nonunion
"	**	400.00		350.00	75.00	125.00	Panlibhonco
	*	65.00	200.00	60.00	25.00	50.00	Nonunion (Teco)
"	*	118.00	50.00	116.00	87.00	87.00	Local 333
	*	126.00	112.00	88.00	88.00		Local 333—24-hour
1975	*	102.14	123.00				MM&P Seattle
"	*	82.00		77.00	48.00	48.00	SIU Coast
	*	132.00	63.00	132.00	100.00	100.00	Local 333 Great Lakes
1983	*	247.00	126.00				SIU Coast
"	*	$150.00		117.71	74.89	76.62	AMO (San Francisco IBU)

Basis of pay period: * equals daily; ** equals monthly.
Note: No overtime, holiday, or weekend premium pay is shown.

Each of these contracts had a variety of benefits, such as travel paid from the member's home to the tug and return. Tug owners contributed a daily amount for hospital, medical, and in some cases, dental coverage as well as into pension funds.

NEGOTIATIONS

At times, this has been the most difficult nonoperational part of towboating. In some ports tug working conditions and wages are negotiated before groups representing owners such as the New York Towing Association and the Northwest Towboat Owners Group in Seattle. Throughout much of the rest of the United States, individual towing companies negotiate directly with union representatives. After a thorny negotiation session, the usual handshakes and good wishes are halfheartedly given. Both management and tugboaters go back to work believing they were victorious. Unfortunately, there are always a few on each side who will belligerently refer to "next time." As the time for contract renewal approaches, rumors begin. Management prepares for total resistance. The membership has a new want list to be presented through its negotiating committee.

The first meetings are, as politicians like to say, candid. They usually consist of testing the waters. If, after several meetings, nothing is resolved at the bargaining table a strike notice may be imminent. Without a court injunction or a cooling-off period, a strike may take place.

This plethora of tug unions has both advantages for owners and disadvantages for members. If a strike occurs, a mixture of unions within a port allows tug owners to play games. Boats may be moved to another more favorable port, or sold. For tug crews changes in ownership may mean the loss of jobs and rights unless a new position is available. Frequently, prior to renewal of contracts tugs have been sold to dummy subsidiaries of the original owners. A strike is the tool of last resort. Its much touted clout is almost an insult forced on the tug worker. Strikes are caused by the failure of negotiations. Somewhere, labor-management communications have broken down. Labor cannot expect management to offer the impossible. Management cannot expect labor to accept their financial assertions, which in reality are not fact. Each side should know what the other has to offer and what each expects. Sudden surprises and changes of attitude at the bargaining table lead nowhere.

Strikes are demeaning. They bring out a little of the best in some and the worst in others. All of the negativisms of the opposition flow forth. Positivism is forgotten. Management looks on the tugmen as pirates ready to rob them. Tugmen think of their suddenly unfriendly employers as misers. If the strike lasts more than one pay period, there is seldom a real winner. The longer the strike the more adverse publicity there will be for the employer. For the tug worker, any wage increase he may receive at the end of a long strike may have been outweighed by wages lost during its duration.

There have been several long strikes in the Port of New York. To prevent the port from being tied up, military personnel operated some of the tugs. Shipmasters and pilots attempted to dock ships without tugs. Damages to some tugs and a few pierheads were high.

Ideally, the level of tug labor relations should not be allowed to deteriorate toward an impasse or strike. Avoiding this requires teamwork by both parties. A joint public relations approach should be made toward marketing the skills of both groups. All too often this has been done unilaterally. The results have been a feeling of combative competitiveness. The ultimate desire should be a unified team of management-labor-employee. At times, it seems that the formula for the simple goal of financial success for the towing company has eluded both sides. Like truth in advertising, there must be truth in revelations of operating costs. From these, an understanding should be evident as to what share of the profits should go to the tug employees.

At least one towing line has tried this. They have held seminars for their tug personnel. These included a cross section of every rating. These seminars are very personal and are held in convention hotels away from home and office. Tug crews meet top management through open daytime lecture discussions and social events in the evening.

Much has been accomplished in these sessions. An atmosphere of pride and praise of the employees' part in the company's success was prevalent. Unfortunately, no union representatives were present or invited even though the particular union had trained, provided, and upgraded many of these tugmen.

The SIU has held its own seminars for tug crews of various towing companies. Management representatives have not, as yet, been asked to take part. Until labor and management learn to trust each other and bargain in good faith, there can be no unity of purpose. To run their tugs, the towing company needs the well-trained personnel furnished by the union. The tugmen need the company to produce work for the tug; they need the union for added training and job security. All three need to work together to survive.

Today's union delegates, patrolmen, and officials are educated business administrators who recognize employer's problems. The frequent visits of union port agents to the tug keep any problems from getting out of hand. The agent's access to a company labor coordinator should be a welcome monthly occurrence. With such a readily accessible communications pipeline, problems should easily be solved. A healthy attitude should prevail. Negotiations should continue until solutions rather than strikes are reached. With such cooperation between labor and management, it is hoped that labor relations in the future will arrive at a plateau where the tugman really feels he is a partner of his employer. Both management and tug crews must present any ideas for better operations. Very often, tug crews afloat have seen wasteful practices overlooked by the company.

HIRING AND TRAINING THE CREW

With a few exceptions, towing companies live up to that portion of their labor contract which states that all personnel shall be furnished by the union. Briefly this means calling the local union hall when crew members are needed. If a person should get a position aboard a union-contracted tug by any other means, he or she will be approached to become a member by paying an initiation fee and monthly dues. If there are qualified people available in the union hall to cover this position he or she will be forced to give up the job. In this way, union solidarity protects job security. To balance the scale for tug employers, the union must provide good, qualified men. There have been qualified men who were not so good. Towing companies have the right to reject unqualified workers. The unions quickly weed them out.

To provide good qualified men, the SIU has developed a fine basic training program in towing at its Harry Lundeberg School of Seamanship. There, at Piney Point, Maryland, they have a standard harbor tug and an inland towboat. On these, the school offers hands-on maneuvering and towing practice with the school's fleet of tank barges and deck barges. Classroom study on scholastic and navigational problems is a part of the program. By having tug crew members return periodically, the school offers them a chance to upgrade themselves from the entry level of ordinary seaman to that of licensed officer. A radar collision-avoidance course is also offered. All levels of instruction will prepare a person for the current Coast Guard examinations.

Another school that offers instruction for towing is the Maine Maritime Academy at Castine, Maine. Many of its graduates have put in their sea time as cadets on tugs and later have returned as licensed mates and engineers. There are other maritime schools in the United States, which annually graduate personnel at the third mate and third assistant engineer level. Some of these graduates eventually enter towing. These schools are The Great Lakes Maritime Academy; The Massachusetts Maritime Academy at Buzzards Bay; The State of New York Maritime College at Fort Schuyler; and the United States Merchant Marine Academy at Kings Point, Long Island. The academy at Kings Point has a former U.S. Navy rescue tug as a training vessel. In the Gulf of Mexico area, Texas A & M has a maritime college in Galveston. For the river towboat men, there is an industry-supported towboat school, the National River Academy, in Helena, Arkansas. On the U.S. West Coast, the California Maritime Academy at Vallejo, California, has provided officers and other ratings for some tug operations on the Pacific.

13. Rules, Rulings, Regulations, and Other Considerations Affecting Towing

'72 COLREGS AND THE NAVIGATION RULES

ALL of these definitely affect towing on international waters. On U.S. inland waters, a great change in the Inland Rules of the Road occurred in the 1980s. To clarify what has happened, particularly to the navigation lights required on tugs and barges, a brief background is offered. Following the 1972 International Convention of the United Nations Inter-Governmental Maritime Consultative Organization (IMCO), a new set of international regulations for preventing collisions at sea was developed. They replaced the International Regulations for preventing collisions of 1960 and are known as the '72 COLREGS.

This was the beginning of changes to and the unification of the International Rules of the Road. These '72 COLREGS apply to all public and private vessels navigating the high seas. Later updated, they went into effect on July 15, 1977, and remain unchanged.

As the '72 COLREGS apply only to the high seas, countries such as the United States which have established lines of demarcation between the high seas and inland waters, were left with some confusing sets of rules governing inland navigation. Within the United States, a new set of unified inland navigation rules "were enacted by law on 24 December, 1980," effective one year from that date.

To further unify these Inland Navigation Rules so that they would closely follow those of the '72 COLREGS, the Rules for Western Rivers were canceled as of December 24, 1981, and those on the Great Lakes were canceled on March 1, 1983. On that date, these new Inland Navigation Rules became effective for all vessels on U.S. inland waters.

The following U.S. Coast Guard publications were canceled:

CG-169-Navigation Rules—International and Inland
CG-169-1 COLREGS Demarcation Lines
CG-184-Rules of the Road—Western Rivers
CG-172-Rules of the Road—Great Lakes

RULES AFFECTING TUGS

All of these changes to the navigation rules affect tugs and towing. As they are explicitly presented in the publication (COMDTINST-M166 72.2) "Department of Transportation U.S. Coast Guard Commandant's Instructions M166 72.2," only those parts affecting tugs and tows will be paraphrased here.

Duty dictates that tug masters, pilots, and those others seeking a deck license should try to memorize the contents of these rules. Wheelhouse personnel on U.S. vessels use or refer to these rules every time they are under way. Questions pertaining to COLREGS appear on every U.S. Coast Guard deck license examinations.

In paraphrasing what follows, "power-driven" vessel has been replaced by tow or towing vessel and is used to indicate both tugs and inland towboats. In placing emphasis on rules that apply to tugs, towboats, or towing, there is no intention of lessening the importance of all of the International and Inland Navigation Rules.

Here are some comments on the latest rules that apply to tugs. Unless stated otherwise, one comment covers both International and Inland Rules:

Rule 2. This was formerly known as the "Precautionary Rule" and remains basically the same. Its impact remains regarding "neglect to comply with these rules" and the necessity to depart from them "to avoid immediate danger."

Rule 3(g)(vi). "Vessel restricted in her ability to move" includes "a towing operation such as severely restricts towing vessel and her tow in her ability to deviate from her course" [see Lights and Shapes as required by Rule 27(c)].

Rule 5. Lookout Rule. This is most important. Note the words "sight and hearing." This means a person, not just radar, which is included in "as well as by all available means."

Rule 6. It is obvious and always applies when running without a tow.

Rule 7. This rule does not give tugs with tows any right-of-way over other vessels.

Rule 9(a). Holding to the starboard of a narrow channel and (a)(i) should normally be adhered to by tugs with tows.

However under (a)(i), on the Mississippi River, all downbound traffic takes the deep part of the bend, which means staying in the outside of curving banks. Upbound traffic holds to the point. As the river has bends both to the right and to the left, a downbound tow can be in the righthand bend and meet an upbound vessel or tow on her starboard. This would be a two-whistle meeting—starboard to starboard. When downbound in a lefthand bend, the meeting would be one whistle and port to port.

This type of meeting with ships is customary below Baton Rouge. It should be noted that in Inland Rule 9(a)(ii), the Great Lakes and Western Rivers are mentioned for the first time. As the limits of these waters are not described within these rules, the old boundaries are still acceptable. They are for the Great Lakes and their connecting waters and tributaries as far east as the lower entrance to St. Lambert's Lock in Montreal; the Thomas J. O'Brien Lock and controlling works at Mile 325-327 on the Calumet River; and the eastern side of the Ashland Avenue Bridge of the Chicago River.

The waterway beyond and below the O'Brien Lock and Ashland Bridge is considered a part of the Western Rivers, including "the waters of the Mississippi River above the Huey P. Long Bridge, at New Orleans, and all of the tributaries emptying thereinto and their tributaries and that part of the Atchafalaya River above its junction with the Plaquemine-Morgan City alternative waterway; and the Red River of the North."

Rule 9(b) also introduces the metric measurement for the first time. As feet and inch measurements are standardly used by the U.S. Maritime industry, metric notations will be reduced to feet and tenths of a foot.* A 20-meter vessel as mentioned is 65.6 feet in length.

Rule 9(e)(i) requires the "bend signal" of "one prolonged blast." Many tugs fail to give this signal. The use of a VHF announcement *does not satisfy the law.*

The traffic separation schemes and vessel traffic services mentioned under Rule 10 have already been discussed in chapter 10.

Rules 11 through 18 need no further comments.

Rule 20 defines when lights and shapes are to be shown.

Rule 21 defines and describes the characteristics of all navigation lights including the "towing light."

Rule 22 lists a required visibility for each navigation light according to the vessel's size: under 12 meters (39.4 feet), between 12 meters and 50 meters (39.4 and 164 feet), and over 50 meters (over 164 feet).

Rule 24. Both the international and inland portion of this rule pertain to towing lights and signals. In particular, this portion applies when towing astern. While the colored sidelight and white stern light remain the same, an added amber towing light that shows over the same arc as the stern light must be located above and vertically in line with the stern light.

Rule 24(a). It is important to note that for the first time the size of the tug (50 meters or 164 feet) is mentioned. If the tug is under 164 feet overall, and towing astern and the length of the tow is over 200 meters

---

*See conversion tables in appendix XII.

(656.2 feet) from the stern of the tug to the after end of the tow, three masthead lights in vertical line must be shown. These masthead lights will show from right ahead to two points abaft the beam on either side as defined in Rule 21.

As each sidelight shows within the same arc on its respective side and both the stern and tow light cover from astern to this two points abaft the beam on either side, the properly lighted tug should always be visible within a distance of 2 miles.

In addition, on a tow of over 200 meters (656.2 feet), the tug must show a black diamond shape "where it best can be seen." When the tow astern is under 200 meters only two masthead lights are required.

Rule 24(b). This refers to the Integrated Tug Barge Unit or ITB. This is evident through the use of the wording "a pushing vessel and vessel being pushed ahead are rigidly connected in a composite unit." This type of unit shows the same navigation lights as a ship ("power driven vessel"), unless under 50 meters (164 feet).

Rule 24(c). When a tug of less than 50 meters (164 feet) is pushing or towing alongside on international waters, she must show, in addition to her sidelights, two masthead lights and a stern light. When the same tug is pushing or towing alongside on inland waters, she shows two yellow towing lights aft in place of the white stern light. All other masthead and sidelights remain the same.

Rule 24(d). Both international and inland covers tugs towing on the hawser astern and the unit being towed, when the length of the tow is 200 meters (656.2 feet) or less. The tug must show sidelights, two masthead lights, a stern light, and a yellow towing light. A second or after masthead light (formerly called a range light) is optional if the tug is less than 50 meters (164 feet).

When the white masthead lights are shown from an aftermast while towing astern with the length of the tow 200 meters or less, or pushing, a forward masthead light is also required. All other lights, side, stern, and towing remain the same.

The unit being towed must show sidelights and a stern light, and if the length of the tow exceeds 200 meters, a black diamond shape must be shown "where it best can be seen."

Rule 24(e). Both international and inland versions of this rule list the required lights for barges being pushed or towed alongside. The one outstanding difference for a unit being pushed on inland waters is the special flashing light as described in Rule 21(f). This light should be displayed at the forward end near the center of the head barge.

Rule 24(i). Under inland waters, this rule retains the basic lights shown on towboats on Western Rivers above the Huey P. Long Bridge at

New Orleans. These are the sidelights and two vertically mounted amber towlights aft.

Rules 25 through 27(a) describe lights and shapes for small vessels, sail vessels, fishing vessels, and vessels not under command, and vessels other than tugs restricted in their ability to maneuver.

Rule 27(c) involves towing operations, lights, and shapes when the towing vessel is severely restricted "to deviate from her course with her tow."

When the tug is restricted in international waters, she must show in addition to her regular running lights three all-round lights arranged vertically. The top and bottom lights shall be red and the middle light white. For the same condition in inland waters, the tug shows in addition to her colored sidelights two masthead lights, stern and towlights, and three all-round lights as described above for tows on international waters. The daytime shapes for all waters are two black balls mounted vertically with a black diamond between them. In all cases, these lights and shapes apply to tows of up to 200 meters and tugs of less than 50 meters (164 feet).

Rule 28. As tugs with heavy or deep-draft tows usually stay within the confines of marked channels, their masters and pilots should recognize the meaning of three red all-round lights on approaching another vessel. These lights in addition to the vessel's regular running lights signify that the vessel is constrained in her draft.

Rule 29. As tugs with tows enter and exit many harbors of the world, their watch standers should be familiar with the lights carried on pilot vessels.

Rule 30. As the tug master may be sent to assist a vessel aground, he should memorize the basic additional lights of two all-round red and the shapes, which are three black balls in a vertical line.

Rule 33. As most tugs are over 12 meters (39.4 feet), they must have a whistle and bell.

Rule 34. The maneuvering and warning signals, also known as meeting or passing signals, need not be explained as all tug wheelhouse personnel should be aware of them and their meaning.

Rule 34(b). Some tugs have had installed an all-round white maneuvering light that will come on simultaneously when the whistle is blown. Towboats on the inland rivers frequently have this maneuvering light. The main reason seems to be that, in addition to the absence of steam escaping as whistles are blown, the new noise of diesel engine exhaust lessens the audibility of sound.

Tugs towing into the St. Lawrence River should not be surprised at night when meeting a Canadian Great Lakes vessel. Frequently, a large alphabetical letter may light up on her foremast. For example, some Great Lakes vessels of Canada Steamship Lines had a large letter C, and those of the Misener fleet had a large letter M, which coincided with the ship's whistle blast.

Rule 34(e). The bend signal should always be given day or night regardless of whether the tug pilot assumes no traffic is approaching from the opposite direction. Unofficially, if another vessel heading in the same direction has just cleared the bend, and advises that there is nothing approaching, the bend signal can be omitted.

Rule 35(c). This rule simplifies the whistle signal for tugs with tows in restricted visibility. Restricted visibility includes fog, heavy rain, falling snow, or other vapors. The signal for tugs with tows both on international and inland waters is one prolonged blast followed by two short blasts at intervals of not more than two minutes.

Rule 35(e). This rule states that in all waters, the last vessel in a tow, *if manned,* shall give one prolonged and three short blasts immediately after the tug's signal is given.

In the Annex of Navigation Rules will be found the technical details for positioning masthead and other navigational lights. One of the weaknesses in the required visibility for the luminosity of navigation lights is that although they may have USCG, ABS, or other approval, no one ever checks the bulbs, cleans the dirty lenses, or corrects the low power, which may often weaken visibility, particularly on towed units.

It should be noted that the new inland rules have omitted mention of lights formerly required on barges, scows, canal boats, and tows on the Hudson River, Lake Champlain, and the inland waters as far east as Narragansett Bay. Also omitted are the former regulations for the length of towing hawsers between barges and those points where seagoing tows should shorten hawser.

U.S. Coast Guard Ruling Regarding Length of Towing Hawser
The following reply was received from the chief of marine information and rules branch in July 1986:

The Navigation Rules do not address hawser lengths except to identify what light or sound signals are required. There is a difference in signals between vessels towing with a length of tow over 200 meters and that of a shorter tow.

Towing hawser lengths are regulated in a number of geographical locations within our inland waters; some places by Federal Regulations and possibly some by State or local requirements. As an example, Title 22 of the Code of Federal Regulations, parts 161-163 and 165, all contain towing hawser limitations.

LAWS REGARDING TUGS AND TOWS

It would be impractical to cover all of the laws and decisions rendered in Admiralty regarding tugs and tows. These have already been compiled in an excellent tome.* It would do well for the interested tug captain to read at least parts of this. In it, he would find information on the obligation his company undertakes when it contracts for towage, including the requirement that the tug have sufficient power and equipment.

Tugs have been found unseaworthy for many seemingly minor deficiencies, such as inadequately equipped life rafts, defective towing pendants, towing capstan inoperable, lack of charts or spare towing hawser, and many more possible shortcomings.

What constitutes unseaworthiness? Tug captains, mates, and engineers should be aware of the full context of the law in order to protect their vessel, their owners, and themselves. Here is what the law states:

> Under maritime law, every shipowner or operator owes to every member of the crew employed aboard the vessel, the non-delegable duty to keep and maintain the ship, and all decks and passageways, appliances, gear, tools and equipment of the vessel, in a seaworthy condition at all times.
>
> To be in a seaworthy condition means to be in a condition reasonably suitable and fit to be used, for the purpose or use for which provided or intended. An unseaworthy condition may result from the lack of an adequate crew, the lack of adequate manpower to perform a particular task on the ship, or an improper use of otherwise seaworthy equipment.
>
> Liability for an unseaworthy condition does not in any way depend upon negligence or fault or blame. This is to say, the shipowner or operator is liable for all injuries and consequent damage proximately caused by an unseaworthy condition existing at any time, even though the owner or operator may have exercised due care under the circumstances, and may have had no notice or knowledge of the unseaworthy condition which proximately caused the injury or damage.

Reading chapter 4 of *The Law of Tug, Tow, and Pilotage* is highly recommended. It points out many duties of the tug, which are of course, duties of the owners and operators and in many cases, duties of the tug master. Such interesting titles appear as: "Seaworthiness and Exercise

---

*The Law of Tug, Tow, and Pilotage by Alex Parks (Centreville, Md.: Cornell Maritime Press, 2nd ed., 1982).

of Due Care," "Tugs Must Have Sufficient Power and Proper Equipment," "Duty of Tug to Make Up the Tow," "Proper Position in Tow and Makeup," "Proper Handling during Makeup," "Duty of Tug as to Towlines," "Duty of Tug as to Lights on Tow," "Meeting Swells," errors in navigation in "Leaving the Channel," "Duty of Tug After Disaster," and many more subjects dealing with fog, sound signals, anchoring, and mooring of barges.

LOOKOUTS AND THE TUG

Case after case, whether settled in court, or through rulings by the USCG has emphasized that the "Duty of the Look-Out is of the highest importance. All navigating vessels are subject to this requirement, tugs and barges no less than the largest vessel afloat."

Rule 5 of the Navigation Rules—International and Inland of 1983 orders that "every vessel shall at all times maintain a proper lookout by sight and hearing and will as well use all available means appropriate in prevailing circumstances and conditions as to make a full appraisal of the situation and the risk of collision." Some of these rulings involving tugs point out a lack of understanding by the watch stander/operator of the full concept of what constitutes maintaining the proper lookout.

Several USCG decisions of the late 1980s point out these errors by tug operators and tug captains. In one case, a tug was pushing a flotilla of high-sided barges down a West Coast river. The weather was clear. Although the tug operator's view was obstructed for 600 feet ahead, and 200 feet on each side, no lookout was posted. All deckhands were in the galley. At 4 knots, the tow was approaching a popular recreational fishing area. Many boats were drifting around in the channel. The tug operator sounded a danger signal. One boat, which was unable to get its engines started, was run down. One person was drowned. The tug operator's license was suspended for two months.

In another case, on the East Coast, a similar situation occurred with a tug pushing a barge that obscured the operator's vision for a little over one-quarter of a nautical mile ahead. In running through a popular fishing ground on a holiday, the tug operator had not posted a lookout or sounded a danger signal and subsequently ran over a small boat, casting three people into the water, one of whom drowned. The tug operator's weak excuse was that customarily, pleasure craft get out of the way. His license was revoked.

The author has overheard rather serious-sounding conversations between other tug captains when eastbound from Execution Rock into Long Island Sound when flotillas of sailboats seemed to cover the course ahead. "I'll get one of the b - - - - - - - -." "Yeh, go ahead, I'll get two," was the reply. This is neither the proper attitude nor a prudent action. A reduction in speed and alteration of course in accordance with the Inland Rules would be proper.

One such conversation did not go unnoticed. It was overheard by a well-known shipping person who was on a regatta committee boat. He called the tug owner later and notices were sent to this tug fleet instructing them to avoid or gingerly pick a way through in these situations.

Back to failure to post a lookout: A sailboat that had become becalmed outside a vessel traffic lane was run down in San Francisco Bay. The cause was a tug and barge that after receiving acknowledgment of its intention from VTS crossed out of the traffic lane and struck and sank the disabled sailboat. While there was no loss of life, the tug master was ruled against for failure to post a "dedicated" lookout. His license was suspended for one month.

Finally, a U.S. tug with a barge in tow on the hawser was proceeding up the East Coast in the fog. The captain sent the able seaman on watch down to the galley. In that instant, the tug came upon and struck a fishing vessel, which for reasons unexplained apparently did not appear on the tug's radar as a target. The commandant of the Coast Guard's ruling was that as the lookout's eyes and ears as required in Rule 5 had been removed, even though the tug operator was using the radar as "all appropriate means," a proper lookout was not posted at that moment. The operator's license was suspended.

This ruling was apparently based on the *Congressional Record,* which regarding the U.S. Inland Rules that became effective 1 March, 1983, reads in part, "The duty of keeping a proper look-out is often termed the first Rule of Seamanship. That duty cannot be discharged by deckhands not specifically assigned as look-outs."

WHO IS THE TUG MASTER?

There are several answers to this question. Throughout the chapters of this book, the terms tug master, tug captain, tug pilot, skipper, or watch stander have been used capriciously. Except when otherwise noted, they refer to the person on watch and in charge of handling the tug. (The term "tug operator" has been avoided.)

The legal master of a U.S. tug is the person whose name appears in the tug's log as such, provided he or she holds an applicable license, or the person whose name appears in the register if the vessel is in foreign trade. In other countries, the tug master is regulated by the vessel's tonnage and routes. Due to the manner in which U.S. tugs are admeasured, other complications arise, mainly in the responsibilities of the tug master. Presently, on U.S. motor tugs of up to 200 gross tons the captain need only have a first-class tug operator's license covering the route within which the tug operates (see "Licensing" in chapter 12). This is a new form and classification of license invoked by the U.S. Coast Guard in the 1970s. This certificate is different in appearance from that of the regular Coast Guard license to merchant marine officers. Prior to the issuance of this new license, a skipper on a U.S. tug of under 200 gross tons needed only a Z card to act as captain. Had the same tug been steam-powered, he would have had to hold a license as master of steam vessels for whatever tonnage he was qualified for. The tug would also have had a certificate of inspection.

While the motor tug of under 200 gross tons is termed an uninspected vessel, those between 200 and 300 gross tons are also classed as uninspected. There is a difference, however. In the latter case, the Coast Guard, which may board at any time, does not inspect or set up certain manning and other requirements on a certificate of inspection. Another difference is that the person designated as master on this tug of between 200 and 300 gross tons must have a regular master's license for uninspected vessels (of whatever tonnage).

A tug operator's license does not cover these vessels. The mate must hold a license as chief mate of uninspected vessels, whereas the mate of a motor tug under 200 gross tons is required to hold only a second-class operator's license. The classification of tug operator may be changed if the Coast Guard's proposed rule making of 1985 is approved and placed in effect (see "U.S.Coast Guard Ruling Regarding Length of Towing Hawser," following).

The differences do not end here. U.S. tugs of over 300 gross tons are considered inspected vessels. All U.S. Coast Guard regulations pertaining thereto must be upheld. This includes a fully licensed master covering the applicable tonnage and waters. All other crew members must be certified for the capacity they hold.

RESPONSIBILITIES—TUG OPERATOR VS. TUG MASTER

It has been ruled that the person designated as tug master—whether tug operator or licensed master—is, when on duty and/or on board, responsible for everything that goes on aboard or happens with the tug. This includes all conditions that could conceivably contribute to a marine casualty, such as a barge's running lights dropping below the minimum luminosity as prescribed in Annex I to the Inland Rules.

How many times has a tug in a narrow channel met another tug with a barge on the hawser whose lights on her tow were not visible until the tugs were abeam and about 150 feet off? These luminosity problems should not be the responsibility of the tug master. These lights provided for tows are usually furnished ashore. The tug captain has only limited control. His sole influence may be an inspection prior to sailing. If the lights pass, there is no guarantee on an unmanned unit that these lights will work throughout the voyage. On long tows, batteries weaken; salt spray, snow, or ice can coat lenses. With such conditions and fog, the tug master has a potential casualty situation that he must overcome.

Another Coast Guard decision is that it is the master's "non-delegable" duty to see that a Form 2692 is filed within five days of any casualty for which such a report is required. If his company insists on filing this form, they should advise him in writing that it has been done. Failure to file will cause the Coast Guard to look toward suspension of the master's license.

Here is how the Coast Guard has ruled regarding the liability of tug masters for the general safety of their vessels. A person licensed as "tug operator" cannot be held liable when he is not on watch. However, if the tug's skipper is a "licensed master," the Coast Guard looks at it differently. As such, he is responsible for all that goes on whether on watch or not.

Frequently, tugs will have two complete crews that relieve each other at regular intervals. If the relief captain does not enter in the tug's log the date and time he relieved the regular captain, the regular captain as master may be called to account for incidents that occur when he is not on board.

A consensus among many New York Harbor tug captains on 24-hour tugs has been, "I'll stand my watch, you stand yours. I learned the hard way. Why should I stay up to train you?" Most of the captains possessing this attitude do not realize that regardless of the mate's qualifications, the master will be held responsible for any mistakes.

This was made very clear by the *Ocean Prince* litigation often referred to by Admiralty attorneys. The *Ocean Prince* was a New York tug bound up the Hudson River with a loaded barge. She had been operating around the harbor as a single-crew day boat. When this out-of-town towage came up, the dispatcher or person in charge of personnel had to round up a second crew quickly, including a licensed mate. Although the *Ocean Prince* was only 198 gross tons, the traditional New York habit left over from steam days remained. All officers had to be licensed to cover their position. The "tug operator" license had not yet been introduced. Thus, both the master and mate of *Ocean Prince* were fully licensed. Whether or not each had first-class pilotage for the Hudson River is questionable, although most New York tug operators demanded this additional endorsement for waters in which the tug operated.

The mate the office assigned to *Ocean Prince* for this voyage was later deemed inexperienced. On the trip upriver, he had relieved the tug captain, who was the legal master, at midnight. Several hours later, the tug passed under Bear Mountain Bridge and approached Con Hook, a small, low, wooded island in the middle of the river. The island was joined to the west shore by a reed-covered mud flat. From the tug's pilothouse a pilot could look over these flats and see the bright lights of Highland Falls and West Point.

There is a lighted buoy (#25) marking a rocky ledge one-eighth of a mile downstream from Con Hook. A lighted beacon on the island itself is not visible from midstream when upbound until a vessel has passed buoy 25. *Ocean Prince*'s mate, seeing the shore lights ahead over the reeds, presumed the lighted buoy was to be left on his starboard; he ran *Ocean Prince* and tow hard aground on the rocky ledge. Considerable damage resulted. Because of several different claimants, the case caused much legal attention. When the tug master, even though he was off watch and asleep, was found guilty of negligence in navigation, a precedent was established. Since then, the *Ocean Prince* case has often been referred to in litigating other cases. The point is that the master was responsible and his license was suspended.

In another instance, which did not go to trial, a fully inspected tug of over 400 gross tons departed from an oil terminal in Port Arthur, Texas. She was towing a deeply loaded oil barge on a short hawser that was connected to a single tow pendant running from the tug's bullnose chock. In such a mode, the barge sheered considerably.

The day before the tow's departure, the tug's regular master had been called home due to an emergency. The towing company had hurriedly hired a man they felt was competent and licensed, as he had served on several larger tugs. These tugs were, however, under 200 gross tons.

He was now at the conn of this 400-gross-ton tug with the radically sheering barge. As the tow approached School House Reach, the barge took off to starboard, striking several barges in the head tier of an eastbound tow. The master of its towboat called the outbound tug. The relief captain acknowledged the casualty but kept on going. After a cursory inspection in Sabine Anchorage, he put to sea with his barge on the hawser. Meanwhile, the owners of the damaged barges reported this pseudo hit-and-run collision to the Coast Guard. The next day, the regular master of the tug was contacted at his home on the East Coast and accused of leaving the scene of a collision and other charges, including failure to file Form 2692.

When the correct details were revealed, it was found that the relief captain had not had his name endorsed on the vessel's enrollment and license because he did not hold a master's license of the grade required to cover the tug. Eventually, the regular master was exonerated, and the relief captain's license was suspended.

PILOTS AND THE TUG AND TOW

In such areas as the Sounds of Denmark, the Baltic, Norway's Vestfjorden, the Kiel Canal, the Hellespont and the Bosphorus, the Suez, the Panama Canal, and the St. Lawrence Seaway, pilotage is compulsory. As in many harbors of the world, this pilotage is regulated by that country's pilotage authority and extends to all vessels, including tugs and tows. In the United States, there are five classes of pilots. A requirement for both the master and mates on any U.S. tug of over 400 gross tons, or any other inspected vessel, is an endorsement as a first-class pilot for certain inland waters. In certain states regardless of whether the captain holds an endorsement as a pilot for any of the local waters, he may be legally forced to take a bar and river pilot, if he has a tow.

If the tug is returning with or without a tow from a foreign voyage upon which the tug has been placed in registry, he will be required to take an authorized state bar pilot just as would be required for any foreign ship or U.S. vessel under registry.

The authorized state pilot is one licensed by the state for service upon the waters he serves. He may, and usually does, hold a U.S. Coast Guard license as master with an endorsement as first-class pilot for the waters his pilot association covers. Pilot groups such as the Pilots Association for the Bay and River Delaware, San Francisco Bar Pilots, or Combined New York-New Jersey Pilots, and others are all regulated by their state pilot commission. Other than in those states where pilotage is mandatory for tugs with tows, these state or bar pilots are seldom called upon to assist the tug master unless the tug is over 400 gross tons and the master does not have a pilotage endorsement.

Bar pilots board tows at the sea buoy in some ports and inside jetties in others. If the distance to the tow's destination is great, two pilots may board so that they may relieve each other. This is common for tows bound upriver at Pilottown on the Mississippi, a few miles above Head of Passes, where the bar pilots debark and the Crescent River pilots take over. If the tow is bound above Avondale Shipyard, a NOBRA or New Orleans Baton Rouge pilot will take over. On other bars such as the Columbia River, pilots also change at a convenient inside station. These river pilots who take over are an extension of the state's pilotage system.

At some major U.S. ports, there exists in addition to these state bar pilots an independent pilot group known as Interport Pilots. These are experienced masters who serve under the authority of their federal Coast Guard license and pilotage endorsements. Normally, they handle only U.S. ships in the coastwise trade which are under enrollment. They cannot handle vessels under registry. They are seldom called upon to handle tugs and tows. This is due to an amendment #157.30-40, effective January 7, 1972, which states, "A person holding a license as master or mate is a licensed pilot of a vessel of not more than 1,000 gross tons within the restriction of his license, on which he is employed as master or mate." As most U.S. tugs are under 1,000 gross tons, no additional pilot is normally required unless the master feels the need for assistance.

The remaining group of pilots is known as docking pilots. For years in such ports as New York, Philadelphia, Baltimore, Hampton Roads, and San Francisco, tug captains boarded ships that were making use of their propelling power. These docking pilots took charge using a tug or tugs of the company that employed them. It has always been understood that when the tug captain boarded the ship to take over the docking operation, that he became an independent contractor and servant of the vessel and her owners. If any damage was done, neither the tugs nor their owners or agents would be liable.

Most docking pilots were and still are excellent shiphandlers. Occasionally, damage occurs. The aggrieved shipowner normally looks to the pilot and the towing company from whose tugs he came for relief from the damage. If the towing company has a contract to provide tug assistance and that contract was signed by the steamship company and included a pilotage clause, the tug company was clear. Obviously, the individual pilot does not and could not afford the necessary protective insurance.

This pilotage clause, as an "exculpatory clause," includes provisions that clear the tug company from an alleged fault or guilt. All towing bills repeat this pilotage clause in various forms on their reverse side. The important thing is that the shipping company or its authorized agents have signed this pilotage clause prior to towage and docking pilot service. In some cases, individual ships have arrived and no such agreement had been signed. The docking pilot often submitted to the captain a pilot ticket at the completion of service. The pilotage clause was spelled out on the ticket's reverse. If any damage had occurred, litigation often took place and frequently did not favor the towing company. In 1967, in order to eliminate loopholes, attorneys on both U.S. coasts created two separate pilotage clauses, one for New York and the other for San Francisco.

Meanwhile, in New York, Philadelphia, Baltimore, and Hampton Roads docking pilots who board from tugs continue to be used. In the balance of East Coast ports south of Cape Hatteras, the bar pilots do the docking as they do in Boca Grande, Tampa, and around the Gulf and Mississipi River wharves and terminals.

San Francisco's docking pilots, many of whom were employed on Red Stack tugs, were terminated by that company in the late 1970s.

With other docking pilots they formed the California Inland Pilots Association. They took over from the bar pilots at various spots in the bays and the Carquinez Strait. In 1984, Governor Deukmejian allowed Assembly Bill 1768 to pass. This disbanded the California Inland Pilots and allowed them to become San Francisco Bar Pilots. Now all of these bar pilots must dock ships with or without tugs at the Bay Area's various terminals.

The so-called riding pilot, when towing a dead ship, barge, or other unit, may be covered by an exculpatory pilotage clause if one is signed prior to service. Often, he is an employee of the towing company.

Any tug captain or mate who is suddenly called upon to furnish his services in docking or undocking a vessel with or without its own power should present some form of exculpatory clause for signature by the vessel's authorized representative prior to commencement of service.

In direct connection with any type of pilot and the tug and tow, there are certain things the tug master must understand. Whatever happens, he as the tug captain is responsible. The pilot is a servant accepted to assist and not to take over the handling of the tow. While many bar and some river pilots have never served as a tug captain, their guidance should be accepted. The steering and running of the tug should not be turned over to a pilot. The captain or a competent watch officer should steer and handle the tow under the pilot's direction.

Any and all deficiencies or unusual characteristics of the tug or tow should be pointed out to the pilot immediately upon his boarding. Any suggestion by the pilot regarding extreme current, the ability to negotiate any section, the need for assistance, the need to anchor and await better visibility or tidal conditions, or other matters that might jeopardize the safe negotiation of the tug and tow should be adhered to. If the tug captain should proceed against the advice of the pilot and run into difficulties, the chain of blame will hang around the tug captain's neck.

The only possible exception to proceeding against a pilot's advice might be that the same tug captain with the same tow under the same conditions had successfully negotiated the waterway when assisted by a different pilot. This has happened.

At no time should a pilot be left alone in the pilothouse except for a few moments. Modern electronic technology does fail on occasions. A pilot cannot be expected to know how to change from autopilot to manual steering or to cope with any other wheelhouse failure. During docking or undocking operations, the tug captain should be wary of any pilot who uses his hand-held VHF in such a manner that the orders he gives to assisting tugs cannot be heard by the tug captain. The result may be two minds operating in different directions with dire results.

Nearly all bar pilots in the U.S. belong to the American Pilots Association headquartered in Washington, D.C. Since 1937, the American

Pilots Association has tried to maintain a frank and cordial relationship with the U.S. Coast Guard. It has been amended several times. The clinker in it was that most state bar pilots must have a U.S. Coast Guard license with pilotage endorsement in addition to their state pilot's license. Whenever the Coast Guard after an investigation of a casualty felt that the pilot was at fault, a report and recommendation for action was sent to the APA. Sometimes, this did not result in what the Coast Guard would have liked. More recently, the Coast Guard has suspended the offender's Coast Guard license. Any suspension automatically removes the basic qualification to operate as a state pilot.

All pilots are subject to a civil suit for damages to property or from pollution their action may have caused. Pilots frequently are unsuspecting offenders. Their worst offense occurs when tugs are working on a line from the ship's stern; it is the thrashing of the ship's propeller in a manner that throws a strong current against the tug. It is easy to visualize the force of the tug already listing as she pulls from a high point on the ship's stern; add to this the opposing force of a jet stream of water hitting low on the tug's hull. Unless quick action is taken, girding is inevitable. When this is about to happen while the tug is towing from the hook, a deckhand should give a quick yank or whack on the release mechanism to relieve the tug. It also relieves the pilot and the ship of the tug's service.

On a tug with a line from a ship to a slip line at her after bitts, quick release depends on the ability of a seaman to cut it with an axe. At no time should the eye of a towline from a ship be dropped over a bitt when towing from the tug's stern. Due to just such an error, the tug *Mr. Gus* sank in an anchorage when a pilot was moving a ship to a berth. The tug got in irons and the ship's lines could not be cut soon enough.

So-called telegraph-happy pilots have existed in various ports from Port Arthur, Texas, to Piraeus, Greece, where a pilot was blamed for sinking a harbor tug as he excitedly increased speed. This has been a common occurrence that tug captains have had to face. Another rather strange pilot-related type of incident occurs while docking a ship when after receiving orders to work half speed head-on, the tug master discovers that all lines are out and the pilot has gone ashore without dismissing the tug.

U.S. COAST GUARD'S PROPOSED RULE MAKING

Many of these proposals would affect tugs and towing. One of the first under consideration concerns pilots aboard tugs towing tank barges of up to 10,000 gross tons. Pilotage services would be dispensed with if the person on watch possessed a pilotage endorsement for those waters. Another suggested change would raise the tonnage for avoiding outside pilotage services to 20,000 gross tons. APA does not approve of either of these proposals.

Other proposed rule changes would raise the minimum age for a pilotage endorsement to 21 and also require pilots to undergo an annual physical examination. Applicants for renewal of a license with pilotage would also be required to have made a minimum of one trip over the route within the last five years.

Tug watchers may soon have difficulty establishing a tug's home port. The hailing port will no longer be required on the tug's stern. As on all U.S. vessels, it may be replaced by the letters "U.S.," followed by the tug's official number.

There is a massive list of proposed changes in C.G.D.-81-059 entitled "Licensing of Marine Personnel." A few of the suggested changes will affect towing personnel. One of the first is a new dividing boundary for inland and "near coastal licenses." Another involves a complete description of requirements for license as "operator of Uninspected Towing Vessels." Additional requirements state that on every uninspected vessel "it is the master's obligation to ensure that the appropriate personnel are carrried to comply with the law and regulations"; that individuals licensed to operate an uninspected vessel of less than 200 gross tons may not work more than 12 hours "except in a consecutive 24-hour period except in an emergency"; and that these licensed operators may be divided into two watches regardless of the length of the voyage. Why safety of the vessel, watch-standing, and maintaining a proper lookout is any less arduous on a two-watch tug of 199.9 gross tons than it would be on board an uninspected tug of 201 gross tons is difficult to understand.

USE OF FOREIGN-FLAG TUGS IN U.S. HARBORS

The U.S. House of Representatives introduced H.R. 2466, which if passed by both houses would have closed the loophole that could allow foreign-flag tugs to assist foreign-flag vessels in U.S. ports. President Reagan vetoed the bill.

IMO CLARIFICATIONS OF COLREGS

Vessel constrained by her draft is clarified as follows:

Rule 3(h). "Not only the depth of the water but also the navigable width should be used as a factor—due account should be taken of the effect of small underkeel clearance on maneuverability of the vessel and her ability to deviate from the course she is following. However, if there is sufficient area in which to take avoiding action, the vessel should not be considered as constricted by her draft."

Rule 3(i)-underway. "A vessel may be underway but stopped and making no way through the water."

Rule 9(b)(c)(d)-10(i)(j) and 18(d)-implication of the words "not to impede." "When a vessel is required not to impede another vessel's passage she shall if practical navigate to avoid risk of collision. If such risk exists the relevant steering and sailing rules shall be complied with."

Rule 10(b)(i). "A vessel using a traffic lane may transfer from one side to another at as small an angle to the general direction of traffic flow as is practicable."

Rule 10(d). "Use of inshore traffic zones by tugs of 65 feet—20 meters, or less is permissible for reasons of safe navigation and to comply with Rule 10(j)."

POLLUTION AND THE TUG

In U.S. waters, tug bilges should be kept clean and as dry as possible. Those that are not must be pumped into a slop barge or shore disposal installation. Similar rules apply in almost every other country. Any oil discharge in U.S. waters, which is traced to a vessel, can result in a fine of $10,000 and suspension of the master's license.

VHF RADIO AND THE TUG

Several questions have arisen about the use of the VHF radio. Most of the procedures on use of VHF radio are covered by international rule. On U.S. vessels each installation requires an FCC station license. Each person who uses the VHF must have a restricted radiotelephone operator's permit (RP). This RP is good for a lifetime. The applicant must be 14 years of age. To apply, use FCC Form 753, which may be obtained from the Federal Communications Commission, Box 1030, Gettysburg, Pennsylvania 17325.

USE OF VHF BRIDGE-TO-BRIDGE TO AVOID COLLISIONS

The International Chamber of Shipping (ICS), which represents shipowners' associations in 30 countries responsible for two-thirds of the world's merchant shipping, has shown legitimate concern over certain uses of the VHF to avoid collision. As this concern would apply to tugs and tows, here are some of the reasons ICS has given to IMO and the U.S. National Transportation Board. They refer primarily to the use of the VHF bridge-to-bridge conversations in open-sea areas and the approaches to ports and pilot stations.

1. Difficulty in establishing identification of the other vessel, particularly if more than one is present.
2. Language problems.

3. Misunderstanding of intentions.
4. Distraction of the bridge officer from his regular duties when he is trying to establish bridge-to-bridge contact.

This danger is unusual on U.S. or other inland waters where pilots, masters, and tug captains speak the same language. Even here, though, we are confronted with local accents. One of the first questions the U.S. Coast Guard investigator will ask after a collision is, "Was the VHF used to avoid this casualty?"

However, in coastal and other open areas such as the English Channel, Straits of Gibraltar, and so on, there is much evidence that collisions have occurred many minutes after radar detection was established, because the bridge officer was making a fruitless effort to establish VHF communication.

PART IV
Appendices

List of Appendices

I. Tugs and Barges: Dimensions and Classifications   505

    Largest Ocean Tugs, 505
    Largest Seagoing Rudderpropeller Tugs, 505
    Recommended Draft for Ocean Towage by
        Lloyd's Register, 506
    Classification of Canadian Tugs, 506
    Barge Dimensions, 506
    Ice Classification of Tugs, 507

II. Towing Equipment   508

    Tug Power Versus Towing Wire Size, 508
    Towing Winch Data, 509
    Towing Winch Drums—Calculating Capacity, 510
    Bollard Pull Certificates and Testing, 511
    Tensile Stength of 6×37 Plow Steel Towing Hawsers, 512
    Breaking Strength for Wire Rope, 512
    Shackles, 513
    Cordage, 513

III. Whistle Signals, Lights, and Buoyage   520

    Whistle Signals, Great Lakes, 520
    Whistle Signals Between Pilot and Tug, 520
    Modification of U.S. Buoys, Beacons, and Day Markers,
        to Conform to IALA System, 521
    Buoyage on the New York State Barge Canal System, 522

IV. VHF/VTS Frequencies   523

    Weather Broadcasts for Tugs, 523
    VHF Channel Assignments in the United States, 524
    VHF and VTS Channels Frequently Used by Tugs in
        North America, 524
    Location of Marine Radiotelephone Terminals, 528
    San Francisco Offshore Vessel Movement Reporting
        System, 530

LIST OF APPENDICES

V. Wind and Ice 531
Wind Velocities for Ocean Towing, 531
Wind Speed, Air Temperature, and Icing Conditions, 533

VI. Canals, Locks, and Narrows: Regulations 537
Dimensions and Limitations of Major Locks
  and Canals, 537
Navigation and Tug Assistance,
  Valdez Narrows, Alaska, 537
Closed Chock and Line Requirements
  for St. Lawrence Seaway and the Panama Canal, 538
Dimensions and Restrictions
  for the St. Lawrence Seaway, 539
Mooring Table for St. Lawrence Seaway
  and Welland Canal Lock Walls, 543
Dimensions and Restrictions
  for Other Canals and Locks, 544

VII. Towing Arrangements and Procedures 548
Towing Arrangements Required
  of Ships on the Great Lakes, 548

VIII. Tugs and Fishing Areas 551
Tugboat/Fishing Lane Negotiations, 551
Gill Net Fishing—Coast Guard Notice, 552
Navigating Through a Gillnet Fleet, 554

IX. Mississippi River Data 557
Towing Procedures and Traffic Lights, 557
Ferry Crossings and Hawser Length, 559
Gauge Boards and Relation to Bridges, 559
Vessel Traffic Service (VTS) Sectors, 562
VTS Reporting Points, 562

X. Harbors, Terminals, and Anchorages 564
New York, 564
San Francisco, 565
Oakland, 566
Redwood City, 567
Richmond, 567
San Pablo Bay, 568
Sacramento, 569
Stockton, 569

XI. Health and Diet for Tug Crews   570
        Substitutions for Excess Calories, 570
        Words Vital to Your Health, 570
        Desirable Weight Ranges for Adults, 572
        Cooking Hints, 572
        Guidelines for Low-Cholesterol,
            Low-Triglyceride Diets, 573

XII. General Information   575
        International VHF Voice Pronunciation and Spelling, 575
        Conversion Tables, 576
        Public Maritime Schools in the United States, 579

Appendix I. Tugs and Barges: Dimensions and Classifications

LARGEST OCEAN TUGS

John Ross: Safmarine, 26,000 HP, twin Mirliss controllable-pitch propellers, 20 knots running free

Wolraad Waltenmade: Safmarine, 26,000 HP, twin Mirliss controllable-pitch propellers, 20 knots running free

Abeille Flandre: les abeilles international, 23,000 HP, 1,600 tons, 208′ × 48′ × 21′

Abeille Languedoc: les abeilles international, 23,000 HP, (four MAK 8M coupled), 17 knots with thrusters

Smit Singapore: Smit Tak International, 22,000 HP, 246′ × 49.2′ × 23′, twin Stork-Werkspoor with four-blade variable-pitch propellers

Arctic and *Oceanic:* Bugsier, 2,087 tons, 20,000 HP, 287′ × 47′ × 24′, two Deutz thrusters, 170 tons of bollard pull

Amsterdam: Bureau Wijsmuller, 7,500 HP, two MAK 9M 452AK

Neptun and *Poseidon:* Neptun Salvage, 7,000 HP, two Nohab, 147.9′ × 36′ × 20′, 490 tons, thrusters

Ulstein 714: Ulstein Group, 4,000 to 8,000 HP, 192′ × 42.6′ × 21′, controllable-pitch propellers in nozzle, 70 tons of bollard pull

Irving Miami: Irving Oil, Saint John, N.B., 7,000 HP

Seaspan Commodore: Seaspan, Vancouver, B.C., 5,750 HP, 132.6′ × 37.2′ × 20.3′, two 16-cylinder GMs

Captain Ioannis S.: Quebec Tugs, 5,600 HP, 136.8′ × 38.5′, 722 tons, 73 metric tons of bollard pull

LARGEST SEAGOING RUDDERPROPELLER TUGS

Bugsier #8: Bugsier, 7,200 HP twin Klochner-Humboldt Deutz, Kort-nozzled, 41 tons of bollard pull

Bugsier #9: Bugsier, range 5,000 miles at 12 knots, 5″ circumference tow wire, rigged for salvage

RECOMMENDED DRAFT FOR OCEAN TOWAGE BY LLOYD'S REGISTER

Length of barge	Recommended draft Forward	Aft
100'	3'	4'
200'	6'	7'9"
300'	8'	10'6"
400'	10'	13'

CLASSIFICATION OF CANADIAN TUGS

Transport Canada classifies Canadian tugs as follows:
    Oceangoing Tug—Approximately 5,000+ HP, 140' × 32' × 19', 650 gross tons
    Coastal-Home Trade Class I—3,600-3,900 HP, 112' × 30' × 19'
    Coastal-Home Trade Class II—1,800 HP, 82' × 24' × 13', 150 gross tons
    Harbor—Up to 1,400 HP at various lengths and tonnages

BARGE DIMENSIONS

The size of ocean barges varies with the area and type of usage. Petroleum and dry-bulk cargo barges used on the United States East and Gulf coasts are considerably larger than similar ones on the West Coast. Here are a few statistics of recently built barges.

East and Gulf Coasts

Name	In feet: Length	Beam	Loaded draft	Tonnage	Cargo
Thoroughbred Topper	550	78	34		Coal: 36,000 st
Ocean 70	350	70	25	5,312 gt	Oil: 73,000 bbl
Ocean 90	400	66	27	6,409	Oil: 96,491 bbl
Ocean 135	435	74	29.5	16,200 dwt	Oil: 135,000 bbl
Ocean 250/255	546	85	40	14,678 gt	Oil: 250,000 bbl
I.O.S. 3301/3302	583.5	87	46		Oil: 285,000 bbl
Union Carbide	500	68	33.5		Chemicals
Corpus Christi Sun	405	72	29.9		Oil: 115,000 bbl
Saint John Carrier	362	82	22	10,826 gt	Newsprint: 8,000 st
La Reina	730	105			Trailers
Marie Flood					Coal: 38,000 t

Pacific Coast and Ocean

Name	In feet: Length	Beam	Loaded draft	Tonnage	Cargo
City of Seward	487	104			Trailers: 330-340 TEU
500-3 (Alaska Hydro Train)	400	100			Railcars
Sause Bros	230	55	12		Cargo: 2,600 st
Seaspan 310	200	50	12		Chip barge: 2,500 st
Seaspan 250					Open deck: 13,000 st
450-10 (Crowley)	450	100			Cargo
Dillingham-Inter-Island-Hawaii	280	60			Trailer and cargo
Columbia-Snake River	207-273				

    In order to interchange routes and ports, many West Coast barges are beamier and have less draft than those built for the East and Gulf coast trade. Because of their shallower draft their speed under tow is greater than that of deeper loaded barges.

ICE CLASSIFICATION OF TUGS

The American Bureau of Shipping (ABS) issues ice classification for icebreakers, tugs, and other vessels as follows:

Class 1AA—good for extreme ice conditions
Class 1A—good for severe ice conditions
Class B—good for medium ice conditions
Class C—good for light ice conditions

Transport Canada's classification for icebreaking tugs and vessels serving in ice runs from:

Class 1—good for light ice
through
Class 8—good for heavy arctic ice as in their northwest passages

Icing in Alaskan Waters. Vessel icing is one of the most serious marine meteorological hazards in high-latitude waters. Icing requires the presence of subfreezing air temperatures, strong winds, and cool sea temperatures. Actual icing potential is a characteristic of each vessel, depending on its design and seakeeping ability. Differences in vessel type, combined with difficulties in making observations during operations, result in great variability in vessel icing observations for similar meteorological conditions.

Observations consisted primarily of fishing vessels, fish processors, towboats, and Coast Guard vessels that operate in Alaskan waters. Most vessels ranged in length from 65.6 feet to 246.1 feet. For towboats, the operating logs of individual vessels were consulted. Towboats were a particularly good source of data because their schedules were not affected by the fishing season or bad weather.*

---

*C. H. Pease, Pacific Marine Environmental Laboratory, 7600 Sand Point Way N.E., Seattle, WA 98115-0070. Also A. L. Comiskey, Northern Technical Services, 750 W. Second Avenue, Anchorage, AK 99501.

Appendix II. Towing Equipment

TUG POWER VERSUS TOWING WIRE SIZE

(Based upon past selections and recommendations, and subject to the individual vessel's configuration and service requirements . . .)

Main engine nominal power (horsepower)	Wire diameter (inches)	Main engine nominal power (horsepower)	Wire diameter (inches)
1,000	1	6,000	$2^{1}/_{4}$
1,500	$1^{1}/_{4}$	6,500	$2^{1}/_{4}$
2,000	$1^{1}/_{2}$	7,000	$2^{1}/_{4}$
2,500	$1^{5}/_{8}$	7,500	$2^{1}/_{4}$
3,000	$1^{3}/_{4}$	8,000	$2^{1}/_{4}$
3,500	$1^{7}/_{8}$	8,500	$2^{1}/_{2}$
4,000	2	9,000	$2^{1}/_{2}$
4,500	2	10,000	$2^{1}/_{2}$
5,000	$2^{1}/_{8}$	12,500	$2^{3}/_{4}$
5,500	$2^{1}/_{8}$	15,000	3

Relationship between 12-inch Nylon Pendant and 6 × 37 Wire Rope Regarding Breakage. The safe working load of 11 percent of the tensile strength (practice safety of 9) is a general figure that will take care of all uses. There are some exceptions to this rule, particularly where actual strain is known and where the nylon is entirely outboard so that there will be no chafing or sharp bending. In cases like this a safety factor of 5 or 6 can be used with confidence.

Wire rope manufacturers recommend a practice safety of 5; this would correspond favorably using 12-inch circumference nylon with a 2-inch diameter 6 × 37 wire rope.

The elasticity of stretch and recovery in the nylon pendant in a particular application will occur throughout its entire length; however, the 200-foot length will absorb about 810,000 foot-pounds of energy whereas the 150-foot length will absorb about 610,000 foot-pounds at 67,000 pounds tension. In other words, the longer the length, the more energy will be absorbed.

Splicing does weaken the rope somewhat; however, this is figured in the factor of safety. In laboratory testing of rope, nearly all breaks occur in the splice; this is, of course, under controlled ideal conditions. Cuts, abrasions, or bending a sharp corner will usually have a greater effect on the rope strength.

Thimbles should be used wherever possible and the bending radius of rope should not be less than four times the diameter of the rope.

Additional Information for Nylon Astern Towlines.* Nylon lines "work in" fairly quickly when subject to wetting and loading, but until then the line should be handled so as to avoid kinks and backturns. Damage from these occur when the lay of the rope is altered appreciably by permitting turn to be thrown into or out of the line in localized places. Normally, good handling methods (described below) will prevent this.

> When first removing the rope from the reel, set the reel up to revolve at least 50 feet (farther if possible) from where the line is to be coiled (clockwise) or, better still, faked down. Keep the line straight and taut as possible from reel to where it is coiled or faked down. Do not let the line rub against sharp edges or abrasive surfaces. If it cannot be faked down, use a long oval flemish coil. For every turn of the coil, a turn will be added to the line back to the reel. Be sure the distance to the reel is long enough to absorb these without kinking. If a kink forms, straighten it by unwinding it. Do not pull it straight, or let it bend sharply. Try to get more distance from reel to coil.
>
> When first taking line from deck coil (counterclockwise) to put to work, lay out so that no kinks or backturns form. Try to keep the line straight until it is wet and load applied gradually. If load must be applied before the line is thoroughly wet, do so gradually and release slowly, if at all possible. Take in and recoil with the same precautions as above.
>
> When removing line from bitts or winches, particularly before it is stable, try always to run it off the winch. If at all possible, do not pull the turns off over the end of the bitt or winch. This concentrates turn in a short distance and can so alter the lay as to lead to backturns. If a backturn or kink tends to form, rotate rope on its own axis so as to restore the lay approximately to where it was when received. It is useful to measure the distance of 20 to 30 crowns when rope is received, and make note of it for this purpose, and also to check the stretch when heavily loading.
>
> To help prevent buildup of turn and prevent backturning, reverse line end for end as conditions indicate.

In some areas, it is necessary to tow "shortened up" a good deal of the time (more than 25 percent of total towing time) in exposed waters or under normal towing loads. At these times, only 250 to 400 feet of line may be out. This is not only working the lines harder because there is less line to absorb the surging, but it works the rope structure more severely and can result in backturns which permanently damage the line. The only answer to this is a larger line which, in addition to withstanding this hard working, will provide longer wear life. For these conditions, the circumference of the line should be approximately 20 percent larger than normally calculated for 1,200-foot line. If the line is used shorter than this for appreciable periods (more than 25 percent of total towing time) under normal towing conditions and loads, it should be larger still.

TOWING WINCH DATA

The following, used as an example, is based on a Markey electric hydraulic towing winch with a drum with a 30-inch core. This size drum will take 2,200 feet of $2^{1}/_{4}$-inch, 6×37 improved plow steel Warrington hemp core cable (such a

---

*Prepared by Plymouth Cordage Company

cable should meet ABS "A Maltese Cross B" approval). The 30-inch winch drum will have received when full $10^{1}/_{2}$ layers of cable at $17^{1}/_{2}$ complete turns per layer. A new wire when led onto the drum should be tight with no space between turns.

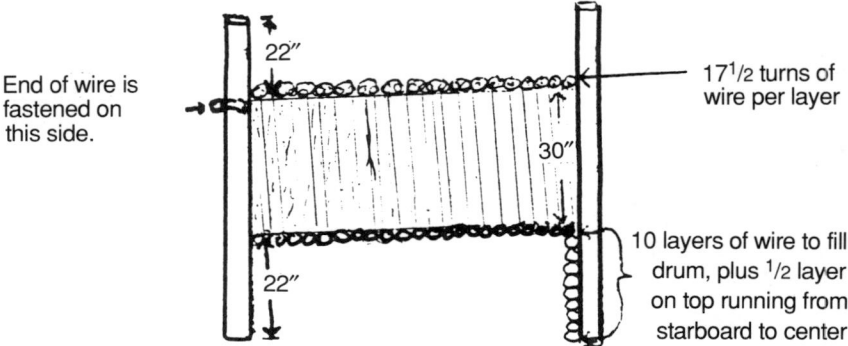

End of wire is fastened on this side.

$17^{1}/_{2}$ turns of wire per layer

10 layers of wire to fill drum, plus $1/2$ layer on top running from starboard to center

Approximate Length of $2^{1}/_{4}$-Inch Cable Out by Layer

Number of layers On drum	Out	Length of single layer	Direction wire is running toward	Amount of hawser out in relation to position on drum at top (in feet)		
				Center	Extreme port	Extreme starboard
$10^{1}/_{2}$	0	18	port	6	171	—
$9^{1}/_{2}$	1	17	starboard	285	438	319
$8^{1}/_{2}$	2	16	port	582	734	726
$7^{1}/_{2}$	3	15	starboard	861	996	741
$6^{1}/_{2}$	4	14	port	1,122	1,010	1,248
$5^{1}/_{2}$	5	13	starboard	1,365	1,482	1,248
$4^{1}/_{2}$	6	12	port	1,590	1,494	1,698
$3^{1}/_{2}$	7	11	starboard	1,797	1,896	1,709
$2^{1}/_{2}$	8	10	port	1,986	1,906	2,076
$1^{1}/_{2}$	9	9	starboard	2,157	2,258	2,085
1	10	8	port			
$1/2$	10	8	port	2,330—Stop		

Six feet of cable runs from top layer at center to outside of threader.

TOWING WINCH DRUMS—CALCULATING CAPACITY

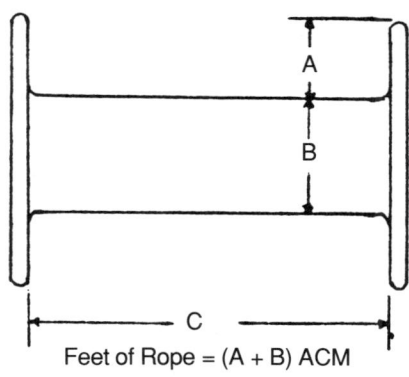

Feet of Rope = (A + B) ACM

Example: How much ½-inch wire rope can be spooled on a drum having a 36-inch diameter flange, 24-inch diameter drum, and 30-inch width between flanges?
Answer: $6 + 24 \times 6 \times 30 \times 1.05$

Values for M

Wire rope diameter (inches)	M	Wire rope diameter (inches)	M	Wire rope diameter (inches)	M
1/4	4.19	7/8	.342	1 3/4	.085
3/8	1.86	1	.262	1 7/8	.074
7/16	1.37	1 1/8	.207	2	.065
1/2	1.05	1 1/4	.168	2 1/8	.058
9/16	.827	1 3/8	.138	2 1/4	.052
5/8	.670	1 1/2	.116	2 3/8	.046
3/4	.465	1 5/8	.099	2 1/2	.042

BOLLARD PULL CERTIFICATES AND TESTING

Bollard pull certificates were first issued by ABS. They are also issued by Lloyd's and may be by other classification societies. The following is provided as a guide in testing for bollard pull. It was chosen without prejudice toward any classification society.

Det norske Veritas Recommendations for Bollard Pull Testing

1. A proposed test programme together with the relevant drawings and specifications for the vessel shall be submitted to Det norske Veritas prior to the testing.
2. The main engine(s) shall be run at the manufacturer's recommended maximum continuous output.
3. The propeller(s) fitted when performing the test shall be the propeller(s) used when the vessel is in normal operation.
4. All auxiliary equipment such as pumps, generators and other equipment which are driven from the main engine(s) or propeller shaft(s) in normal operation of the vessel shall be connected during the test.
5. The length of the tow wire shall not be less than 300 m, measured between the stern of the vessel and the shoreline.
6. The water depth at the test location shall not be less than 20 m within a a radius of 100 m of the vessel.
7. The test shall be carried out with the vessel's displacement corresponding to full ballast and full fuel capacity.
8. The vessel shall be trimmed at even keel or at a trim by the stern not exceeding 1% of the vessel's length.
9. The vessel shall be able to maintain a fixed course for not less than five minutes while pulling as specified in items (2) and (5) above.
10. The test shall be performed with a wind speed not exceeding 5 m/sec.
11. The current at the test location shall not exceed one knot in any direction.
12. The strain cell used for the test shall be approved by Det norske Veritas and be calibrated at least once a year. The accuracy of the strain cell shall be ±2% within a temperature range of −10 °C and +40 °C.
13. An instrument giving a continuous read-out and also a recording instrument recording the bollard pull graphically as a function of the time shall both be connected to the strain cell. The

instruments shall be placed and monitored ashore.

14. The strain cell shall be fitted between the eye of the towing wire and the bollard.

15. The figure certified as the vessel's continuous bollard pull shall be the towing force recorded as being maintained without any tendency to decline for a duration of not less than five minutes.

16. Certification of bollard pull figures recorded when running the engine(s) at overload, reduced RPM or with a reduced number of main engines or propellers operating can be given and noted on the certificate.

17. A communication system shall be established between the vessel and the person(s) monitoring the strain cell and the recording instruments ashore, by means of VHF or telephone connection, for the duration of the test.

18. Should it not be possible to comply with one or more of these recommendations a notation of this fact shall be made on the certificate.

TENSILE STRENGTH* OF 6 × 37 PLOW STEEL TOWING HAWSERS

Size (inches)	Weight per foot (pounds)	Improved plow steel (pounds)	Standard plow† (pounds)	
3/4	.87	21,000	18,200	
7/8	1.2	28,400	24,700	
1	1.5	36,900	32,100	
1 1/8	2.0	46,500	46,400	
1 1/4	2.4	57,100	49,700	
1 1/2	3.5	81,500	70,900	
1 5/8	4.1	96,300	82,900	Swede wire 70,000
1 3/4	4.7	110,000	96,700	
1 7/8	5.5	126,000	109,000	
2	6.2	143,000	124,000	
2 1/4	7.8	179,000	156,000	

*In the above table tensile strength refers to the safe, steady, working load.
†Some towing companies have started to return to galvanized towing tables. They feel that the price, although higher, is compensated for by the extended life of the cable. Others turn the cable end for end after 18 months. On tugs where the work requires only half or less of the tow cable to be out, its use will not be extended.

BREAKING STRENGTH TABLE FOR WIRE ROPE

Typical 6 × 37 Fiber Core Wire Rope

Wire diameter (inches)	Breaking strength (pounds)	Wire diameter (inches)	Breaking strength (pounds)
1	79,600	1 7/8	272,000
1 1/8	100,200	2	308,000
1 1/4	123,000	2 1/8	346,000
1 3/8	148,200	2 1/4	386,000
1 1/2	175,800	2 1/2	472,000
1 5/8	206,000	2 3/4	568,000
1 3/4	238,000	3	670,000

Note: This table refers to the pounds at which the cable should part or break. (Courtesy U.S. Steel.)

SHACKLES

Shackle failure has often been the cause of tows going adrift. A shackle may lose a pin if it is not properly secured, or it may break. It is wise to use the next size larger than required and, to make sure the nut will not come off, insert a stainless steel cotter key through a hole drilled in the pin just outside the nut.

6053

Size Inches	Approximate Weight Each	DIMENSIONS Inches Length in Clear	Jaw Opening	Pin Diam.
1½	25	10	4½	1⅝
1¾	45	12	5	2
2	60	12	6	2¼
2½	90	14½	7	2¾
3	170	16	8	3¼

- ▲ With toggle pin and hand grip
- ▲ Can be supplied in any size and with screw pin if required

Towing Shackles

- ▲ With screw pins
- ▲ May be furnished in other sizes

Size Inches	Approximate Weight Each	DIMENSIONS Inches Length in Clear	Jaw Opening	Pin Diam.
1⅜	12	8	4	1⅝
1½	15	9	4½	1¾
1¾	22	10	4½	2
2	35	10	5	2¼

6054

DATA FOR SHACKLES

Size Inches	Approximate Weight Each	Approx. Safe Working Load Tons	DIMENSIONS Inches Length in Clear Chain	Anchor	Jaw Opening	Pin Diameter
¼	2 oz.	¼	⅞	1⅛	7/16	5/16
5/16	3	⅜	1	1¼	½	⅜
⅜	4	¾	1¼	1 7/16	11/16	7/16
½	¾ lb.	1¼	1⅝	1⅞	⅞	⅝
⅝	1¼	2	2	2⅜	1⅛	¾
¾	2⅛	3	2¼	2⅝	1¼	⅞
⅞	3	4	2¾	3 3/16	1⅜	1
1	4⅜	5	3⅛	3⅝	1⅝	1⅛
1⅛	6¼	6	3⅜	4	1⅞	1¼
1¼	8¾	8	3¾	4⅝	2	1⅜
1⅜	11¼	10	4¼	5¼	2⅛	1½
1½	14	12	4¾	5½	2⅜	1⅝
1⅝	18	14	5¼	6¼	2⅝	1¾
1¾	22	16	5¾	7	3	2
2	35	20	6¾	7¾	3¼	2¼
2¼	48	25	7	9¼	3¾	2½
2½	60	30	7½	10½	4	2¾
2¾	93	35	8	11½	4½	3
3	120	50	8⅜	13⅛	5	3¼

Anchor

Chain

6051

6052

Bosloc Safety Shackles

Full pin bearing. Used for cargo fall, topping lift and guy (vang) blocks. Available in sizes ⅜″ to 3½″. For data see above. Recommended for safety aloft and on deck.

CORDAGE

Traditionally, rope manufacturers have worked to make cordage for the marine industry lighter, stronger, and more flexible. As a result there are new, small-diameter synthetic ropes with exceedingly high tensile strength equal to stainless steel cable of the same size. KARAT, a DuPont fiber, is one. It is so strong that it is laid flat. If twisted it will cut and break itself. It is very good as a replacement for wire standing rigging. As rope it was developed in Holland, but the manufacturing rights have been sold to Norwegian interests. In the United States, the Tubbs Cordage Company has taken this new extraordinary fiber as a core and encased it with 100 percent plied-yarn- filament polyester covering. The result, as claimed by Tubbs, is lightweight, easy-to-handle line with about half the weight, and 90 percent more strength than a polypropylene line of equal diameter. This new line is sold as KARAT Combo.

Weights and Wet Strengths

	KARAT 3 or 8* Strand			KARAT COMBO 3 Strand		
DIAMETER INCHES	CIRCUMFERENCE INCHES	APPROX. WT./100 FT.**	BREAK STRG. LB.***	APPROX. WT./100 FT.**	BREAK STRG. LB.	DIAMETER INCHES
1½	4½	43.5	52.920	—	—	1½
1⅝	5	53.3	63.505	—	—	1⅝
1¾	5½	64.0	75.630	74.0	75.630	1¾
2	6	75.9	88.860	87.0	88.860	2
2⅛	6½	89.4	103.194	103.0	103.194	2⅛
2¼	7	103.5	118.630	120.0	118.630	2¼
2½	7½	118.8	134.945	138.0	134.945	2½
2⅝	8	135.3	152.365	157.0	152.365	2⅝
2⅞	8½	152.8	170.885	177.0	170.885	2⅞
3	9	171.2	190.510	199.0	190.510	3
3¼	10	211.3	233.730	—	—	3¼
3½	11	255.7	280.035	—	—	3½
4	12	304.4	330.750	—	—	4

* 8-STRAND NOT AVAILABLE IN 1½" AND 1⅝" DIAMETERS
** WEIGHTS MAY VARY BY 5%
*** TENSILE STRENGTHS ARE AVERAGES AND MAY VARY BY 10%. BREAK STRENGTH FIGURES INCLUDE SPLICES. CERTIFIED BREAK TEST NUMBERS AVAILABLE UPON REQUEST. SAFE WORKING LOAD IS APPROXIMATELY 17% OF BREAK STRENGTH.

Samson Braided Rope

U.S. MEASURE		2-in-1 NYLON (Nylon-Nylon)		2-in-1 POWER BRAID (Nylon-Polypropylene)		2-in-1 STABLE BRAID (Polyester-Polyester)		12-STRAND BRAID BLUE STREAK (Polyester-Polypropylene)	
Diameter inches	Circumference inches	Strength pounds	Weight lbs. 100 ft.	Strength pounds	Weight lbs. 100 ft.	Strength pounds	Weight lbs. 100 ft.	Strength pounds	Weight lbs. 100 ft.
1/4"	3/4"	2,200	1.65	2,200*	1.65*	2,000	2.0	2,000	2.0
5/16"	1"	3,400	2.6	3,400*	2.6*	3,000	3.0	2,500	2.5
3/8"	1-1/8"	4,800	3.7	4,300	3.3	4,300	4.4	4,000	4.0
7/16"	1-1/4"	6,500	5.1	5,800	4.5	5,800	6.0	5,000	5.0
1/2"	1-1/2"	8,300	6.6	7,500	5.9	7,500	7.9	7,000	6.7
9/16"	1-3/4"	11,200	9.0	10,000	8.0	10,000	10.7	9,000	9.0
5/8"	2"	14,500	12.0	13,000	10.5	13,000	14.0	12,000	11.0
3/4"	2-1/4"	18,000	15.0	16,000	13.0	16,000	18.0	15,000	15.0
13/16"	2-1/2"	22,000	18.0	20,000	16.0	20,000	22.0	—	—
7/8"	2-3/4"	26,500	22.0	24,000	20.0	24,000	27.0	22,000	23.0
1"	3"	31,300	26.0	28,400	24.0	28,400	32.0	25,000	27.0
1-1/16"	3-1/4"	36,500	31.0	33,200	28.0	33,200	37.0	—	—
1-1/8"	3-1/2"	42,000	36.0	38,000	32.0	38,000	43.0	33,000	36.0
1-1/4"	3-3/4"	47,700	41.0	43,600	37.0	43,600	49.0	38,000	42.0
1-5/16"	4"	54,000	47.0	49,000	42.0	49,000	56.0	43,000	48.0

(continued)

APPENDIX II

U.S. MEASURE		2-in-1 NYLON (Nylon-Nylon)		2-in-1 POWER BRAID (Nylon-Polypropylene)		2-in-1 STABLE BRAID (Polyester-Polyester)		12-STRAND BRAID BLUE STREAK (Polyester-Polypropylene)	
Diameter inches	Circumference inches	Strength pounds	Weight lbs. 100 ft.	Strength pounds	Weight lbs. 100 ft.	Strength pounds	Weight lbs. 100 ft.	Strength pounds	Weight lbs. 100 ft.
1-1/2"	4-1/2"	67,500	60.0	61,400	53.0	61,400	71.0	54,000	60.0
1-5/8"	5"	82,600	74.0	75,000	66.0	75,000	88.0	64,000	72.0
1-3/4"	5-1/2"	99,000	89.0	90,000	80.0	90,000	106.0	80,000	88.0
2"	6"	117,000	106.0	106,000	95.0	106,000	126.0	94,000	108.0
2-1/8"	6-1/2"	136,000	124.0	124,000	111.0	124,000	148.0	108,000	126.0
2-1/4"	7"	156,000	144.0	142,000	129.0	142,000	172.0	124,000	147.0
2-1/2"	7-1/2"	178,000	165.0	162,000	148.0	162,000	197.0	140,000	168.0
2-5/8"	8"	202,000	188.0	183,000	168.0	183,000	224.0	160,000	192.0
2-3/4"	8-1/2"	226,000	212.0	205,000	190.0	205,000	253.0	180,000	216.0
3"	9"	252,000	238.0	229,000	213.0	229,000	284.0	200,000	243.0
3-1/4"	10"	308,000	294.0	280,000	263.0	280,000	350.0	250,000	300.0
3-5/8"	11"	369,000	356.0	336,000	318.0	336,000	424.0	300,000	360.0
4"	12"	436,000	423.0	396,000	379.0	396,000	504.0	350,000	432.0
4-1/4"	13"	504,000	497.0	461,000	444.0	461,000	592.0	400,000	504.0
4-5/8"	14"	586,000	576.0	531,000	515.0	531,000	686.0	450,000	576.0
5"	15"	666,000	662.0	606,000	592.0	606,000	788.0	500,000	648.0
5-1/4"	16"	753,000	753.0	685,000	673.0	685,000	896.0	—	—
5-1/2"	17"	846,000	850.0	768,000	760.0	768,000	1,012.0	—	—
6"	18"	940,000	953.0	857,000	852.0	857,000	1,134.0	—	—
6-1/4"	19"	1,040,000	1,061.0	949,000	949.0	949,000	1,264.0	—	—
6-1/2"	20"	1,150,000	1,176.0	1,050,000	1,052.0	1,050,000	1,400.0	—	—
7"	21"	1,260,000	1,297.0	1,150,000	1,160.0	1,150,000	1,544.0	—	—

Strengths and weights are Approximate Averages   °Nylon Cover — Nylon Core

METRIC MEASURE		2-in-1 NYLON (Nylon-Nylon)		2-in-1 POWER BRAID (Nylon-Polypropylene)		2-in-1 STABLE BRAID (Polyester-Polyester)		12-STRAND BRAID BLUE STREAK (Polyester-Polypropylene)	
Diameter MM	Circumference inches	Strength Kg.	Weight Kg./100M	Strength Kg.	Weight Kg./100M	Strength Kg.	Weight Kg./100M	Strength Kg.	Weight Kg./100M
6	3/4"	1,000	2.45	1,000	2.45°	908	2.98	910	2.69
8	1"	1,544	3.87	1,544	3.87°	1,362	4.47	1,140	3.73
9	1-1/8"	2,179	5.51	1,952	4.91	1,952	6.55	1,820	5.37
10	1-1/4"	2,951	7.59	2,633	6.71	2,633	8.94	2,275	7.46
12	1-1/2"	3,768	9.83	3,405	8.79	3,405	11.8	3,180	9.70
14	1-3/4"	5,085	13.4	4,540	11.9	4,540	15.9	4,090	13.3
16	2"	6,583	17.9	5,902	15.6	5,902	20.9	5,450	17.9
18	2-1/4"	8,172	22.4	7,264	19.4	7,264	26.8	6,820	22.4
20	2-1/2"	9,988	26.8	9,080	23.8	9,080	32.8	—	—
22	2-3/4"	12,031	32.8	10,896	29.8	10,896	40.2	10,000	32.8
24	3"	14,210	38.7	12,894	35.8	12,894	47.7	11,365	38.8
26	3-1/4"	16,571	46.2	15,053	41.8	15,073	55.1	—	—
28	3-1/2"	19,023	53.6	17,252	47.7	17,252	64.1	15,000	52.2
30	3-3/4"	21,656	61.1	19,794	55.1	19,794	73.0	17,275	61.1
32	4"	24,471	70.0	22,246	62.6	22,246	83.4	19,500	68.7
36	4-1/2"	30,645	89.4	27,876	79.0	27,876	105.8	24,500	88.1
40	5"	37,500	110.0	34,050	98.3	34,050	131.0	29,000	107.0
44	5-1/2"	44,901	133.0	40,860	119.0	40,860	158.0	36,400	131.0
48	6"	53,118	158.0	48,124	142.0	48,124	188.0	42,700	155.0
52	6-1/2"	61,774	185.0	56,296	164.0	56,296	221.0	49,100	184.0
56	7"	70,824	215.0	64,468	192.0	64,468	256.0	56,400	212.0
60	7-1/2"	80,812	246.0	73,548	221.0	73,548	294.0	63,600	243.0
64	8"	91,708	280.0	83,082	250.0	83,082	334.0	72,700	278.0
68	8-1/2"	102,604	316.0	93,070	283.0	93,070	377.0	81,800	313.0
72	9"	114,408	355.0	103,966	317.0	103,966	423.0	90,900	351.0
80	10"	139,832	438.0	127,120	392.0	127,120	522.0	114,000	433.0
88	11"	167,526	530.0	152,544	474.0	152,544	632.0	136,000	522.0
96	12"	197,944	630.0	179,784	565.0	179,784	751.0	159,000	624.0
104	13"	228,816	741.0	209,291	662.0	209,291	882.0	182,000	731.0
112	14"	266,044	858.0	241,074	767.0	241,074	1,022.0	204,000	848.0

(*continued*)

METRIC MEASURE		2-in-1 NYLON (Nylon-Nylon)		2-in-1 POWER BRAID (Nylon-Polypropylene)		2-in-1 STABLE BRAID (Polyester-Polyester)		12-STRAND BRAID BLUE STREAK (Polyester-Polypropylene)	
Diameter MM	Circumference inches	Strength Kg.	Weight Kg./100M	Strength Kg.	Weight Kg./100M	Strength Kg.	Weight Kg./100M	Strength Kg.	Weight Kg./100M
120	15"	302,364	986.0	275,124	882.0	275,124	1,174.0	227,000	973.0
128	16"	341,862	1,122.0	310,990	1,003.0	310,990	1,335.0	—	—
136	17"	384,084	1,267.0	346,402	1,132.0	346,402	1,508.0	—	—
144	18"	426,760	1,420.0	389,078	1,269.0	389,078	1,690.0	—	—
152	19"	472,160	1,581.0	430,846	1,414.0	430,846	1,883.0	—	—
160	20"	522,100	1,752.0	476,700	1,567.0	467,700	2,086.0	—	—
168	21"	572,040	1,933.0	522,100	1,782.0	522,100	2,301.0	—	—

Application Data:

Supply Vessel Mooring AR-1
Single Point Mooring — Java Sea AR-2
Single Point Mooring — Humber River AR-3
Single Point Mooring — Ekofisk AR-5
Single vs. Double Leg Hawsers AR-6
Utility Winch Lines AR-7
Utility Stringing Lines AR-8
Utility Hand and Block Lines AR-9

Engineering Reports:

Deep Sea Mooring Systems ER-1
Towing Systems ER-2
Pierside Mooring Systems ER-3

Samson Braided Rope Systems for the Ocean Industry OC-1A

Elongation and Elasticity on Repeat Loading

(Based on tests made at machine speed of 4"/minute.)

	Permanent Elongation[1]						Working Elasticity[2]					
Rope Lay	Standard		Heavy Marine				Standard		Heavy Marine			
% of Break Strength	20%	30%	10%	20%	30%	50%	20%	30%	10%	20%	30%	50%
Manila	1.5	2.0	4.0	7.0	9.0	11.0	7.0	8.0	7.0	7.0	8.0	8.0
Nylon/GoldLine	4.0	7.0	8.0	12.0	14.0	19.0	22.0	24.0	18.0	22.0	24.0	25.0
Dacron	0.5	0.5	6.0	10.0	14.0	16.0	9.0	9.0	8.0	9.0	9.0	10.0
P/D	—	—	6.0	10.0	12.0	14.0	—	—	6.5	8.0	8.5	9.0
Polyethylene	5.0	6.0	6.0	8.0	9.0	13.0	8.0	10.0	7.0	8.0	10.0	11.0
Monofilament Polypropylene	2.5	3.0	7.0	9.0	12.0	17.0	12.0	13.0	10.0	12.0	13.0	14.5
Multifilament Polypropylene	3.0	4.0	8.5	13.0	16.0	20.0	14.0	16.0	13.5	14.0	14.0	16.0

(1) Permanent elongation is the permanent increase in length after several loadings to equilibrium and dynamic relaxation at the indicated percentages of breaking strength.

(2) Working elasticity is the recoverable stretch after several loadings to equilibrium and dynamic relaxation at the loads indicated and is based on the new rope length (original length plus permanent elongation).

Plymouth Rope Weight and Strength Specifications for Standard Construction

Nominal Size In.		NET WEIGHT IN POUNDS PER 100 FEET						TENSILE STRENGTH IN POUNDS FOR STANDARD CONSTRUCTION					
		Maximum Fed. Spec.	Approximate Average					Minimum Fed. Spec.	Approximate Average				
Cir.	Dia.	Manila	Polyethylene	Polypropylene Monofilament**	P/D	Dacron	GoldLine and Nylon	Manila	Polyethylene	Polypropylene Monofilament**	P/D	Dacron	GoldLine and Nylon
5/8	3/16	1.47	.71	.71	—	1.50	0.97	450	700	800	—	1,300	1,100
3/4	1/4	1.96	1.25	1.25	—	2.45	1.74	600	1,200	1,350	—	2,150	1,950
1	5/16	2.84	1.88	1.90	—	3.60	2.65	1,000	1,750	1,960	—	3,300	2,960
1 1/8	3/8	4.02	2.94	2.75	5.0	5.00	3.85	1,350	2,500	2,650	3050	4,500	4,200
1 1/4	7/16	5.15	4.00	3.70	6.5	6.60	5.25	1,750	3,400	3,350	3800	6,000	5,550
1 3/8	15/32	6.13	—	—	—	—	—	2,250	—	—	—	—	—
1 1/2	1/2	7.35	5.00	4.75	8.5	8.40	6.95	2,650	4,100	4,200	4800	7,600	7,200
1 3/4	9/16	10.20	6.30	6.10	9.5	10.50	8.65	3,450	4,600	5,000	6100	9,250	9,000
2	5/8	13.1	8.1	7.7	10.5	12.8	10.6	4,400	5,700	5,700	6700	11,250	11,000
2 1/4	3/4	16.3	11.5	11.0	15.0	18.0	15.5	5,400	7,800	8,200	8900	15,600	15,300
2 1/2	13/16	19.1	—	—	—	—	—	6,500	—	—	—	—	—
2 3/4	7/8	22.0	16.2	15.0	20.5	23.5	20.8	7,700	11,000	11,200	11,500	20,000	21,000
3	1	26.5	19.6	18.7	26.5	30.0	27.5	9,000	13,300	14,000	14,000	25,000	26,500
3 1/4	1 1/16	30.7	—	—	—	—	—	10,500	—	—	—	—	—
3 1/2	1 1/8	35.2	24.9	23.7	34.0	40.0	35.0	12,000	16,500	18,300	21,000	26,900	36,000
3 3/4	1 1/4	40.8	28.6	27.2	39.0	47.0	38.0	13,500	18,600	20,700	24,000	30,600	41,500
4	1 5/16	46.9	32.4	30.8	44.0	53.0	44.0	15,000	21,200	23,500	27,000	35,000	47,500
4 1/2	1 1/2	58.8	41.0	39.0	55.0	68.0	55.0	18,500	26,700	29,700	34,000	45,000	60,000
5	1 5/8	73.0	50.3	47.9	67.0	84.0	68.0	22,500	32,700	36,300	42,000	56,300	74,000
5 1/2	1 3/4	87.7	61.2	58.3	80.0	98.0	84.0	26,500	39,500	43,900	50,000	66,300	90,000
6	2	105.0	72.5	69.0	95.0	117.0	100.0	31,000	47,700	53,000	60,000	77,000	107,000
6 1/2	2 1/8	123.0	86.0	81.7	112.0	139.0	118.0	36,000	55,800	62,000	70,000	91,000	125,000
7	2 1/4	143.0	98.1	93.5	127.0	163.0	138.0	41,000	63,000	70,000	80,000	106,900	145,000
7 1/2	2 1/2	163.0	113.5	108.0	147.0	187.0	160.0	46,500	72,500	80,500	92,000	120,000	166,000
8	2 5/8	187.0	127.0	121.0	165.0	216.0	183.0	52,000	81,000	90,000	105,000	137,500	190,000
8 1/2	2 7/8	211.0	144.0	137.0	—	—	—	58,000	91,800	102,000	—	—	—
9	3	237.0	161.0	153.0	208.0	274.0	233.0	64,000	103,000	114,000	130,000	180,000	240,000
9 1/2	3 1/8	264.0	—	—	—	—	—	71,000	—	—	—	—	—
10	3 1/4	292.0	200.0	191.0	253.0	336.0	288.0	77,000	123,000	137,000	163,000	221,000	300,000
11	3 1/2	360.0	244.0	232.0	—	412.0	350.0	91,000	146,000	162,000	—	270,000	360,000
12	4	426.0	289.0	275.0	—	490.0	420.0	105,000	171,000	190,000	—	324,000	430,000

**Figures for M/M Polypropylene Ropes are approximately the same as figures given for Polypropylene Monofilament Ropes.

Handling Rope

Removing Rope from Reel or Coil. Synthetic fiber ropes normally are shipped on reels for maximum protection while in transit. The rope should be removed from the reel by pulling it off the top while the reel is free to rotate. This can be accomplished by passing a pipe through the center of the reel and jacking it up until the reel is free from the deck. Rope should never be taken from a reel lying on its side. If the rope is supplied in a coil, it should always be uncoiled from the inside so that the first turn comes off the bottom.

Avoid Kinking and Hockling. The continuous use of a line on one side of a winch or windlass is a common abuse that can render a line useless in a comparatively short time. Repeated hauling of a line over a winch in a counterclockwise direction will extend the lay of the rope and simultaneously shorten the twist of each strand. As this action continues, strand kinks (or hockles) will develop. Once these hockles appear, they cannot be removed, and the rope is permanently damaged at the point of hockling.

If, on the other hand, the line is continuously hauled over a winch in a clockwise direction, the rope lay is shortened, the rope becomes stiff and will kink readily.

To avoid these detrimental conditions, the direction of turns over the winch should be alternated regularly. Clockwise turns are recommended for the initial use of a new line. If this practice is observed, the original rope balance will be maintained and the ship's lines will have a longer useful life.

Avoid Overheating. When synthetic ropes are used on a capstan or winch, care should be exercised to avoid surging while the winch head or capstan is rotating. The friction from this slippage causes a localized overheating which melts and fuses the fibers of the outer yarns. In extreme cases of fusion, the damaged section should be removed and the ends joined with a short splice. Such a condition is less likely to occur with PNX Floterope because the fibers are treated with Resistex, a lubricant developed and used exclusively by American because it is highly resistant to heat, water, and abrasion.

Avoid Cutting. Whenever possible, synthetic lines should be passed through roller chocks that are free to rotate and free of burrs and rust. *Do not* use manila, wire, or spring lay in conjunction with synthetic ropes on the same chocks or bitts. In ports where there is a heavy undertow and a ship is moored to buoys, synthetic lines should be parceled with canvas strips or other protective material at the sections that come into contact with the chocks.

Don't Drag Rope. Dragging rope over rough surfaces and sharp edges or permitting one rope to chafe against another causes wear and fraying. Dirt and grit picked up by the rope gradually work into the strands, cutting the inside fibers.

Don't Overload Rope. To make allowance for worn or weakened rope, figure the working load of the rope at not more than one-fifth of its breaking strength. Never use a nylon line, which has a high stretch factor, with a polypropylene, polyethylene, or Dacron line having low stretch factors. The nylon line will not take its proportionate share of the load, thus putting additional strain on the other line or lines.

Splicing. Before unlaying the rope, put a whipping of friction tape around the line far enough from the end to provide sufficient lengths of strand for the required tucks. When unlaying the rope, unlay each strand separately, taking care to hold the twist, and whip with tape at intervals along the strand. After each strand is properly whipped, the procedure is the same as splicing any fiber rope. Four tucks should be used when splicing synthetic fiber ropes. Short splices should be used instead of long splices when splicing end to end.

Safety Precautions. Check the line regularly for frayed strands and broken yarns. Twist open the strands of rope slightly and look for powdered fiber which warns of internal wear. Careful attention to the condition of the rope will determine when it should be replaced and help prevent accidents. When handling any rope take care to stay out of the direct line of pull. This is a normal safety precaution but is especially significant when working with lines of synthetic fibers.

Storage. Synthetic fiber rope can be stored below while wet but should not be exposed to sunlight for extended periods.

Appendix III. Whistle Signals, Lights, and Buoyage

WHISTLE SIGNALS, GREAT LAKES

Ship to Tug
One long and one short blast (— -): ship ready to proceed, O.K., or all right.
One long and two short blasts (— - -): let go of the bow towline.
One long and three short blasts (— - - -): let go the stern towline.

Both Ship to Tug and Tug to Ship
One short blast (-): work propeller ahead when propeller is stopped; stop propeller when propeller is working either ahead or astern.
Two short blasts (- -): work propeller astern.
Three short blasts (- - -): "check" or slow down.
Four short blasts (- - - -): work propeller strong in the same direction it is working at the time the signal is blown.
Five or more short blasts (- - - - -): the standard danger signal.

WHISTLE SIGNALS BETWEEN PILOT AND TUG

These signals, which may be sounded by ship's whistle, police type mouth whistle, or voice on VHF, are based on the old New York Harbor bell signals from wheelhouse to engine room.

Status	*Signal*	*Instruction*
Stopped	One blast	Slow ahead
Slow ahead	One blast	Stop
Slow ahead	Lots of short blasts	Full speed
Full ahead or astern	One blast	Slow
	Two blasts	Stop
Stopped	One long and two short	Let go—dismissed

MODIFICATION OF U.S. BUOYS, BEACONS, AND DAY MARKERS TO CONFORM TO IALA SYSTEM

Lateral aids marking the sides of channels as seen when entering from seaward

Port Side / Odd Numbers — Lighted Buoy (Green Light Only), Can Buoy (Unlighted), Daymark

Starboard Side / Even Numbers — Lighted Buoy (Red Light Only), Nun Buoy (Unlighted), Daymark

Light Rhythms: FIXED, FLASHING, OCCULTING, QUICK FLASHING, EQ INT

Modifications: Port hand aids will be green with green lights. All starboard hand aids will have red lights.*

Safe water aids marking midchannels and fairways

(No Numbers - May Be Lettered)

Light Rhythm: Morse Code Mo(A)

Lighted (White Light), Spherical Buoy (Unlighted), Daymark

Modifications: Red and white will replace black and white. Buoys will be spherical or will have a red spherical topmark.†

Preferred channel aids (mark bifurcations)

(No numbers—May be lettered)

Preferred Channel to Starboard — Lighted Buoy (Green Light Only), Can Buoy (Unlighted), Daymark

Preferred Channel to Port — Lighted Buoy (Red Light Only), Nun Buoy (Unlighted), Daymark

Light Rhythm: Composite Group Flashing (2 + 1)

Modifications: Green will replace black. Light rhythm will be changed to Composite Gp Fl (2 + 1).

---

*White lights will no longer be used on port and starboard hand aids in waters used by international mariners. However, in the Western Rivers including the Mississippi River system above Baton Rouge, white lights may still be used on crossing aids.

†Lighted and/or sound safe water buoys will have a red spherical topmark.

BUOYAGE ON THE NEW YORK STATE BARGE CANAL SYSTEM

The navigation buoys provided for the Northern and Mohawk/Lake Oneida to Three Rivers sections are based on entry from sea at Waterford. As in other harbor and river or breakwater entrances the red even-numbered buoys are on the right or starboard, green buoys and markers to the left or port. However, at Three Rivers at the head of the Oswego River, the buoyage is reversed as you are exiting to Lake Ontario at Oswego, because at that point the St. Lawrence River is also regarded as an entrance from the sea.

Appendix IV. VHF/VTS Frequencies

WEATHER BROADCASTS FOR TUGS

In the United States tugs with VHF channels WX 1, 2, or 3 may get continuous weather forecasts on either coast and are seldom beyond range of these stations. In Canada all coast stations give regular weather broadcasts.

National Weather Service, NOAA Weather Radio, West Coast Marine Network

WEATHER RADIO FREQUENCY 162.550 MHz
- KIH-36 Neah Bay, Wa.
- KHB-60 Seattle, Wa.
- KIG-98 Portland, Or.
- KIH-33 Newport, Or.
- KIH-37 Crescent City/Brookings
- KHB-49 San Francisco, Ca.
- KIH-31 San Luis Obispo, Ca.
- KWO-37 Los Angeles, Ca.

WEATHER RADIO FREQUENCY 162.400 MHz
- KEC-91 Astoria, Or.
- KIH-32 Coos Bay, Or.
- KEC-82 Eureka, Ca.
- KIH-30 Point Arena, Ca.
- KEC-49 Monterey, Ca.
- KIH-34 Santa Barbara, Ca.
- KEC-62 San Diego, Ca.

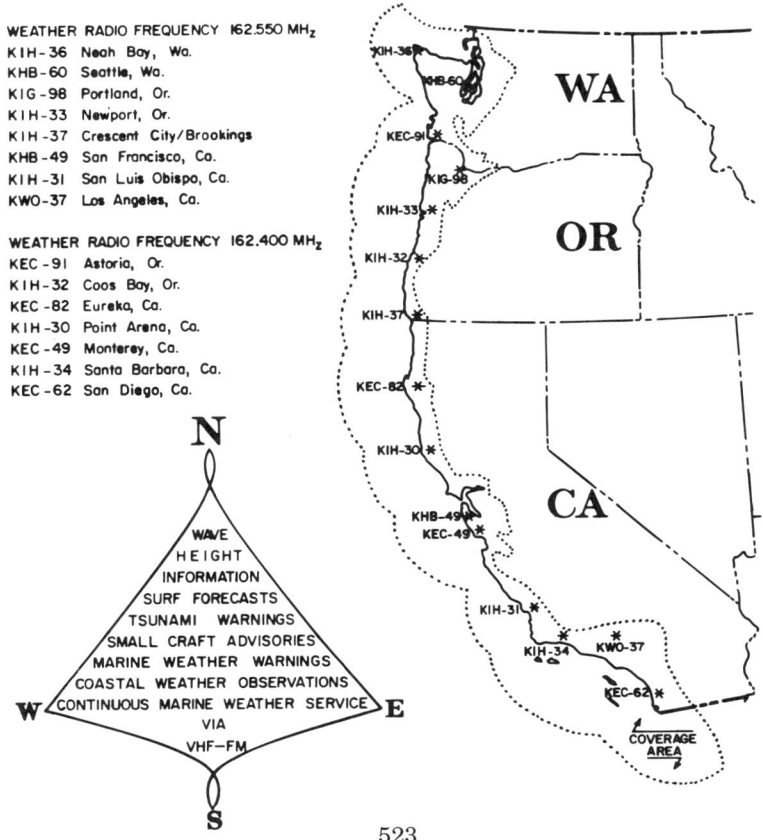

VHF CHANNEL ASSIGNMENTS IN THE UNITED STATES

Channel numbers	Type of communication	Suggested channel selection Recreational vessels 6 ch.	Suggested channel selection Recreational vessels 12 ch.	Suggested channel selection Commercial vessels 12 ch.
16	DISTRESS, SAFETY & CALLING Intership & ship to coast	*	*	*
6	INTERSHIP SAFETY Intership. NOT to be used for non-safety intership communications	*	*	*
22A	Communications with U.S. Coast Guard ship, coast, or aircraft stations	1	1	1
65A, 66A, 12, 73, 14, 74, 20	PORT OPERATIONS Intership & ship to coast		1	2
13	NAVIGATIONAL Intership & ship to coast		1	1
68, 9, 69, 71, 78A	NON-COMMERCIAL Intership & ship to coast	1	3	
70, 72	NON-COMMERCIAL Intership		2	
7A, 9, 10, 11, 18A, 19A, 79A, 80A	COMMERCIAL Intership & ship to coast			3
67, 8, 77, 88A	COMMERCIAL Intership			1
24, 84, 25, 85, 26, 86, 27, 87, 28	PUBLIC CORRESPONDENCE Ship to public coast	2	2	2
162.400 (WX-2) 162.550 (WX-1) 162.475 (WX-3)	NOAA WEATHER SERVICE Ship receive only	†	†	†

*These channels are REQUIRED to be installed in every ship station equipped with a VHF radio.

†The weather receive channels are half-channels (receive only) one or more of which are recommended to be installed in each ship station. Many manufacturers include one or more of these channels in their sets in addition to the normal six or twelve channel capacity.

VHF AND VTS CHANNELS FREQUENTLY USED BY TUGS IN NORTH AMERICA

The Gulf of St. Lawrence and River to Montreal has been divided into six sectors. Each sector uses a different channel frequency on VTS. These sectors and channels are as follows:

Sector One	from Sept-Iles to Pte. Manicouagan	Channel 14	(156.70)
Sector One-A	from Pte. Manicouagan to Ile Blanche (this includes pilot station at Les Escoumins and Saguenay River to Chicoutimi)	Channel 9	(156.45)

Sector Two	from Ile Blanche to between St. Laurent Wharf and Ste. Petronille Wharf, Ile d'Orleans	Channel 12	(156.60)
Quebec Harbour uses		Channel 6	(156.3)
Sector Three	from Ile d'Orleans to Grondines rear light	Channel 11	(156.55)
Sector Four	Grondines rear light to Yamachiche Light	Channel 13	(156.65)
Sector Five	Yamachiche to Cap St. Michel	Channel 9	(156.45)
Sector Six	Cap St. Michel to Montreal	Channel 10	(156.5)
Montreal Harbour uses		Channel 6	(156.3)

St. Lawrence Seaway (Montreal to Lake Erie) Radio Communications

Listening Watch and Notice of Arrival. (1) Vessels shall be on radio listening watch on the applicable assigned frequency while within a Seaway traffic control sector as shown on the General Seaway Plan and shall give notice of arrival in the manner prescribed in section 64 upon reaching any designated calling in point; and (2) notice of arrival shall be deemed to have been given when it is acknowledged by a Seaway station.

Assigned Frequencies. The Seaway stations operate on the following assigned VHF frequencies:

Distress and Calling	Channel 16	(156.6)
Working (Canadian Stations other than Lake Ontario and Erie)	Channel 14	(156.7)
Working (U.S. Stations, Lake Ontario, and Sector 4 of the river)	Channel 13	(156.65)
Working (U.S. Stations other than Lake Ontario and Sector 4 of the river)	Channel 12	(156.6)
Working (Canadian Stations, Sector 3, Lake Ontario and Lake Erie)	Channel 11	(156.55)

Calling in Points. Each sector of the St. Lawrence River and Seaway has calling in points (CIP) from Sept-Iles (Seven Islands) to the exit to Lake Erie at Port Colborne. As pilotage is compulsory in most of these waters, the pilot usually makes the report as the tug passes each of the CIPs.

Straits of Belle Isle, and Cabot to Bay of Fundy. The Canadian government requires that all vessels about to enter, pass through, or depart from their waters or harbors shall request traffic clearance through a Coast Guard Canada Radio Station. Vessels of over 500 tons gross, which will include many tugs, will be cleared by Eastern Canada Traffic System, referred to in publications as ECCAREG. Tugs under 500 gross tons wishing to enter Canso may call Canso Traffic on Channel 16 for further advice.

Halifax Traffic will answer on Channel 16. If a tug is running without tow and taking shelter in bad weather without a pilot on board, the calling in points are when passing the following buoys: Outer Central and Neverfail and also at Meaghers Beach and Ives.

Vessels entering the Bay of Fundy must call Fundy Traffic. The Saint John pilots monitor Channels 16 and 14. The Irving Tugs use Channel 10.

New England Ports. There are TSS lanes for the approaches to Portland and Boston. There is *no* VTS. Contact with local pilots or towing companies may be established through the local marine operator on the appropriate channel. Providence Steamboat Company in that port monitors Channel 16.

New York Harbor. The Coast Guard monitors a voluntary VTS for ships and some tows entering from sea. Bridge-to-bridge is overwhelmingly used for moving and location announcements by tugs and docking pilots. Here are a few other well-used channels with the dispatcher's identification: 616-Turecamo-18A; 234-Red Star/Spentonbush-12; 672-Berman Tank Cleaning-18A; Texaco, tugs and dock-18A.

Mid-Atlantic
Cape May Ferry	Channel 7A
Delaware Bay and River Pilots	Channel 14
Chesapeake and Delaware Canal	Channels 16-13-14-18A
Port of Philadelphia and Wilmington	
Curtis Bay Towing	Channel 12
McAllister Towing	Channel 10
Taylor & Anderson Tugs	Channel 7A
Maritrans KSK 241	Channel 18A
Texaco Dock, Eagle Point & Tugs KSK 322	Channel 18A
Philadelphia Marine Operator	Channel 26
Wilmington Marine Operator (Delaware)	Channel 28
Baltimore and Hampton Roads	
Curtis Bay Towing	Channel 12
McAllister Towing	Channel 10

There are marine operators at various ports from Baltimore to Corpus Christi, Texas. Each is operated by the area telephone company using Channels 25-26-28, etc. The exception is the excellent private radio VHF and SSB station, WLO MOBILE, which can usually handle traffic anywhere in the Gulf of Mexico if a vessel has a good transmitter and receiver.

The Mississippi, the Passes to Baton Rouge, Voluntary Vessel Traffic Service (VTS). There is a very good voluntary traffic control system in place on the river from outside the passes to the end of deep-draft navigation at Baton Rouge.

    Vessel Traffic Service Area
      a. The entire Mississippi River from Southwest Pass Entrance Mid-channel Lighted Buoy to Devil's Swamp Light, mile 242.4 AHP.
      b. The Mississippi River—Gulf Outlet from the Mississippi River Gulf Outlet Lighted Bell Buoy "2" to mile 60 MR-GO, thence westerly along the Gulf Intracoastal Waterway (GIWW) including the Inner Harbor Navigational Canal (IHNC), from the turning basin to the IHNC Locks, but excluding the IHNC from the turning basin north to Lake Pontchartrain.

Galveston Bay to Houston. This busy waterway, with its narrow dredged cut in upper Galveston Bay, is controlled voluntarily by pilots and tug masters who

APPENDIX IV 527

give frequent "Securité" calls and bridge-to-bridge meeting and passing arrangements.

San Francisco Bay. South to Redwood City and north to The Brothers Light is included in the mandatory USCG VTS. The local activities monitor the following frequencies:

Marine operator	Channels 26, 84, 87
San Francisco bar pilots	Channels 10, 16
American Navigation Tugs	Channel 18A
Bridges, rivers, and delta	Channels 9, 13, 16
Crowley Maritime (Red Stack) Tugs	Channel 10
Harbor Launch	Channel 10
Marine exchange	Channels 10, 18A
Oscar Niemeth Tugs	Channels 14, 18A
Port of Sacramento	Channels 16, 18A
Port of Stockton	Channels 7, 16, 18A
Sacramento Locks	Channels 9, 13, 16
Saunders Towboat	Channel 18A
Smith Rice	Channel 14

Marine Operators along the northern California coast include:

San Luis Obispo	Channel 26
Monterey	Channel 28
Santa Cruz	Channel 27
Bodega Bay	Channel 25
Fort Bragg	Channel 28

Straits of Juan de Fuca, Haro, Rosario, Georgia, and Puget Sound. These areas include the Canadian ports of Victoria, Sydney, Nanaimo, New Westminster, and Vancouver; and the U.S. ports of Bellingham, Port Townsend, Everett, Seattle, and Tacoma. Each of the main channels off these ports or their river entrance are served by traffic separation schemes and voluntary vessel traffic services. Canadian waters are divided into four sectors numbered "2 to 5." Sector 1 was formerly the Strait of Juan de Fuca. This is now served by the USCG as Seattle Traffic, which covers all of Puget Sound on Channel 14. Canadian Sectors 2-5 are controlled by Vancouver Traffic.

Sector 2	Channel 11	(156.55)
Sector 3	Channel 12	(156.6)
Sector 4	Channel 74	(156.725)
Sector 5	Channel 71	(156.575)

West Coast of Vancouver Island. This area is in the Tofino Traffic Sector 1 operating on Channel 74 (156.725) and is voluntary.

Prince Rupert Traffic serves the waters of northern British Columbia, north of a line from Triangle Island to Cape Caution, and monitors on a voluntary participation Channels 11 and 71.

Note that both the Puget Sound VTS and all of those in Canadian waters have calling in points. They are listed in the U.S. Coast Guard and Canadian *Notice to Mariners.*

LOCATION OF MARINE RADIOTELEPHONE TERMINALS

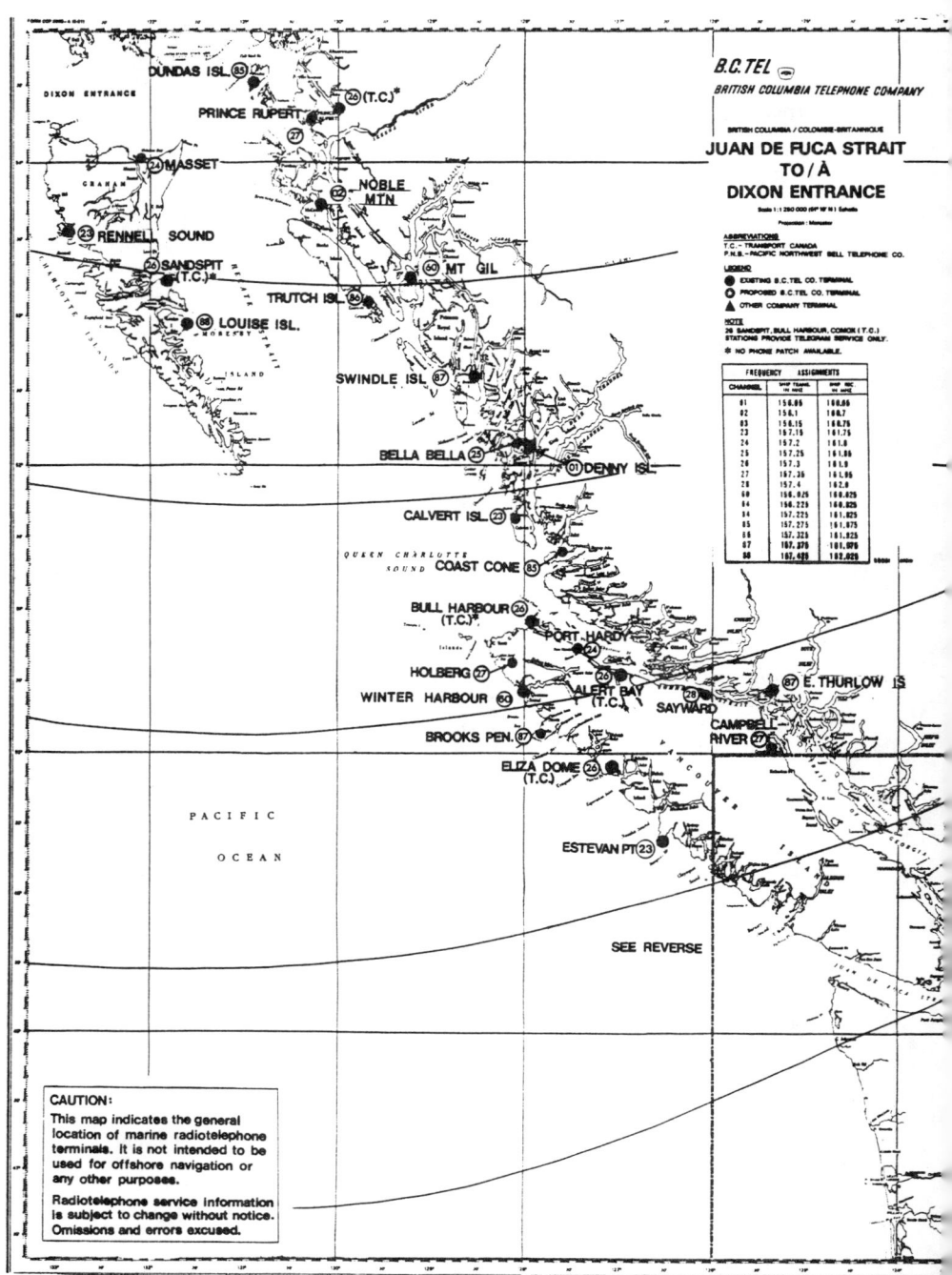

APPENDIX IV

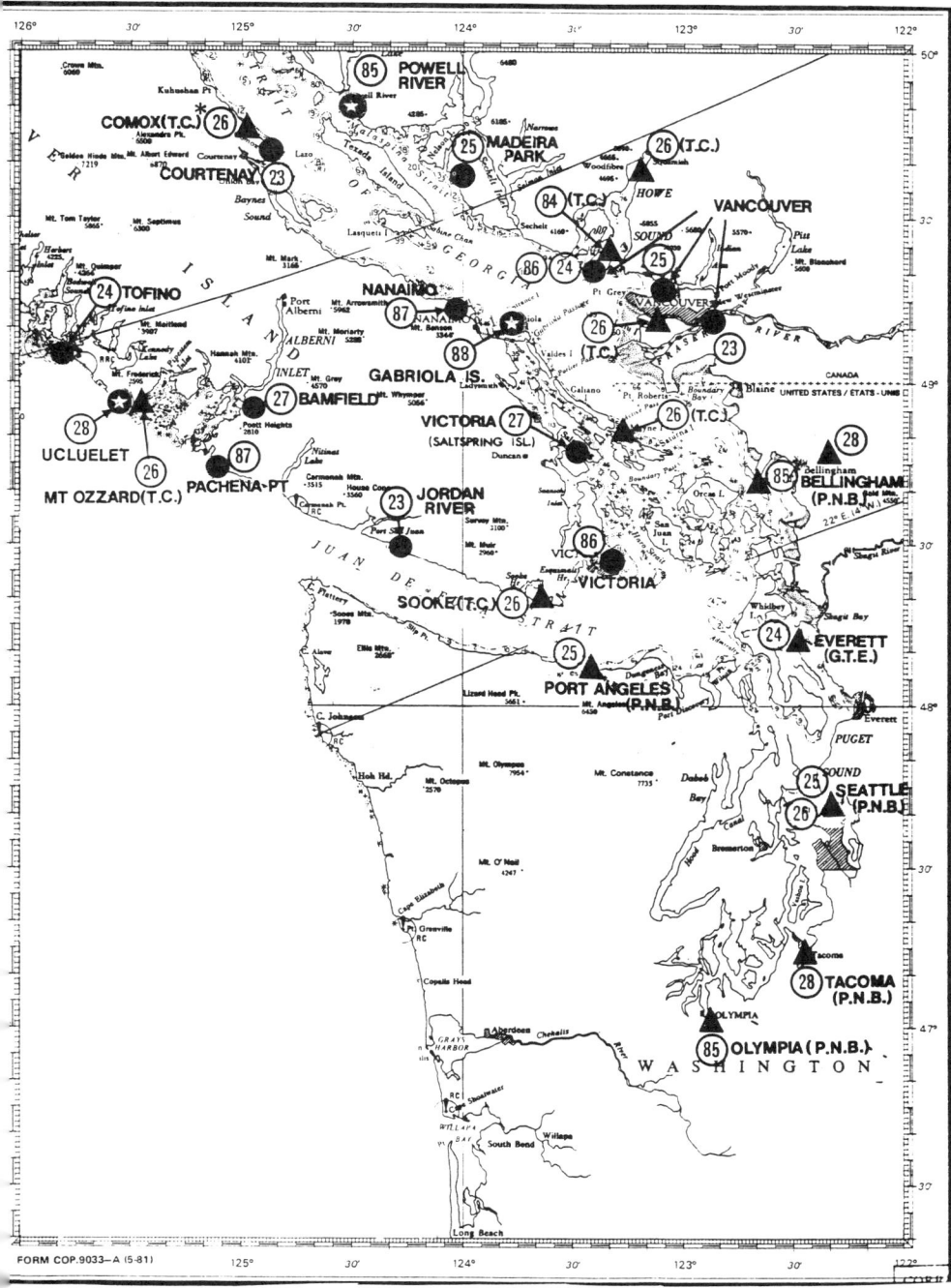

SAN FRANCISCO OFFSHORE VESSEL MOVEMENT REPORTING SYSTEM

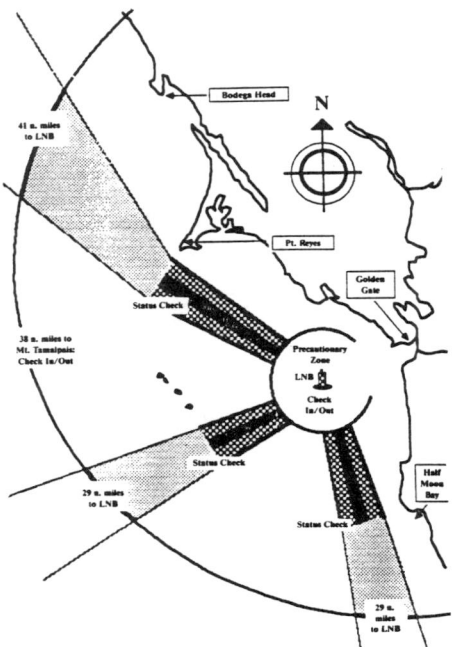

Beginning December 1, 1986, the U.S. Coast Guard will operate an Offshore Vessel Movement Reporting System (OVMRS) as a service to all vessels transiting the ocean approaches to San Franscisco Bay, California, U.S.A. The OVMRS will aid in detecting potential close encounters and improve communications by providing the names of transiting deep-draft vessels. Although participation is voluntary, the U.S. Coast Guard urges all vessels transiting the ocean within the OVMRS area to participate as follows:

- Vessels over 300 gross registered tons and vessels with tows (referred to as "deep-draft" vessels) should actively participate by reporting their movements to the Coast Guard's Vessel Traffic Service (VTS).
- All vessels should listen to VTS broadcasts of deep-draft vessel transits in their area, and navigate and communicate appropriately; identify your vessel and your intentions to approaching vessels.

OVMRS Area. The OVMRS applies to vessels transiting ocean waters within a 38 n.m. radius of Mount Tamalpais (37° 55.8' N, 122° 34.6' W). The distances of this radius from the San Francisco Approach Lighted Horn Buoy (or LNB for Large Navigation Buoy; U.S. Coast Guard Light List No 360; 37° 45' N, 122° 41.5' W) measured seaward beyond each of the three charted Traffic Separation System lanes are: Northern, 41 n.m.; Western, 29 n.m.; Southern, 29 n.m.

Responsibilities of Deep-Draft Vessels. Before entering the OVMRS area from sea or from San Francisco Bay, deep-draft vessels will call the Coast Guard's San Francisco Vessel Traffic Service on VHF-FM channel 16. Switching to channel 12, the vessel reports its type, name, position, route, speed, and estimated time of arrival at either the LNB if the vessel is inbound, or the seaward radius of the OVMRS if the vessel is outbound. VTS will broadcast this report. Upon reaching the seaward end of one of the Traffic Separation System lanes, approximately halfway in the transit, the vessel will report an update of its progress to VTS. After an initial call on channel 16, VTS will broadcast the status of all participating deep-draft vessels on channel 12 on the hour and half hour.

Responsibilities of All Vessels. By monitoring the radio transmissions of vessels reporting into the OVMRS, as well as scheduled broadcasts, listeners will be informed of participating deep-draft vessel movements. All vessels should communicate their identity and intentions to other vessels operating in their vicinity. The OVMRS is only an information service. The conduct of mariners will still be governed by: agreements reached by communicating between vessels; good seamanship; and the international Rules of the Road.

Appendix V. Wind and Ice

WIND VELOCITIES FOR OCEAN TOWING

Many ocean tugs with tows have been caught at sea with tows in wind velocity of from 9 to 11 and occasionally Force 12 on the Beaufort Scale. However, the tug master who departs with a tow when winds of Force 8 or above are forecast is not prudent. A dispatcher for one of the world's largest towing lines has stated he would not allow a tow to sail in Force 8 weather.

Wind Speed in knots (33' above sea level)	Wind Condition	Sea Conditions	Approx. Wave ht. (in ft.)	Force
0-1	Calm	Smooth, mirror-like sea.	–	0
1-3	Light Air	Scale-like ripples; no foam crests.	¼	1
4-6	Light Breeze	Short wavelets; glassy crests; non-breaking.	½	2
7-10	Gentle Breeze	Large wavelets; glassy crests, some breaking; occasional white foam.	2	3
11-16	Moderate Breeze	Small waves becoming longer; frequent white foam crests.	3½	4
17-21	Fresh Breeze	Moderate waves with more pronounced long form; many white foam crests; some spray.	6	5
22-27	Strong Breeze	Large waves form; white foam crests everywhere; probably more spray.	9½	6
28-33	Near Gale	Sea heaps up; white foam from breaking waves is blown in streaks with the wind.	13½	7
34-40	Gale	Moderately high waves with greater length; edges of crests break into spindrift; foam is blown in well-defined streaks with the wind.	18	8

(continued)

Wind Speed in knots (33' above sea level)	Wind Condition	Sea Conditions	Approx. Wave ht. (in ft.)	Force
41-47	Strong Gale	High waves; crests of waves start to topple and roll over; spray may affect visibility.	23	9
48-55	Storm	Very high waves; overhanging crests; resulting foam is blown in dense white patches with the wind; sea surface takes on a whiter look; the tumbling of the sea becomes heavy and shock-like; visibility affected.	29	10
56-63	Violent Storm	Exceptionally high waves; sea covered with long white patches of foam blown in direction of wind; all wave crests are blown into froth; visibility affected.	37	11
64+	Hurricane	Air is filled with spray and foam; sea is completely white with driving spray; visibility seriously affected.	45	12

Note: In enclosed waters or near land with an offshore wind, wave heights will be less, the waves steeper and not so long. In many tidal waters wave heights are apt to increase considerably in a very short time and conditions can be more dangerous near land than in the open sea.

WIND CHILL FACTOR

Wind Speed (mph)	Temperature in Fahrenheit					
	+30	+20	+10	0	-10	-20
10	+16	+2	-9	-22	-31	-45
20	+3	-9	-24	-40	-52	-68
30	-2	-18	-33	-49	-63	-78
40	-4	-22	-36	-54	-69	-87
50	-7	-24	-38	-56	-70	-88

WIND SPEED, AIR TEMPERATURE, AND ICING CONDITIONS

Icing Conditions for Vessels Heading into or Abeam of the Wind for Water Temperatures of +1°C (34°F).

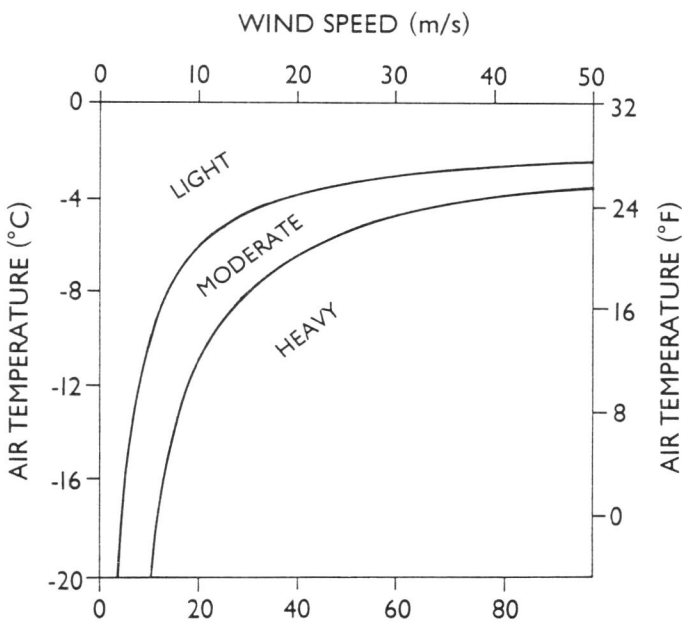

Light Icing—Less than 0.7 cm/hr (0.3 in/hr)
Moderate Icing—0.7 cm/hr (0.3 in/hr) to 2.0 cm/hr (0.8 in/hr)
Heavy Icing—Greater than 2.0 cm/hr (0.8 in/hr)

Icing Conditions for Vessels Heading into or Abeam of the Wind for Water Temperatures of +3°C (37°F).

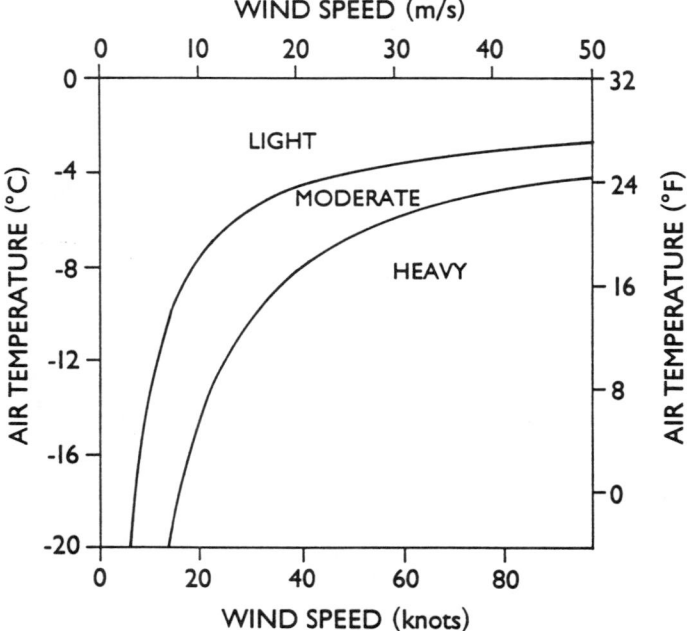

Light Icing—Less than 0.7 cm/hr (0.3 in/hr)
Moderate Icing—0.7 cm/hr (0.3 in/hr) to 2.0 cm/hr (0.8 in/hr)
Heavy Icing—Greater than 2.0 cm/hr (0.8 in/hr)

APPENDIX V

Icing Conditions for Vessels Heading into or Abeam of the Wind for Water Temperatures of +5°C (41°F).

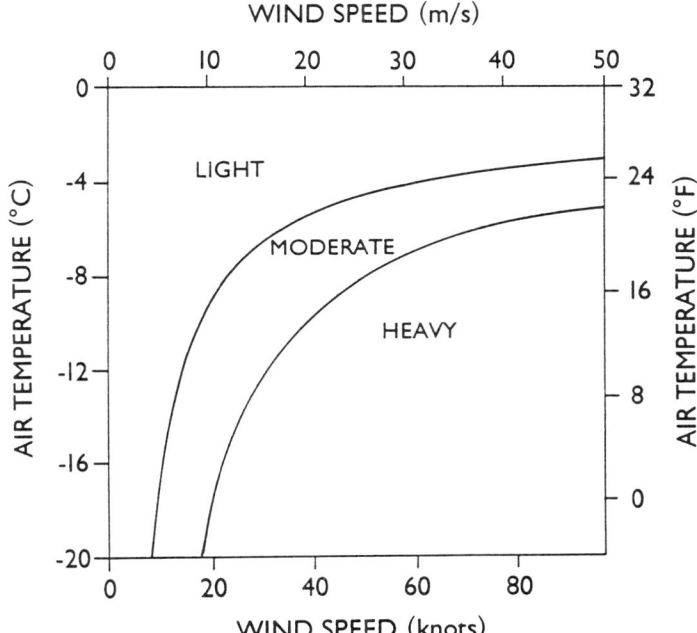

Light Icing—Less than 0.7 cm/hr (0.3 in/hr)
Moderate Icing—0.7 cm/hr (0.3 in/hr) to 2.0 cm/hr (0.8 in/hr)
Heavy Icing—Greater than 2.0 cm/hr (0.8 in/hr)

Icing Conditions for Vessels Heading into or Abeam of the Wind for Water Temperatures of +7°C (45°F).

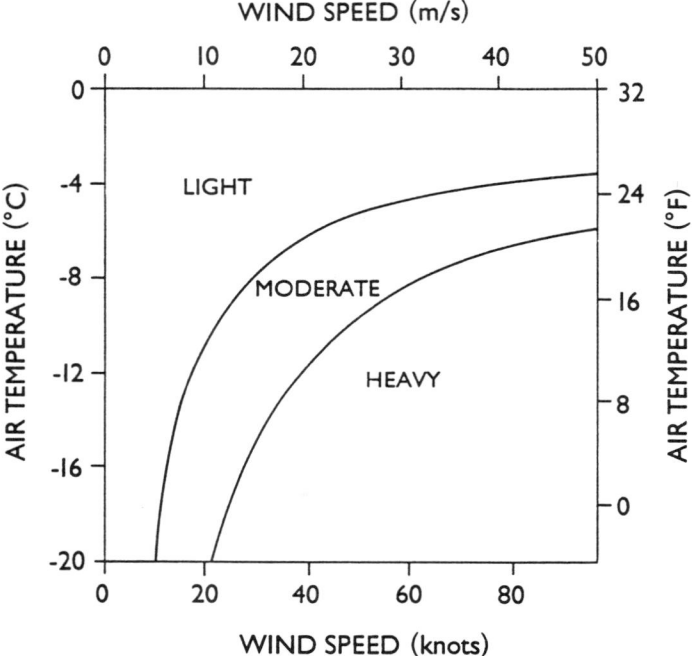

Light Icing—Less than 0.7 cm/hr (0.3 in/hr)
Moderate Icing—0.7 cm/hr (0.3 in/hr) to 2.0 cm/hr (0.8 in/hr)
Heavy Icing—Greater than 2.0 cm/hr (0.8 in/hr)

Appendix VI. Canals, Locks, and Narrows: Regulations

DIMENSIONS AND LIMITATIONS OF MAJOR LOCKS AND CANALS

The Suez. The maximum-size vessel to pass through as of 1985 was the 423,642 dwt *Buyul Selcuklu,* 1,240' × 226', with a 75' loaded draft.
The Panama. "Panamax Size" is 948' × 106' × 34.8' draft.
St. Lawrence Seaway and Welland Canal. Montreal to Lake Erie (including Welland), 13 locks all 830' × 80' × 27', maximum vessel length 730'.
Sault Ste. Marie Locks. Lake Huron to Lake Superior. Weitzel Lock, 550' × 80' × 18'; Canada Lock, 900' × 60' × 22'; McArthur Lock, 800' × 80' × 31'; Poe Lock, 1,200' × 110' × 32'; Davis and Sabin Locks, 1,350' × 80' × 23'.

NAVIGATION AND TUG ASSISTANCE, VALDEZ NARROWS, ALASKA

161.387 Valdez Narrows One-Way Traffic Area. Valdez Narrows One-Way Traffic Area consists of the navigable waters of the United States in Valdez Arm, Valdez Narrows, and Port Valdez northeast of a line bearing 307° true from Tongue Point at 61°02'06" N, 146°40'00" W, and southwest of a line bearing 307° true from Entrance Island Light at 61°05'06"N, 146°36'42" W.

161.378 Tug Assistance for Tank Vessels. For the purpose of this section, tug assistance means the use of a sufficient number of tugs properly manned and positioned, with enough power and maneuverability to enable the vessel to accomplish the intended maneuvers safely.

Factors to Be Considered:

1) Existing and expected conditions of wind, tide, and current
2) Size, displacement, and maneuvering capability of the vessel
3) No laden tank vessel of 20,000 dwt or more may transit the Valdez Narrows One-Way Traffic Area unless: a) a sufficient number of tugs, as determined by the VTC, are standing by at the northern entrance to Valdez Narrows, and b) tug assistance is utilized when directed by the VTC.

CLOSED CHOCK AND LINE REQUIREMENTS

THE PANAMA CANAL

Table of Minimum Requirements

Vessel Type	Length Overall	Mooring Lines Required each side of vessel	Power-Operated Winches	Power-Operated Capstan or Windlass	Fairleads or Chocks at Each Side of Vessel – Closed Fairleads of Acceptable Type	Fairleads or Chocks at Each Side of Vessel – Closed Chocks
All Vessels	90′ or less	2	—	—	—	2
Non-Self-Propelled Vessels	More than 90′	4	—	—	—	4
Self-Propelled Vessels	More than 90′ and up to 125′	4	—	2	—	4
Self-Propelled Vessels	More than 125′ and up to 225′	4	2	2	2	2
Self-Propelled Vessels	More than 225′	4	4	—	4 single or 2 double	—

Table of Location of Fairleads

Overall Length of Vessel in Feet	For Mooring Lines Nos. 1 and 2	For Mooring Lines Nos. 3 and 4
200′—300′	Between 30′ and 80′ from the stem	Between 30′ and 80′ from the stern
Over 300′—400′	Between 40′ and 100′ from the stem	Between 50′ and 110′ from the stern
Over 400′—500′	Between 40′ and 110′ from the stem	Between 50′ and 130′ from the stern
Over 500′—600′	Between 50′ and 130′ from the stem	Between 60′ and 150′ from the stern
Over 600′—730′	Between 60′ and 160′ from the stem	Between 70′ and 170′ from the stern

Hand Lines

Hand lines must be of Manila or other acceptable material and must have a minimum diameter of one-half inch and a minimum length of one hundred feet and must not be knotted or weighted when they are to be used in the chamber of a lock.

Anchor Marking Buoys

An orange coloured anchor marking buoy, of an approved type and fitted with seventy-five feet of suitable line, shall be secured directly to each anchor so that it will mark the location of the anchor when it is dropped.

APPENDIX VI

ST. LAWRENCE SEAWAY

For requirements concerning lines, any tow exceeding 80 meters in length transiting the St. Lawrence Seaway should apply in writing to:

    The St. Lawrence Seaway Authority
    202 Pitt Street
    Cornwall, Ontario
    K6J 3P7
    Canada

Tows of up to 80 meters in length should apply in writing to:

The Niagara Region:

The Maisonneuve Region:

The St. Lawrence Seaway Authority
P.O. Box 370
St. Catharine's, Ontario
L2R 6V8
Canada

The St. Lawrence Seaway Authority
P.O. Box 97
St. Lambert, Quebec
J4P 3N7
Canada

DIMENSIONS AND RESTRICTIONS FOR THE ST. LAWRENCE SEAWAY

Montreal-to-Maitland Section

Lock Data

LOCK	NORMAL LIFT FT.	LGTH. BREAST WALL TO FENDER	WIDTH OF CHAMBER FT.	LENGTH OF UPPER ENT. FT.	LENGTH OF LOWER ENT FT.
IROQUOIS	0.5 - 6.0	766'	80	3000	1500
EISENHOWER	38 - 42	766'	80	2043	1650
SNELL	45 - 49	766'	80	1650	2394
UPPER BEAUHARNOIS	36 - 40	766'	80	2265	1023
LOWER BEAUHARNOIS	38 - 42	766'	80	1012	2124
COTE STE. CATHERINE	33 - 35	766'	80	1590	1603
ST. LAMBERT	13 - 20	766'	80	2268	2954

NOTE - MINIMUM DEPTHS ON LOCK GATE SILLS ---- 30 FT.
CONTROLLING CHANNEL DEPTHS -------- 27 FT.
MILEAGES ALONG ₵ OF SAILING COURSE SHOWN THUS ⊥ 54
ZERO MILEAGE TAKEN AT INTERSECTION OF SEAWAY
CHANNEL ₵ AND BELLERIVE PARK RANGE OF SHIP CHANNEL

BRIDGE & NAVIGATION LEGEND

HL.   HIGH LEVEL BRIDGE
VL.   VERTICAL LIFT BRIDGE
RL.   ROLLING LIFT BRIDGE
SW.   SWING BRIDGE

- WHISTLE SIGN
- LIMIT OF APPROACH SIGN
- SEAWAY RADIO - TELEPHONE DESPATCH STATION
- C.I.P. CALLING - IN POINT
- ANCHORAGE AREA

REFER TO CANADIAN HYDROGRAPHIC SERVICE CHARTS 1340, 1409, 1410, 1411, 1412, 1413, 1414, 1415, 1416, 1417

St. Lambert Lock

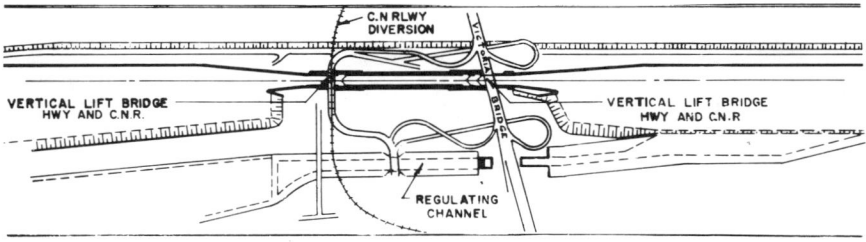

Cote Ste. Catherine Lock Through Upper Beauharnois Lock

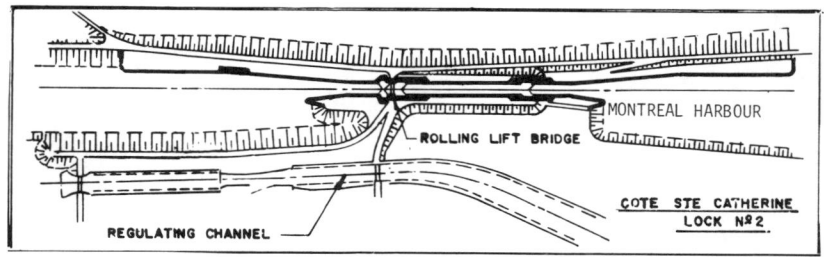

Lower Beauharnois Lock

Upbound this lock may be difficult to enter. Its entrance is at right angle to the main river current and prevailing wind.

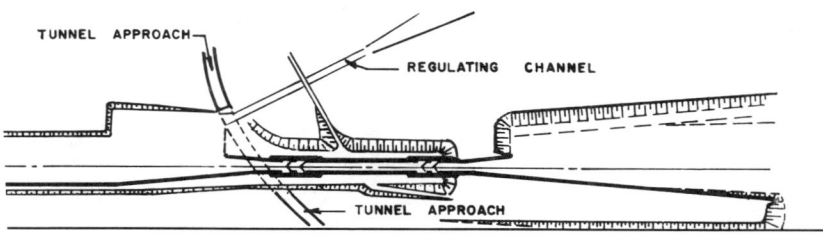

Upper Beauharnois Lock

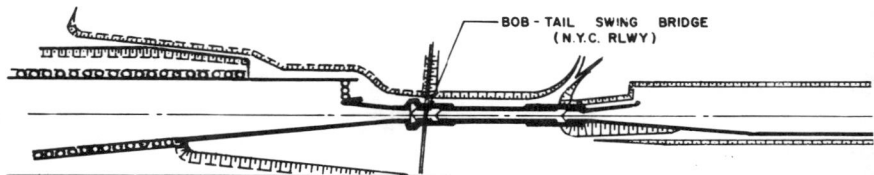

Snell Lock

Snell Lock is the first U.S. lock and is connected to Eisenhower Lock by the short, straight Wiley-Dondero Canal. Entering Snell Lock is difficult with a tow as the main river current will change from dead ahead to the starboard bow and beam as the approach is made. If the Massena Canal to port is in freshet it may act as a countercurrent, with eddies running toward the lock gates.

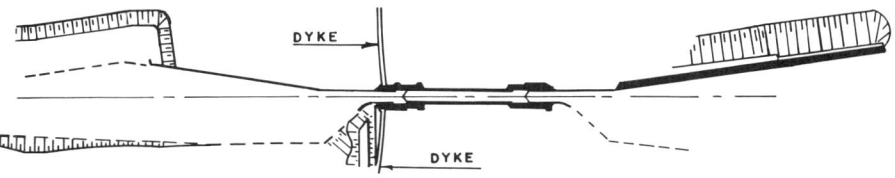

Eisenhower Lock

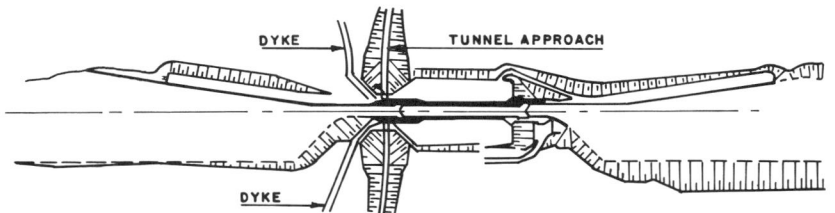

Iroquois Lock

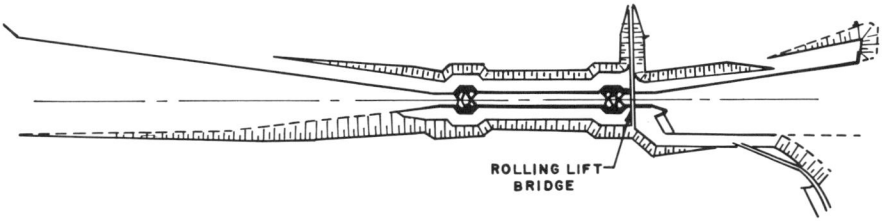

Welland Canal Section, Lake Ontario to Lake Erie

LOCKS	NORMAL LIFT m.	USABLE LENGTH m.	WIDTH OF CHAMBER m.	LENGTH-L/A 2 TO END OF WALL	
				UPPER ENT. m.	LOWER ENT. m.
LOCK 1	14	222.50	24.38	370	840
LOCK 2	14	222.50	24.38	510	459
LOCK 3	14	222.50	24.38	453	441
LOCK 4	15	222.50	24.38		291
LOCK 5	15	222.50	24.38		
LOCK 6	13	222.50	24.38		
LOCK 7	14	222.50	24.38	604	305
LOCK 8	0.5–3.5	350.0	24.38	E 469 W 380	E 240 W 345

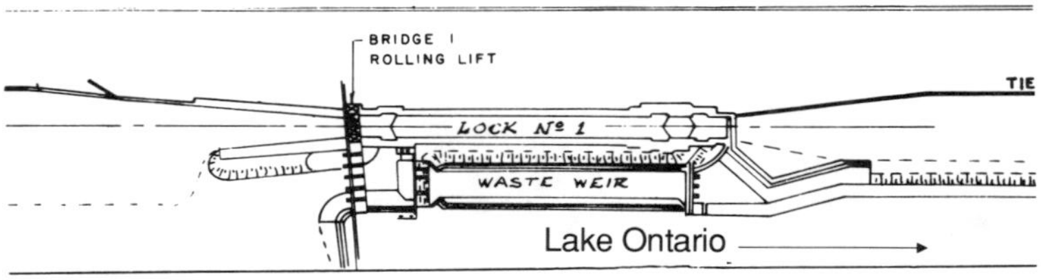

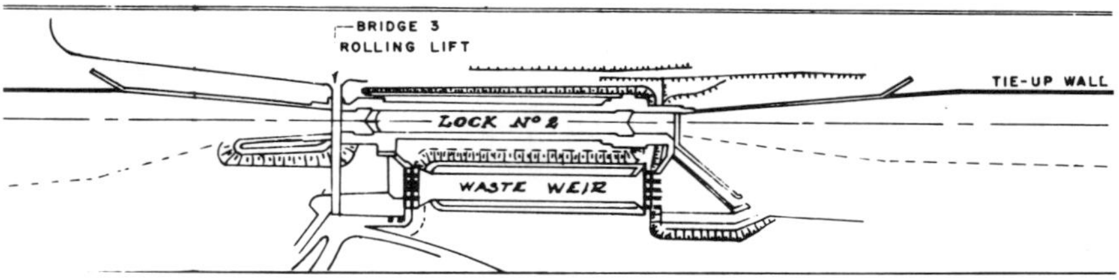

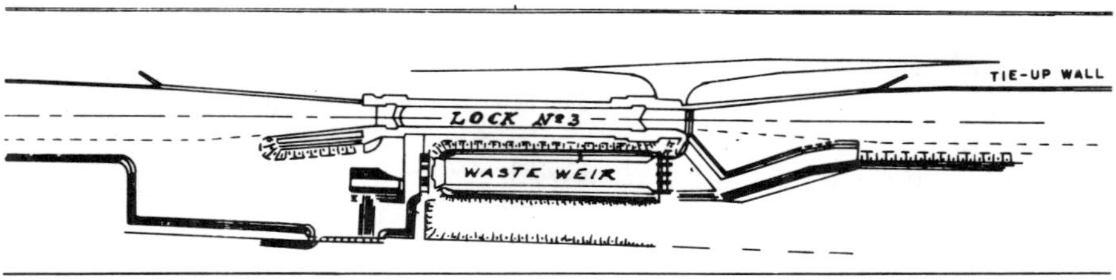

APPENDIX VI

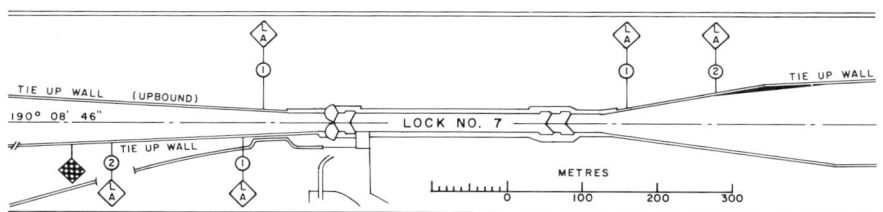

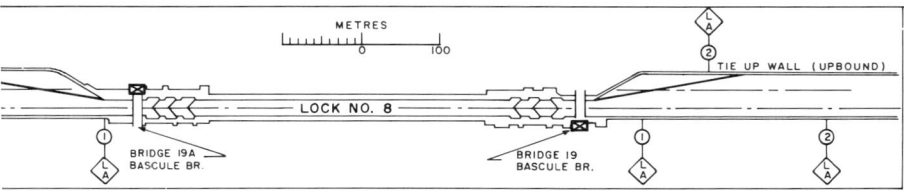

◄──── To Port Colborne and Lake Erie

MOORING TABLE FOR ST. LAWRENCE SEAWAY AND WELLAND CANAL LOCK WALLS

S = Starboard   Upb = Upbound
P = Port   Dnb = Downbound

MONTREAL TO IROQUOIS

	South Shore		Beauharnois			Wiley-Dondero		Iroquois
	St. Lambert	Cote Ste. Catherine	Lower	Pool	Upper	Snell	Eisen-hower	
Locks								
Upb	P	P	S		S	S	S	P
Dnb	S	S	P		P	P	P	S
Tie-up Walls								
Upb	S	S	P	P		S	S	S
Dnb	P	P		S	S	P	P	P

WELLAND CANAL

Locks	1	2	3	4	5	6	7	Guard Gate Cut	8
Upb	S	S	P	P	P	P	P		S
Dnb	P	P	S	P	P	P	S		P
Tie-up Walls									
Upb	S	S	S	S			S	S	P or S
Dnb	P	P	P				S	P	P or S

Mooring Procedure in Locks

Mooring lines shall only be placed on the mooring posts as directed by the Lockmaster, the lines leading astern normally being placed on the posts first, and winches from which the mooring lines run shall not be operated until the Lockmaster or a linesman has signalled that the line has been placed on the post.

All of the St. Lawrence Seaway and Welland Canal and approaches to the Sault have speed restrictions for all vessels. These are, when upbound, 7 to 8 MPH (over the bottom); when downbound, 6 MPH.

DIMENSIONS AND RESTRICTIONS FOR OTHER CANALS AND LOCKS

New York State Barge Canal Locks. 230′ × 40′ × 12′. There are some restrictions as far as the time of lockage for pleasure craft.

Chesapeake and Delaware Canal. Restricts ships to maximum length of 675′ and minimum safe speed in Chesapeake City area.

Tenn-Tom Canal. This 234-mile canal opened in 1985. It is dredged to 12′ and has ten locks, all of which are 600′ × 100′ in the lock chamber. An eight-barge tow with units of 195′ × 35′ drawing $8^{1}/_{2}$′ arranged in three tiers would be about 585′ × 105′. The Tenn-Tom Canal Authority advises that a towboat of between 1,800 to 2,400 HP could safely push such a tow at 6 MPH.

The canal is divided into three sections. The lower or river section is 149 miles of the dredged Tombigbee River. It has the following locks and dams: Gainesville, Aliceville, Columbus, and Aberdeen. The middle or canal section is known as the Chain of Lakes and is dredged to 12′ for a width of 300′. There are five locks, designated as A, B, C, D, and E. The northernmost section or divide section is 39 miles long, with only one lock, at Beaver Springs. The width in this section is 280′ with the same 12′ depth available. There is a maximum vertical clearance of 52′. For vessels and tows bound for the Tennessee, Kentucky, Ohio, or Mississippi above Cairo, and which have originated in the Gulf Intracoastal Waterway, it is a great shortcut, with little or no current until arriving in the Tennessee River, where it is favorable to the Ohio.

APPENDIX VI   545

Mississippi River Locks. From its mouth at the passes in the Gulf of Mexico to above Baton Rouge, the river is held by levees to keep it from flooding the lower land beyond on either side. As there are several places where exits from the main river are necessary, individual locks have been provided. Upbound they are: Ostrica Lock, a small exit for fishermen, oyster boats, and small tugs to Quarantine Bay and beyond. It is on the right ascending bank below the Ostrica Oil Terminal. On the left ascending bank at Empire there is a larger lock, which is privately maintained for fishing vessels and supply craft wishing to get into the many bayous beyond. It is $200' \times 40'$.

New Orleans and Mississippi

HARVEY CANAL

0.2M S HARVEY LOCK	T&P RR CO.	BASCULE	75	12	9
0.2M S HARVEY LOCK	4TH ST. (HWY. 18)	BASCULE	75	10	7
0.8M S HARVEY LOCK	HWY 90 (WESTBANK)	FIXED	300	100	97
2.8M S HARVEY LOCK	JEFF. PAR. (LAPALCO)	BASCULE	150	47	45

PORT ALLEN LOCK

0.2M W PORT ALLEN LOCK	STATE HWY. 1	FIXED	84	73	65
0.2M W PORT ALLEN LOCK	T&P RAILROAD	VERT LIFT	84	23	14
			OPEN	81	73

 # Less than 10 feet of clearance
+ Information is not available

Algiers Lock opposite Meraux on left ascending bank	$425' \times 75' \times 12'$	Mile 88 AHP
Inner Harbor Navigational Canal on right ascending bank above Chalmette (also known as Industrial Canal)	$640' \times 75' \times 35'$	Mile 92.4 AHP
Harvey Canal and Lock on left ascending bank above Jackson Avenue ferry, joins with Algiers Canal	$425' \times 75' \times 12'$	
Pearl River Locks #1, 2, 3	$310' \times 65' \times 10'$	
Warrior River Locks #13 and 16	$285.5' \times 53' \times 9.5'$	
Warrior River Locks #14 and 15	$282' \times 52' \times 9.5'$	
Columbia and Snake River Locks		
Bonneville	$500' \times 76' \times 24'$	
The Dalles	$675' \times 86'$	on
John Day	$675' \times 86'$	Columbia
McNary	$675' \times 86'$	
Ice Harbor	$664' \times 86'$	Restriction to
Lower Monument	$675' \times 86'$	$650' \times$
Little Goose	$675' \times 86'$	84' on
Lower Granite	$675' \times 86'$	Snake River

Hiram M. Chittenden Locks connect Puget Sound at Shilshole north of Seattle to Salmon Bay, Union Bay, Lake Union, and Lake Washington.

        Large Lock   760' × 80' × 29'
        Small Lock   123' × 28' × 16'

North East Sea Canal and Lock (Kiel, FRG). This 62-mile (98.6- kilometer) canal runs from Kiel Bay south of Denmark to the Elbe River at Brunsbüttel and is well used by ships and tugs and tows. Its locks are restricted to ships over 61,000 metric tons.

North Sea Locks and Canal. One of the largest and most unique set of locks allows shipping to enter through the North Sea dyke at IJmuiden. These locks lower vessels to Holland's below-sea-level 14-mile canal to Amsterdam. There are four sets of locks:

	Length	*Width*	*Depth*
The Big Lock	1,314.4' (400m)	164' (50m)	49.2' (15m)
The Middle Lock	738.2' (225m)	78.8' (24m)	32.8' (10m)
The Southern Lock	390.4' (119m)	56' (18m)	26.2' (8m)
The Small Lock	226.3' (69m)	39.4' (12m)	16.4' (5m)

The approach to the IJmuiden Locks can be very rough and dangerous until well inside the jetties. Regulations for transit are contained in the *North Sea Pilots* issued by the United States, the United Kingdom, The Netherlands, and other countries.

Small Boat Basins on Either End of Cape Cod Canal. On east end, 13 feet mean low water, on south side of Sandwich, available for mooring small boat traffic; on west end, channel 15 feet at mean low water, 100 feet wide leads from northeast side of Hog Island Channel abreast of Hog Island to harbor in Onset Bay. Fuel, supplies, and phone service at both locations.

Cape Cod Canal Regulations. Signal or traffic lights on eastern end at Sandwich (Cape Cod Bay Terminal); western end near Wings Neck Lighthouse (Buzzards Bay entrance)

Entering from eastern end: Lights south side of entrance to canal
    Red light: Vessels must stop clear of outer end of channel. Small vessels under 65 feet may proceed against red light to east mooring basin or east boat basin at Sandwich.
    Amber light: Proceed as far as east mooring basin dolphins opposite Sandwich terminal.
    Green light: Canal open to westbound traffic.

Entering from western end: Lights near Wings Neck Lighthouse
    Red light: Vessels must stop clear of channel. Small vessels under 65 feet may proceed against red light to the west mooring basin at Hog Island, or the State Pier, Buzzards Bay, and wait there until given clearance by radiotelephone, or Corps of Engineers patrol boat.
    Amber light: Proceed as far as the west mooring basin at Hog Island, and wait there for clearance as above.
    Green light: Canal open to eastbound traffic.
    When traffic lights are extinguished: all vessels over 65 feet cautioned not to enter canal until clearance given, as above.

APPENDIX VI 547

Two-way traffic through canal (vessels 65 feet and over) allowed when dispatcher on duty controlling traffic considers conditions suitable.

Clearance under highway bridges: 135 feet at mean high water. Buzzards Bay railroad bridge maintained in up, or open, position except when lowered for trains or maintenance purposes.

Minimum time limits of passage:

	Station 35 to 388	*Station 388 to 661*
Against head tide	60 minutes	46 minutes
With fair tide	30 "	23 "
Slack water	45 "	33 "

Special Note. Vessels that are to transit the Cape Cod Canal will monitor channel 16 continuously to establish contact with traffic controllers. The vessels will be asked to switch to channel 12 or 14 as a working channel to pass information between the traffic controller and the vessel. However, channel 13 may be used only when the above channels are not available. This is in accordance with Public Law 92-63, effective January 1, 1973.

As in previous years, the following lines of communication are still available: phone service at all hours, 759-4431, 4432; ship-to-shore, call letters W O U, Scituate, Massachusetts; radio telegraph, call letters W I M, W C C Chatham, Massachusetts.

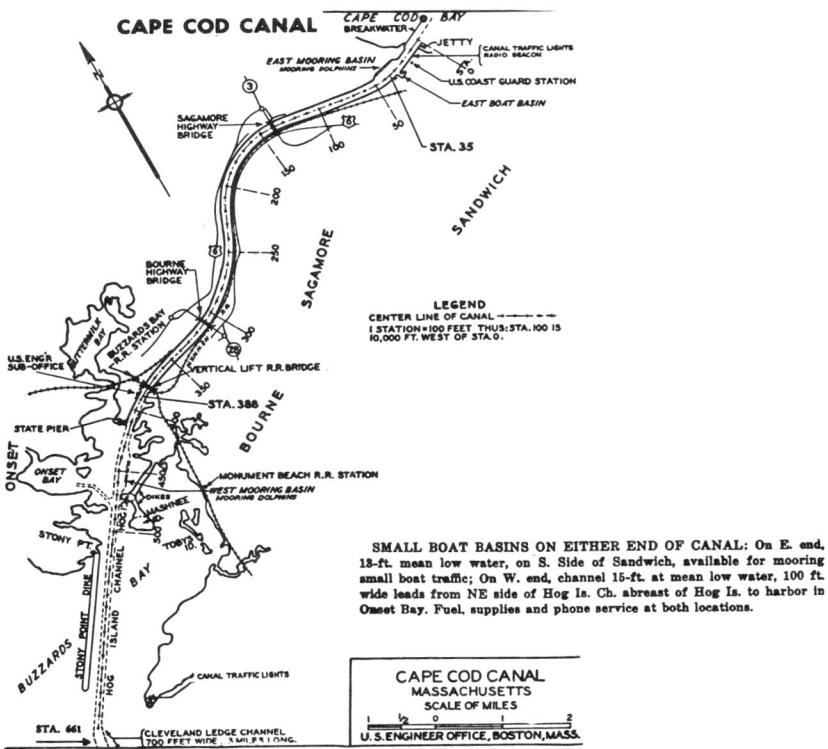

Appendix VII. Towing Arrangements and Procedures

TOWING ARRANGEMENTS REQUIRED OF SHIPS ON THE GREAT LAKES*

The diagrams below show the material a ship must furnish and how it must be rigged. Each of these towing pendants and fittings are passed to the tug and after completion of towage are hauled back aboard the ship. This is a standard method introduced years ago by Great Lakes Towing Company.

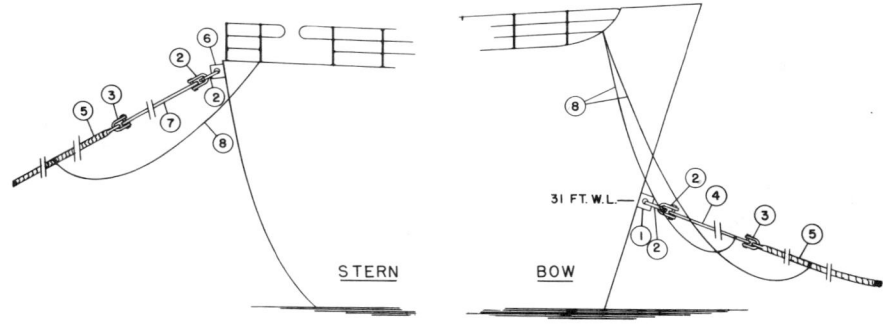

STERN   BOW

BILL OF MATERIAL		
ITEM	QUANTITY	DESCRIPTION
1	1	TOWING PAD (DETAILS WILL BE SUPPLIED UPON REQUEST).
2	4	SHACKLE - 2" FORGED STEEL, BOLT & NUT TYPE PIN.
3	2	SHACKLE - 1-1/4" FORGED STEEL, BOLT & NUT TYPE PIN.
4	1	PENDANT - 1" DIAM. 6 x 19 IMPROVED PLOW STEEL WITH WIRE CORE. LENGTH TO BE NOT LONGER THAN REQUIRED TO REACH CLOSEST POINT ON SHIP FOR TOWLINE ATTACHMENT.
5	2	TOWLINE - 8" CIRCUMFERENCE MANILA OR EQUIVALENT 20 FATHOMS IN LENGTH WITH A THIMBLE ON THE SHIP END AND BITTER SEIZING AT TUG END.
6	1	TOWING PAD - ON CENTERLINE OF VESSEL (DETAILS WILL BE SUPPLIED UPON REQUEST).
7	1	PENDANT - 1" DIAM. 6 x 19 IMPROVED PLOW STEEL WITH WIRE CORE. LENGTH TO BE 20 FT. MAXIMUM.
8	3	TRIP LINE - 3/4" DIAM. MANILA. FOR HAULING TOWLINES AND PENDANTS ABOARD.

*Courtesy Great Lakes Towing Company, 1800 Terminal Tower, Cleveland, Ohio 44113.

Towing Tackle. The arrangement of towing tackle for use in Great Lakes harbors necessarily differs from that found in many other areas due to the types of maneuvers which the nature of the harbors requires of tug and tow. Our early experience with Seaway shipping has indicated that some special provision for towing tackle on your ship is necessary for its safe handling in Great Lakes ports. We have therefore included a schematic drawing of the towing tackle arrangement which we believe is best designed for this purpose. We strongly urge that your ship be equipped with this tackle.

Due to the fact that the available channels in some Great Lakes harbor areas are quite narrow and further restricted by bridge draws, protruding piers and moored and moving vessels, tugs are required to make frequent and rapid changes in position relative to their tow. These are often coupled with equally rapid changes in belayment of the towline aboard the tug. Positioning of the tug alongside the ship as is done in many other areas is impossible, as are long towline leads.

The following are examples of the maneuvers to which we refer. When a single tug is assisting a loaded vessel in a left turn, the tug must change position from ahead of the vessel with the towline on its after bitts to alongside the vessel's port bow with the towline on the tug's starboard-breast cleat. It is impossible to properly assist the vessel with a towline leading from a high bow chock on the vessel. When a ship is being towed through narrow bridge draws or locks into narrow slips, the tug cannot exert the necessary power at acute angles to the vessel unless there is a short and nearly horizontal lead to the towline. Thus, again, lines leading from high bow chocks are not acceptable. When tugs are moving vessels stern-first into narrow slips, two 180° changes of position of the tugs, relative to the towed vessel, are required with four changes in belayment of the towlines aboard the tugs. This maneuvering requires the type of towing tackle described in our drawing. There are many other maneuvers which, not only in our own experience but in the experience of those masters who, since the opening of the Seaway, have become familiar with Great Lakes harbors, indicate the advisability of using our recommended tackle in substitution for towlines or bridles running from ships' chocks.

Wires of any kind are completely unacceptable for towing, as are towlines with eyes in the tug end. The use of tackle heavier than that indicated in our drawing is not desirable because of the difficulty inherent in handling.

Understanding the Maneuver. As you know, towing practices differ greatly from port to port, and even in our own operation local conditions require different procedures in various ports. This makes it of the utmost importance that there be a clear understanding between the vessel and tug as to all aspects of the maneuver that is to be undertaken.

We do not need to remind you that the safety of any maneuver depends upon this understanding between the tug and vessel. So long as it is not the practice to have a tug officer on your vessel during the towing maneuver, your understanding with the tug must be had by conversation with its captain before the maneuver commences. You will find our tug captain willing to give you the benefit of his knowledge of local conditions and customs, although in the last analysis he will rely upon your decision as to the manner in which you are to be towed or assisted. Once the tow is undertaken, it is a cooperative venture. Both tug and tow must be alert for the safety of both. In every case it is the intention of this Company to furnish you safe and efficient service.

Radio Equipment. Contact may be had with our tug dispatchers in the Buffalo, Cleveland, Duluth, and Chicago areas by means of VHF (FM) radiotelephone, and the same equipment can be used to communicate directly with our tugs in those areas for purposes other than ordering service. Our VHF equipment operates on the following frequencies and channels.

Safety and Calling—Channel 1—156.8 megacycles (Tug offices and tugs)
Ship-to-Shore—Channel X—156.6 megacycles (Tug offices)
First Intership—Channel 2—156.3 megacycles (Tugs)
Second Intership—Channel 4—156.6 megacycles (Tugs)

In the ports of Erie, Pennsylvania; Conneaut, Ashtabula, Fairport, Lorain, Huron, Sandusky, and Toledo, Ohio; Detroit and Sault Ste. Marie, Michigan; and Milwaukee, Wisconsin, direct contact for ordering service and for other communication may be had on MF (AM) radiotelephone. Our tugs in those ports carry MF (AM) equipment which operates on the following frequencies and channels.

Safety and Calling—Channel 51—2182 kilocycles (Tugs only)
Intership—Channel 40—2003 kilocycles (Tugs only)
Ship-to-Shore—Channel 39—Duplex frequency (2514 and 2118) for public cor-
 respondence

Appendix VIII. Tugs and Fishing Areas

TUGBOAT/FISHING LANE NEGOTIATIONS*

Background on This Agreement

As a result of mutual interference between West Coast crab fishermen and towboats with tows, a "gentlemen's agreement" was reached in 1971 to provide towing lanes for towboats along a major portion of the West Coast. Almost every year since, a meeting has been held to review and revise these towboat lanes; some significant changes have been made.

The general agreement is that crab fishermen will not put crab pots in the designated lanes. If they choose to do so, they forfeit the right to complain if tugs and tows destroy their pots.

The towboaters agree to stay within the designated lanes, or well outside the fishing areas, as long as weather and ship safety will allow.

The agreement has worked well over the years, saving many dollars for both fishing and towing industries. Regulatory authorities recognize the existence of this *voluntary* agreement; they have publicly stated they see no reason to involve themselves as long as the two industries—fishing and towing—are working to solve their own problems.

The OSU Extension/Sea Grant staff has acted as negotiator, administrator, and producer of charts for this effort. The mutual agreement is between the two industries; OSU Sea Grant provides a forum for the discussion and handles the printing and distribution of reports and chartlets that result from annual meetings.

List of Effective Charts Which Are Reviewed and Revised Annually

18500	Columbia River to Destruction Island	Revised 1987
18520	Yaquina Head to Columbia River	Revised 1987
18521	Columbia River, Pacific Ocean to Harrington Pt.	Revised 1987
18580	Cape Blanco to Yaquina Head	Revised 1987
18600	Trinidad Head to Cape Blanco	Revised 1987
18620	Point Arena to Trinidad Head	Revised 1987
18640	San Francisco to Point Arena	Revised 1987

If you're holding charts with revised dates earlier than these, please destroy them. Additional sets of charts are available from: Captain Gib Carter, OSU Extension/Sea Grant, 211 SE 80th Ave., Portland, OR 92715 (Tel. 503-254-1500) or Captain Boone Taylor, Executive Director, NW Towboat Association, 115 North 85th St., Seattle, WA 98103 (Tel. 206-782-3860).

*This 1987 agreement is reprinted with permission from the Oregon State University Extension/Sea Grant Program, Corvallis, Oregon 97331.

Incident Reporting Format

Please include this primary information when you report an incident:

Date _____

Position _____

Course of tug _____
  (and, generally, N or S bound)

Description of vessel _____
_____
  (name, captain, company, etc.)

Other (any information that may assist in identification)
_____

Note: The underlying purpose is to locate, and communicate with, the vessel or its owner. Many times, the captain is unaware of the lane agreement and/or does not have current, effective charts on board.

Radio Communications

Towboaters can be contacted on the following channels or frequencies:
  VHF—Channels 7A, 13, 16
  Medium Frequency—2182 kHz SSB
  Sideband—Foss 4143.6, 2429
      Sause Bros. 4125
      Crowley VHF channels 13 and 16; 4419 SSB
      Seaspan VHF channels 16 and 22

Fishermen can be contacted by VHF on the following channels:

  Fort Bragg—66   Coos Bay—88
  Crescent City—80, 72   Astoria—88A
  Brookings—88A   Grays Harbor—88
  Port Orford—9

Notice to Mariners. The U.S. Coast Guard will publish a *Notice to Mariners* near the start of the winter crab fishing season, reminding mariners of this agreement. In addition, timely notice will be broadcast on the USCG safety net at periodic intervals during the fishing season.

For further information, contact Captain W. G. Carter, Extension Marine Agent, Multnomah County Office, OSU Extension Service, 211 S.E. 80th, Portland, OR 97215, telephone 503-254-1500.

GILL NET FISHING—COAST GUARD NOTICE

Special Notice. Washington—Strait of Juan de Fuca—Admiralty Inlet—Puget Sound—Puget Sound Vessel Traffic Service—Information

The purpose of this notice is to announce a procedure intended to minimize conflict between commercial gillnet fishing activities and vessels using the Puget Sound Vessel Traffic Service (PSVTS) during All Citizen Gillnet Fishing periods. As a result of a Fall 1982 meeting with the Puget Sound Gillnetters Association, the Purse Seine Vessel Owners Association, the Northwest Towboat Association, the Washington State Ferry System, the Puget Sound Steamship Operators Association, and the Port Angeles Pilots, and the successful operation of this same procedure during the 1982 gillnet fishing season, the Commander, Thirteenth Coast Guard District has authorized temporary deviations to the Traffic Separation Scheme (TSS) Rules.

A temporary special traffic lane will be made available commencing at TSS buoy "SE" (Point No Point), and extending to TSS buoy "TC" (Browns Point). A line connecting traffic separation buoys "SE", "SF", "SG", "T", "TA", "TB", and "TC" will form the center line of the temporary special traffic lane. A line parallel to and one-quarter mile east of this center line will form the eastern boundary and a line parallel to and one-quarter mile west of this center line will form the western boundary of this special lane. Additionally, a temporary special one-quarter mile wide ferry traffic lane will be made available along that route normally used by the Edmonds-Kingston ferry. The lanes will be available on only those nights when All Citizen Gillnet Fishing is authorized.

Use of this special lane will be authorized for through traffic by the Vessel Traffic Service on a case-by-case basis as a deviation to the existing rules. Gillnetters are to avoid fishing within these relatively narrow special traffic lanes. Direct communications between all vessels involved are encouraged on VHF/FM Channel 13.

This temporary deviation to the Traffic Separation Scheme Rules is being implemented on a trial basis and does not modify any existing rules of the road. Continued direct radiotelephone communication between fishing vessels and VTS participants is encouraged. As always, good seamanship and safe navigation are of the utmost importance during periods of heavy congestion.

Washington—Strait of Juan de Fuca—San Juan Archipelago—Georgia Strait—Puget Sound—Hood Canal—Information

1. *Restricted Fishing Area:* The following restrictions apply to fishing within the waters of the COTP Seattle zone including Puget Sound, Commencement Bay, Elliott Bay, Possession Sound, the Strait of Juan de Fuca, San Juan Archipelago, Georgia Strait, Rosario Strait, and adjacent waters:

a) A tug with tow, whose intended course will take it through waters occupied by gill net gear, shall sound one long blast, followed by one short blast, of a whistle or horn, and during a darkness or fog shall, in addition, indicate its intended course by directing a searchlight beam on such course. Gill net fishermen operating within the indicated course of the tug shall draw in their gear or otherwise maneuver to permit passage of the tug and its tow without hindrance or unreasonable delay.

b) A tug without tow or any other vessel, if unable to determine the lay of the nets and doubt exists aboard the tug or vessel as to the best course to take, may request assistance of the nearest gill net boat which shall, without delay, drop its net and pilot the tug or vessel through. If assistance of a pilot boat is not obtainable or if nets are so concentrated as to make it impracticable to lay a course through the nets, the tug or vessel shall proceed as indicated in paragraph (1)(a) of this section for a tug with tow, and nets shall be lifted or maneuvered out of the way to permit passage of the tug or vessel without hindrance or unreasonable delay.

c) A boat with at least one man in it capable of controlling the net shall be in constant attendance upon each net while it is laid out, except when providing pilot service as provided in paragraph (1)(b) of this section.
2. *Prohibited fishing area in Possession Sound between Mukilteo and Columbia Beach.* Fishing with gill nets is prohibited within 440 yards on each side of a straight line connecting the ferry landings at Mukilteo and Columbia Beach.

NAVIGATING THROUGH A GILLNET FLEET

By Jim Humphreys (Washington Sea Grant Marine Advisory Services)

From mid-summer through Thanksgiving, a group of commercial fishermen in Puget Sound and the San Juan Islands are usually very busy trying to catch salmon with commercial gillnets. Because these fishermen work mainly at night, the boater must exercise extreme caution when navigating these waters during the gillnet fishing season.

Hitting a gillnet can cause extensive damage to a vessel as well as the net, and if a boat damages a legally marked gillnet, the boater may be financially liable for the damage. New gillnets cost up to $8000!

What is a gillnet?

A gillnet is a type of net that is designed to float in the water from the surface to a depth of 80 feet. Gillnets that are fished in Puget Sound waters are usually 1,800 feet long. The top of the net consists of a rope with corks attached every 3 to 5 feet. This rope is called the corkline. The bottom of the net is also a rope, but it is lead filled so that it will sink. This rope is called the leadline. In between the corkline and the leadline is very lightweight nylon webbing that has been manufactured into diamond-shaped openings that are 4 to 8 inches long. When salmon swim into these openings, they cannot back out because their gills act like barbs on fishing hooks.

What does a gillnet look like in water?

To recognize a gillnet while it is being fished requires patience, good eyesight, and cooperation between the boater and the fishermen. A *legally marked* gillnet will be 1,800 feet long; the end of the net will be marked with a white light, called a *jacklight;* and the boat fishing the net will display a red light over a white light. The boat will not display other lights while fishing, but usually will have several bright spotlights turned on to illuminate the working area in either the stern or bow of the boat. The jacklight end of the gillnet will also have a large colored buoy attached, but that buoy is not legally required.

Generally, a gillnet will be set by the fishermen in somewhat of a straight line perpendicular to the tidal flow, and the boat will be attached to the downwind end of the net. The nets usually are set in water deeper than 50 feet. Once the net is in the water, the fisherman will allow it to drift for 1 to 2 hours. Occasionally, the fisherman may check his net by running alongside it, and when he is doing that, both ends of the net will be marked with white-lighted jacklights. Under normal circumstances only one end will be marked with the jacklight, and the other end will be attached to the boat. Under these conditions you can expect to see the red over white light on the boat, the white jacklight at the end

APPENDIX VIII 555

of the net, and depending upon light conditions, the corkline which is floating at the surface. Sometimes these markers will be difficult to see because of poor light conditions at dawn or dusk, rough seas, or rainy and foggy weather. *At night the corkline is virtually impossible to see without the aid of a spotlight.*

How can a boater avoid a gillnet?
To avoid hitting a gillnet, a boater should follow a few basic guidelines.
  If you think nets are present, *slow down* and then try to identify the jacklight associated with the fishing vessel.
  Approach the fishing vessel directly. This greatly reduces the chance of hitting a net since the nets usually are set in a straight line.
  Carry a spotlight, as this will allow you to find the corkline when you get close to it. Without a spotlight, the corkline will be virtually impossible to see at night.
  When you get close enough to the boat to see the corkline and the lay of the net from the boat, simply run around the other end of the boat. If there are a considerable number of gillnet boats, you can avoid the nets by running from boat to boat.
  Position someone on the bow to watch for the corkline.
  Operate the boat from the flybridge if you have one.

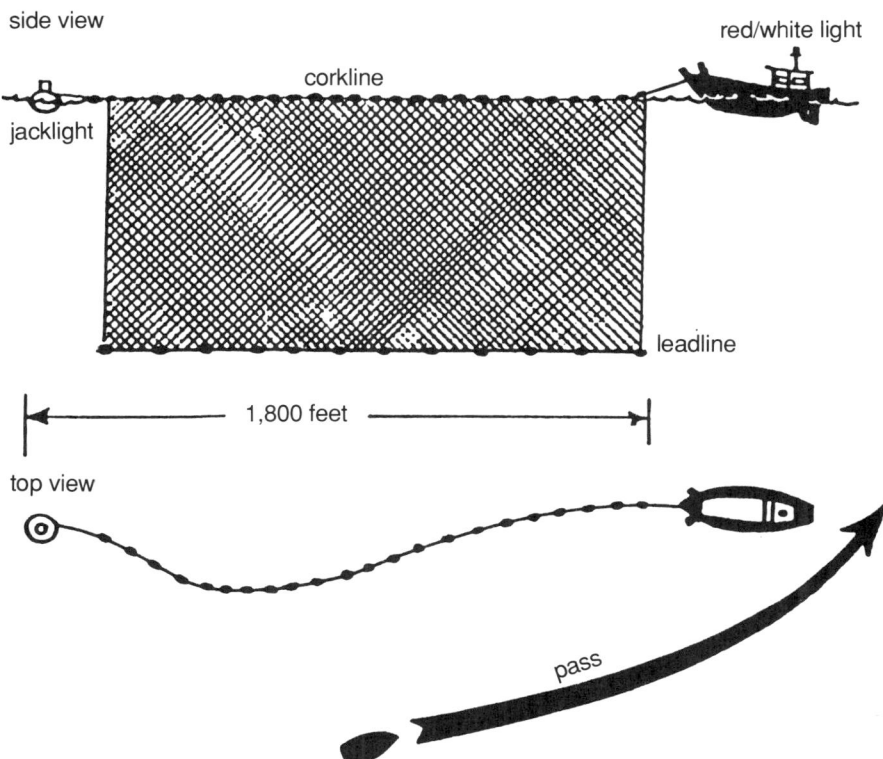

How can a gillnetter help a boater avoid a gillnet?
A gillnetter can follow several basic guidelines to help other boaters to avoid his nets.
- Mark your nets and boat properly while fishing. This not only helps the boater to see the net, but also puts the gillnetter legally in the right if his net is hit.
- As a boat approaches your net, tighten your net to pull the corks to the surface and flash a spotlight down the net in the direction of the jacklight. This will enable the boater to find the corkline. Once the boater sees the corkline, he should signal the fisherman with a single acknowledging flash of a spotlight, running lights, or a flashlight, and then easily and safely pass around the other end of the gillnet boat.
- If there are several nets in the area try to make radio contact with the boat to help the boater to pass safely through the fleet.
- Use frantic or rapid flashing to tell the boater to stop immediately if his boat is about to hit or has just hit the net. A blast with a horn or siren will also help to get the boater's attention.

What if a boat hits a gillnet?
If you should hit a gillnet, remain calm and make the best of a bad situation.
- Immediately try to get your vessel stopped by shifting into neutral. Continued running will only worsen the situation.
- If you have net and corkline wound in your propeller, wait for the fisherman to arrive to assist you in untangling. Most fishermen have some experience in how to free net from the wheel.
- Never enter the water to free net from your wheel, as you can easily become entangled in the net.
- Never "blast" your way out of the net. This will only do more damage to the net, your boat and engine.
- If damage is done, be sure to exchange names and insurance companies as that information will expedite any claims.

Adapted from a Puget Sound Gillnetter's Association fact sheet by Bill Sibbett.

Appendix IX. Mississippi River Data

TOWING PROCEDURES AND TRAFFIC LIGHTS

NEW ORLEANS VESSEL TRAFFIC SERVICE

GENERAL RULES

§ 161.401 Purpose and applicability.

(a) Sections 161.401 and 161.402 prescribe rules for vessel operation in the New Orleans Vessel Traffic Service Area (VTS Area) to prevent collisions and groundings and to protect the navigable waters of the VTS Area from environmental harm resulting from collisions and groundings.

(Sec. 104, Pub. L. 92-340, 86 Stat. 424 (33 U.S.C. 1224); 49 CFR 1.46(n)(4))

[CGD 75-082, 42 FR 51759, Sept. 29, 1977]

§ 161.402 Vessel operation.

(a) Mississippi River below Baton Rouge, LA, including South and Southwest Passes:

(1) *Supervision.* The use, administration, and navigation of the waterways to which this paragraph applies shall be under the supervision of the District Commander, Eighth Coast Guard District.

(2) *Speed; high-water precautions.* When passing another vessel (in motion, anchored, or tied up), a wharf or other structure, work under construction, plant engaged in river and harbor improvement, levees withstanding flood waters, building partially or wholly submerged by high water, or any other structure liable to damage by collision, suction or wave action, vessels shall give as much leeway as circumstances permit and reduce their speed sufficiently to preclude causing damages to the vessel or structure being passed. Since this subparagraph pertains directly to the manner in which vessels are operated, masters of vessels shall be held responsible for strict observance and full compliance therewith. During high river stages, floods, or other emergencies, the District Commander may prescribe by navigation bulletins or other means the limiting speed in land miles per hour deemed necessary for the public safety for the entire section or any part of the waterways covered by this paragraph, and such limiting speed shall be strictly observed.

(3) *Towing.* Towing in any formation by a vessel with insufficient power to permit ready maneuverability and safe handling is prohibited.

(b) Movement of vessels in vicinity of Algiers Point, New Orleans Harbor:

(1) *Control lights.* When the Mississippi River reaches 8 feet on the Carrollton Gage on a rising stage, and until the gage reads 9 feet on a falling stage, the movement of all tugs with tows and all ships, whether under their own power or in tow, but excluding tugs or towboats without tows or river craft of comparable size and maneuverability operating under their own power, in the vicinity of Algiers Point shall be governed by red and green lights designated and located as follows: Governor Nicholls Light located on the left descending bank on the wharf shed at the upstream end of Esplanade Avenue Wharf, New Orleans, approximately 94.3 miles above Head of Passes; and Gretna Light located on the right descending bank on top of the levee at the foot of Ocean Avenue, Gretna, approximately 96.6 miles above Head of Passes. Governor Nicholls Light has lights visible from both upstream and downstream, and Gretna Light has lights visible from upstream, all indicating by proper color the direction of traffic around

Algiers Point. From downstream, Gretna Light always shows green. All lights are visible throughout the entire width of the river and flash once every second. A green light displayed ahead of a vessel (in the direction of travel) indicates that Algiers Point is clear and the vessel may proceed. A red light displayed ahead of a vessel (in the direction of travel) indicates that Algiers Point is not clear and the vessel shall not proceed. Absence of lights shall be considered a danger signal and no attempt shall be made to navigate through the restricted area.

(2) *Ascending vessels.* Ascending vessels shall not proceed farther up the river than a line connecting the upper end of Atlantic Street Discharge Light (on right descending bank) with the lower end of Desire Street Wharf (on left descending bank) when a red light is displayed. Vessels waiting for a change of signal shall keep clear of descending vessels.

(3) *Descending vessels.* (i) Descending vessels shall not proceed farther down the river than a line connecting the lower end of Julia Street Wharf (on left descending bank) with the vertical flagpole at Eastern Associated Terminals (on right descending bank) when a red light is displayed. Vessels shall round to and be headed upstream before they reach that line, if the signal remains against the vessel. Vessels waiting for a change of signal shall keep clear of ascending vessels.

(ii) Vessels destined to a wharf above the lower end of Julia Street Wharf shall signal the Gretna towerman three long blasts and one short blast of a whistle or horn to indicate that the vessel is not bound below the Julia Street Wharf.

(iii) The master, pilot, or authorized representative of any vessel scheduled to depart from a wharf between Governor Nicholls Light and Louisiana Avenue, bound downstream around Algiers Point, shall communicate with the Governor Nicholls Light towerman by telephone to determine whether the channel at Algiers Point is clear before departure. When the point is clear, vessels shall then proceed promptly so that other traffic will not be unnecessarily delayed.

NOTE: Telephone numbers of both signal towers will be published in navigation bulletins in advance of each operating period.

(4) *Minor changes.* The District Commander is authorized to waive operation or suspension of the lights whenever prospective river stages make it appear that the operation or suspension will be required for only a brief period of time or when river stages will rise or fall below the critical stage which is established for operation or suspension by only a few tenths on the Carrollton Gage.

(5) *Underpowered vessels.* When the Carrollton Gage reads 12 feet or higher, any vessel which is considered by the master or pilot as being underpowered or a poor handler shall not navigate around Algiers Point without the assistance of a tug or tugs.

(6) *Towing.* When the Carrollton Gage reads 12 feet or higher, towing on a hawser in a downstream direction between Julia Street and Desire Street is prohibited except by special permission of the District Commander.

(c) Navigation of South and Southwest Passes.

(1) No vessel, except small craft and towboats and tugs without tows, shall enter either South Pass or Southwest Pass from the Gulf until after any descending vessel which has approached within two and one-half (2½) miles of the outer end of the jetties and visible to the ascending vessel shall have passed to sea.

(2) No vessel having a speed of less than ten mph shall enter South Pass from the Gulf when the stage of the Mississippi River exceeds 15 feet on the Carrollton Gage at New Orleans. This paragraph does not apply when Southwest Pass is closed to navigation.

(3) No vessel, except small craft and towboats and tugs without tows, ascending South Pass shall pass Franks Crossing Light until after a descending vessel shall have passed Depot Point Light.

(4) No vessel, except small craft and towboats and tugs without tows, shall enter the channel at the head of South Pass until after an ascending vessel which has reached Franks Crossing Light shall have passed through into the river.

(5) When navigating South Pass during periods of darkness no tow shall consist of more than one towed vessel other than small craft, and during daylight hours no tow shall consist of more than two towed vessels other than small craft. Tows may be

in any formation. When towing on a hawser, the hawser shall be as short as practicable to provide full control at all times.

(6) When towing in Southwest Pass during periods of darkness no tow shall consist of more than two towed vessels other than small craft, and during daylight hours no tow shall consist of more than three towed vessels other than small craft.

(Sec. 104, Pub. L. 92-340, 86 Stat. 424 (33 U.S.C. 1224); sec. 2, Pub. L. 95-474, 92 Stat. 1471 (33 U.S.C. 1223); 49 CFR 1.46(n)(4))
[CGD 75-082, 42 FR 51759, Sept. 29, 1977, as amended by CGD 78-080, 44 FR 47933, Aug. 16, 1979]

FERRY CROSSINGS AND HAWSER LENGTH

Ferry Crossings. In addition to ascending and descending river traffic, many ferries cross the Mississippi River above the Head of Passes. This is the point where South and Southwest Pass join the main river a few miles below Pilottown. Head of Passes is used as a reference in statute miles, for points upriver from there. As an example, Pointe a La Hache Ferry is 48.6 AHP or 48.6 statute miles above Head of Passes.

MM (AHP)	Name
35.4	Bass Enterprises Private Ferry
48.6	Pointe a La Hache Ferry
75.7	Belle Chasse/Scarsdale Ferry
88.8	Chalmette Ferry
92.8	Navy Shuttle Boat
95.0	Canal Street/Algiers Ferry
97.2	Jackson Ave./Gretna Ferry
137.9	Reserve/Edgard Ferry
147.6	Lutcher/Vacherie Ferry
191.2	White Castle/Carville Ferry
207.7	Plaquemine Ferry

Hawser Length. Towing hawser—amount used in entering and exiting passes with an empty or loaded 20,000- to 30,000-ton dry or liquid cargo barge:

Empty barge entering either pass	One full layer of 6×37, $2\ 1/2''$ wire off winch drum
Loaded barge passing in SW Pass	One full layer
Loaded barge passing out SW Pass	One-half layer or less until out clear of Sea Buoy

GAUGE BOARDS AND RELATION TO BRIDGES

BRIDGES —VS— GAGE LOCATIONS

LOCATION MILE	NAME	GAGE	LOCATION MILE
MISSISSIPPI RIVER			
95.6AHP	GREATER NEW ORLEANS	CARROLLTON	102.8AHP LDB COE DOCK
95.7AHP	GREATER NEW ORLEANS	CARROLLTON	102.8AHP LDB COE DOCK

(continued)

LOCATION MILE	NAME	GAGE	LOCATION MILE
MISSISSIPPI RIVER			
106.0AHP	HUEY P. LONG	CARROLLTON	102.8AHP LDB COE DOCK
121.6AHP	HALE BOGGS	RESERVE	138.7AHP LDB WHARF
145.9AHP	GRAMERCY	RESERVE	138.7AHP LDB WHARF
167.5AHP	SUNSHINE	DONALDSONVILLE	
			175.4AHP RDB BAYOU LAFOURCHE WATER INTAKE
229.2AHP	I-10	BATON ROUGE	228.4AHP RDB SOUTH WALL PORT ALLEN LOCK

BRIDGES

LOCATION MILE	NAME	TYPE	CLEARANCE HORZ	VERT LW	HW
MISSISSIPPI RIVER GULF OUTLET					
60.8 MR-GO	PARIS ROAD	FIXED	500	140	135

Note: National Ocean Survey Charts 11367 and 11369 show a vertical clearance of 135 feet for the Paris Road Bridge over the Mississippi River Gulf Outlet at mile 13.0 east of the Harvey Canal Lock. This is the minimum vertical clearance, including hurricane effects, and was taken from the Coast Guard bridge permit drawing. A more normal measurement of the vertical clearance, excluding hurricane effects, is as follows:

Vertical Clearance Across 500 ft. Channel (in feet)	Above Water Level (Excludes Hurricane Effects)	Elevation NGVD = MSL (in feet)
137.61	Average Annual High Tide	3.03
142.31	Average Annual Low Tide	−1.67
140.01	Average Daily Tide	0.63

LOCATION	NAME	TYPE	HORZ	VERT LW	HW
INDUSTRIAL CANAL					
8.5M N IHNC LOCK	DANZIGER	BASCULE	100	+	9
8.3M N IHNC LOCK	I-10	FIXED	250	+	115
1.0M N IHNC LOCK	FLORIDA AVE.	BASCULE	91	___	___#
AT IHNC LOCK	CLAIBORNE AVE.	VERT LIFT	305	45	40
			OPEN	160	156
AT IHNC LOCK	ST. CLAUDE AVE.	BASCULE	75	+	16
ALGIERS CANAL					
1.0M S ALGIERS LOCK	STATE HWY. 407	VERT LIFT	125	___	___#
			OPEN	103	100
3.7M S ALGIERS LOCK	MOPAC RR	VERT LIFT	125	___	___#
			OPEN	103	100

APPENDIX IX

BRIDGES

Note: Low water reference plane does not necessarily correspond with "O" readings on the gages.

Vertical Clearances under the Greater New Orleans and the Huey P. Long Bridge. Information extracted from Local Notice to Mariners Number 51-82. All figures are for midspan.

CARROLLTON GAGE	GREATER NEW ORLEANS BRIDGE		HUEY P. LONG BRIDGE	
	MID 750 FEET MAIN SPAN		MID 500 FEET MAIN SPAN	
FEET	FEET	METERS	FEET	METERS
0.0	170.0	51.82	153.0	46.64
2.0	168.0	51.21	151.0	46.03
4.0	166.0	50.60	149.0	45.42
6.0	164.0	49.99	147.0	44.81
8.0	162.0	49.39	145.0	44.20
10.0	160.0	48.78	143.0	43.59
12.0	158.0	48.17	141.0	42.98
14.0	156.0	47.56	139.0	42.37
16.0	154.0	46.95	137.0	41.76
18.0	152.0	46.34	135.0	41.15
20.0	150.0	45.73	133.0	40.54

Vertical Clearances under the Hale Boggs Bridge and Gramercy Bridge. Information extracted from Local Notice to Mariners Number 51-82. All figures are for midspan.

RESERVE GAGE	HALE BOGGS BRIDGE		GRAMERCY BRIDGE	
	Full 1200 Feet Main Span			
FEET	FEET	METERS	FEET	METERS
0.0	158.4	48.28	164.9	50.26
2.0	156.6	47.73	162.9	49.66
4.0	154.8	47.19	160.8	49.01
6.0	153.0	46.64	158.7	48.37
8.0	151.2	46.09	156.5	47.70
10.0	149.4	45.54	154.4	47.06
12.0	147.6	44.99	152.3	46.42
14.0	145.8	44.44	150.2	45.78
16.0	144.0	43.89	148.1	45.14
18.0	142.2	43.34	146.0	44.50
20.0	140.4	42.80	143.8	43.83
22.0	138.6	42.25	141.7	43.19
24.0	136.8	41.70	139.6	42.55
26.0	135.0	41.15	137.5	41.91

VESSEL TRAFFIC SERVICE (VTS) SECTORS

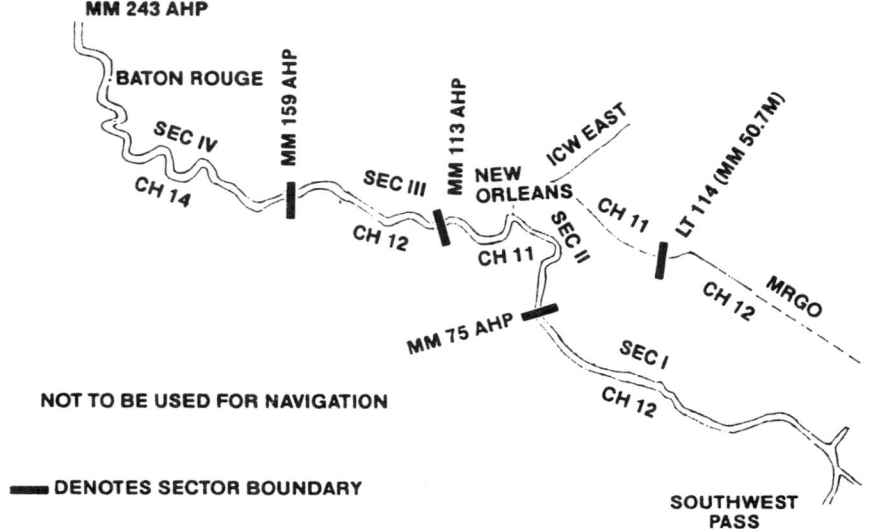

NOT TO BE USED FOR NAVIGATION

▬▬ DENOTES SECTOR BOUNDARY

a. Sector I, (Channel 12)—The Mississippi River from Southwest Pass Entrance Midchannel Lighted Buoy to the Belle Chasse Ferry Landing (LDB), mile 75.5 AHP, and from the Mississippi River—Gulf Outlet Lighted Bell Buoy "2" to MR-GO Light "114," mile 50.7 MR-GO.

b. Sector II, (Channel 11)—the Mississippi River from mile 75.5 AHP to Guy Mallory Light, mile 113.0 AHP, and from Light "114" to mile 60.0 MR-GO and westerly along the GIWW and the IHNC.

c. Sector III, (Channel 12)—The Mississippi River from mile 113.0 AHP to Shell Pipe Line Corp. Dock Lights, mile 159.5 AHP.

d. Sector IV, (Channel 14)—The Mississippi River from mile 159.5 AHP to Devil's Swamp Light, mile 242.4 AHP.

VTS REPORTING POINTS

a) Mississippi River (LDB—left descending Bank; RDB right descending bank).

NAME	LOCATION	SECTOR	CHANNEL
Southwest Pass Entrance Mid-Channel Lighted Whistle Buoy	28-52.7N 89-25.9W	I	12
Scott's Canal Lt "23"	RDB Mile 4.9 BHP	I	12

APPENDIX IX 563

a) **Mississippi River** (LDB—left descending Bank; RDB right descending bank).

NAME	LOCATION	SECTOR	CHANNEL
Old Quarantine Station	LDB Mile 4.0 AHP	I	12
Baptiste Collette Bayou Light	LDB Mile 11.3 AHP	I	12
Bolivar Point Light	LDB Mile 22.0 AHP	I	12
Nestor Canal Light	LDB Mile 39.7 AHP	I	12
Phoenix Tower	LDB Mile 57.6 AHP	I	12
Belle Chasse Ferry Landing (Sector Boundary)	LDB Mile 75.5 AHP	I/II	12/11
Inner Harbor Navigational Canal Light "2"	LDB Mile 92.6 AHP	II	11
Harvey Lock	RDB Mile 98.3 AHP	II	11
Guy Mallory Light (Sector Boundary)	LDB Mile 113.0 AHP	II/III	11/12
Taft Light	RDB Mile 127.3 AHP	III	12
Cargill Grain	LDB Mile 139.4 AHP	III	12
Crescent Light	RDB Mile 148.8 AHP	III	12
Shell Pipe Line Dock Light (Sector Boundary)	RDB Mile 159.5 AHP	III/IV	12/14
Sunshine Bridge	Mile 167.5 AHP	IV	14
Bayou Lafourche Intake	RDB Mile 175.4 AHP	IV	14
Alhambra Range Light	LDB Mile 189.4 AHP	IV	14
Georgia Pacific Wharf	RDB Mile 205.5 AHP	IV	14
Mulberry Grove Light	LDB Mile 217.3 AHP	IV	14
Port Allen Lock Upper Forebay	RDB Mile 228.4 AHP	IV	14
Devil's Swamp Light	LDB Mile 242.4 AHP	IV	14

b) **Mississippi River—Gulf Outlet (ML-MRGO Lower, M-MRGO Upper)**

NAME	LOCATION	SECTOR	CHANNEL
MRGO Lighted Bell Buoy "2"	Mile 9.4 ML	I	12
MRGO Lighted Bell Buoy "20"	Mile 0.2 ML	I	12
MRGO Light "62"	Mile 20.1 M	I	12
MRGO Light "94"	Mile 36.4 M	I	12
MRGO Light "114" (Sector Boundary)	Mile 50.7 M	I/II	12/11
Junction MRGO & GIWW	Mile 60.0 M	II	11
Junction GIWW & Inner Harbor Navigational Canal	Mile 66.0 M	II	11
Industrial Locks	Mile 68.4 M	II	11

Appendix X. Harbors, Terminals, and Anchorages

The following detailed information about selected terminals and anchorages is representative of the kind of data tug masters will want to have available on their tugs. Similar information is available for many other areas of the world.

NEW YORK

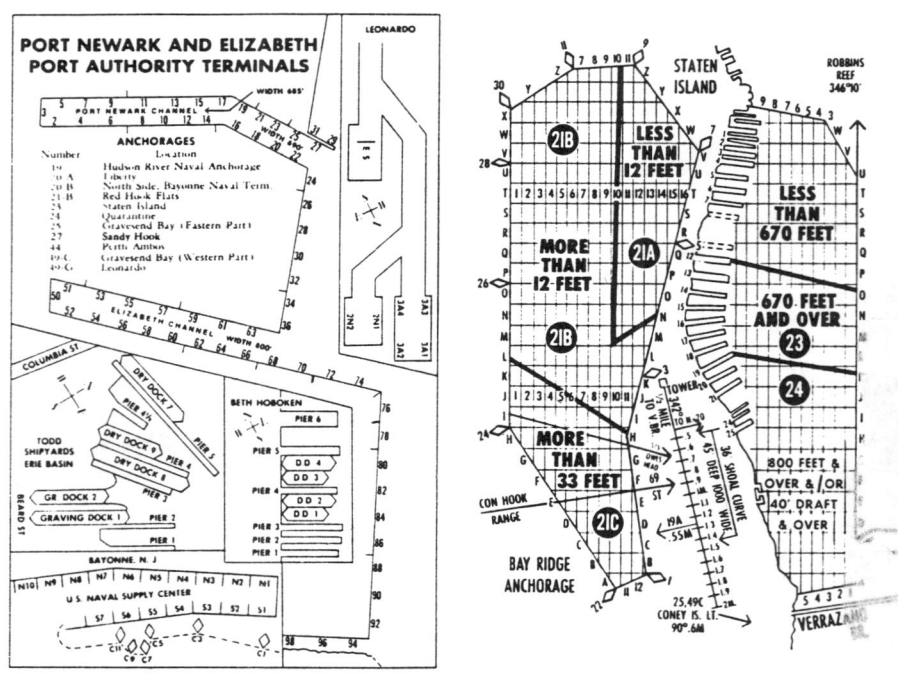

APPENDIX X

SAN FRANCISCO

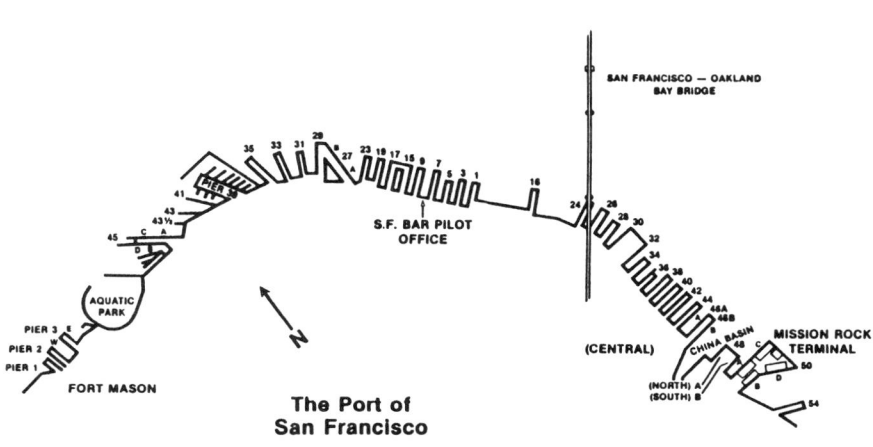

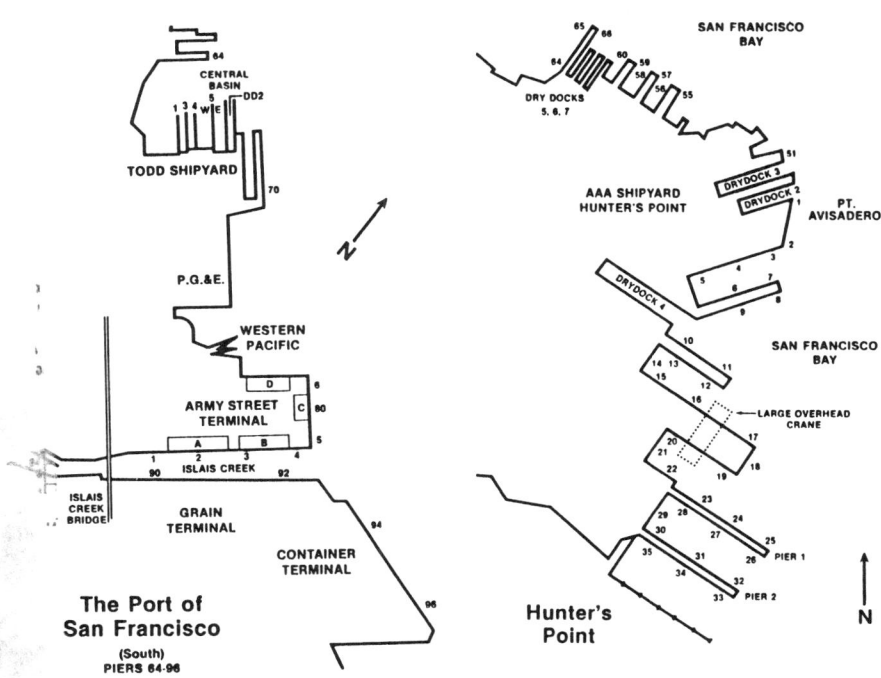

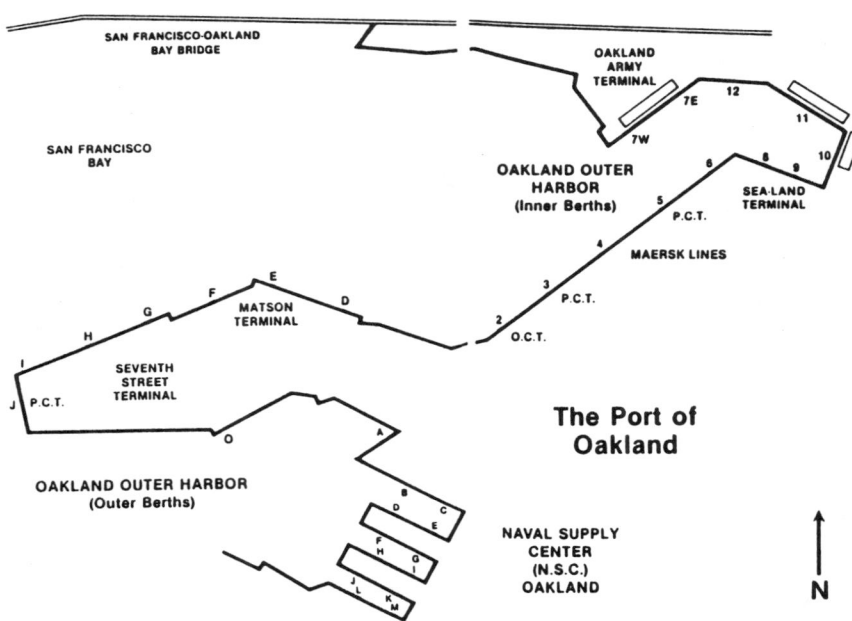

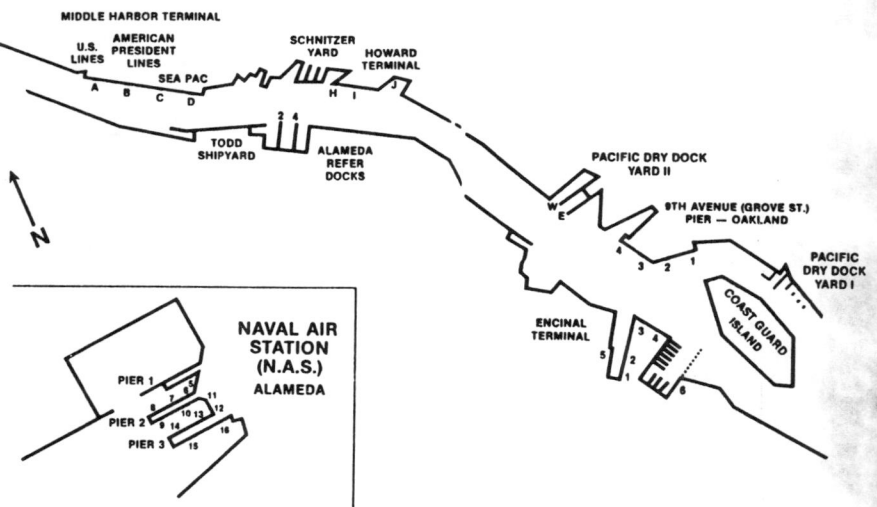

APPENDIX X 567

REDWOOD CITY

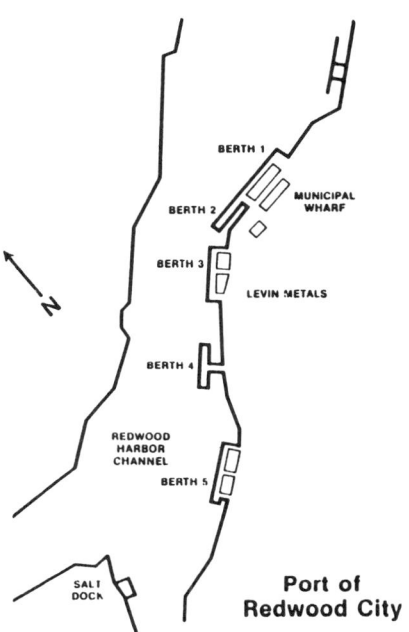

RICHMOND

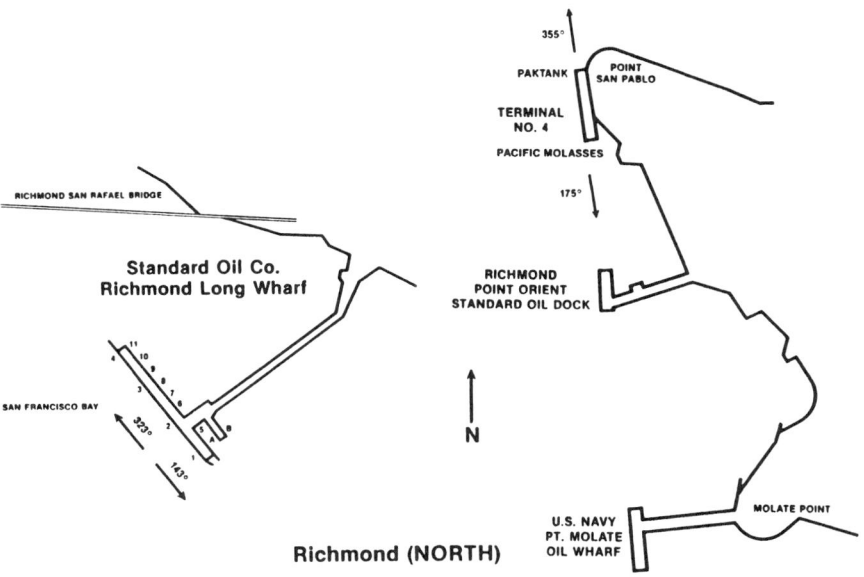

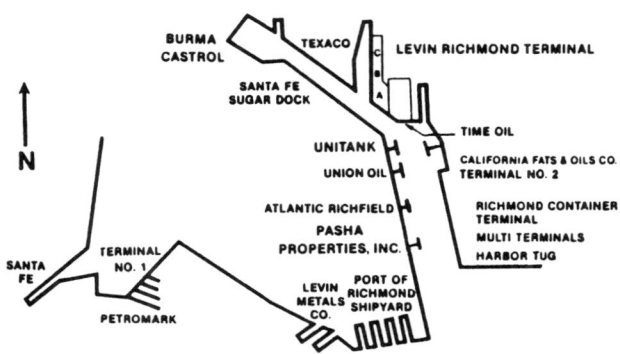

Richmond Inner Harbor

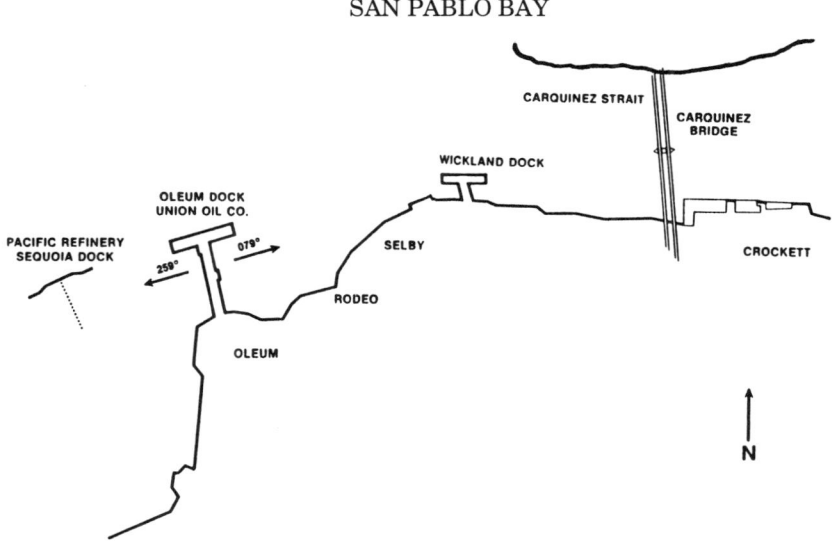

APPENDIX X 569

SACRAMENTO

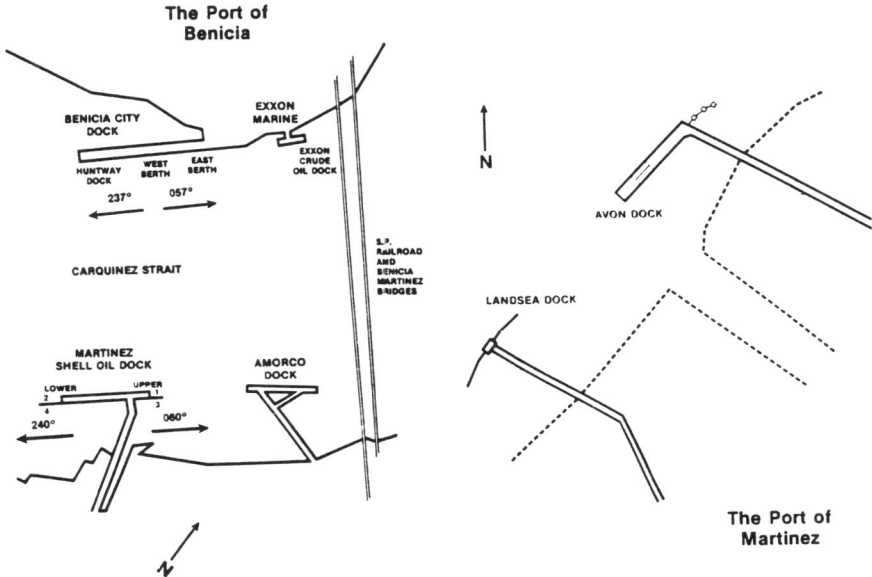

Port Chicago

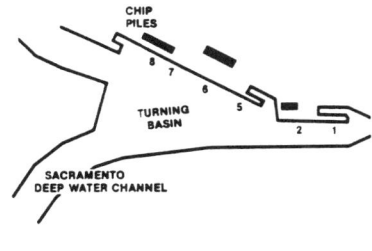

THE PORT OF SACRAMENTO

STOCKTON

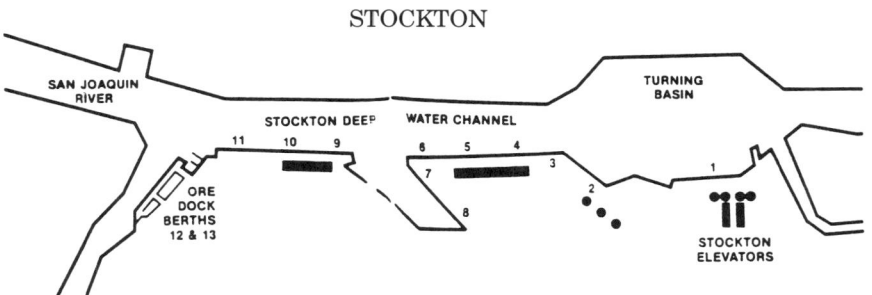

Appendix XI. Health and Diet for Tug Crews

SUBSTITUTIONS FOR EXCESS CALORIES

Cutting excess calories is the first step in slimming the body.

For this item	*Calories*	*Substitute this item*	*Calories*
1 cup whole milk	170	1 cup skim milk or buttermilk	80
1 scrambled egg	115	1 poached or boiled egg	70
1 cup apple juice	120	1 cup tomato juice	35
12 oz. can soft drink	150	12 oz. can diet soda drink	1
1 piece of apple pie	345	Baked apple with cinnamon	80
5 fish sticks	200	4 oz. broiled fish	150
3 oz. average hamburger	240	3 oz. lean hamburger	150
Fried potatoes, 1 cup	480	Baked potato	70
10 potato chips	115	10 pretzels	35
1 cup cream soup	210	1 cup chicken noodle soup	110
1 cup corn	185	1 cup green beans	25
12 oz. beer	150	12 oz. light beer	100
1 cup shelled peanuts	1,375	1 cup plain popcorn	25
3 oz. pork chop	340	3 oz. roast chicken	160
1 teaspoon blue cheese dressing	120	1 teaspoon vinegar and oil	60
1 cup coffee, 2 lumps	50	1 cup coffee black	5

Many of the items in the left-hand column are also high in fat and cholesterol.

WORDS VITAL TO YOUR HEALTH

Cholesterol. This is a waxy material used in many of the body's chemical processes. Everyone requires it in correct amounts for good health, but too much cholesterol in the blood encourages the development of heart and blood vessel diseases. The body obtains cholesterol in two ways: It is manufactured in the body, and it is absorbed directly from foods of animal origin. Egg yolks and organ meats are very high in cholesterol, and shrimp is moderately high. These foods should be restricted in a modified fat, low-cholesterol diet. There is no cholesterol in foods of plant origin, such as fruits, vegetables, grains, cereals, and nuts, and these foods are recommended.

Saturated Fats. These fats, which harden at room temperature, tend to raise the level of cholesterol in the blood and are therefore restricted in this diet. They are found in most animal products and some hydrogenated vegetable products and in coconut and palm oil (used in commercially prepared cookies, pie fillings, and in nondairy milk and cream substitutes). Saturated animal fats are found in beef, lamb, pork, ham, butter, cream, and whole milk.

Polyunsaturated Fats. These are recommended in this meal plan and are usually liquid oils of vegetable origin. Oils such as corn, cottonseed, safflower, sesame seed, soybean, and sunflower seed are high in polyunsaturated fat. The fat in fish and walnuts is polyunsaturated. These polyunsaturated fats and oils tend to lower the level of cholesterol in the blood. Your daily use of salad dressing, cooking fats, and margarines should emphasize the recommended polyunsaturated vegetable oils for their cholesterol-lowering effect.

Monounsaturated Fats. These fats neither contribute to atherosclerosis nor help prevent it. These are fats such as those found in olive oil, avocado, peanuts, and most other nuts. If you do not need to lose weight, you may want to use them for flavor occasionally, but they should not take the place of polyunsaturated vegetable oils.

A fat is never completely saturated, polyunsaturated, or monounsaturated. For practical purposes, however, it is one of these. This distinction depends on the chemical makeup of the particular fat. If it is largely composed of polyunsaturated fatty acids, it is called a polyunsaturated fat.

Triglyceride. This is the chemical term for fat. When physicians say that the triglycerides are too high, they are referring to the level of triglycerides in the blood. There are laboratory tests for the triglyceride levels as well as the cholesterol levels in the blood. If both, or either, are excessive, the doctor will probably prescribe a specific low-saturated-fat, low-cholesterol diet. With some patterns of high triglyceride levels, controlling the sugar in the diet is necessary along with the low-saturated-fat, low-cholesterol diet.

Hydrogenation. This changes liquid fats to solid fats. Completely hydrogenated (hardened) oils resemble saturated fats and should be avoided or used in moderation.

Fat-Controlled Diet. This is a cholesterol-lowering diet in which polyunsaturated fat intake is increased to equal that of saturated fats and total fat and cholesterol are moderately reduced.

Atherosclerosis. A disease in which there is a thickening and narrowing of the arteries (major blood vessels). The mushy deposit of fat, cholesterol, and other materials in the inner layer of the arterial wall interferes with the normal blood flow and the nourishment of the tissues. Atherosclerosis is often called "hardening of the arteries" and is the basic cause of most heart attacks and strokes.

Coronary Heart Disease. The most common form of adult heart disease, in which the main arteries of the heart (the coronary arteries) have atherosclerotic deposits and the normal blood flow to the heart is impaired.

DESIRABLE WEIGHT RANGES FOR ADULTS

	Height *(without shoes)* *(inches)*	*Weight* *(without clothes)* *(pounds)*
Men:	64	122–144
	66	130–154
	68	137–165
	70	145–173
	72	152–182
	74	160–190
Women:	60	100–118
	62	106–124
	64	112–132
	66	119–139
	68	126–146
	70	133–155

COOKING HINTS

Pam spray and other brands all work very well for baking and browning meats and vegetables.

Sauté vegetables using water instead of fat.

Make your own bread, leaving out the fat and limiting the amount of salt by using less salt than standard recipes call for. Commercial breads are still low in fat.

Make pancakes, waffles, breads with yeast instead of baking powder and baking soda.

Buy low-sodium baking powder.

Stretch servings of meat with noodles, beans, vegetables, and tofu.

Use less meat than recipes suggest.

Salad dressings: Use more vinegar and less oil; make dressing from vinegar, fruit juice, and spices; try adding 1 tablespoon of catsup to vinegar/oil/spice mixture—it adds flavor and smoothness and allows you to use less oil (or no oil). If you like a "cheesy" salad dressing try making one with buttermilk or yogurt and herbs.

Use non-stick cookware!!! The new products are very durable and effective. They are considered much better than the old-style coated pans. Saucepans are now available with Teflon coating. SilverStone is a cook's dream.

Use water-packed tuna instead of oil-packed tuna. To make lunch spreads without mayonnaise, thicken meat/vegetable mixture with a cooked cornstarch paste. Spend a little extra on unsalted tuna and add a small amount of salt if you want it.

Make spaghetti sauce without meat or sausage; sauce made with tomatoes and tomato paste and herbs, spices, and mushrooms is excellent. Use grated dry Monterey jack cheese instead of the higher-fat parmesan.

In baking, use just the egg whites instead of the whole egg. Make frosting from egg whites instead of butter when you make a cake for a special occasion.

Use cocoa instead of chocolate squares and don't add additional fat.

Use more spices in your cooking to help make low-salt/salt-free food taste better—garlic, rosemary, sage, oregano, dill; cinnamon, cloves, ginger.

Serve meals with many courses—the objective being that you will fill up on foods other than meat and fat. Start with soup—homemade broth with vegetables (low-salt and low-fat); add a little macaroni for variety; make vegetable cream soups. Serve salad, entrée, lots of bread (no butter or margarine), two vegetables. For dessert serve fresh fruit; on special occasions serve homemade gelatin desserts or angel food cake. Canned broth and packaged gelatin mixes are very high in sodium.

GUIDELINES FOR LOW-CHOLESTEROL, LOW-TRIGLYCERIDE DIETS*

Foods to Use

Meats, Fish. Choose lean meats (chicken, turkey, veal, and nonfatty cuts of beef with excess fat trimmed). (One serving = 3 oz. of cooked meat.) Also, fresh or frozen fish and canned fish packed in water. Meats and fish should be broiled (pan or oven) or baked on a rack.

Eggs. Egg whites (use freely). Egg yolks (limit three per week).

Fruit. Eat three servings of fresh fruit per day (1 serving = $1/2$ cup). Be sure to have at least one citrus fruit daily. Frozen or canned fruit with no sugar or syrup added may be used.

Vegetables. Most vegetables are not limited. One dark-green (string beans, escarole) or one deep-yellow (squash) vegetable is recommended daily. Vegetables may be boiled, steamed, strained, or braised with polyunsaturated vegetable oil (see below).

Beans. Dried peas or beans (1 serving = $1/2$ cup) may be used as a bread substitute.

Nuts. Pecans, walnuts, and peanuts may be used sparingly. 1 serving = 1 tablespoonful.

Breads, Grains. One roll or one slice of whole-grain or enriched bread may be used, or three soda crackers or four pieces of melba toast as a substitute. Spaghetti, rice, or noodles ($1/2$ cup) or $1/2$ large ear of corn may be used as a bread substitute. In preparing these foods, do not use butter or shortening; use soft margarine. Also use egg and sugar substitutes.

Cereals. Use $1/2$ cup of hot cereal or $3/4$ cup of cold cereal per day. Add a sugar substitute if desired.

Milk products. Always use skim milk or skim milk products such as low-fat cheeses (farmer's, uncreamed cottage, mozzarella), low-fat yogurt, and powdered skim milk.

Fats, Oils. Soft margarine and polyunsaturated vegetable oils derived from safflower, soybean, sunflower, corn, or sesame seeds.

Desserts/Snacks. Limit to two servings per day; substitute each serving for a bread/cereal serving: ice milk, water sherbet ($1/4$ cup); unflavored gelatin or gelatin flavored with sugar substitute ($1/3$ cup); pudding prepared with skim milk ($1/2$ cup); egg white soufflés; unbuttered popcorn ($1 1/2$ cups).

Beverages. Fresh fruit juices (limit 4 oz. per day); black coffee, plain or herbal teas; soft drinks with sugar substitutes; club soda, preferably salt-free; cocoa made with skim milk or nonfat dried milk and water (sugar substitute added if desired); clear broth. Alcohol: limit two servings per day.

Miscellaneous. You may use the following freely: Vinegar, spices, herbs, nonfat bouillon, mustard, Worcestershire sauce, soy sauce, flavoring essence.

*Courtesy Kentfield Rehab Center, Kentfield, California.

Foods to Avoid

Meats, Fish. Pork, bacon, sausage, and other pork products; fatty fowl (duck, goose); skin and fat of turkey and chicken; processed meats; luncheon meats (salami, bologna); frankfurters and fast-food hamburgers (they're loaded with fat); organ meats (kidneys, liver); canned fish packed in oil. Shellfish (lobster, shrimp, crab, oysters) should be used sparingly.

Eggs. Limit egg yolks to three per week.

Fruits. Coconuts.

Vegetables. Avoid avocados and olives. Starchy vegetables (potatoes, corn, lima beans, dried peas, beans) may be used *only* if substituted for a serving of bread or cereal.

Beans. Commercial baked beans with sugar and/or pork.

Nuts. Avoid nuts. Limit pecans, walnuts, and peanuts to one tablespoonful per day.

Breads, Grains. Any baked goods with shortening and/or sugar. Commercial mixes with dried eggs and whole milk. Avoid sweet rolls, doughnuts, and breakfast pastries (Danish).

Milk Products. Whole milk and whole-milk packaged goods; cream, ice cream; whole milk puddings, yogurt, or cheeses; nondairy cream substitutes.

Fats, Oils. Butter, saturated fats (olive, peanut, and coconut oil), lard, solid margarine, commercial salad dressings, gravies, bacon drippings, cream sauces.

Desserts, Snacks. Fried snack foods like potato chips; chocolate; candies in general; jams, jellies, syrups; whole milk puddings; ice cream and milk sherbets. Hydrogenated peanut butter.

Beverages. Sugared fruit juices and soft drinks; cocoa made with whole milk and/or sugar. When using alcohol (1 oz. liquor, 5 oz. beer, or $2^{1}/_{2}$ oz. dry table wine per serving), one serving must be substituted for one bread or cereal serving (limit, two servings of alcohol per day).

Special Notes:
1. Remember that even nonlimited foods should be used in moderation.
2. While on a cholesterol-lowering diet, be sure to avoid animal fats and marbled meats.
3. While on triglyceride-lowering diet, be sure to avoid sweets and to control the amount of carbohydrates you eat (starchy foods such as flour, bread, potatoes).
4. Consult your physician if you have any questions.

Appendix XII. General Information

INTERNATIONAL VHF VOICE PRONUNCIATION AND SPELLING

Phonetic Spelling Alphabet. The spelling alphabet is to be used to identify letters when spelling out words, names, abbreviations, and call signs in voice communications.

Letter to be identified	Identifying word	Spoken as:
A	Alfa	**AL** FAH
B	Bravo	**BRAH** VOH
C	Charlie	**CHAR** LEE (OR **SHAR** LEE)
D	Delta	**DELL** TAH
E	Echo	**ECK** OH
F	Foxtrot	**FOKS** TROT
G	Golf	**GOLF**
H	Hotel	HOH **TELL**
I	India	**IN** DEE AH
J	Juliett	**JEW** LEE **ETT**
K	Kilo	**KEY** LOH
L	Lima	**LEE** MAH
M	Mike	**MIKE**
N	November	NO **VEM** BER
O	Oscar	**OSS** CAH
P	Papa	PAH **PAH**
Q	Quebec	KEH **BECK**
R	Romeo	**ROW** ME OH
S	Sierra	SEE **AIR** RAH
T	Tango	**TANG** GO
U	Uniform	**YOU** NEE FORM (or **OO** NEE FORM)
V	Victor	**VIK** TAH
W	Whiskey	**WISS** KEY
X	X-ray	**ECKS** RAY
Y	Yankee	**YANG** KEY
Z	Zulu	**ZOO** LOO

CONVERSION TABLES
British—U.S.—International and Metric Units

British − U.S. − International Units				Metric Units (SI)
LENGTH				
1	nautical mile (U.S.-Int.)		=1.852	km
0.53996	nautical mile (U.S.-Int.)		=1	km
1	nautical mile (Br)		=1.853 184	km
0.539612	nautical mile (Br)		=1	km
1	mile		=1.609 344	km
0.621371	mile		=1	km
1	inch		=2.540	cm
0.39370	inch		=1	cm
1	foot		=0.304 8	m
3.2808	feet		=1	m
MASS				
1	ton	(long)	=1 016.046 908 8	kg
0.9842065	ton	(long)	=1000.0	kg
1	ton	(short)	=907.184 74	kg
1.10231	tons	(short)	=1000.0	kg
1	pound		=0.453 592 37	kg
2.20462262	pounds		=1	kg
CAPACITY				
1	gallon (Br)		=4.546 092	dm³
0.219969	gallon (Br)		=1	dm³
1	gallon (U.S.)		=3.785 412	dm³
0.264172	gallon (U.S.)		=1	dm³

Note: Relevant units used with the S1
    a) 1 tonne (t) or 1 metric ton = 1000 kg
    b) 1 litre = 1 dm³

APPENDIX XII 577

Conversion of Metric to U.S. Customary/Imperial Units

The table below refers to the metric measurements used in the COLREGS and Inland Rules and Regulations.

Metric Measure	U.S. Customary/ Imperial Measure (approx.)
1000 Meters (M)	3280.8 ft.
500 M	1640.4 ft.
200 M	656.2 ft.
150 M	492.1 ft.
100 M	328.1 ft.
75 M	246.1 ft.
60 M	196.8 ft.
50 M	164.0 ft.
25 M	82.0 ft.
20 M	65.6 ft.
12 M	39.4 ft.
10 M	32.8 ft.
8 M	26.2 ft.
7 M	23.0 ft.
6 M	19.7 ft.
5 M	16.4 ft.
4.5 M	14.8 ft.
4.0 M	13.1 ft.
3.5 M	11.5 ft.
2.5 M	8.2 ft.
2.0 M	6.6 ft.
1.5 M	4.9 ft.
1 M	3.3 ft.
.9 M	35.4 in.
.6 M	23.6 in.
.5 M	19.7 in.
300 Millimeters (mm)	11.8 in.
200 mm	7.9 in.

Feet to Meters

Feet	0	1	2	3	4	5	6	7	8	9
0	0.00	0.30	0.61	0.91	1.22	1.52	1.83	2.13	2.44	2.74
10	3.05	3.35	3.66	3.96	4.27	4.57	4.88	5.18	5.49	5.79
20	6.10	6.40	6.71	7.01	7.32	7.62	7.92	8.23	8.53	8.84
30	9.14	9.45	9.75	10.06	10.36	10.67	10.97	11.28	11.58	11.89
40	12.19	12.50	12.80	13.11	13.41	13.72	14.02	14.33	14.63	14.93
50	15.24	15.54	15.85	16.15	16.46	16.76	17.07	17.37	17.68	17.98
60	18.29	18.59	18.90	19.20	19.51	19.81	20.12	20.42	20.73	21.03
70	21.34	21.64	21.95	22.25	22.55	22.86	23.16	23.47	23.77	24.08
80	24.38	24.69	24.99	25.30	25.60	25.91	26.21	26.52	26.82	27.13
90	27.43	27.74	28.04	28.35	28.65	28.96	29.26	29.57	29.87	30.17

Fathoms to Meters

Fathoms	0	1	2	3	4	5	6	7	8	9
0	0.00	1.83	3.66	5.49	7.32	9.14	10.97	12.80	14.63	16.46
10	18.29	20.12	21.95	23.77	25.60	27.43	29.26	31.09	32.92	34.75
20	36.58	38.40	40.23	42.06	43.89	45.72	47.55	49.38	51.21	53.03
30	54.86	56.69	58.52	60.35	62.18	64.01	65.84	67.67	69.49	71.32
40	73.15	74.98	76.81	78.64	80.47	82.30	84.12	85.95	87.78	89.61
50	91.44	93.27	95.10	96.93	98.75	100.58	102.41	104.24	106.07	107.90
60	109.73	111.56	113.39	115.21	117.04	118.87	120.70	122.53	124.36	126.19
70	128.02	129.85	131.67	133.50	135.33	137.16	138.99	140.82	142.65	144.47
80	146.30	148.13	149.96	151.79	153.62	155.45	157.28	159.11	160.93	162.76
90	164.59	166.42	168.25	170.08	171.91	173.74	175.56	177.39	179.22	181.05

Meters to Feet

Meters	0	1	2	3	4	5	6	7	8	9
0	0.00	3.28	6.56	9.84	13.12	16.40	19.68	22.97	26.25	29.53
10	32.81	36.09	39.37	42.65	45.93	49.21	52.49	55.77	59.06	62.34
20	65.62	68.90	72.18	75.46	78.74	82.02	85.30	88.58	91.86	95.14
30	98.42	101.71	104.99	108.27	111.55	114.83	118.11	121.39	124.67	127.95
40	131.23	134.51	137.80	141.08	144.36	147.64	150.92	154.20	157.48	160.76
50	164.04	167.32	170.60	173.88	177.16	180.45	183.73	187.01	190.29	193.57
60	196.85	200.13	203.41	206.69	209.97	213.25	216.54	219.82	223.10	226.38
70	229.66	232.94	236.22	239.50	242.78	246.06	249.34	252.62	255.90	259.19
80	262.47	265.75	269.03	272.31	275.59	278.87	282.15	285.43	288.71	291.99
90	295.28	298.56	301.84	305.12	308.40	311.68	314.96	318.24	321.52	324.80

Meters to Fathoms

Meters	0	1	2	3	4	5	6	7	8	9
0	0.00	0.55	1.09	1.64	2.19	2.73	3.28	3.83	4.37	4.92
10	5.47	6.01	6.56	7.11	7.66	8.20	8.75	9.30	9.84	10.39
20	10.94	11.48	12.03	12.58	13.12	13.67	14.22	14.76	15.31	15.86
30	16.40	16.95	17.50	18.04	18.59	19.14	19.68	20.23	20.78	21.33
40	21.87	22.42	22.97	23.51	24.06	24.61	25.15	25.70	26.25	26.79
50	27.34	27.89	28.43	28.98	29.53	30.07	30.62	31.17	31.71	32.26
60	32.81	33.36	33.90	34.45	35.00	35.54	36.09	36.64	37.18	37.73
70	38.28	38.82	39.37	39.92	40.46	41.01	41.56	42.10	42.65	43.20
80	43.74	44.29	44.84	45.38	45.93	46.48	47.03	47.57	48.12	48.67
90	49.21	49.76	50.31	50.85	51.40	51.95	52.49	53.04	53.59	54.13

APPENDIX XII

PUBLIC MARITIME SCHOOLS IN THE UNITED STATES

While not the sole educational and training facilities in the country, the state and federal maritime academies have furnished the towing industry with well-trained personnel.

California Maritime Academy, Vallejo, California 94590
Great Lakes Maritime Academy, Traverse City, Michigan 49684
Maine Maritime Academy, Castine, Maine 04420
Massachusetts Maritime Academy, Buzzards Bay, Massachusetts 02530
State University of New York Maritime College, Fort Schuyler, Bronx, New York 10465
Texas A&M Maritime College, Galveston, Texas 77553
United States Merchant Marine Academy, Kings Point, New York 11024

References

Armitage, Captain B. F. *Tug and Towing Operations.* London: Maritime Press, Ltd., 1962

Brown, R. Ingram, and Cotter, Charles H. *Brown's Nautical Almanac and Daily Tide Tables.* Glasgow: Brown, Son & Ferguson Ltd., 1984

The Canadian Maritime Law Association, The Maritime Law Section of the Canadian Bar Association, and The Continuing Legal Education Society of British Columbia. *The Law of Tug and Tow.* Vancouver: University of British Columbia, 1979.

Crowley Maritime Corporation. *Hauls of Fame.* Seattle: Crowley Maritime Corporation, 1983

de Kerchove, René. *International Maritime Dictionary.* 2d ed. New York: Van Nostrand Reinhold, 1983.

Duffy, Francis J. *Tow Line.* Greenwich, Conn.: Moran Towing & Transportation Company, Winter 1986.

Lloyd's. *100A1.* London: Lloyd's Register, July 1985.

MacDiarmid, Ian. *Marine Pilotage.* Ottawa, Ont.: Transport Canada, 1986.

Office of Federal Register, National Archives and Records Administration. *Code of Federal Regulations,* Titles 33 and 46, As Revised October 1, 1987. Washington, D.C.: U.S. Government Printing Office, 1987.

Parks, Alex L. *The Law of Tug, Tow, and Pilotage.* 2d ed. Centreville, Md.: Cornell Maritime Press, 1982.

Plummer, Carlyle J. *Ship Handling in Narrow Channels.* 2d ed. Cambridge, Md.: Cornell Maritime Press, 1966.

White, Robert Eldridge. *Eldridge Tide and Pilot Book.* Boston: R. E. White, 1986

U.S. Department of Transportation, Coast Guard. *Navigation Rules: International—Inland.* Washington, D.C.: U.S. Government Printing Office, 1983.

Index

Abandon-ship drill, 263
Abeille Languedoc, 330, 340
Able lookout, 470
Able seaman, 470
Able seaperson, 470
Able watch, 470
Accident
  reporting, 316
  VHF bridge-to-bridge, 499-500
AFL-CIO, Maritime Trades Department, 476
Afterdeck engine controls, 99
After steering, 99
Air conditioning, 144
Air temperature, 533-36
Ajax, 329
A. J. Hoyle tow winch, 337-39
Alaska, 24-26
  North Slope, 10-12
  tug-barge carriers, 9, 10
Albert, 192
Alcohol, 459
Allison C, 81
Almon-Johnson winch, 337
Alongside, defined, 29
Alongside towing, 74-76
  advantages, 74-75
  disadvantages, 75
  weather, 76
America, 58, 59, 132
American Bureau of Shipping, 124
Amidships, defined, 29
Anchor
  design, 138-39
  drill rig moving, 255
  oil barge, 421, 422
Anchor-handling supply vessel, fire fighting, 415
Anchor-handling tug, 362-63
Anchor-handling vessel, ice-breaking, 400-402

Anchoring
  barge with tug, 425-26
  docking on hawser, 423-24
  getting under way, 426
  with tow astern, 424-25
Anchor shackle
  defined, 29
  illustrated, 30
Andrew Foss, 12
Annex of Navigation Rules, 487
AquaMaster bow thruster unit, 125
AquaMaster unit, 131
Arthur Foss, 197
Artubar tug, 48, 148
Astern towing, 71-74
  assisting tug, 73-74
  current, 73
  landing-barge lines, 73
  landing on hawser alone, 74
  weather, 72
  wind, 73
ATF fleet tug, 148
Athwartship, defined, 29
Atlantic Intracostal Waterway, 18
Atomic, 410
Automatic gear, defined, 29
Automatic Radar Plotting Aids, 346

Bar, defined, 29
Bar pilot, 189
  defined, 36
Barge
  anchoring, getting under way, 426
  big notched, 231-33
  chafing gear, 293-94
  classifications, 505-7
  deeper-notched, 303
  dimensions, 505-7
  getting under way, 216-21
  head-and-tail, 325-26
  IOT notch arrangement, 303

landing in river currents, 274
letting go, 324-25
lookout, 294
making up alongside, 213-16
mooring buoy, 428-30
oil, harbor towing, 221-24
partially loaded, 371
picking up light, 325-26
pinhole system, 301-2
power application, 294
pushing, 224-36, 304-5; advantages, 224; barge notch approach, 227; big notch, 231-33; getting into notch, 227; making up notch, 227-30; rapid connecting, 233-36; rapid disconnecting, 233-36; safety lines in notch, 233; sea dangers, 305
shallow-notch, 301
swift river exit, 294
taking charge, 293-94
towing in rough weather, 428
tug, 212-36
unmanned, crew training, 276
wire-rope system, 301
work on hawser, 199-202
Barge flotilla, short hawser towing, 68
Barge notch approach, 227
Barge pickup line, 293-94
Barge retrieving line, 277
Barge-steering system, 308
Bar-tight, defined, 29
Bay of Fundy, 376
Bay of the Seine, 376
Beacon marker, 521
Beaufort scale, defined, 29
Becket, defined, 29
Bend signal, 484, 487
Bentol, 43
Bergy bit, defined, 29
Bilge keel, defined, 31
Bitt, 117. *See also* Specific type
Bitter end, defined, 31
Boat drill, 263
Bollard pull, 124
  certificates, 511
  testing, 511
Bootheel pilothouse, 91
Bow bitt, 117
Bowditch Navigation System, 348

Bow eyelet, 118
Bow pudding, 177
Bridge, electronics, 97-98
Bridge protocol, 454
Bridle, 182
  defined, 31
  Seaspan 251's, 299
Bridle leg, defined, 31
British Columbia, 24-26
  problem areas, 394-96
Brynn Foss, 92
Bulbous bow mark, 188
Bullnose, defined, 31
Bulwark
  defined, 31
  design, 102
Buzzards Bay, 377-78

Cable, design, 138-39
Cable hold-down, 115
California, 21-22
Canada, inland waters, 55-57
Canal, 242-53
  approaches, 245-53
  ascending, 242
  dimensions, 537, 544-47
  limitations, 537
  narrow dredged, 317
  regulations, 537-47
  restrictions, 544-47
  size, 243-45
Cape Cod Canal, 377
Capt. Ioannis S., 194
Captain. *See also* Tug master
  as supervisor, 458-59
  background, 456-57
  defined, 31
  duties, 459-60
  legal issues, 490-91
  log book, 459-60
  office control, 458
  personal characteristics, 457-58
  promotion, 457
  relief, 460-61
  responsibilities, 491-93
  selection, 457
  substance abuse, 459
Captain Bob, 226
Car-barge towing, 240-42
Catenary, defined, 31
Catherine Foss, 397

Celestial navigation, 348
Central America, 21
Chafing board, 309, 310
Chafing gear, 293-94
  defined, 31
  ocean tow, 337
  rigging, 309-12
  types, 309
Chain shackle
  defined, 31
  illustrated, 32
Challenger, 195
Channel, narrow dredged, 317
Charles H. Cates II, 187
Chart, 258
  sea tow, 313-14
Chateau, 63-66
Chesapeake Bay, 15-17
Chine, defined, 31
Chock
  closed, illustrated, 32
  defined, 31
  open, illustrated, 32
CIDS, defined, 31
Clamshell dredge, 236-40
Cleat
  defined, 31
  illustrated, 32
Cleveland, 92
Clevis
  defined, 31
  illustrated, 32
Clontarf Monarch, 138, 238
Coal, 15-17
Coast Guard notice, gillnet fishing, 552-54
Coastwise, defined, 31
Coastwise tow
  chart, 276
  connection selection, 278-79
  on hawser, 84-87
  safety equipment, 276
  weather, 277-78
Coastwise tug, 47, 147
Code of Federal Regulations, 467
Coleraine, 238
Collision avoidance training, 454
COLREGS, 1972, 259, 482
  IMO clarifications, 498-99
Columbia River system, 57
Columbia-Snake River, 22-24

COMDTINST-M166 72.2, 483
Communications, 171
  ocean tow, 346
Company seminar, 444
Connecticut River, 380, 381
Connecting up, defined, 31
Containership, ship assisting, 187
Controllable-pitch propeller, 133-34
Control station, 99
Conversion table, 576-78
Cook, 351-52, 449-51, 466
Coos Bay, Oregon, 22
Cordage, 513-17
Counter, defined, 31
Covered barge, 264, 265
Crew
  hiring, 480-81
  relief, 460-61
  size, 467-73
  training, 480-81
Crusader, 404
Culver, 192
Cycloidal propulsion system, 132

Damage, reporting, 316
Day marker, 521
Day signal, 326-27
Dead ship, defined, 33
Dead ship towing, 202-6
Dead-in-the-water, defined, 33
Dead-reckoning sensor, 348
Decca, 348
Deck capstan, 109-10
Deck light, 121
Deck line, 178-79
Deck winch, 120
Deckhouse, 99-100
  design, 102-4
Deep-notched towing, sea, 82
Deep-sea salvage tug, 41
Deep-sea towing, 27, 28
Deep-sea tug, 365
  mess room, 331
  quarters, 331
Delaware, 210
Delaware Bay, 14-15
Department of Transportation U.S. Coast Guard Commandant's Instructions M166 72.2, 483
Departure, 156-57
  from slip, 156-57

Det norske Veritas, 124
Diesel engine, 123
Diet, 351-52, 449-51, 570-74
Dip, defined, 33
Dipper dredge, 236-40
Dispatching, 442-43
Diving equipment, 184
Docking, 60-66
  European tow hook method, 66
  in tideway, 63-66
  on hawser, anchor, 423-24
  towing hook, 60, 62
  tug pair positioning, 60
  U-bolt, 60, 66, 67
  U.S. alongside mode, 66
  wire cable winch, 66
Docking method, 191-96
Docking pilot, 188-89
  defined, 36
Double bow bitt, illustrated, 30
Downbound river tow, 225
Downbound tow, 272
Down-by-the-head, defined, 33
Draft, recommended, ocean
  towage, 506
Drag, defined, 33
Drawbridge, 256-58, 372-75
  alignment, 373
  clearance gauge, 374
  regulations, 257
  special hazard locations, 257
Dredge, 365-68
Dredge towing, 236-40
Drift buoy, 225
Drill rig, 28
  connecting to, 279
Drill rig towing, 253-56
  anchor handling, 255
  ocean towing, 363
Drugs, 459
Dry-docking, 139
Dump scow, 365-68

Electrolysis, 140
Electronic bridge, navigation, 346-50
Electronic navigation, ocean tow, 348
Electronic Position-Indicating Radio
  Beacon, 141
Electronics, 97-98
Electronic viewing equipment, 97
El Gaucho, 306

Ellena Hicks, 50, 297
Engine, 122-24
  automation, 123
  auxiliaries, 122
  diesel, 123
  horsepower, 123-24
  valves, 123
Engine controls
  standard tugs, 135-36
  tractor tug, 136
Engineer, 465-66
  license, 466
Engine room, 121-22
Engine warm-up, 155
Englishman, 91
Entertainment system, 352
Esteff S. DeFelice, 106
European Tug Owners' Association, 5
Exercise, 351-52, 451-52

Fair current, defined, 33
Fairlead, 110-13, 117
  retractable, 116
Fender, 171
  bow, 177
  design, 100-102
  handmade rope, 102
  long line, 172
  manufactured rubber tire side, 102
  purpose, 101-102
  rope hung, 172
  rubber bow, 172
  rubber fender rail, 101
  side, 171-77
  stern, 177
  tires, 101
  vertical single bow, 173
Fid, defined, 33
Fire Boss, 414
Fire drill, 263
Fire equipment, 140
Fire fighting
  anchor-handling supply vessel, 415
  oil rig, 414
  oil tanker, 414-15
  salvage, 360, 362
  salvage tug, 415
  supply vessel, 414
  tug, 413-17
Fire fighting drill, ocean tow, 350-51
Fire monitor, 48

INDEX 585

Fishing area, 434-35, 551-56
Fishing vessel, 434-35
Fish plate, defined, 33
Florida
  East Coast, 18-19
  Gulf Coast, 19-20
  Mobile Bay, 20
Flounder plate
  defined, 33
  illustrated, 30
Fog, approaching port, 426
Form CG-735(T), 155
Forward bitt, 117-18
  defined, 31
  illustrated, 30
Foss water tractor, 198-99
Frank B. White, 368
Free-surface effect, defined, 33

Galley, 143-44
Gate line towing, 69
Gayle B, 343
Gear, defined, 33
George T. Horton, 409
Getting under way, 155-56
Getting way off, defined, 33
Getting way on, defined, 33
Gillnet
  defined, 554
  legally marked, 554
Gillnet fishing, 434-35
  Coast Guard notice, 552-54
  navigating through fleet, 554-56
Girded, defined, 33
Glomar Beaufort Sea #1, 26
GM, defined, 33
Gobrope, defined, 33
Great Lakes, 13
  towing arrangements required, 548-50
Grounding, 315-16
  accidental, 316
  aid for, 316
  intentional, 315-16
  reporting, 316
Ground tackle, defined, 33
Grub, 448-49
Grub check, 448
Guard, design, 100-102
Gulf Bayou Coast, Louisiana, 20
Gulf of Mexico, lightering, 391-92

Gunnel, defined, 33
Gunwale, defined, 33
Gypsy, defined, 33
Gyrocompass, 97

H bitt
  defined, 31
  illustrated, 30
Hand signal, 156
Harbor of Possible Refuge, 385
Harbor pilot, 189
  defined, 36
Harbor tug, 41-42, 43, 48, 147, 148, 150, 193, 194, 403
  anchor, 138
  cleat sheath, 69
  design, 41-42
  fire fighting, 413
  line-handling arrangement, 68
  power, 58
  removable sheath, 68
  wire and line, 237
Hard over, defined, 33
Haul towboat, 54
Hawser
  connecting at transition points, 288-92
  connecting head-and-tail, 284; getting away from, 284-87
  connecting in port, 283-84
  defined, 34
  frequently used areas, 280-82
  installation, 342-45
  jackstay, 295
  length, U.S. Coast Guard ruling, 487
  letting go, 324-25
  lying on, 320-21
  maintenance, 342-45
  parting the, 318
  tensile strength, 512
  to notch from, 325
Hawser board, defined, 34
Hawser length, inland water, 279-80
Hawser tension damper, 107-8
Hawser territory, 282-83
Head-and-tail, defined, 34
Head current, defined, 34
Health, 449-51, 570-74
Heating, 144
Heave in, defined, 34

Heave to, defined, 34
Heaving line, defined, 34
Hell Gate, 381-85
Holland, 7
Hooking up, defined, 34
Hopper barge, 264, 265
Horizontal roller, 114
Horsepower, 123-24
Hurricane readiness, 430-32
Hydralift skeg, 308, 309
Hydropad, 302

IALA, defined, 34
IALA buoyage marker, 521
Ice, towing, 399-403
Icebreaking tug, 334, 399, 402
Icing, 533-36
Ikaluk, 400
Illumination, 121
IMCO, defined, 34
IMO, defined, 34
Improved plow steel tow cable, 344-45
Indiana, 210
Ingram ITB, 306
Inland Boatmen's Union, 475-76
Inland river towboat, 54-57
  design, 54-55
  flanking rudders, 55
  pilothouse equipment, 55
  thruster, 55
Inland water
  hawser length, 279-80
  push tow, 273
Inland water towing, 264
INMARSAT, defined, 34
Integrated river tow, 265-71
  barge coupling, 267-71
  making up, 266-67
Intercom system, 121
International Chamber of Shipping, 499-500
International Maritime Satellite Organization, 346
Interport container feeder barge, 28
ITB, 485

Jacksonville, Florida, 18-19
Jackstay, hawser, 295
John Ross, 9
Joystick, 162-64

Joystick steering, 125

Karl, 103
Kodiak II, 415
Kort nozzle, defined, 34
Kuanza, 43

Labor-management relations, 456-81
  negotiations, 478-80
Ladder, outside, 144
Lake Charles, Louisiana, 21
La Reina, 18
Lash barge, 371-72
Lash up, defined, 34
LCD, defined, 34
LED, defined, 34
Left-hand turn, defined, 34
Let go, defined, 34
Letting go, 156-57
  from slip, 156-57
Libby Black, 291
Licensing, 469-70
Life jacket, 141
Life raft, 142
Life ring, 142
Lifesaving equipment, 140-42
Light, 485-86
Light tug, defined, 34
Lightering, 16, 17
  oil barge, 418-19
Line, small, 180-81
Line-handling vessel, 194
List, defined, 35
Lloyd's Open Form of Salvage Agreement, 358
Load-cell, 298
Lock, 242-53
  approaches, 245-53
  ascending, 242
  dimensions, 537, 544-47
  entering, 271-72
  limitations, 537
  making, 271-72
  regulations, 537-47
  restrictions, 544-47
  size, 243-45
Log
  captain's duties, 459-60
  defined, 35
  sea tow, 313-14
Log raft, 396-99

INDEX 587

Log sheet, 260
Lookout, rulings, 489-90
Lookout Rule, 483
Loran, 97, 348
Lounge, 144
Lubec Narrows, 377
Lubrication, 183

Main bitt, illustrated, 30
Making up, defined, 35
Management, 5-6
Manning, 467-73
Manoir, 63-66
Marine Engineers Beneficial Association, 466
Marine insurance, ocean tow, 332
Mariner, 94, 288
Marine radiotelephone terminal, 528
Marine Safety Reporting Program, 454
Marine weather report, 277-78
Maritime provinces, 13-14
Maritime school, 579
Markey TDSD-36 tow winch, 106
Markey winch, 339, 341
Mark K, 150
Master
  defined, 35
  responsibility, 154-55
Masters, Mates and Pilots Union, 476
Mate, 461-62
Matilda, 60
Mess room, deep-sea tug, 331
Metacentric height, defined, 35
Mexico, 21
Miscaroo, 399, 401
Mississippi River
  course, 56
  data, 557-63
  ferry crossings, 559
  flanking, 56
  gauge boards, 559-61
  hawser length, 559
  lower, 20
  passes, 387-91
  relation to bridges, 559-61
  rudder action, 56
  steering, 56
  towing procedures, 557-59
  traffic lights, 557-59

  vessel traffic service sector, 562-63
Mohawk, 150
Molly gogger, 117
  defined, 35
Monkey fist, 180
  defined, 35
Mooring fittings, 117
Mud ball, defined, 35
Mud towing, 365-68
Multipurpose tug, defined, 35
Murdock tension damper, 107, 108, 339-41

Narrow passage, 316-17
Narrows, regulations, 537-47
Navigation
  electronic bridge, 346-50
  ocean tow, 346-50
  piloting, 348
  sea tow, 313
Navigation light, 142
Navigation rules, 482
New Brunswick, 13-14
New York
  anchorage, 564
  harbor, 564
  terminal, 564
Niigata Z-drive, 132
Noordzee, 52
Norman pin, 112
  defined, 35
  hydraulic, 114
North Slope, 10-12
Notched ocean barge, 80-83

Oakland, California
  anchorage, 566
  harbor, 566
  terminal, 566
Ocean
  current, 352-55
  routes, 352-55
  weather, 352-55
Ocean barge, deep-notched tug pushing, 70
Ocean Hawk, 196
Oceanic, 365
Ocean salvage and rescue tug, 149
Ocean tow, 44
  approaching port, 322-26
  chafing gear, 337

chart, 276, 313-14
communications, 346
computer, 351
connecting to, 336
connection selection, 278-79
departure, 336-37
drill rig, 363
electronic devices, 348-49
electronic navigation, 348
fire-fighting drill, 350-51
inventory, 351
log, 313-14
maintenance, 314
marine insurance, 332
navigation, 313, 346-50
on hawser, 84-87
preparing, 335-36
radio, 346
radio watch, 313
rigging, 335-36
safety drill, 350-51
safety equipment, 276
short, routine, 312-14
tow lights, 313
towing winch, 337-42
waste disposal, 314
watch stander, 349-50
weather, 277-78
wind velocity, 531-32
Ocean tug, 49, 51-53
defined, 51
futuristic, 27
historical aspects, 51-53
largest, 505
Offshore terminal, ship-assist tug, 47
Oil barge
anchor, 421, 422
empty, turning into wind, 417-20
harbor towing, 221-24
lightering, 418-19
Oil rig, fire fighting, 414
Oil tanker, fire fighting, 414-15
Oil terminal docking, 211-12
One by four, defined, 35
Open bridge, 91
Operations, 442-43
Oregon, 13
Orville hook, 322, 323
Otto Candies, 45
Out of shape, defined, 35
Outboard bearings, 140

Outside ladder, 144

Pacific Coast, 393
  problem areas, 393-94
Pad eye
  defined, 35
  illustrated, 32
Painting, 139
Panama Canal, 21, 57-58, 538
Panamax, defined, 35
Parted, defined, 35
Parting the hawser, 318
Pascagoula, Mississippi, 20
Passageways, 143
Patriarch, 49
Pay out, defined, 35
Pelican hook, defined, 35
Pendant, 182
  defined, 35
Pennant, 182
  defined, 36
Personnel, 5, 456-81
  manning, 467-73
  safety, 262
  transferring boat to boat, 375
Peter Moran, 89
Petit Passage, Nova Scotia, 377
Petroleum barge, head-and-tail, 285
Petroleum industry, 7-8
  anchor handling, 362-63
Petroleum industry tug, 53, 54
  design, 53
Pigtail, defined, 36
Pilot, 191
  defined, 36
  legal aspects, 494-97
Pilothouse
  deck, 98-99
  design, 95-97
  electronics, 97-98
  river towboat, 95
  visibility, 96, 97, 98-99
Piloting, navigation, 348
Pollution, 499
Portland, 190
Portsmouth, New Hampshire, 377
Power, 4
  historical aspects, 4
Power-barge, integrated, 305-6
Precautionary Rule, 483
President Lincoln, 197

Profit sharing, 445
Propeller. *See also* Specific type
  fixed, 130
  inboard turning, 159
  left-hand turning, 158
  outboard turning, 159
  right-hand turning, 158
Propeller wash, 203, 204
Protection and indemnity coverage, 440-41
Public Law #96-380, 453
Puget Sound, 24-26
Push cable, 182
Pushing
  advantages, 79
  coastwise, 80-83
  historical aspects, 81-83
  hookup, 83
  notched-stern barges, 80
  ocean, 80-83
Push towing, 76-83, 79-83
  advantages, 300
  deep notch, 78
  historical aspects, 77
  hydrodynamic efficiency, 300
  inland river, 77-79
  inland water, 273
  shallow notch, 78
  types, 76-77
Put out, defined, 36

Quarter bitt
  defined, 31
  illustrated, 30
Quarters, 144
  deep-sea tug, 331

Rabigh I, 96
Radar, 97
Radio, 97
  international rule, 499
  ocean tow, 346
Radio navigational aid, 348
Radio operator, 466-67
Radiotelephone terminal, 528
Radio watch, sea tow, 313
Railroad car float towing, 240-42
Railroad tug, 91-95
Rake, defined, 36
Rapids, 316-17
Recreation, 352

Redwood City
  anchorage, 567
  harbor, 567
  terminal, 567
Regulations, 482-500
Reporting
  accident, 316
  damage, 316
  grounding, 316
Reporting aboard, 153-54
Restricted visibility, 487
Richmond, Virginia
  anchorage, 567
  harbor, 567
  inner harbor, 568
  terminal, 567
Riding pilot, defined, 36
Riding sheave, 115-17
Rig, defined, 36
River entrance, 259
River pilot, 189
  defined, 36
Rope, defined, 36
Rope handling, 518-19
Rope stopper
  defined, 36
  illustrated, 32
Rouse out, defined, 36
Rudder, 126-30
  angle degree, 128-30
Rudderpropeller, control wheel, 160
Rules, 482-500
Rulings, 482-500
Rust prevention, 139

Sacramento
  anchorage, 569
  harbor, 569
  terminal, 569
Safety, 261-63, 453-55
  personnel, 262
  towing company, 453-55
Safety drill, ocean tow, 350-51
Safety shackle
  defined, 36
  illustrated, 37
Saint John, New Brunswick, 376-77
St. Lawrence River, 13
St. Lawrence Seaway, 538
  dimensions, 539-43
  lock wall mooring table, 543-44

restrictions, 539-43
Sales, 445-46
Salvage, 356-62
　fire fighting, 360, 362
Salvage and rescue tug, 53
Salvage tug, 359, 403
　fire fighting, 415
San Francisco, California
　anchorage, 565
　harbor, 565
　terminal, 565
San Francisco Offshore Vessel Movement Reporting System, 530
Sandy Hook Channel, 385
Sarah Hays, 231, 290, 291
Satellite receiver, 97
Schottel joystick steering system, 162-64
Schottel propelled water tractor, 149
Schottel rudderpropeller system, 127
Schottel tractor tug, handling, 161
Screw pin shackle
　defined, 36
　illustrated, 37
Sea King, 329
Sea pilot, defined, 36
Sea tow. *See* Ocean tow
SeaBee barge, 371-72
Seafarers International Union, 475-76
　substance abuse, 459
Seagoing rudderpropeller tug, largest, 505
Seaman, 462-64
　alertness, 462-63
　document requirements, 468-69
　duties, 463-64
　linehandling, 464
Seamen's Union of the Pacific, 475-76
Searchlight, 142
Seaspan Regent, 8
Seaspan tug, 213
Seine net, 434, 435
Seminole Chief, 42
Sewage, 124
Shackle, 182, 183, 513
Sheave, 115-17, 120
Sheering, 314-15
Shelley Foss, 68, 69
Sheltered water towing, 67-83

alongside towing, 74-76; advantages, 74-75; disadvantages, 75; weather, 76
astern towing, 71-74; assisting tug, 73-74; current, 73; landing on hawser alone, 74; landing-barge lines, 73; weather, 72; wind, 73
push towing, 76-83, 79-83; advantages, 79; coastwise, 80-83; deep notch, 78; historical aspects, 77, 81-83; hookup, 83; inland river, 77-79; notched-stern barges, 80; ocean, 80-83; shallow notch, 78; types, 76-77
type selection, 67-71
Shifting, 71
Ship, work on hawser, 199-202
Ship assisting, 185-89
　containership, 187
　defensive action, 207
　defined, 36
　Great Lakes ports, 208-10
　pulling to pushing, 192
　tractor tug, 189-91
Ship-assist tug, 41-42, 43
　offshore terminal, 47
　ULCC, 47
　VLCC, 47
Ship chandler, 449
Ship hull, 188
Shock line, 181-82
　defined, 36
　pickup, 290
Shore gang, defined, 36
Short hawser towing, 68
　deep-draft units, 236
　tow winch, 70
Short stay, defined, 36
Shot, defined, 38
Side bitt, 117
　defined, 31
　illustrated, 30
Simpson, 403
Sisu, 410
Skeg, 306-9
　defined, 38
　Hydralift skeg, 308, 309
　movable, 307-9; maintenance, 307
Small line, 180-81
Smit Houston, 365

INDEX 591

Smit London, 4, 26
Smit New York, 254, 365
Smit Rotterdam, 26, 254
Smit Singapore, 338
Snotter, defined, 38
Socket, 345
Soft rope
  defined, 38
  illustrated, 37
Sounder, 97
Soundproofing, 144
South America, East Coast, 393
Spider, defined, 38
Spud, defined, 38
Standing part, defined, 38
Steamboat ratchet
  defined, 38
  illustrated, 37
Steel tow hawser
  defined, 38
  illustrated, 37
Steering, without tow, 160
Steering system, 125-26
Stemhead, 118-19
Stern
  design, 104-5
  general layout, 104
  protective measures, 104-5
  towing arch, 105
Stern bitt, 117
  illustrated, 30
Stern H bitt, 117
Stern roller
  horizontal, 113
  vertical, 113
Stick out, defined, 38
Stockton
  anchorage, 569
  harbor, 569
  terminal, 569
Stop off, defined, 38
Stowage, 120-21
Straight up and down, defined, 38
Stream, defined, 38
Strike, 479
Substance abuse, SIU facility, 459
Sun XVIII, 42, 96
Superstructure, 103
Supply, 448-49
Supply vessel, 44, 53, 54

  design, 53
  fire fighting, 414
  hydraulic cable controls, 117
  icebreaking, 400-402
Surge chain, 294-99
Surge gear, defined, 38
Survival suit, 141
Swedish wire rope, defined, 38
Synthetic fiber rope
  defined, 38
  illustrated, 37

Tail shaft, 140
Tandem tow, 85, 86, 299-300
  defined, 38
  icing, 404
  triple, 302
  underrunners, 302
Tandem tug, towing ahead, 363-65
Tank barge, 264, 265
Telephone, 121
Terry Fox, 334
Texas, 21
Thimble, 181
Thruster, 124-25, 188
Thumb chock, 113
Tidewater Virginia, 15-17
Tier, defined, 38
Titan, 6, 359, 403
TNT Active Flap Rudder and Barge
    Steering System, 308
Tornado readiness, 430-32
Tow
  departure for sea, 275-76
  helicopter reconnection, 319-20
  leaving river terminals, 274-75
  live ship, 427-28
  piloting, 258
  reconnecting, 318-19
  retrieving, 318-19
  sinking, 321-22
  turning, 274-75
  underway inspection, 355, 356
Towage contract, 332
Towage market, 332
Towage planning, 332-34
Towage rates, 447-48
Tow attempts to pass, 314-15
Tow bitt, work on hawser, 199-201
Towboat, 42, 148. *See also* Tug

defined, 41
historical aspects, 3-4
terminology, 3, 54
Towboat design, 88-150. *See also* Tug design
  historical aspects, 88-95
  hull form, 100
  safety, 88
  stability, 100
  strength, 88
Tow cable
  catenary, 345
  hold-down, 338
  improved plow steel, 344-45
  pay-out, 345
  tension, 337
Tow hook, 119-20
  defined, 38
  illustrated, 39
Towing. *See also* Specific type
  equipment, 508-19
  from hook, 236
  historical aspects, 6-7
  ice, 399-403, 407-13
  log raft, 396-99
  long-distance, 328-30
  special problem locations, 376-99
  unit towed's condition, 264
Towing Advisory Committee, 453
Towing arch, 105, 238
  free-running sheave, 309
  rainbow, 310
  single, 115-17, 310
Towing company, 5, 439-55
  company seminars, 444
  computer management, 442
  corporate structure, 443
  dispatching, 442-43
  grub, 448-49
  maintenance, 446-47
  management, 440
  marketing, 445-46
  operating cost, 439
  operations, 442-43
  profit sharing, 445
  protection and indemnity coverage, 440-41
  relationships, 443
  repairs, 446-47
  safety, 453-55
  sales, 445-46
  supervisors afloat, 444-45
  supply, 448-49
  vessel readiness, 442-43
  vessel supervisor, 444
Towing industry, 3
Towing master, defined, 38
Towing orders, 170
Towing vessel, terminology, 3
Tow light, 313, 326-27, 484
Towline placement, 427
Tow Master Rudder System, 128-30
Tow winch, 84
  cable guide, 109
  chafing, 85-87
  controls, 99, 105-6
  data, 509-10
  defined, 40
  double-drum, 84-85
  dual-drum, 339
  fastening, 105-6
  illustrated, 39
  line retrieval, 109
  location, 105-6, 339
  maintenance, 341
  ocean tow, 337-42
  power, 108
  single-drum, 84-85
  synchronized, 340
  towline drum amount, 341-42
  types, 84, 106-7
  warping drum, 109
  waterfall, 339
Tow wire
  maintenance, 341
  size, 508-9
Toxic-waste barge towing, 369-70
Tractor tug, 11, 12, 47, 187
  all-joystick controls, 92
  automation, 92
  defined, 40
  dry-docking, 139
  engine controls, 136
  Schottel rudderpropeller type, 47
  ship assisting, 189-91
Traffic separation scheme, 433-34
Trailer barge, 298
Trial maneuver, 157-60
  new tug, 168
Triple bow bitt, illustrated, 30
TSS, defined, 40
Tug. *See also* Towboat

INDEX 593

advance, 166-68
Artubar type, 48
barge, 212-36
Canadian, classification, 506
classification, 146, 505-7
coming alongside moving object, 170-71
coming alongside stationary object, 168-70
crew size, 145-46
delayed en route, anchoring, 425-26
dimensions, 505-7
equipment, 171-84; spares, 183-84
escort, 210-11
fire fighting, 413-17
fixed-propeller, turning, 166-68
foreign-flag, in U.S. harbors, 498
Gulf of Mexico, 47
historical aspects, 4
ice classification, 507
icing on, 404-6
long distance towing, 328-30
maintenance, 260-61
piloting, 258
selection for ship assisting, 186-87
sinking, 321-22
size, 145-46
small, towing routes, 386-87
smallest type multipurpose, 329
stopped, 320-21
terminology, 3
tonnage, 145-46
tow astern, anchoring, 424-25
transfer, 166-68
turning, 166-68
types, 41-59
Tug-barge, integrated, 305-6
Tug-barge unit, 83-84
 defined, 83
 historical aspects, 84
 integrated, 83
Tugboat/fishing lane negotiations, 551-52
Tug design, 88-150. *See also* Towboat design
 historical aspects, 88-95
 hull form, 100
 safety, 88
 stability, 100
 strength, 88
Tug interior, 142-44

Tug log, 260
Tug master, 456-60. *See also* Captain
 as supervisor, 458-59
 background, 456-57
 captain, 456-60
 duties, 459-60
 legal issues, 490-91
 log book, 459-60
 office control, 458
 personal characteristics, 457-58
 promotion, 457
 relief, 460-61
 responsibilities, 491-93
 selection, 457
 substance abuse, 459
Tug operator, responsibilities, 491-93
Tug pilot, defined, 36
Tug power, 508-9
Tumble home, design, 102-4
Turmoil, 51, 110, 111
Turning circle, single-screw tug vs. automobile, 167
Twin-screw steerable rudderpropeller, 131-33
 efficiency, 133
Typhoon readiness, 430-32

U.S. Coast Guard, proposed rule making, 497-98
Ultra-large crude carrier
 oil terminal docking, 211-12
 ship-assist tug, 47
Under run, defined, 40
Undocking, 60-66
 European tow hook method, 66
 towing hook, 60, 62
 tug pair positioning, 60
 U-bolt, 60, 66, 67
 U.S. alongside mode, 66
 wire cable winch, 66
Union, 473-76
 historical aspects, 473-75
 negotiations, 478-80
Unseaworthiness, 488-89
 law, 488
Upbound tow, 272

Valdez Narrows, Alaska, 537
Valiant, 50
Valour, 82
Vanguard, 101

Vent, 120
Very large crude carrier
  oil terminal docking, 211-12
  ship-assist tug, 47
Vessel, defined, 40
Vessel readiness, 442-43
Vessel supervisor, 444
Vessel traffic service, 432-33
VHF radio
  bridge-to-bridge, 499-500
  defined, 40
  international voice pronunciation and spelling, 575
VHF/VTS frequencies, 523-30
Victory, 235
Viewnav Electronic Chart Navigation System, 258
Vineyard Sound, 377-78
Visibility, 484
  pilothouse, 98-99
  restricted, 487
Voith-Schneider cycloidal propeller, 134-35
Voith-Schneider propeller, 165
Voyage, length, 471-73
VTS, defined, 40

Wage, 477-78
Warping drum, defined, 40
Wastewater, 124
Watch stander, ocean tow, 349-50
Water tractor, 101, 198-99
  defined, 40
Watercom, 346
Wave height, 278
Way, defined, 40
Weather, 277

marine weather reports, 277-78
  ocean, 352-55
  rough, 428
  wave height, 278
Weather broadcast, 523
Weather chart display, 97
Weather recorder, 347
Wedell Foss, 67, 197
West Coast surge chain, 294-99
Wheelhouse, electronics, 97-98
Whip, defined, 40
Whistle signal
  between pilot and tug, 520
  Great Lakes, 520
  pilot-tug, 206-7
Wildcat, defined, 40
Wilfred M. Cohen, 209
Williwaw, defined, 40
Winch, single-drum, 342
Winch drum, calculating capacity, 510-11
Wind speed, 533-36
Wind velocity, ocean towing, 531-32
Wire rope, breaking strength table, 512
Wisconsin, 210
Work arrangement, 477-78
Working area floodlight, 121
Work vest, 141
World War II, 7

Yvon Dupre, 91

Z card, 468
Zinc, 140
Z-peller tug, 45, 46
Zwarte Zee, 358

About the Author

John S. Blank, 3rd, now a towing consultant, went to sea immediately out of high school and was a licensed third mate at 20. He currently holds the eleventh issue of his marine licenses as master and first class pilot. At the end of several years of active duty during World War II he had achieved the rank of Lieutenant Commander and been awarded the Distinguished Service Medal. In addition to his many years as a tug captain, he has been a port captain, marine superintendent, and shipbuilding inspector; he frequently appears as an expert witness in legal cases involving towing and piloting.

Captain Blank and his wife Constance, who were married in 1950, lived aboard ship with their six children for many years. Their current home in California is close to the sea on which the captain worked for more than 50 years.

ISBN 0-87033-372-0